Understanding the way in which large-scale struc
remains the most challenging problem in cosmc
provides an up-to-date and pedagogical introducti
of research. Part 1 deals with the Friedmann model, the thermal history of the universe, and includes a description of observed structures in the universe. Part 2 describes the theory of gravitational instability in the linear regime and the statistics of density fields. This part also includes chapters on the microwave background radiation and large scale velocity fields. Part 3 of the book covers nonlinear instability, high redshift objects, inflation, cosmic strings and dark matter. Each chapter is accompanied by a comprehensive set of exercises to help the reader in self-study.

The book will be of interest to research and graduate students in cosmology, relativity, theoretical physics, astrophysics and astronomy. It is also suitable for use as a graduate textbook for introductory graduate level courses.

Structure formation in the universe

Structure formation in the universe

T. Padmanabhan

Inter-University Centre for Astronomy and Astrophysics,
Pune, India

Published by the Press Syndicate of the University of Cambridge
The Pitt Building, Trumpington Street, Cambridge CB2 1RP
40 West 20th Street, New York NY 10011-4211, USA
10 Stamford Road, Oakleigh, Victoria 3166, Australia

First published 1993

Printed in Great Britain at the University Press, Cambridge

A catalogue record for this book is available from the British Library

Library of Congress cataloguing in publication data

Padmanabhan, T. (Thanu)
Structure formation in the universe / T. Padmanabhan.
p. cm.
Includes bibliographical references and index.
ISBN 0 521 41448 2 – ISBN 0 521 42486 0 (pbk)
1. Cosmology. 2. Astrophysics. I. Title.
QB981.P245 1993
523.1–dc20 92-14397 CIP

ISBN 0 521 41448 2 hardback
ISBN 0 521 42486 0 paperback

UP TAG

To that special person in my life, but for whom much of this would have been impossible

Contents

Preface

"kasmin nu, Bhagavo, vignaate
sarvamidam vignaatam bhavatiti?"
("What is That, Lord, which being known,
all these become known?")
– Mundako Upanishad

How do large scale structures like galaxies form? This remains the major unsolved problem in cosmology in spite of an impressive increase in observations and computing capabilities. Structure formation was one of the primary research areas in the eighties and is likely to attract more attention in coming years.

This book is aimed at helping physics students to learn this subject in a systematic manner. It is, by and large, a *text book* intended for the *non-expert.* Most of the book will be intelligible to a graduate student in physics who has completed first courses in quantum mechanics, electrodynamics and statistical physics. Parts of the book use concepts from general relativity and field theory; but I have provided alternative routes ('Track 1') for readers who are not familiar with these topics. The 'advanced' sections can be skipped during the first reading without losing a sense of continuity.

The real difficulty in making the book self-contained was the following: the target reader is not expected to be familiar with astrophysical concepts and processes at a level beyond, say, articles in *Scientific American.* However, the subject of structure formation requires inputs from several areas of astrophysics. I have attempted to tackle this difficulty in three different ways:

Firstly, I have put together in chapter 1 several pieces of astrophysical information which are needed later. (This makes chapter 1 very loosely structured; reviewers should start reading the book from chapter 2 and read chapter 1 only at the end!)

Secondly, I have included several exercises which develop some of the necessary astrophysical concepts *ab initio.* Thomson and Compton scattering, bremsstrahlung, synchrotron radiation, free–free absorption, collisional and violent relaxation, polytropes, etc. are some of the physical processes which are presented as exercises. Sufficient help is provided in these exercises to enable a serious reader to work out these processes on his (her) own.

Finally, on a few occasions, I am forced to state the necessary result without any justification. On those occasions, I have tried to keep the statement free of astrophysical (or field theoretical) jargon and have provided references which the reader can look up.

The emphasis, throughout the book, is on theoretical ideas and general physical principles. Nevertheless, I have devoted sufficient space to discussion of the results of crucial observations and their implications. There is, for example, one chapter each on microwave background radiation and large scale streaming velocities.

The emphasis on physical principles was essential for another reason: Fashions change very rapidly in this area. (The half-life for the 'best' models is about four years!) Hence I do not expect much of the *details* of the models discussed in this book to be relevant after, say, five years; but the physical principles on which these models are designed are likely to remain useful for a much longer time.

I have kept citations to the original literature at a level appropriate for a text book rather than a monograph. These references (with occasional annotations) appear at the end of the book. I offer my apologies to anyone whose priority or contribution has been overlooked inadvertently.

The detailed structure of the book is as follows. Chapter 1 contains several items of astrophysical information which are needed at various stages of the book. It is probably best if the reader glances through this material initially but returns to the relevant sections later whenever they are needed. Chapters 2 and 3 provide a description of the standard big bang model and form the foundation for the chapters to come. Chapters 4 and 5 develop the full machinery of linear perturbation theory and, in some sense, form the core of the book. The derivation of the relativistic perturbation equation in chapter 4 uses fairly advanced concepts from general relativity. However, the reader can skip the derivation (and take the final result on faith) during the first reading. Since the results from perturbation theory can be somewhat confusing when encountered for the first time (merely because of the number of distinct cases which should be taken care of), I have also provided a simplified derivation of the results in one of the sections. This discussion, I hope, will help the reader keep track of the physics while wading through the mathematics in the remaining sections. Chapters 6 and 7 apply the results of linear perturbation theory to understanding observations related to the microwave background radiation and large scale streaming velocities. These are probably the two most important observations which can be analyzed within the context of the linear theory and I have tried to present a fairly detailed discussion. Chapter 8 continues with the theoretical framework from chapters 4 and 5, developing analytical approximations for studying the nonlinear evolution. The spherical top-hat model and the Zeldovich approximation are discussed in some detail and are used to estimate the observed characteristics (like mass, angular momentum etc.) of galaxies. Chapter 9 compares the predictions from the nonlinear theory with observations about quasars and other high-redshift objects. This chapter is similar in spirit to chapters 5 and 7 but involves more complicated astrophysics. If the reader prefers, he (she) can cover the theoretical developments in chapters 4, 5 and 8 first and then read chapters 6, 7 and 9 together. Chapters 10 and 11 discuss the ideas which are important in providing a more complete view on structure formation. The origin of seed perturbations and the nature of dark matter are discussed in these chapters. Some of the sections in these chapters use concepts from field theory and particle physics; once again, I have arranged the material in such a manner that the reader who is not familiar with these concepts can skip these sections with minimum loss of continuity. The last chapter is a summary and, to a great extent, reflects the personal bias of the author.

There are, of course, several topics which I could not cover due to lack of space. I particularly regret the omission of the following topics: isocurvature perturbations, formation and physics of clusters of galaxies, textures and other cosmological defects.

It is possible to adapt parts of the text for shorter courses. For example, chapters 2 and 3 (with some material from chapter 1) can make up a one-semester course in standard cosmology. Similarly, chapters 4, 8 and 10 can be used in designing advanced level courses on structure formation.

Several people have contributed to the making of this book.

I thank Martin Rees for emphasizing the need for such a book and encouraging me to take up the venture. Rufus Neal of Cambridge University Press was extremely efficient and helpful in sorting out the logistics and in clarifying my own ideas as to what the book should be like; it was a pleasure working with him.

Many of my colleagues read the earlier versions of the manuscript and offered detailed comments. I thank Sai Iyer, Rajaram Nityananda, Vasanthi Padmanabhan, Somak Raychaudhury (especially for his detailed comments on chapter 7), Tejinder Singh, Ramesh Sinha and Kandaswamy Subramanian for this help.

I am grateful to D. Burstein, A. Dressler, J. Silk and the editors and publishers of *Nature* and *Reports on Progress of Physics* for giving me permission to reproduce the figures which appear in chapter 7. Some of the material is based on a review article which I co-authored with K. Subramanian and I thank him for permitting me to draw upon our joint work.

I am grateful to H.M. Antia for much help in devising the original TeX macros for this book and to Sunita Nair and D. Narasimha for their assistance with the numerical work and graphs.

F. Fernandez and P. Joseph have helped me considerably in getting the manuscript into final shape. The members of the Drawing and Photography sections of TIFR have done an excellent job in producing the diagrams used in this book. I thank them for their cooperation and help.

My style of thinking about this subject has been significantly influenced by my interaction with Jayant Narlikar and Donald Lynden-Bell. I thank both of them for conveying to me the sense of excitement they feel about this subject.

T. Padmanabhan

Part one

The smooth universe

1

Introducing the universe

1.1 Introduction

This chapter contains a discussion of several observational and theoretical topics which form the background for the rest of the book. After a brief theoretical discussion (in section 1.2) of the physical processes which determine the masses and sizes of various structures, sections 1.3 and 1.4 review the known properties of stars and galaxies. The mathematical description of a galaxy, based on the solutions to the collisionless Boltzmann equation, is covered in section 1.5. The next two sections, 1.6 and 1.7, review the observed properties of the larger structures – groups, clusters and superclusters – and the expansion of the universe. Section 1.8 contains a brief outline of the properties of quasars. The extragalactic background radiation – including the microwave background – is described in section 1.9. The last two sections review the procedures used in determining the distance and timescales in the universe.

The material discussed in sections 1.3, 1.4, 1.6, 1.7, 1.10 and 1.11 will be useful in several later chapters. The material in section 1.5 will be required in chapters 8 and 11 while that in sections 1.8 and 1.9 will be needed in chapter 9.

Units in which $\hbar = 1$, $c = 1$ and $k_B = 1$ are used throughout this book. These factors will be indicated in the equations only when some special purpose needs to be served by such a display. Numbers appearing as superscripts to the text refer to the literature cited, which appears at the end of the book.

1.2 Existence of large scale structures

In studying the behaviour of large scale structures, it is convenient to use the galaxy as the primary unit. A typical galaxy is a gravitationally bound system made of about 10^{11} stars with a total mass of about 10^{45} g and size of about 10^{23} cm. Each star has a mass of about 10^{33} g and radius of about 10^{11} cm. We will begin by discussing the physical principles which play a crucial role in determining the mass and length scales of stars and galaxies.

To do this, let us consider the forces which govern the behaviour of bulk matter, viz. the forces of electromagnetism and gravity. These two long-range forces differ considerably in their strength. The ratio of the electromagnetic and gravitational forces acting between two protons, for example, is

$$(F_e/F_g) = [(e^2/r^2)/(Gm_p^2/r^2)] = \frac{e^2}{Gm_p^2} = \left(\frac{e^2}{\hbar c}\right)\left(\frac{\hbar c}{Gm_p^2}\right) \simeq 10^{36}. \quad (1.1)$$

The largeness of this ratio is due to the small value of the 'gravitational fine-structure' constant $\alpha_G \equiv (Gm_p^2/\hbar c) \simeq 10^{-38}$; the corresponding dimensionless number for electromagnetism is $\alpha \equiv (e^2/\hbar c) \simeq 10^{-2}$.

Because of this fact, the properties of matter at small scales are completely determined by the electromagnetic processes, with gravity playing virtually no role. The atomic size ($a_0 \simeq \alpha^{-1} m_e^{-1}$), binding energy of atomic systems ($E_b \simeq \alpha^2 m_e$), typical densities of matter ($\rho_0 \simeq (3Am_p/4\pi a_0^3) \simeq (A\alpha^3/4)\, m_p m_e^3$ where A is the atomic weight) are all decided by electromagnetic processes. However, it should be noted that bulk neutral bodies do not exert electrical forces on each other, whereas the gravitational interaction exists between any two bodies. It follows that, as the body becomes bigger, the effect of gravity will become increasingly important; at sufficiently large scales, gravity will be the dominant force governing the structure of the system.

If the mass of a body is larger than some minimum mass M_C, then the self-gravity of matter will start affecting the structure of the body significantly. This minimum mass can be estimated in the following way. A spherical solid body of mass $M = Nm_p$ and radius $R = N^{1/3} a_0$, made of N atoms, will have a gravitational potential energy $E_g \simeq GM^2/R \simeq N^2(Gm_p^2/R)$. The internal (atomic) binding energy of such a body will be $E_0 \simeq N\alpha^2 m_e$. The body can exist in a stable configuration, resisting the tendency of the gravitational force to crush it, provided $E_0 > E_g$. Writing

$$E_g \simeq N^2 \frac{Gm_p^2}{R} \simeq N^{5/3}\left(\frac{Gm_p^2}{a_0}\right) \simeq N^{5/3}\left(\frac{Gm_p^2}{\alpha}\right)\alpha^2 m_e, \quad (1.2)$$

we see that the condition $E_g < E_0$ implies the bound $N < N_{\max}$ where

$$N_{\max} = \left(\alpha/Gm_p^2\right)^{3/2} = \alpha^{3/2}\alpha_G^{-3/2} \simeq 10^{54}. \quad (1.3)$$

Such an object can have a maximum radius $R = N_{\max}^{1/3} a_0 \simeq \alpha^{1/2}\alpha_G^{-1/2} a_0 \simeq 10^{10}$ cm and maximum mass $M \simeq N_{\max} m_p \simeq 10^{30}$ g. These correspond to the mass and radius of a typical large planet. (One can obtain similar results by computing the central pressure on a spherical body and equating it to the maximum pressure a solid can sustain without undergoing significant deformation. Such a procedure will lead to the same order of magnitude estimate). Objects which are less massive are dominated by solid state forces and have approximately constant density; for these objects, $M \propto R^3$.

The evolution of systems with $M > M_C$ is complicated and depends on the relative importance of several physical processes. To illustrate these processes, we shall consider two simple situations. As a first example, let us consider the evolution of a gaseous sphere of hydrogen, with $N > N_{\rm max}$. For a body which is dominated by gravity and is (at least approximately) in steady state, the virial theorem implies that the magnitudes of kinetic energy (K) and potential energy (U) must be comparable. Since $|U| \simeq (GM^2/R)$ and $K \simeq NT$ where T is the mean temperature of the body, it follows that $T \simeq (GMm_p/R)$. Notice that, for $N > N_{\rm max}$, this temperature is higher than the typical atomic binding energy of solids; it is, therefore, reasonable to consider the matter to be in a plasma state, with the electrons moving freely between the nuclei. As the body contracts under the action of its own gravitational force, its temperature will increase according to the relation $NT \simeq GM^2/R$. Writing the inter-nucleon separation as $d \simeq RN^{-1/3}$, we can express the temperature as $T \simeq N^{2/3}(Gm_p^2/d)$. From this relation, it might appear that T can increase without bound as R – and hence, d – is decreased. This conclusion, however, is not quite correct. As d is decreased, the electrons in the atoms are confined to progressively smaller regions and their degeneracy pressure will soon begin to contribute. Non-relativistic electrons, confined within a region of size d, will have kinetic energies of the order of $(p^2/m) \simeq (\hbar^2/md^2) \simeq (md^2)^{-1}$. We may, therefore, assume that the dependence of T on d is given by the relation $[T + (m_e d^2)^{-1}] \simeq N^{2/3}(Gm_p^2/d)$, or

$$T(d) \simeq N^{2/3}\left(\frac{Gm_p^2}{d}\right) - \left(\frac{1}{m_e d^2}\right). \tag{1.4}$$

The second term, due to degeneracy pressure of electrons, is important at high densities. (This relation is not exact but has the correct limiting behaviour. A more precise analysis is given in exercise 1.5). The function $T(d)$ reaches its maximum value at $d = d_\star = 2N^{-2/3}(m_e\alpha_G)^{-1}$; the corresponding maximum central temperature is $T = T_\star \simeq N^{4/3}\alpha_G^2 m_e$.

The evolutionof the gas cloud depends crucially on the value of $T_\star$. If this value is high enough to trigger nuclear reactions at the centre of the cloud, then further contraction can be halted. If not, the body cools and solidifies in a very short timescale, forming a planet-like system. The temperature needed to trigger nuclear reactions is about $T_N \simeq \eta\alpha^2 m_p$ where η depends on the details of the reaction. (The electrostatic potential barrier between the positively charged protons can be higher than 1 MeV at nuclear separations. Nuclear reactions, however, can be triggered even at a lower energy of about 1 keV because sufficient number of protons can tunnel quantum mechanically through the Coulomb barrier; see exercise 1.1). The condition $T_\star > T_N$ implies that $N > \eta^{3/4}(m_p/m_e)^{3/4}(\alpha/\alpha_G)^{3/2} \simeq (0.1\eta)^{3/4}10^{57}$; that is, $M > M_\star$ where

$$M_\star = (\eta)^{3/4}\left(\frac{m_p}{m_e}\right)^{3/4}\left(\frac{\alpha}{\alpha_G}\right)^{3/2} m_p \simeq 10^{32}\,\mathrm{g} \tag{1.5}$$

for $\eta \simeq 0.1$. Once the nuclear reaction starts, the gravitational contraction will be halted; the body will become luminous, in the sense that the energy generated by nuclear reactions will be radiated away by the body.

Thus, elementary physical considerations suggest the possible existence of bodies, with large masses and radii, which can be luminous because of nuclear energy generation. Such objects are identified with the stars. One can also show, by comparing the radiation pressure and gas pressure in a star, that it will be stable only if M is less than about $100M_\star$ (see exercise 1.2). Hence most stars are confined to a fairly narrow range in mass. For comparison, note that the mass and radius of the Sun are $\mathrm{M}_\odot = 2\times 10^{33}\,\mathrm{g}$ and $\mathrm{R}_\odot \simeq 7\times 10^{10}\,\mathrm{cm}$.

A more careful calculation[1] suggests that the critical mass for igniting nuclear fuel is about $0.08\,\mathrm{M}_\odot$. Gas clouds which are less massive will end up as objects conventionally called brown dwarfs.

As a second example of bodies dominated by gravity, consider a gas cloud with a mass which is significantly larger than that of a typical star: $M \gg \mathrm{M}_\odot$. In this case, we have to take into account the possibility that the body may fragment into smaller objects. A region of size λ in the body has thermal energy $E_{\rm th} \sim (\rho\lambda^3 T/m_p)$ and gravitational energy $E_g \sim G(\rho\lambda^3)^2\lambda^{-1}$. Such a region can be stable against gravitational contraction only if the thermal energy dominates over the gravitational energy: $E_{\rm th} > E_g$; i.e., only if $T \gtrsim G\rho\lambda^2$. If the body cools rapidly, thereby decreasing T, this condition can be violated. The body will then have a tendency to fragment into smaller bodies (with $\lambda \ll R$), each of which can satisfy the condition $E_{\rm th} > E_g$ and evolve separately. Hence, to study the dynamics of the gas cloud, we need to know the dominant cooling processes which operate in it. For systems with temperature $T \simeq (GMm_p/R)$ which is much higher than the ionization potential $\alpha^2 m_e$, the dominant cooling mechanism is bremsstrahlung. The cooling time for the bremsstrahlung process is

$$t_{\rm cool} \simeq (n\alpha\sigma_T)^{-1}\left(\frac{T}{m_e}\right)^{1/2} = \frac{m_e^2}{\alpha^3 n}\left(\frac{T}{m_e}\right)^{1/2} \tag{1.6}$$

where $\sigma_T \simeq \alpha^2\, m_e^{-2}$ is the Thomson scattering cross section and n is the number density of charged particles (see exercise 1.3). This timescale should be compared with the timescale for gravitational collapse

$$t_{\rm grav} \simeq \left(\frac{GM}{R^3}\right)^{-1/2}. \tag{1.7}$$

The condition for efficient cooling ($t_{\rm cool} < t_{\rm grav}$) leads to the constraint $R < R_g$ with

$$R_g \simeq \alpha^3 \alpha_G^{-1} m_e^{-1}\left(\frac{m_p}{m_e}\right)^{1/2} \simeq 2.23\times 10^{23}\,{\rm cm} \cong 74\,{\rm kpc}. \tag{1.8}$$

In arriving at the last equality, we have introduced the unit, 'kiloparsec' with $1\,{\rm kpc} \cong 3\times 10^{21}$ cm. This unit, along with parsec ($1\,{\rm pc} = 10^{-3}\,{\rm kpc}$) and megaparsec ($1\,{\rm Mpc} = 10^3\,{\rm kpc}$) will be extensively used in future discussions. The condition that $T > \alpha^2 m_e$ at $R \simeq R_g$ (which was assumed in the above analysis) implies that M must be larger than the critical mass

$$M_g \simeq \alpha_G^{-2}\alpha^5\left(\frac{m_p}{m_e}\right)^{1/2} m_p \simeq 3\times 10^{11}\, M_\star. \tag{1.9}$$

Clouds of gas with $M \gtrsim M_g$ and $R \lesssim R_g$ can cool by the bremsstrahlung process and fragment into smaller bodies. The smaller fragments evolve as independent units and can end up as stars or planets. Systems with $R > R_g$ will evolve quasistatically with $T \propto R^{-1}$ until the condition $R < R_g$ is satisfied. We will study these processes in greater detail in chapter 8.

The above analysis suggests that one may expect to find gravitationally bound systems which have about 10^{12} or more stars. Such systems with 10^{11} to 10^{12} stars may be identified with galaxies. Thus fundamental considerations suggest the possible existence of large scale objects like stars and galaxies, which are dominated by gravity[2].

The analysis presented above should be thought of as a feasibility argument rather than as a rigorous derivation. There are several uncertain factors which could affect the formation and evolution of these structures. Notice, for example, that the above discussion offers no insight into the initial conditions which lead to the formation of these structures. Neither does it explain the variety and complexity which exist among stars and galaxies. The limited nature of the argument also prevents us from determining the fate of systems in the intermediate mass range.

Most of the discussion in the later chapters will concentrate on these issues and on scenarios leading to the formation of galaxies and bigger systems. In the remaining parts of this chapter, we will briefly review some of the observed features of these large scale structures.

1.3 Stars

The primary unit of study in the formation of large scale structures will be a galaxy. But since galaxies are made of stars, it is preferable to begin with an overview of some of the properties of the stars[3] which will be of importance in future discussion.

We have seen in the last section that stars are essentially fuelled by nuclear reactions. The luminosity of a typical star (which is the amount of energy radiated per second by the star) can be estimated as follows: The nuclear energy generated at the core of a star is transported to the surface, and radiated away, by photons. In this process, the photons are repeatedly scattered by the charged particles that make up the plasma in the star. Let the mean free path of the photon be l. If the photon travels a distance R (from the core to the surface) after Q collisions, then $R \simeq Q^{1/2} l$; i.e., $Q = (R/l)^2$. Since the mean time between collisions is also l (in the units with $c = 1$), the photon takes a time $t_{\rm esc} \simeq Ql \simeq (R^2/l)$ to escape from the core. The luminosity, L, of a star is the ratio between the total radiant energy available and the escape time $t_{\rm esc}$. Since the energy available in the star in the form of photons is about $(aT^4)R^3$, where $a = (\pi^2 k^4/15\hbar^3 c^3)$ is the radiation constant (which, in the units we are using, is $a = (\pi^2/15) \simeq 1$), it follows that

$$L = \frac{aR^3T^4}{R^2} l \simeq RT^4 l. \tag{1.10}$$

In general, mean free path l can depend on the temperature and density of the star in a complicated manner. In the simplest case, applicable to high mass stars, $l \simeq (\sigma_T n_e)^{-1}$ where $\sigma_T \simeq \alpha^2 m_e^{-2}$ is the Thomson scattering cross section and $n_e \simeq NR^{-3}$ is the number density of electrons. In that case,

$$L \simeq \frac{T^4 R^4}{\sigma_T N} \simeq \frac{G^4 M^3}{\alpha^2 m_e^{-2}} m_p^5 \simeq 10^{34}\,{\rm erg\,s^{-1}} \left(\frac{M}{M_\star}\right)^3. \tag{1.11}$$

We have also used the relation $T \simeq GMm_p/R$ in arriving at the above result. This is a typical luminosity of a *high* mass star; for comparison, note that the luminosity of the sun is $L_\odot \cong 4 \times 10^{33}\,{\rm erg\,s^{-1}}$. This is the total luminosity – called the 'bolometric' luminosity – obtained by integrating the flux over all wavelengths.

Before proceeding further, we shall comment on the units used to measure luminosity and related quantities. Since the luminosity of stars varies with the wavelength of the emitted radiation, it is usual to quote the luminosity in a specified waveband. The 'visual' (or V) band is centred at the wavelength $\lambda = 5500\,\text{Å}$, the blue (B) band is centred at $4400\,\text{Å}$ and the ultraviolet (U) band is centred at $3650\,\text{Å}$. These bands have a width of $(\Delta\lambda/\lambda) \simeq 0.2$. For example, the luminosities of Sirius in the B and V bands are $L_B = 42.5\,\mathrm{L}_{\odot B}$ and $L_V = 23.3\,\mathrm{L}_{\odot V}$.

For historical reasons, luminosity is often measured in terms of a logarithmic unit called 'magnitude'. The absolute magnitude, M, of an object is defined to be $M = (-2.5\log L +\text{ constant})$ where the value of the constant depends on the waveband chosen. Note that the magnitude *decreases* when the luminosity increases. It is conventional to choose these constants so that $\mathrm{M}_{\odot B} = 5.48$ and $\mathrm{M}_{\odot V} = 4.83$. One can also define a bolometric magnitude, M_b, from the total luminosity; for the Sun, $\mathrm{M}_{\odot b} = 4.72$.

The quantity which is directly accessible to measurement is not the luminosity of the object but the flux of radiation received from this object at earth, $f = L/(4\pi d^2)$, where d is the distance to the object. The 'apparent magnitude' of the object is defined as $m = (-2.5\log f +\text{ constant})$ where the constant is chosen such that $m = 0$ for an object with the flux $f = 2.52 \times 10^{-5}\,\mathrm{erg\,cm^{-2}s^{-1}}$. From these definitions, one can easily show that

$$m = M + 5\log(d/3\times 10^{19}\,\mathrm{cm}) = M + 5\log(d/10\,\mathrm{pc}), \qquad (1.12)$$

where we have used the unit 1 parsec $= 1\,\mathrm{pc} = 3\times 10^{18}\,\mathrm{cm}$ introduced earlier. Equation (1.12) shows that $m = M$ for an object at 10 pc. The conversion from magnitude to physical units can be achieved using the following formulas:

$$\begin{aligned} L &= 3.02\times 10^{35-0.4M_b}\ \mathrm{erg\,s^{-1}}, \\ f &= 2.52\times 10^{-5-0.4m_b}\ \mathrm{erg\,cm^{-2}\,s^{-1}}. \end{aligned} \qquad (1.13)$$

Sometimes the luminosity and the magnitude are quoted for unit frequency range; in that case, the conversion factor is

$$\begin{aligned} \left(\frac{df}{d\lambda}\right)_{\mathrm{at}\ \lambda\simeq 4400\,\text{Å}} &= 6.76\times 10^{-9-0.4m_B}\mathrm{erg\,cm^{-2}\,s^{-1}\,\text{Å}^{-1}} \\ &= 4.36\times 10^{-20-0.4m_B}\mathrm{erg\,cm^{-2}\,s^{-1}\,Hz^{-1}}, \end{aligned} \qquad (1.14)$$

where m_B is the apparent blue magnitude.

From the luminosity of the star one can estimate its lifetime. The nuclear reactions in a star convert the rest mass of the nuclei into energy with an efficiency which varies from reaction to reaction. The simplest reaction, producing a helium nucleus from four hydrogen nuclei, will liberate an energy equivalent to $0.03m_p$, which is about 0.7 per cent of the original mass $4m_p$. If the average efficiency of nuclear reactions is ϵ, then the amount of fuel available for nuclear burning is (ϵM). Hence the total duration for which nuclear reactions can support the star against gravitational contraction is:

$$t_\star = \frac{\epsilon M}{L} \simeq 3 \times 10^9 \,\text{yr} \left(\frac{\epsilon}{0.01}\right) \left(\frac{M_\star}{M}\right)^2 . \tag{1.15}$$

This gives the typical lifetime of an average star. Notice that stars with larger masses have shorter lifetimes.

The result $L \propto M^3$ derived above depends on the specific form of the scattering (viz., the Thomson scattering) which was assumed. In general, depending on the form of the scattering, one gets[3] a relation of the form $L \propto M^n$ with $n \simeq$ (3–5). For example, in lower mass stars, the mean free path varies as $l \propto T^{7/2}\, n_e^{-2} \propto T^{7/2} R^6 M^{-2}$. In this case, $L \propto R^7 T^{15/2} M^{-2} \propto M^{5.5} R^{-1/2}$; if we further take $M \propto R$, then $L \propto M^5$. For stars fuelled by H–He conversion, we may assume a relatively constant temperature so that $R \propto M$; more generally, it turns out that $R \propto M^p$ with $p \approx 0.6$ to 1. Since the surface temperature of the star, T_s, is related to L by $L \propto R^2 T_\text{s}^4$, we find that

$$T_s \propto L^{1/4} R^{-1/2} \propto L^{1/4} M^{-p/2} \propto L^{1/4 - p/2n}. \tag{1.16}$$

For $n \simeq 3, p \simeq 1$, we get $T_s \propto L^{1/12}$ while for $n = 5, p = 1$, we get $T_s \propto L^{3/20}$. If the stars are indicated as points in a $\ln T_s$–$\ln L$ plot (called the Hertzsprung–Russell diagram, or H–R diagram for short), then we would expect most of the stars to lie on a straight line with slope in the range of 1/12 to 3/20. Observations suggest that this is indeed true; the observed slope is about 0.13.

The time evolution of the star, which can be depicted as a path in the H–R diagram, is quite complicated because of the several physical processes which need to be taken into account. Detailed calculations, based on the numerical integration of the relevant equations, have provided us with a fairly comprehensive picture of stellar evolution[4]. We summarize below a few of the relevant details.

One of the primary sources of stellar energy is a series of nuclear reactions converting four protons into a helium nucleus. Since the simultaneous collision of four particles is extremely improbable, this process of converting hydrogen into helium proceeds though two different sequences of intermediate reactions, one called the *p–p* chain and the other called the CNO cycle. In the *p–p* chain, helium is formed through deuterium and ^{3}He in the intermediate steps; this reaction is the dominant mechanism for H–He conversion at temperatures below about 2×10^7 K. In the CNO cycle, hydrogen is converted into helium through a sequence of steps involving ^{12}C as a catalyst (i.e., the amount of ^{12}C remains the same at the end of the cycle of reaction). Since the Coulomb barrier for carbon nuclei is quite high, the CNO cycle is dominant only at higher temperature.

The evolution of a star – like the Sun – during the phase of H–He conversion (called the 'main sequence' phase) is fairly stable and uneventful. The stability is essentially due to the following regulatory mechanism: Suppose the temperature decreases slightly causing the nuclear reaction rate to decrease. This will make gravity slightly more dominant, causing a contraction. Once the star contracts, the temperature will again increase, thereby increasing the rate of nuclear reactions and the pressure support. This will restore the balance.

After the burning of core hydrogen ends, the core will undergo a contraction, increasing its temperature; if the star now heats up beyond the helium ignition temperature, then the burning of helium will start and stabilize the star. In principle, such a process can continue with the building up of heavier and heavier elements. But to synthesize elements heavier than He is not easy because He has very high binding energy per nucleon among the light elements. Stars achieve synthesis of post-He elements through a process known as 'triple-alpha reaction' which proceeds as $^4\mathrm{He}(^4\mathrm{He}, ^8\mathrm{Be})\gamma$; $^8\mathrm{Be}(^4\mathrm{He}, ^{12}\mathrm{C})\gamma$. (Here, we are using the notation $A(B,C)D$ to denote the reaction $A + B \to C + D$). Once $^{12}\mathrm{C}$ has been synthesized, production of heavier elements like $^{16}\mathrm{O}$, $^{20}\mathrm{Ne}$, $^{24}\mathrm{Mg}$ etc. can occur through various normal channels (provided temperatures are high enough) and the star can evolve through successive stages of nuclear burning. The ashes of one stage can become the fuel for the next stage as long as each ignition temperature is crossed. Such stars will evolve into structures which contain concentric shells of elements. For example, a $15\,\mathrm{M}_\odot$ star, during its last phase can have layers of iron, silicon, oxygen, neon, carbon, helium and hydrogen all burning at their inner edges.

The details of the above process, which occurs after the exhaustion of most of the fuel in the core, depend sensitively on its mass. Consider, for example, a star with $M \gtrsim 1\,\mathrm{M}_\odot$. Its evolution proceeds in the following manner: Once the hydrogen is exhausted in the core, the core (containing predominantly He) undergoes gravitational contraction. This increases the temperature of the material just beyond the core and causes renewed burning of hydrogen in a shell-like region. Soon, the core contracts rapidly, increasing the energy production – and the pressure – in the shell, thereby causing the outer envelope to expand. Such an expansion leads to the cooling of the surface of the star. About this time, convection becomes the dominant mechanism for energy transport in the envelope and the luminosity of the star increases due to the convective mixing. This is usually called the 'red-giant' phase.

During the core contraction, the matter gets compressed to very high densities (about $10^5\,\mathrm{g\,cm^{-3}}$) so that it behaves like a *degenerate* gas and not as an ideal gas. Once the core temperature is high enough to initiate the triple-alpha reaction, helium burning occurs at the core. Since degenerate gas has a high thermal conductivity, this process occurs very rapidly (called 'helium flash'). If the core was dominated by *gas pressure*, such an explosive ignition would have increased the pressure and led to an expansion; but since the *degeneracy pressure* is reasonably independent of temperature, this does not happen. Instead the evolution proceeds as a run-away process: the increase in temperature causes an increase in triple-alpha reaction rate, causing further increase in temperature etc. Finally, when the temperature becomes about 3.5×10^8 K, the electrons become non-degenerate; the core then expands and cools.

The star has now reached a stage with He burning in the core and H burning in a shell around the core. Soon the core is mostly converted into carbon and the reaction again stops. The process described above occurs once again, this time with a carbon-rich, degenerate core, and a He-burning shell. This situation, however, turns out to be unstable because the triple-alpha reaction is highly sensitive to temperature. This reaction can overrespond to any fluctuations in pressure or temperature thereby causing pulsations of the star with increasing amplitude. It is believed that one such violent pulsation can eject the cool, outer layers of the star leaving behind a hot core. The ejected envelope becomes what is known as a 'planetary nebula'.

The above discussion assumes that the core could contract sufficiently to reach the ignition temperature for carbon burning. In low mass stars, degeneracy pressure stops the star from reaching this phase and it ends up as a 'white dwarf' supported by the degeneracy pressure of electrons against gravity.

A more complicated sequence of nuclear burning is possible in stars with $M \gg 1\,\mathrm{M}_\odot$. After the exhaustion of carbon burning in the core, one can have successive phases with neon, oxygen and silicon burning in the core with successive shells of lighter elements around it. This process can go on until ^{56}Fe is produced in the core. The binding energy per nucleon is maximum for the ^{56}Fe nucleus; hence it will not be energetically feasible for heavier elements to be synthesized by nuclear fusion. The core now collapses catastrophically reaching very high (about 10^{10} K) temperatures. The ^{56}Fe photo-disintegrates into alpha particles, and then even the alpha particles disintegrate in the heat to become protons. The collapse of the core squeezes together protons and electrons to form neutrons and the material reaches near nuclear densities forming a 'neutron star'. There exist several physical processes which can transfer the gravitational energy from core collapse to the envelope, thereby leading to the forceful

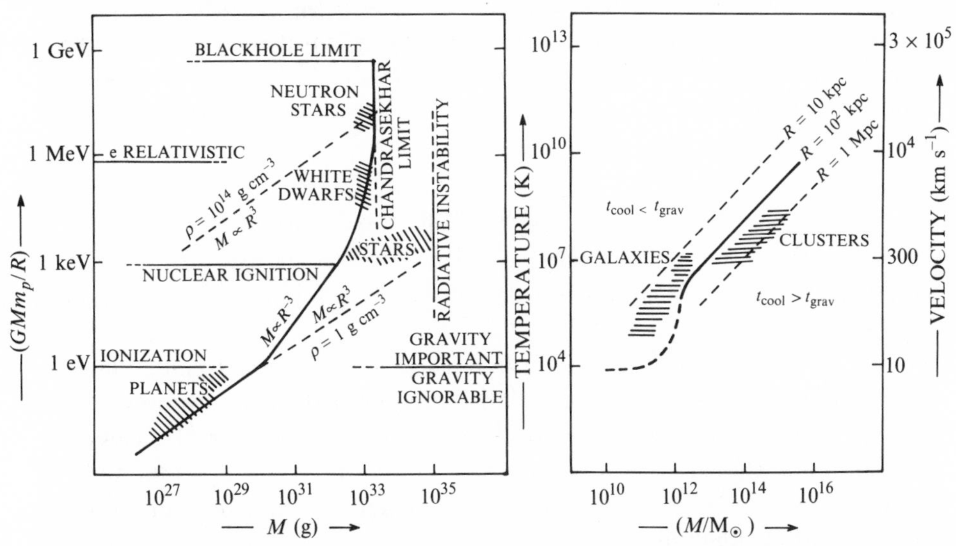

Fig. 1.1. The mass–radius relationship for various structures discussed in this section is summarized here. The left part of the figure describes the stellar mass objects while the right half (with a break of scale) covers galaxies and clusters. The vertical axis is the gravitational potential energy per baryon; equivalent kinetic temperature and the velocity dispersion are also indicated. At the smallest mass scale are the objects with constant density dominated by solid state forces. (Lines of constant density have a slope of (2/3) in the logarithmic scale which is used). Objects supported by degeneracy pressure lie on the curve $M \propto R^{-3}$ which has a slope of (4/3). This curve joins with the vertical line indicating the Chandrasekhar limit for systems supported by relativistic degeneracy pressure. Several other regions of physical interest and the regions in the M–R space occupied by various structures are also indicated. The right half of the diagram shows the curve $t_{\rm cool} = t_{\rm grav}$ by a thick line. For the bremsstrahlung cooling discussed in the text, this line has unit slope for $M \gg 10^{12}\,{\rm M}_\odot$. The behaviour of this curve for $M < 10^{12}\,{\rm M}_\odot$ is indicated by a broken, thick line. (This curve will be discussed in detail in chapter 8). Also shown are the lines corresponding to the radius of 10 kpc and 1 Mpc. Galaxies fall in the region $t_{\rm cool} < t_{\rm grav}$ while the clusters occupy the region $t_{\rm cool} > t_{\rm grav}$.

ejection of the outer envelope causing a 'supernova explosion'. A remnant with smaller mass is left behind.

The mass–radius relation for the stellar remnant of such an explosion, supported by the degeneracy pressure, can be easily determined (see exercise 1.5). If the inter-particle separation is $d \simeq n^{-1/3}$, then the degeneracy momentum is $k \simeq \hbar/d \simeq \hbar n^{1/3}$. The corresponding energy (of each particle) is $\epsilon_{\rm NR} \simeq (k^2/m) \simeq (\hbar^2 n^{2/3}/m)$ in the non-relativistic case and $\epsilon_{\rm R} \simeq k \simeq \hbar n^{1/3}$ in the extreme relativistic case. The degeneracy pressure is $p \simeq n\epsilon$ and becomes $p_{\rm NR} \simeq (\hbar^2 n^{5/3}/m)$ and $p_{\rm R} \simeq \hbar n^{4/3}$ in the two cases. The pressure due to gravity (i.e., the gravitational force per unit area) will be $p_g \simeq (GM^2/R^2)\ (1/R^2) \simeq (GM^2/R^4)$. In the case of a white dwarf supported by non-relativistic, electron-degeneracy pressure, the condition $p_g = p_{\rm NR}$ gives

$$M^{1/3}R = \frac{\hbar^2}{Gm_e^{8/3}}\left(\frac{m_e}{m_p}\right)^{5/3} = \text{constant}, \tag{1.17}$$

so that $M \propto R^{-3}$. A similar result will hold for a neutron star with m_e replaced by $m_n \simeq m_p$.

The situation is quite different in the case of an *extreme relativistic,* degenerate gas of nucleons. The maximum degeneracy pressure of such a neutron star is $p_{\rm R} \simeq \hbar n^{4/3} \simeq \hbar (M/m_p)^{4/3}\ (1/R^4)$. Hence it can support the gravitational pressure only if $p_{\rm R} > p_g$; i.e., only if $(GM^2/R^4) < \hbar(M/m_p)^{4/3}\ (1/R^4)$. This leads to a limiting mass (called the 'Chandrasekhar mass'):

$$M_{\rm Ch} \simeq \alpha_G^{-3/2} m_p \simeq 1\,{\rm M}_\odot \tag{1.18}$$

(a more precise calculation gives the value 1.4 $\rm M_\odot$; see exercise 1.5). Stellar remnants which are more massive than $M_{\rm Ch}$ will be unable to resist gravity and will end up as blackholes. Figure 1.1 summarizes the mass–radius relation for the various objects.

Numerical studies[4] show that stars with $M > 8\,{\rm M}_\odot$ burn hydrogen, helium and carbon and evolve rather smoothly. During the final phase, such a star explodes as a supernova leaving behind a remnant which could be a white dwarf, neutron star or black hole. It is generally believed that stars with masses in the intermediate range, $2.3\,{\rm M}_\odot < M < 8\,{\rm M}_\odot$, do not burn hydrogen and helium in degenerate cores but evolve through carbon burning in degenerate matter (for $M > 4\,{\rm M}_\odot$) ending again in a supernova explosion. Stars with lower mass do not explode but end up as planetary nebulae. Low mass stars with $M < 2.3\,{\rm M}_\odot$ ignite helium in degenerate cores at the tip of the red giant branch and then evolve in a complicated manner.

The synthesis of elements as described above proceeds smoothly up to ^{56}Fe. Heavier elements are formed[4] by nuclei absorbing the free neutrons (produced in earlier reactions) by two different processes called the 'r-process' (rapid process) and the 's-process' (slow process). During the supernova explosion, a significant part of the heavy elements synthesized in the star will be thrown out into the interstellar space. A second generation of stars can form from these gaseous remnants. The initial composition of material in this second generation will contain a higher proportion of heavier elements (collectively called 'metals') compared to the first generation stars. Both these types of stars are observed in the universe; because of historical reasons, stars in the second generation are called population I stars while those in the first generation are called population II stars.

The above discussion shows how stars could synthesize heavier elements, even if they originally start out as gaseous spheres of hydrogen. The study of the spectra of stars allows us to determine the relative proportion of various elements present in the stars. Such studies show that population II stars are made of about 75 per cent hydrogen and 25 per cent helium; even population I stars consist of an almost similar proportions of hydrogen and helium with a small percentage of heavier elements. It is possible to show, using the stellar evolution calculations, that it is difficult for the stars to have synthesized elements in such a proportion, if they originally had only hydrogen. Hence, such a universal composition leads us to conjecture that the primordial gas from which population II stars have formed must have been a mixture of hydrogen and helium in the ratio 3:1 by weight. Heavier elements synthesized by these population II stars would have been dispersed in the interstellar medium by supernova explosions. The population I stars are supposed to have been formed from this medium, containing a trace of heavier elements. The helium present in the primordial gas should have been synthesized at a still earlier epoch and cannot be accounted for by the stellar evolution. We will see in chapter 3 how cosmological models can provide an explanation for the presence of this primordial helium.

1.4 Galaxies

Galaxies range widely in their sizes, shapes and masses; nevertheless, one may talk of a typical galaxy as something made out of about 10^{11} stars or so. Taking the average mass of a star to be that of the sun, the luminous mass in a galaxy is about $10^{11}\,\mathrm{M}_{\odot} \simeq 2 \times 10^{44}\,\mathrm{g}$. This mass is distributed in a region with a size of about 20 kpc. Even though most galaxies have a mass of about $(10^{10}\text{–}10^{12})\,\mathrm{M}_{\odot}$ and a size of (10–30) kpc, there are several exceptions at both ends of the spread. For example, 'dwarf galaxies' have masses in the range $(10^{5}\text{–}10^{7})\,\mathrm{M}_{\odot}$ and radii of only about (1–3) kpc. There are also some giant galaxies with masses as high as $10^{13}\,\mathrm{M}_{\odot}$.

Galaxies exhibit a wide variety in their shapes as well and are usually classified according to their morphology[5]. Broadly speaking, one may divide them into 'ellipticals' and 'discs'.

Ellipticals are smooth, featureless, distributions of stars, ranging in mass from $10^{8}\,\mathrm{M}_{\odot}$ to $10^{13}\,\mathrm{M}_{\odot}$. The proportion of elliptical galaxies in a region depends sensitively on the environment. They contribute only about ten per cent of all galaxies in low density regions of the universe but nearly forty per cent in dense clusters of galaxies. The surface brightness of an elliptical galaxy is very well fitted by the de Vaucouleurs formula

$$I(R) = I_0 \exp\left(-kR^{1/4}\right) = I_e \exp\left\{-7.67\left[(R/R_e)^{1/4} - 1\right]\right\}, \quad (1.19)$$

where R_e is the radius containing half the total luminosity and I_e is the brightness at R_e; R_e is about 3 kpc for bright ellipticals.

The luminosities of ellipticals vary over seven orders of magnitude. The relative number of elliptical galaxies with luminosities between L and $L + dL$ is given (approximately) by the empirical formula

$$\phi(L)\,dL = \eta_\star \left(\frac{L}{L_\star}\right)^{-\alpha} \exp\left(-\frac{L}{L_\star}\right)\frac{dL}{L_\star}, \quad (1.20)$$

where $\eta_\star = 1.2 \times 10^{-2} h^3\,\mathrm{Mpc}^{-3}, \alpha = (1.1\text{–}1.25)$, and $L_\star = 1.0 \times 10^{10}$ $h^{-2}\,\mathrm{L}_\odot$ in the V band. (This is called the Schecter luminosity function). The parameter h which occurs here has a value between 0.5 and 1; it is related to the Hubble constant which we will discuss later. The integral of $\phi(L)$ over all L diverges in the lower limit; so clearly, this formula breaks down at very small L and needs to be truncated at some $L_{\rm min}$. However, observations do suggest that the number of galaxies increases significantly at low values of luminosity which is correctly reflected in the formula. From this luminosity function, we can *formally* calculate the mean number density of galaxies

$$n = \int_0^\infty \phi(L)\,dL = \eta_\star \Gamma(1-\alpha) \simeq 1.2 \times 10^{-2} h^3 \Gamma(1-\alpha)\,\mathrm{Mpc}^{-3}, \quad (1.21)$$

which, of course, diverges for $\alpha > 1$; nevertheless, $\eta_\star$ often provides a rough estimate of the number density of bright galaxies. This number density of $\eta_\star$ corresponds to the mean intergalactic separation of $\eta_\star^{-1/3} \simeq$ 4.4 Mpc. The mean luminosity

$$< L > = \int_0^\infty L\phi(L)\,dL = \eta_\star L_\star \Gamma(2-\alpha) \simeq 1.2 \times 10^8 h \Gamma(2-\alpha)\,\mathrm{L}_\odot\,\mathrm{Mpc}^{-3} \quad (1.22)$$

is, however, finite. Taking the mass of a galaxy to be $M_g \simeq 10^{11}\,\mathrm{M}_\odot$, the average mass density in the form of galaxies is

$$\rho_{\rm gal} \simeq n M_g \simeq 10^{-31} h^3\,\mathrm{g\,cm}^{-3}. \quad (1.23)$$

It may be noted that the luminosity function $\phi(L)$ is essentially determined from the density of galaxies in the nearby region. The actual counts of faint galaxies show some crucial disparities with respect to $\phi(L)$; these observations will be discussed in detail in chapter 9.

Another relevant parameter characterising a galaxy is its angular momentum. The angular momentum of any galaxy can be expressed conveniently in the following way: Consider a galaxy with mass M, radius R, angular momentum L and energy $E = -|E| \simeq -(GM^2/R)$. The angular velocity of such a system will be about

$$\omega \simeq (L/MR^2). \tag{1.24}$$

On the other hand, the angular velocity $\omega_{\rm sup}$ needed for the system to be rotationally supported against gravity is determined by the equation

$$\omega_{\rm sup}^2 R \simeq GM/R^2; \tag{1.25}$$

or $\omega_{\rm sup} \simeq (GM/R^3)^{1/2}$. The ratio $(\omega/\omega_{\rm sup})$ between the actual angular velocity ω and the angular velocity $\omega_{\rm sup}$ needed to provide rotational support represents the degree of rotational support available in the system. This ratio can be expressed as

$$\lambda = \frac{\omega}{\omega_{\rm sup}} = \left(\frac{L}{MR^2}\right)\left(\frac{R^{3/2}}{G^{1/2}M^{1/2}}\right) = \frac{L}{G^{1/2}M^{3/2}R^{1/2}} = \frac{L|E|^{1/2}}{GM^{5/2}} \tag{1.26}$$

and serves as a convenient dimensionless parameter characterizing the angular momentum of the system. For elliptical galaxies $\lambda \simeq 0.05$, showing very little systematic rotation and insignificant rotational support. Most ellipticals do show a certain degree of oblateness. The rotation of the ellipticals, however, is too small to account for this feature; it is more likely that the oblateness is due to the anisotropy of the velocity dispersion; see exercise 1.6.

In virial equilibrium, gravitational potential energy of the galaxy must be comparable to the kinetic energy of the constituents. This equality can arise either due to steady rotational motion or due to random motion of the stars. The stars in ellipticals have large random velocities which support them against the mean gravitational pull. (This concept will become clearer in section 1.5, when we discuss the models for the galaxy). The velocity dispersion σ can be measured from the broadening of the spectral lines and is found to be well correlated with the luminosity of the galaxy by the relation (called the Faber–Jackson law):

$$\sigma \simeq 220(L/L_\star)^{1/4}\,\text{km s}^{-1}. \tag{1.27}$$

The second major type of galaxy is the 'spiral' (or 'disc') to which our own galaxy, the Milky Way, belongs. Spirals have a prominent disc, made of Population I stars and contain a significant amount of gas and dust. (The name originates from the distinct spiral arms which exist in many of these galaxies. We shall, however, use the terms 'spirals' and 'discs' interchangeably). In low density regions of the universe, nearly eighty per cent of the galaxies are spirals while only ten per cent of galaxies in dense clusters are spirals. This is complementary to the behaviour of ellipticals. The stars in a disc galaxy are supported against gravity by their systematic rotation velocity with $\lambda \simeq (0.4\text{–}0.5)$. The rotational speed of stars, $v(R)$, at radius R has the remarkable property that it remains constant for large R in almost all spirals; the constant value is typically between 200 and 300 km s^{-1}. This fact is of crucial importance and will be discussed in chapter 11.

Most discs also contain a spheroidal component of Population II stars. The luminosity of the spheroidal component relative to the disc correlates well with several properties of these galaxies. This fact has been used for classifying the disc galaxies into finer divisions.

Our own galaxy, the Milky Way, is a spiral galaxy. Most of the visible stars in our galaxy are found in a flat, axisymmetric, disc-like structure. The Sun is located at a distance of about $R_0 = 2.55 \times 10^{22}\,\mathrm{cm} = 8.5\,\mathrm{kpc}$ from the centre of this disc. The surface brightness of our galaxy (and other spiral galaxies similar to ours) has the form

$$I(R) = I_0 \exp\left(-R/R_D\right), \tag{1.28}$$

where $R_D \cong 3.5 \pm 0.5\,\mathrm{kpc}$ and $I_0 = 114\,\mathrm{L}_\odot\,\mathrm{pc}^{-2}$. The sun lies at a radius from within which nearly seventy per cent of the light in our galaxy is emitted. Most of the stars in the disc travel in (nearly) circular orbits around the galactic centre. At the location of the Sun, the speed of a test star in a circular orbit will be about $v(R_0) \simeq 220 \pm 15\,\mathrm{km\,s}^{-1}$.

The disc of our galaxy has a small but finite thickness. Very massive and young stars (with typical ages less than 10^7 yr) are distributed within a thickness of about 10^2 pc while average stars (like the sun, with an age of about 5×10^9 yr) are distributed within a larger thickness of about 700 pc.

In addition to the disc, our galaxy also contains a spheroidal distribution of stars which are distinctly different from the stars seen in the disc. These stars in the spheroidal part are older than the disc stars, exhibit very little rotation and have large random velocities. By and large, the disc stars are Population I stars and the spheroidal stars are Population II stars. The spheroid is comparatively small and contributes only about 15 to 30 per cent of the total luminosity of the galaxy.

The stars in a galaxy are not distributed in a completely uniform manner. A typical galaxy contains several smaller stellar systems, each containing about $(10^2\text{–}10^6)$ stars. These systems, usually called star clusters, can be broadly divided into two types called 'open clusters' and 'globular clusters'. Open clusters consist of $(10^2\text{–}10^3)$ Population I stars bound within a radius of (1–10) pc. Most of the stars in these clusters are quite young. In contrast, globular clusters are Population II systems with $(10^4\text{–}10^6)$ stars. Our galaxy contains about 200 globular clusters which are distributed in a spherically symmetric manner about the centre of the galaxy. Unlike open clusters, the stars in globular clusters are quite old. (We will discuss the age of these objects in section 1.11.) The number density of stars in the core of a globular cluster ($10^4\,M_\odot\,pc^{-3}$) is much higher than that of a typical galaxy ($0.05\,M_\odot\,pc^{-3}$). The core radius of the globular clusters is about 1.5 pc while the 'tidal radius' (which is the radius at which the density drops nearly to zero) is about 50 pc.

In addition to the stars, our galaxy also contains gas and dust which contribute about 5 to 10 per cent of its mass. This interstellar medium may be roughly divided into a very dense, cold, molecular component (with about 10^4 particles per cubic centimetre and a temperature of about 100 K) made of interstellar clouds, a second component which is atomic but neutral (with $n \simeq 1\,\mathrm{cm}^{-3}$ and $T \simeq 10^3\,\mathrm{K}$) and a third component which is ionized and very hot (with $n \simeq 10^{-3}\,\mathrm{cm}^{-3}$ and $T \simeq 10^6\,\mathrm{K}$). Though the interstellar medium is principally made of hydrogen, it also contains numerous other chemical species (including molecules as complex as HC_9N) and an appreciable quantity of tiny solid particles ('dust'). The spiral arms are concentrations of stars and interstellar gas and are characterized by the presence of ionized hydrogen. This is also the region in which young stars are being formed.

For a more detailed study of galaxies, one can divide ellipticals and spirals into subsets and also add two more classes of galaxies, called 'lenticulars' and 'irregulars'. The ellipticals are subdivided as $(E1, \ldots, En, \ldots)$ where $n = 10(a-b)/a$ with a and b denoting the major and minor axis of the ellipticals. The 'lenticulars' (also called SO) are the galaxies 'in between' ellipticals and spirals. They have a prominent disc which contains no gas, dust, bright young stars or spiral arms. Though they are smooth and featureless like ellipticals, their surface brightness follows the exponential law of the spirals. They are rare in low density regions (less than ten per cent) but constitute nearly half of the galaxies in the high density regions.

The spirals are subdivided into Sa, Sb, Sc, Sd with the relative luminosity of the spheroidal component decreasing along the sequence. The amount of gas increases and the spiral arm becomes more loosely wound as we go from Sa to Sd. Our own galaxy is between the types Sb and Sc. There also exist another class of spirals called 'barred spirals' which exhibit a bar-like structure in the centre. They are classified as SBa, SBb etc.

Finally, irregulars are the galaxies which do not fall in the above mentioned morphological classification. These are low luminosity, gas rich systems with massive young stars and large HII regions (i.e. regions containing ionized hydrogen). More than one third of the galaxies in our neighbourhood are irregulars. They are intrinsically more difficult to detect at larger distances because of their low luminosity.

We have seen earlier that as the stars evolve, their luminosity and hence the colour changes. Since galaxies are made of stars, galaxies will also exhibit colour evolution. Besides, the gas content and elemental abundances of the galaxies will change as the stars are formed and end as planetary nebulae or supernovae. We shall discuss these aspects in chapter 9.

1.5 Models for galaxies

The distribution[6] of stars in a galaxy can be described by a function $f(\mathbf{x}, \mathbf{v}, t)$ which may be interpreted as the (relative) probability of finding a star in the phase space in the interval $(\mathbf{x}, \mathbf{x}+d^3\mathbf{x};\ \mathbf{v}, \mathbf{v}+d^3\mathbf{v})$ at time t. The smoothed-out mass density of the stars at any point $\mathbf{x}$ will be

$$\rho(\mathbf{x}, t) = m \int f(\mathbf{x}, \mathbf{v}, t)\, d^3\mathbf{v}, \tag{1.29}$$

where m is the average mass of the stars. Such a smoothed-out density will produce a gravitational potential $\phi(\mathbf{x}, t)$ where

$$\nabla^2 \phi = 4\pi G \rho. \tag{1.30}$$

We may assume that each star moves in this smooth gravitational field along some specific orbit. The conservation of the total number of stars, expressed by the equation $(df/dt) = 0$, will reduce to

$$\begin{aligned} \frac{df}{dt} &= \frac{\partial f}{\partial \mathbf{x}} + \dot{\mathbf{v}} \cdot \frac{\partial f}{\partial \mathbf{v}} + \dot{\mathbf{x}} \cdot \frac{\partial f}{\partial \mathbf{x}} \\ &= \frac{\partial f}{\partial t} + \mathbf{v} \cdot \frac{\partial f}{\partial \mathbf{x}} - \nabla\phi \cdot \frac{\partial f}{\partial \mathbf{v}} = 0. \end{aligned} \tag{1.31}$$

Models for galaxies are based on the solution to the coupled equations (1.30) and (1.31). Since these equations allow a wide variety of solutions, different classes of models are possible for the galaxies.

It should be noted that the *actual* gravitational potential at any point $\mathbf{x}$, due to the stars is

$$\phi_{\mathrm{act}}(\mathbf{x}, t) = -\sum_i \frac{Gm}{|\mathbf{x} - \mathbf{x}_i(t)|} \tag{1.32}$$

where $\mathbf{x}_i(t)$ is the position of the ith star at time t. This will be different from the ϕ in (1.30) which was produced by the smooth density. Because of this difference, the actual trajectories of the stars will differ appreciably from the orbits of the smooth potential after sufficiently long time intervals. However, it can be easily shown (see exercise 1.7) that this timescale is very large compared with the orbital timescales in galaxies. Hence, equations (1.30) and (1.31) provide an excellent description of galaxies over reasonable timescales.

When the galaxy is in a steady state, $f(t, \mathbf{x}, \mathbf{v}) = f(\mathbf{x}, \mathbf{v})$. To produce such a steady state solution we can proceed as follows: Let $C_i = C_i(\mathbf{x}, \mathbf{v})$, $i = 1, 2, \cdots$ be a set of integrals of motion for the stars moving in the potential ϕ (which, right now, is not known). It is obvious that any function $f(C_i)$ of the C_i will satisfy the steady-state Boltzmann equation; $(df/dt) = (\partial f/\partial C_i)\ \dot{C}_i = 0$ since $\dot{C}_i$ is identically zero. If we can now determine ϕ from f self consistently, and populate the orbits of ϕ with stars, we have solved the problem. Let us consider some specific examples to see how this idea works.

In these calculations, it is convenient to shift the origin of $|\phi|$ by defining a new potential $\psi \equiv -\phi + \phi_0$ where ϕ_0 is a constant. (We will choose the value of ϕ_0 such that ψ vanishes at the 'boundary' of the galaxy). The new potential satisfies the equation

$$\nabla^2 \psi = -4\pi G\rho, \tag{1.33}$$

and the boundary condition $\psi \to \phi_0 \quad \text{as } |\mathbf{x}| \to \infty$. We will also define a 'shifted' energy for the stars $\epsilon = -E + \phi_0$; since $\phi_0 = \psi + \phi$, $\epsilon = -E + \psi + \phi$ $= -(1/2)v^2 + \psi$.

The simplest galactic models are the ones in which $f(\mathbf{x}, \mathbf{v})$ depends on $\mathbf{x}$ and $\mathbf{v}$ only through the quantity ϵ so that $f = f(\epsilon) = f[\psi - (1/2)v^2]$. The density $\rho(\mathbf{x})$ corresponding to this distribution is

$$\rho(\mathbf{x}) = \int_0^{\sqrt{2\psi}} 4\pi v^2 dv\, f\left[\psi - (1/2)v^2\right] = \int_0^{\psi} 4\pi\, d\epsilon\, f(\epsilon)\sqrt{2(\psi - \epsilon)}. \tag{1.34}$$

The limits of integration are chosen in such a way to pick only the stars bound in the galaxy's potential. The right hand side is a known function of ψ, once $f(\epsilon)$ is specified. The Poisson equation

$$\frac{1}{r^2}\frac{d}{dr}\left(r^2\frac{d\psi}{dr}\right) = -4\pi G\rho = -16\pi^2 G\int_0^{\psi} d\epsilon f(\epsilon)\sqrt{2(\psi - \epsilon)} \tag{1.35}$$

can now be solved – with some central value $\psi(0)$ and the boundary condition $\psi'(0) = 0$ – determining $\psi(r)$. Once $\psi(r)$ is known all other variables can be computed. Three different choices for $f(\epsilon)$ have been extensively used in the literature to describe spherically symmetric systems. We will summarize these models briefly.

(a) Polytropes

The simplest form of $f(\epsilon)$ is a power law with $f(\epsilon) = A\epsilon^{n-3/2}$ for $\epsilon > 0$ and zero otherwise. Using (1.34), we see that this corresponds to density distributions of the form $\rho = B\psi^n$ (for $\psi > 0$) with $B = (2\pi)^{3/2} A\Gamma(n - 1/2)[\Gamma(n+1)]^{-1}$. (Clearly, we need $n > 1/2$ to obtain finite density). The Poisson equation now becomes

$$\frac{1}{l^2}\frac{d}{dl}\left(l^2\frac{d\xi}{dl}\right) = \begin{cases} -\xi^n & \xi \geq 0 \\ 0 & \xi \leq 0 \end{cases} \tag{1.36}$$

where we have introduced the variables,

$$L = (4\pi G\psi(0)^{n-1}B)^{-1/2}; \quad l = (r/L); \quad \xi = (\psi/\psi(0)). \tag{1.37}$$

This equation is called the Lane–Emden equation and its properties are well studied. The case which is of interest in galactic modelling arises for $n = 5$. In this case, (1.36) has the simple solution

$$\xi = \left(1 + \frac{1}{3}l^2\right)^{-1/2} \tag{1.38}$$

which corresponds to a density profile of $\rho \propto [1 + (1/3)l^2]^{-5/2}$ and a total mass of $M = (\sqrt{3}L\psi(0)/G)$. This profile provides a reasonable description of some elliptical galaxies.

(b) Isothermal sphere

This model corresponds to the distribution function

$$f(\epsilon) = \frac{\rho_0}{(2\pi\sigma^2)^{3/2}} \exp\left(\frac{\epsilon}{\sigma^2}\right), \tag{1.39}$$

parametrized by two constants ρ_0 and σ. One can easily verify that the mean square velocity $< v^2 >$ is $3\sigma^2$ and that the density distribution is $\rho(r) = \rho_0 \exp(\psi/\sigma^2)$. The central density is $\rho_c = \rho_0 \exp(\psi(0)/\sigma^2)$. It is conventional to define a core radius and a set of dimensionless variables by

$$r_0 = \left(\frac{9\sigma^2}{4\pi G\rho_c}\right)^{1/2}; \quad l = \frac{r}{r_0}; \quad \xi = \frac{\rho}{\rho_c}. \tag{1.40}$$

Then the Poisson equation can be rewritten in the form

$$\frac{1}{l^2}\frac{d}{dl}\left(\frac{l^2}{\xi}\frac{d\xi}{dl}\right) = -9\xi. \tag{1.41}$$

This equation has to be integrated numerically (with the boundary condition $\xi(0) = 1$ and $\xi'(0) = 0$) to determine the density profile. It can be shown that (see exercise 1.8) the solution has the asymptotic limit of $\xi \simeq (2/9l^2)$. (This function, incidentally, is an *exact* solution; but, of course, it is unphysical because of the singular behaviour at the origin). Hence the exact solution has a mass profile with $M(r) \propto r$ at larger r; the model has to be cut off at some radius to provide a finite mass.

For $l \lesssim 2$, the numerical solution is well approximated by the function $\xi(l) = (1 + l^2)^{-3/2}$. The projected two dimensional surface density corresponding to this $\xi(l)$ is $S(R) = 2(1 + R^2)^{-1}$, where R is the projected radial distance. In other words, the (true) central surface density $\Sigma(0) \simeq 2\rho_c r_0$ where ρ_c is the central volume density and r_0 is the core radius.

Since the mass density of the isothermal sphere varies as r^{-2} at large distances, the potential ϕ varies as $\ln r$. It can be easily shown that a disc of stars embedded in such an isothermal sphere will have constant rotational velocities. We shall use these properties of the isothermal sphere in chapter 11.

(c) King models

These are based on the distribution function with

$$f(\epsilon) = \frac{\rho_c}{(2\pi\sigma^2)^{3/2}}\left(e^{\epsilon/\sigma^2} - 1\right); \quad \epsilon \geq 0 \tag{1.42}$$

and $f(\epsilon) = 0$ for $\epsilon < 0$. Since $f(\epsilon)$ vanishes for $\epsilon < 0$, this model may be thought of as a truncated isothermal sphere. The density, when integrated from the origin numerically, will vanish at some radius r_t (called the tidal radius). The quantity $c = \log(r_t/r_0)$ is a measure of how concentrated the system is. Bright elliptical galaxies are well described by such a model with $[\psi(0)/\sigma^2] \simeq 10$ and $c \simeq 2.4$; the density profile of globular clusters can be fitted with $[\psi(0)/\sigma^2] \simeq 3$–$7$.

Incidentally, notice that a given $\rho(r)$ generates a unique $f(\epsilon)$ by the following procedure: Given $\rho(r)$, we can determine $\psi(r)$ and hence, by eliminating r, the function $\rho(\psi)$. Writing (1.34) as

$$\frac{1}{\sqrt{8}\pi}\rho(\psi) = 2\int_0^{\psi} f(\epsilon)\sqrt{\psi - \epsilon}\, d\epsilon \tag{1.43}$$

and differentiating both sides with respect to ψ, we get

$$\frac{1}{\sqrt{8}\pi}\frac{d\rho}{d\psi} = \int_0^{\psi}\frac{f(\epsilon)\, d\epsilon}{\sqrt{\psi - \epsilon}}. \tag{1.44}$$

This equation (called Abel's integral equation) has the solution

$$f(\epsilon) = \frac{1}{\sqrt{8}\pi^2}\frac{d}{d\epsilon}\int_0^{\epsilon}\left(\frac{d\rho}{d\psi}\right)\frac{d\psi}{\sqrt{\epsilon - \psi}}, \tag{1.45}$$

which determines $f(\epsilon)$. Though this procedure gives an $f(\epsilon)$, there is no assurance that it will be positive definite.

If we allow the distribution function to depend on the angular momentum J as well as on ϵ, then a wider class of models can be constructed to fit the same $\rho(r)$. To see this, assume that f depends on ϵ and J^2 only through the combination

$$Q = \epsilon - \frac{J^2}{2R^2}; \quad R = \text{constant}. \tag{1.46}$$

One can easily integrate out the angular variables in the velocity space and obtain the density distribution

$$\rho(r) = \frac{2\pi\sqrt{8}}{(1+r^2/R^2)}\int_0^{\psi} f(Q)\sqrt{\psi - Q}\, dQ \equiv \frac{2\pi\sqrt{8}}{(1+r^2/R^2)}\mu(\psi). \quad (1.47)$$

Comparing (1.47) and (1.34), we see that

$$f(Q) = \frac{1}{\pi^2\sqrt{8}}\frac{d}{dQ}\int_0^Q \left(\frac{d\mu}{d\psi}\right)\frac{d\psi}{\sqrt{Q-\psi}}. \quad (1.48)$$

Thus, given a $\rho(r)$ we determine $\psi(r)$ and another function $\mu(r) \equiv \rho(r)$ $(1+r^2/R^2)$ $(2\pi\sqrt{8})^{-1}$. Eliminating r between $\mu(r)$ and $\psi(r)$ we obtain $\mu(\psi)$ and consequently $f(Q)$. This distribution function will lead to the density distribution $\rho(r)$ we started with. When $f = f(\epsilon)$, the velocity dispersion is isotropic with $< v_r^2 > =< v_\theta^2 > =< v_\phi^2 >$. For distribution function of the form $f(\epsilon, J^2)$, we have $v_\theta^2 = v_\phi^2 \neq v_r^2$; this is the key difference between the two cases.

Similar modelling also works for a disc-like system if the distribution function is taken to be of the form $f = f(\epsilon, J_z)$. We will mention two examples: If the distribution function is taken to be

$$f(\epsilon, J_z) = \begin{cases} AJ_z^n \exp(\epsilon/\sigma^2) & (\text{for } J_z \geq 0) \\ 0 & (\text{for } J_z \leq 0) \end{cases} \quad (1.49)$$

then we obtain a surface density for the disc, which is

$$\Sigma(R) = \frac{\Sigma_0 R_0}{R}. \quad (1.50)$$

The circular velocity of stars $v_c^2 = -R(\partial\psi/\partial R) = 2\pi G\Sigma_0 R_0$ is a constant for this model, called 'Mestel's disc'. These parameters are related to the parameters in the distribution function by

$$n = (v_c^2/\sigma^2) - 1; \quad A = \Sigma_0 R_0 \left[2^{n/2}\sqrt{\pi}\Gamma(n+1/2)\sigma^{n+2}\right]^{-1}. \quad (1.51)$$

The velocity dispersion in the radial direction σ^2 is a free parameter characterizing the disc. The parameter n is a measure of the 'coldness' of the disc.

A more complicated set of disc models (called the 'Kalnajs disc') can be obtained from the distribution function which has the form

$$f(\epsilon, J_z) = A\left[(\Omega_0^2 - \Omega^2)a^2 + 2(\epsilon + \Omega J_z)\right]^{-1/2} \quad (1.52)$$

when the term in square brackets is positive and zero otherwise. This function leads to the following surface density and potential:

$$\Sigma(R) = \Sigma_0\left(1 - \frac{R^2}{a^2}\right)^{1/2}; \quad \phi(R) = \frac{\pi^2 G\Sigma_0}{4a}R^2 \equiv \frac{1}{2}\Omega_0^2 R^2 \quad (1.53)$$

with $\Sigma_0 = 2\pi Aa\sqrt{\Omega_0^2 - \Omega^2}$. The parameter Ω is free and describes the mean (systematic) rotational velocity of the system: $< v_\phi >= \Omega R$ (see exercise 1.9).

1.6 Distribution of matter

The nearest large galaxy to the Milky Way is the spiral, M31, usually called the 'Andromeda galaxy' which is about 710 kpc away. Its mass is about $3 \times 10^{11}\,\mathrm{M}_\odot$ and it has a size of about 50 kpc. Studies show that it is a spiral galaxy which we see from the earth practically 'edge on'. (It may be noted that galaxies are packed in the universe in a manner very different from the way the stars are distributed inside a galaxy: The distance from the Milky Way to the nearest large galaxy is only 20 galactic diameters while the distance from the sun to the nearest star is thirty million times the diameter of these individual stars.)

There is some evidence to suggest that Andromeda and the Milky Way are gravitationally bound to each other. They have a relative velocity – towards each other – of about $300\,\mathrm{km\,s^{-1}}$. In fact, these two are only the two largest members of a bunch of about 30 galaxies all of which together constitute what is known as the 'Local Group'. The entire Local Group can be contained within a spherical volume of about a few Mpc in radius.

This kind of clustering of galaxies into groups is typical in the distribution of the galaxies in the universe. A careful study[7] within a size of about 20 Mpc from our galaxy shows that only (10–20) per cent of the galaxies do *not* belong to any group; they are called 'field' galaxies.

Groups may typically consist of about (5–100) galaxies; a system with more than 100 galaxies is conventionally called a 'cluster.' The sizes of groups range from a few hundred kiloparsecs to 1 or 2 megaparsecs. Clusters have a size of typically a few megaparsecs. Just like galaxies, one may approximate clusters and groups as gravitationally bound systems of effectively point particles. The large gravitational potential energy is counterbalanced by the large kinetic energy of random motion in the system. The line of sight velocity dispersion in groups is typically $200\ \mathrm{km\ s^{-1}}$ while that in clusters can be nearly $1000\ \mathrm{km\ s^{-1}}$. There are several similarities between clusters of galaxies and stars in an elliptical galaxy. For example, the radial distribution of galaxies in a cluster can be adequately fitted by the $R^{1/4}$ law with an effective radius $R_e \simeq (1\text{–}2)h^{-1}$ Mpc. About ten per cent of all galaxies are members of large clusters. In addition to galaxies, clusters also contain very hot intracluster gas at temperatures $(10^7\text{–}10^8)$ K.

The two large clusters nearest to the Milky Way are the Virgo cluster and the Coma cluster. The Virgo cluster, located about 15 Mpc away, has a diameter of about 3 Mpc and contains several thousand galaxies. This is a typical irregular cluster and does not exhibit a central condensation or a discernible shape. The Coma cluster, on the other hand, is almost spherically symmetric with a marked central condensation. It has an overall size of about 3 Mpc while its central core is about 600 kpc in size. The core is populated with elliptical and spheroidal galaxies with a density nearly 30 times larger than our Local Group. These values are typical for large clusters. Coma is located at a distance of about 80 Mpc.

The distribution of galaxies around our Local Group has been studied extensively. It turns out that most of the galaxies nearby lie predominantly in a plane – called the supergalactic plane – which is approximately perpendicular to the plane of our own galaxy. The dense set of galaxies in this plane is called the Local Supercluster and the Virgo cluster is nearly at the centre of this highly flattened disc-like system. (The term 'supercluster' is used to denote structures bigger than clusters). Broadly speaking, the Local Supercluster consists of three components: About 20 per cent of the brightest galaxies, forming the core, is the Virgo cluster itself; another 40 per cent of galaxies lie in a flat disc with two extended, disjoint groups of galaxies; the remaining 40 per cent is confined to a small number of groups scattered around. Nearly 80 per cent of all matter in the Local Supercluster lies in a plane[8]. Studies of distant galaxies show that there are many superclusters in our universe separated by large voids.

To study such clustering of galaxies quantitatively one needs a good survey of the universe giving the coordinates of galaxies in the sky. Of the three coordinates needed to specify the position of the galaxy, the two angular coordinates are easy to obtain. There exist today several galaxy catalogues[9], containing the angular positions of galaxies in particular regions of the sky, complete up to a chosen depth. (The APM galaxy survey has about 5×10^6 galaxies out to a depth of $600h^{-1}$ Mpc; the Lick catalogue has about 1.6×10^6 galaxies and depth of $200h^{-1}$ Mpc; the IRAS catalogue has more than 14000 galaxies which are prominent in the infrared band; these are a few of the major catalogues available today.) If we know the redshift z of these galaxies as well, then we can attribute to it a line-of-sight velocity $v \cong zc$. If we further assume that this velocity is due to cosmic expansion, then we can assign to the galaxy a radial distance of $r \cong H_0^{-1}v$. (We will discuss this relation in the next section). This will provide us with the galaxy position (r, θ, ϕ) in the sky.

The main difficulty in completing the survey lies in obtaining telescope time to make a systematic measurement of redshifts for the galaxies which are members of a catalogue. We know the redshifts of only about 30 000 or so galaxies (out of millions which exist in catalogues) and the largest systematic survey – the CfA survey – has only about 9000 redshifts.

Even the limited amount of data we have today points to an interesting pattern in galaxy distribution[10]. The single most useful function characterizing the galaxy distribution is what is called the 'two-point-correlation function': $\xi_{GG}(r)$. This function is defined via the relation

$$dP = \overline{n}^2(1 + \xi_{GG}(\mathbf{r}_1 - \mathbf{r}_2))\, d^3\mathbf{r}_1\, d^3\mathbf{r}_2, \tag{1.54}$$

where dP is the probability to find two galaxies simultaneously in the regions $(\mathbf{r}_1, \mathbf{r}_1 + d^3\mathbf{r}_1)$ and $(\mathbf{r}_2, \mathbf{r}_2 + d^3\mathbf{r}_2)$ and $\overline{n}$ is the mean number density of galaxies in space. The homogeneity of the background universe guarantees that $\xi_{GG}(\mathbf{r}_1, \mathbf{r}_2) = \xi_{GG}(\mathbf{r}_1 - \mathbf{r}_2)$ and isotropy will further make $\xi_{GG}(\mathbf{r}) = \xi_{GG}(|\mathbf{r}|)$. From (1.54) it follows that $\xi_{GG}(r)$ measures the excess probability (over random) of finding a pair of galaxies separated by a distance r; so if $\xi_{GG}(r) > 0$, we may interpret it as a clustering of galaxies over and above the random Poisson distribution. A considerable amount of effort was spent in the past decades in determining $\xi_{GG}(r)$ from observations. These studies show that

$$\xi_{GG}(r) \simeq \left(\frac{r}{5h^{-1}\,\mathrm{Mpc}}\right)^{-1.8} \tag{1.55}$$

in the range $0.1h^{-1}\,\mathrm{Mpc} \lesssim r \lesssim 20h^{-1}\,\mathrm{Mpc}$. We will discuss the properties of this correlation function in chapter 5.

It is also possible, using catalogues of rich clusters like the Abell catalogue which contains 4076 clusters, to compute the correlation function between galaxy clusters. The result turns out to be

$$\xi_{CC}(r) \simeq \left(\frac{r}{25h^{-1}\,\mathrm{Mpc}}\right)^{-1.8}. \tag{1.56}$$

Comparing with the galaxy–galaxy correlation function ξ_{GG} quoted before, we see that clusters are more strongly correlated than the individual galaxies[11]. If firmly established, this result has important implications for the theories of structure formation. Unfortunately, $\xi_{CC}(r)$ is not as well established as ξ_{GG} and hence one has to be cautious in interpreting results which depend on ξ_{CC}.

In recent years researchers have also resorted to less quantitative – but more appealing – diagnostics to demonstrate clustering of galaxies. Many galaxy surveys – like the slices of the CfA survey – present striking *visual* patterns of galaxy distribution. The patterns are consistent with the interpretation that the universe contains several voids of size about $(20h^{-1}$ to $50h^{-1})$ Mpc. The CfA slices, in fact, suggest that galaxies are concentrated on sheet-like structures surrounding nearly empty voids[12].

The distribution of matter seems to be reasonably uniform when observed at scales bigger than about $100h^{-1}\,\mathrm{Mpc}$ or so. For comparison, note that the size of the observed universe is about $3000h^{-1}\,\mathrm{Mpc}$. Thus, one may treat the matter distribution in the universe to be homogeneous while dealing with the phenomena at scales larger than 100 Mpc or so. The standard cosmological models, which we will discuss in the next chapter, are based on this assumption of 'large scale' homogeneity.

1.7 Expansion of the universe

A system of point particles interacting via Newtonian gravity cannot have a static equilibrium configuration which is stable. For example, consider the situation in which an infinite collection of such particles is distributed on a regular cubic lattice. By symmetry, the force on any one particle will be zero. However, the slightest deviation from equilibrium will trigger an instability and cause the particles to move. It is therefore natural to expect structures in the universe to be moving under the gravitational influence of each other.

We can detect the radial motion of a distant galaxy by comparing its spectrum with a laboratory spectrum. By and large, one would have expected the spectrum to show blueshifts and redshifts depending on whether the galaxies are moving towards us or away from us. However, the observations show a surprising result. It turns out that all the *distant* galaxies exhibit *only* redshifts indicating that the universe, on the large scale, is 'expanding'. This result has been verified for over 25 000 galaxies and is quite universal. This feature appears in various wavebands (optical, infrared, radio) and for a wide class of galaxies[13].

If the distance to the galaxy (d) is known, then one can study the correlation between the redshift (z) and distance. It is usual to expand the (unknown) function $z(d)$ in a Taylor series, in the form:

$$z = \left(\frac{H_0 d}{c}\right) + \frac{1}{2}(q_0 - 1)\left(\frac{H_0 d}{c}\right)^2 + \mathcal{O}(d^3) \tag{1.57}$$

where H_0 (called the Hubble constant) has the dimension of inverse time and q_0 (called the deceleration parameter) is dimensionless. For small redshifts $z \lesssim 1$, the linear term in (1.57) describes the observations quite well. The value of H_0 turns out to be

$$H_0 = 100h \,\mathrm{km\,s^{-1}\,Mpc^{-1}} \tag{1.58}$$

where h represents the uncertainty which arises due to various *systematic* errors in the observations[14]. This quantity h lies *certainly* between 0.4 and 1.0 and is *likely* to be between 0.4 and 0.6. If we associate a velocity $v \equiv cz$ with the redshift z, then (1.58) shows that a distance of 1 Mpc corresponds to a velocity of $100h \,\mathrm{km\,s^{-1}}$.

The most natural interpretation of this redshift is based on an overall expansion of the universe. (This concept will be discussed in detail in chapter 2). In such an interpretation, H_0 should be identified with the present expansion rate of the universe. The time scale H_0^{-1} and the length scale cH_0^{-1} indicate the characteristic scales of our universe. These are given by

$$H_0^{-1} = 9.78h^{-1} \times 10^9 \,\mathrm{yr}$$
$$cH_0^{-1} = 9.25 \times 10^{27} \,\mathrm{cm} \cong 3000h^{-1} \,\mathrm{Mpc}. \tag{1.59}$$

By measuring z and d for larger values of d, and using (1.57), one can – in principle – determine the function $z(d)$; or, at least, the parameter q_0. The systematic errors in distance measurements have so far prevented any useful conclusion about q_0 being reached.

From the Hubble constant H_0, one can also obtain a quantity with the dimensions of density proportional to (H_0^2/G). We define a quantity (called 'critical density') by the equation

$$\rho_c = \frac{3H_0^2}{8\pi G} \cong 1.88 \times 10^{-29} h^2 \,\mathrm{g\,cm^{-3}}. \tag{1.60}$$

The coefficient $(3/8\pi)$ may be understood in the following manner: Consider a spherical region of size l containing mass M. A test particle (or a galaxy) at the edge of this region will be moving outwards with velocity $v = H_0 l$ due to the overall expansion of the universe. This velocity will be larger than the escape velocity at this distance $v_{\rm esc} = (2GM/l)^{1/2} = (8\pi G\overline{\rho} l^2/3)^{1/2}$ if $\overline{\rho}$, the average density inside the sphere, is less than ρ_c. This suggests that the expansion of the universe will go on indefinitely if $\rho < \rho_c$ but will be halted – to be followed by a contracting phase – if $\rho > \rho_c$. (This discussion is based on Newtonian gravity. Strictly speaking, we should perform this analysis using general relativity, which will be done in the next chapter. However, such an analysis leads to the same expression for the critical density.)

It is therefore clear that the ratio $(\overline{\rho}/\rho_c)$ will be of importance in deciding the dynamics of the universe. It is conventional to denote this quantity by the symbol Ω. We have seen earlier that the density of *luminous* matter (which exists in the form of galaxies) is only about $10^{-31}\,\mathrm{g\,cm^{-3}}$, giving $\Omega_{\rm lum}$ to be about 0.01. Observations based on the gravitational effects of matter, however, suggest that the total Ω for our universe is about 0.2 to 1. In other words, a large quantity of matter in the universe must be non-luminous or 'dark'. We will discuss these observations in detail in chapter 11.

Some properties of the large scale structures discussed so far are summarized in table 1.1.

We shall see in the next chapter that the redshift serves two important (related) roles in cosmological description. Since $z \simeq H_0 d$, cosmological redshift serves as a measure of distance; the higher the redshift, the farther the object is. Moreover, notice that the light we receive today from a distant object would have been emitted by that object at an earlier

Table 1.1. Large scale structures

Object	Mass (g)	Radius (cm)	$\sigma = (GM/R)^{1/2}$ km s^{-1}	$\rho = (3M/4\pi R^3)$ g cm^{-3}
jupiter	2×10^{30}	6×10^{9}	47	2.3
sun	2×10^{33}	7×10^{10}	470	1.4
red giant	$(2\text{–}6)\times10^{34}$	10^{14}	(37–63)	$(4.8\text{–}14.3)\times10^{-9}$
white dwarf	2×10^{33}	10^{8}	3×10^{4}	5×10^{8}
neutron star	3×10^{33}	10^{6}	1.4×10^{5}	7×10^{14}
glob. cluster	1.2×10^{39}	1.5×10^{20}	8.4	8.5×10^{-23}
open cluster	5×10^{35}	3×10^{19}	0.3	4.4×10^{-24}
spiral	$2\times(10^{44}\text{–}10^{45})$	$(6\text{–}15)\times10^{22}$	(150–300)	$(14\text{–}22)\times10^{-26}$
elliptical	$2\times(10^{43}\text{–}10^{45})$	$(1.5\text{–}3)\times10^{23}$	(30–210)	$(0.14\text{–}1.8)\times10^{-26}$
group	4×10^{46}	3×10^{24}	300	3.5×10^{-28}
cluster	2×10^{48}	1.2×10^{25}	10^{3}	2.7×10^{-28}
universe	$7.5\times10^{55}\Omega h^{-1}$	$10^{28}h^{-1}$	$2.2\times10^{5}\Omega^{1/2}$	$1.8\times10^{-29}\Omega h^{2}$

instant of time. Thus, observations of high redshift objects probe the universe at earlier epochs; in an evolving universe the physical conditions could have been different at earlier epochs and thus the study of high redshift objects will be quite valuable.

Let the present age of the universe be t_0 as calculated from the instant at which the spatial interval between objects would have been zero. (This notion will be made more precise in the next chapter). The light we receive today from an object at redshift z, would have been emitted at a time $t(z) < t_0$. The quantity $\tau(z) \equiv (t_0 - t(z))/t_0$ is called the 'look-back time' and is a measure of how far 'back in time' we can see using objects of redshift z. The explicit form of $\tau(z)$ depends on the matter content of the universe. For example, in a universe with $\Omega = 1$, we will see (in chapter 2) that,

$$\tau(z) = 1 - \frac{1}{(1+z)^{3/2}}. \tag{1.61}$$

Nearby galaxies have a redshift of about 0.001 and clusters of galaxies are visible up to a redshift of about $z \cong 1$. Certain radio galaxies can be seen at redshifts up to $z \simeq 3$ and a class of objects called quasars have redshifts as high as 4.9. The object with the highest redshift known at present is a quasar at redshift of 4.9. Using objects with redshift of, say,

$z \simeq 4$ we can obtain information about the universe when its age was only 10 per cent of the present age (see figure 1.2).

1.8 Quasars

Most galaxies which we can see have a fairly low redshift ($z \lesssim 0.5$) and an extended appearance on a photographic plate. There exists another important class of objects, called 'quasars', which exhibit large redshifts (up to $z \simeq 5$) and appear as point sources in the photographic plate[15]. Estimating the distance from the redshift and using the observed luminosity, it is found that quasars must have a luminosity of about $L_q \simeq (10^{46}–10^{47})$ erg s^{-1}. It is possible to estimate the size of the region emitting the radiation from the timescale in which the radiation pattern is changing (see exercise 1.10). It turns out that the energy from

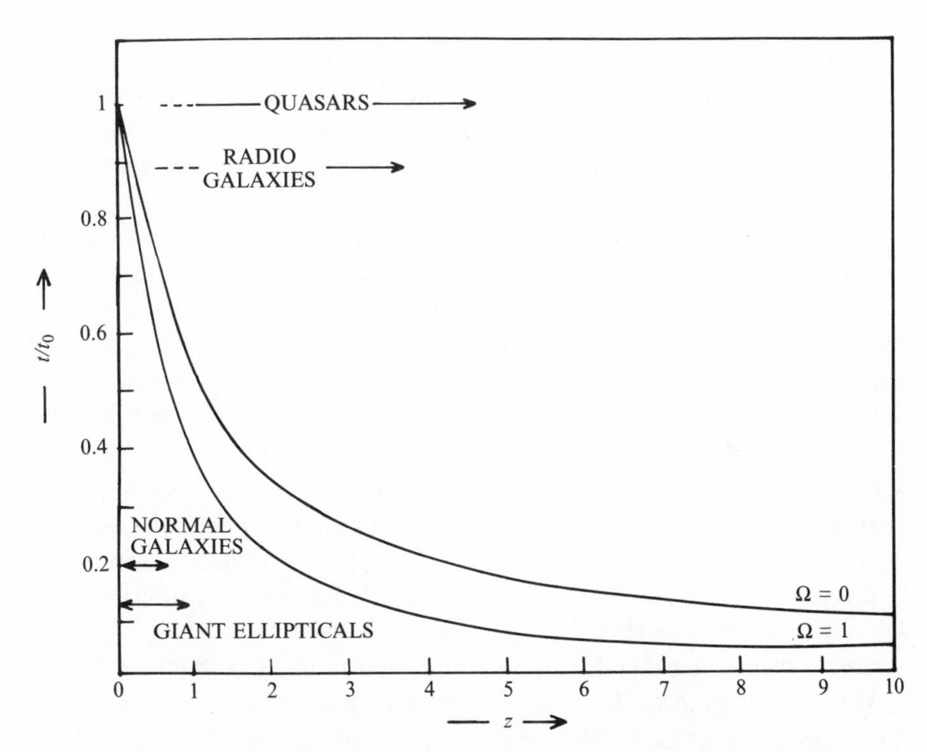

Fig. 1.2. Because of the finite velocity of propagation of the light, high redshift objects allow us to 'look back' in time. The time $t(z)$ at redshift z (in units of the present age of the universe t_0) is plotted against the redshift of the object. The two curves correspond to two different cosmological models containing different amount of matter. Also shown are the typical ranges of redshifts probed by different class of astrophysical objects.

the quasar is emitted from a very compact region. One can easily show that nuclear fusion cannot be a viable energy source for quasars.

It is generally believed that quasars are fuelled by the accretion of matter into a supermassive blackhole ($M \simeq 10^8 \, M_\odot$) in the centre of a host galaxy. The friction in the accretion disc causes the matter to lose angular momentum and hence spiral into the blackhole; the friction also heats up the disc. Several physical processes transform this heat energy into radiation of different wavelengths. Part of this energy can also come out in the form of long, powerful, jets. The innermost regions of the quasar emit X- and γ-rays. Outer shells emit UV, optical and radio continuum radiation in the order of increasing radius.

The typical spectrum of a quasar shows several interesting features. The part of the spectrum at wavelengths longer than that of the Lyman-α line (which has a wavelength of $\lambda_\alpha = 1216$ Å) can be characterized by a power law of the form $P(\nu) \propto \nu^n$ with $n \simeq -1$; this relation remains valid till far infrared or even radio, where a flattening occurs. (Such a power law is interpreted as due to synchrotron emission; see exercise 1.11). In addition to the power law, there is a substantial increase in the continuum radiation near the UV band. The excess intensity is that of a black body with a temperature of $T \simeq 3 \times 10^4$ K and is believed to arise from the accretion disc.

Superposed on this continuum spectrum are strong emission lines due to resonance transitions of common elements like H, C, O and N. These emission lines are quite broad with Doppler velocities corresponding to $(10^8$–$10^9)$ cm s^{-1}. At $\lambda < \lambda_\alpha$, several narrow absorption lines are seen, which are attributed to absorption in clouds of gas with low column density ($\lesssim 10^{16}$ atom cm^{-2}) which may exist along the line-of-sight to the quasar. Most of these lines are due to the Lyman-α *absorption* at some redshift z_{abs}. This region of the spectrum is so thick with lines that it is called the 'Lyman-α forest'; the cumulative effect of absorption by the clouds is a substantial drop in the apparent quasar continuum at $\lambda < \lambda_\alpha$. Further, at $\lambda < \lambda = 912(1 + z_{\text{abs}})$ Å a significant fraction of the UV photons are absorbed by high column density ($\gtrsim 10^{17}$ atom cm^{-2}) systems, producing a sudden spectral discontinuity (see figure 1.3).

Very bright quasars have an apparent magnitude of $m_B \simeq 14$ (which corresponds to a flux of 10^{-25} erg s^{-1} cm^{-2} Hz^{-1}) while the faintest ones have $m_B \simeq 23$. The absolute magnitudes of the quasars are typically $-30 < M_B < -23$; in contrast, galaxies fall in the band $-23 < M_B < -16$. The relation between m and M for any given quasar depends on the estimated distance to the quasar, which in turn depends on the cosmological model and the quasar's redshift.

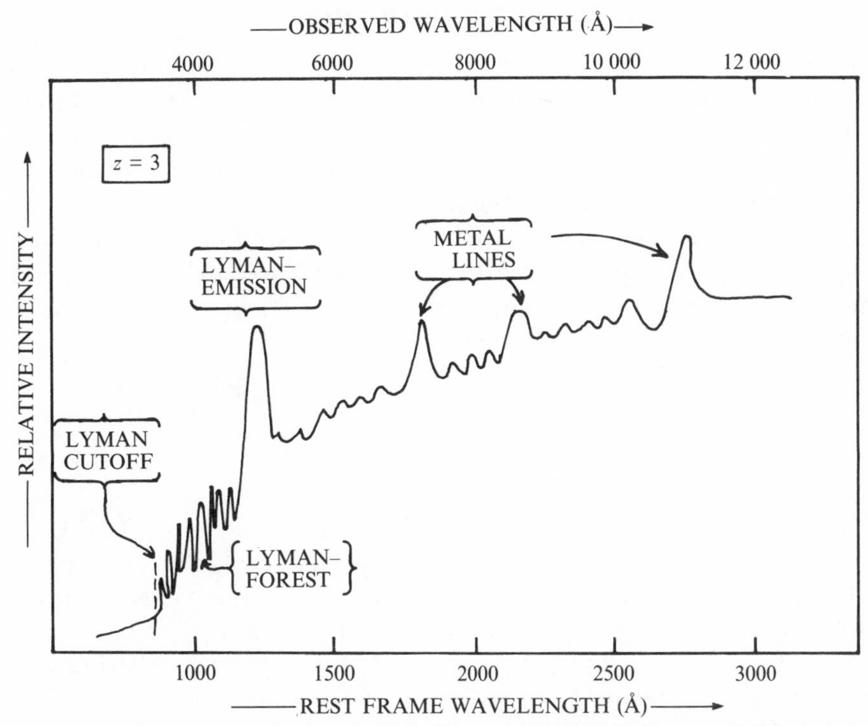

Fig. 1.3. The spectrum of a typical quasar showing some of the features mentioned in the text. These features will be discussed in detail in chapter 9.

The luminosity function of the quasars, as a function of its redshift, has been a subject of extensive study[16]. These investigations show that: (i) the space density of bright galaxies (about $10^{-2}\,\text{Mpc}^{-3}$) is much higher than that of quasars (about $10^{-5}\,\text{Mpc}^{-3}$) with $z < 2$ and (ii) bright quasars were more common in the past ($z \simeq 2$) than today ($z \simeq 0$). We shall see in chapter 9 that quasars serve as an important probe of the high redshift universe.

Quasars are believed to be one extreme example of a wide class of objects called 'active galaxies'. By and large, this term denotes a galaxy which seems to have a very energetic central source of energy. This source is most likely to be a blackhole powered by accretion. One kind of active galaxy which has been studied extensively are radio galaxies. The most interesting feature about these radio galaxies is that the radio emission does not arise from the galaxy itself but from two jets of matter extending from the galaxy in opposite directions. It is generally believed that this emission is caused by the synchrotron radiation of relativistic electrons moving in the jets. The moving electrons generate two elongated clouds containing magnetic fields which, in turn, trap the electrons and lead to the synchrotron radiation.

1.9 Radiation in the universe

The universe has been studied in a wide variety of wavebands from very long waves ($\simeq 10\,\mathrm{m}$) to ultra high frequency γ-ray bands. At most of these wavelengths, one can distinguish two kinds of radiation: (i) the radiation which arises from definite sources in the sky and (ii) a diffuse background radiation which cannot be attributed to any (resolved) source in the sky. The first one carries a considerable amount of information about the particular class of sources while the second one is of more relevance to cosmology. We shall now summarize the main results in various wave bands, concentrating on the background radiation[17].

In what follows, we will use the units 1 watt $= 1\,\mathrm{W} = 10^7$ erg s^{-1} and 1 jansky $= 1\,\mathrm{Jy} = 10^{-26}\,\mathrm{W\,m^{-2}\,Hz^{-1}} = 10^{-23}\,\mathrm{erg\,s^{-1}\,cm^{-2}\,Hz^{-1}}$.

(a) Radio

$$\lambda = 30\,\mathrm{cm} - 10\,\mathrm{m}; \quad \nu \simeq (3 \times 10^7 - 10^9)\,\mathrm{Hz}; \quad T \simeq (10^{-3}\text{–}0.05)\,\mathrm{K}$$

Several discrete sources (supernova remnants, radio galaxies, quasars, ...) emit radio waves. Along the spiral arms of galaxies, there exist clouds of HII regions which are clouds of hot gas (10^4 K) ionized by newly formed massive stars. These clouds emit thermal bremsstrahlung radiation with a relatively flat spectrum ($I_\nu \propto \nu^{-0.1}$). The total power from the galaxy is in the range of (10^{29}–10^{33}) W. Radio galaxies have a more complicated pattern of emission. The radiation usually arises from two 'blobs' placed on either side of the central galaxy with a separation ranging from 3 kpc to 1 Mpc. The power from these radio sources is quite high: (10^{33}–10^{39}) W; special models are needed to explain this emission. The faintest detectable flux in this band is about 10^{-2} Jy; there are about 10^6 discrete sources in the sky up to this level.

Radio observations can also detect the presence of neutral hydrogen in the universe. The two states of the hydrogen atom, in which the spins of the proton and the electron are parallel or antiparallel, differ slightly in energy. The transition radiation between these two levels has a wavelength of 21 cm in the laboratory. Neutral hydrogen at a redshift of z can, therefore, be detected by observations at the wavelength of $21(1+z)$ cm which is in the radio band for sources with the redshift in the range of, say, 1 to 10.

The diffuse background in the radio band arises from our galactic disc, halo and from unresolved extragalactic radio sources. The background flux varies from about 3×10^4 Jy at 100 cm to 6×10^5 Jy at 10^4 cm.

(b) Microwave and submillimetre

$$\lambda = (0.02\text{–}30)\,\mathrm{cm}; \quad \nu \simeq (10^9 - 3 \times 10^{12})\,\mathrm{Hz}; \quad T = (0.05\text{–}300)\,\mathrm{K}$$

The discrete sources in this band are usually dust clouds, hydrogen gas and quasars. The background radiation in this band, however, is of tremendous theoretical importance and has been studied extensively. In fact, this band has the maximum intensity of background radiation. It turns out that the major component of the background radiation in the microwave band can be fitted very accurately by a thermal radiation at a temperature of about 2.75 K. Besides, this radiation shows a remarkable degree of isotropy[18]. Attempts to detect the fractional temperature difference $(\Delta T/T)$, between the microwave radiation received from two different directions (separated by angle θ) in the sky, have drawn a blank so far;* the upper bound on $(\Delta T/T)$ is about 10^{-5}. It seems reasonable to interpret this radiation as a relic arising from the early, hot, phase of the evolving universe. We will study these results in greater detail in chapter 6.

The statements in the above paragraph require one qualification. The microwave radiation does show a clear 'dipole' anisotropy; that is, the observed temperature varies in the sky according to the law

$$T(\theta) \cong T_0(1 + \beta\cos\theta) \tag{1.62}$$

where θ is measured from the direction of maximum temperature. This variation, however, can be eliminated by a Lorentz transformation along this direction with the speed $\beta \simeq 10^{-3}$. In other words, there exists a frame of reference in which microwave radiation is completely isotropic; the motion of the earth with respect to this frame leads to the dipole anisotropy mentioned above.

The temperature of $T \simeq 2.75\,\mathrm{K}$ corresponds to an energy density of radiation $\rho_\gamma = 4.8 \times 10^{-34}\,\mathrm{g\,cm^{-3}}$, which contributes a fraction $\Omega_\gamma = 2.56 \times 10^{-5} h^{-2}$ to the critical density. Since the maximum background energy density of photons exists in the microwave band, Ω_γ may be taken to be the contribution from the electromagnetic radiation to the density parameter.

(c) Infrared

$\lambda \simeq 8000\,\text{Å}$–$0.01\,\mathrm{cm}$; $\nu \simeq (3 \times 10^{12} - 10^{14})\,\mathrm{Hz}$; $T = (300$–$4000)\,\mathrm{K}$

Several interesting astrophysical processes contribute in this band. However this is one of the most difficult range of frequencies to study owing to the enormous opacity of earth's atmosphere and contamination due to emission from interstellar and interplanetary dust. In addition, hydrogen and dust clouds in our galaxy and outside galaxies as well as star-forming regions in our galaxy contribute to this band. The 'near' infrared band of $(1$–$10) \times 10^{-4}$ cm will also receive contributions from the redshifted light

---

* The analysis of COBE data now shows that MBR has an anisotropy of $(\Delta T/T) \simeq (1.1 \pm 0.2) \times 10^{-5}$. This result and its implications are discussed in Appendix C.

associated with the initial epoch of galaxy formation. There have been several attempts to subtract out the galactic contamination and obtain the extragalactic IR flux. Though firm upper bounds are available, there is still substantial uncertainty in the actual shape of the background IR spectrum. The faintest detectable flux is about 1.0 Jy and there are about 10^4 discrete sources up to this limit.

(d) Optical and ultraviolet

$\lambda \simeq (100\text{–}8000)\,\text{Å}$; $\nu \simeq (8 \times 10^{14} - 3 \times 10^{16})\,\text{Hz}$; $T = (4000 - 3 \times 10^4)\,\text{K}$
This band includes visible, UV, far UV and what may even be called soft X-ray. Most of the astronomical observations are still carried out in the optical band in which one can reach up to 10^{-6} Jy. Stars, galaxies and quasars contribute dominantly in this band; there are about 10^{10} discrete sources in the sky. We still do not have conclusive evidence suggesting the existence of a smooth background in this band. Since there is a large amount of contamination from zodiacal light, back scattering of radiation from interstellar gas and hot stars in the field of view, these observations are quite difficult.

(e) X-ray and γ-ray

$\lambda = (3 \times 10^{-3} - 100)\,\text{Å}$; $\epsilon \simeq (0.12\text{–}400)\,\text{keV}$; $T = (3 \times 10^4 - 10^9)\,\text{K}$
Since X-rays and γ-rays are strongly absorbed in earth's atmosphere, observations in this waveband need to be carried out from outside the atmosphere. The X-ray satellites UHURU, HEAO-1, EINSTEIN, EXOSAT and ROSAT have given us a significant amount of information about the X-ray universe. One requires equivalent temperatures of about 10^8 K to produce hard X-rays. Besides in the central cores of stars, such high energies can be usualy found only in binary stars and in supernova remnants. The accretion of matter from one star to a compact companion can lead to the production of X-rays; so can the explosion of a supernova. The main extragalactic sources of X-rays are quasars and hot ionized gas in clusters of galaxies.

In addition, there exists a well defined, diffuse X-ray background in the range of 1 keV to 100 MeV. Part of this background could be due to unresolved point-like sources and another part may be due to hot, ($T \simeq 10^9$ K) diffuse, intergalactic plasma. In the range (1–3) keV, the X-ray spectrum can be fitted by a power law:

$$I \simeq 11(E/1\,\text{keV})^{-0.4}\,\text{erg}\,\text{cm}^{-2}\,\text{s}^{-1}\,\text{erg}^{-1}\,\text{str}^{-1} \tag{1.63}$$

and in the range (3–50) keV it can be fitted by an optically thin thermal bremsstrahlung at the temperature (40 ± 5) keV.

No object in the universe is hot enough to produce high energy γ-rays by thermal radiation. The γ-rays are produced by accretion of matter on compact objects and by the collision of high energy particles in the cosmic rays with the nuclei of atoms in our galaxy.

1.10 Determination of extragalactic distances

The properties of galaxies and other cosmologically interesting structures cannot be studied in detail until the distances to these objects are measured reliably. Most of the uncertainties in cosmology arise from systematic errors in the measurement of extragalactic distances. We shall review briefly how such measurements are made[19].

The usual procedure for measuring distances is the following. Some method of measurement is used for objects which are nearby; a second method of measurement is calibrated by using both the methods to measure distances to objects which are located at an intermediate distance. By repeating this procedure and building up a series of methods, an extragalactic distance 'ladder' can be constructed.

The most direct measurement of distance is possible to stars in our galaxy which are nearby. As the earth goes around the sun, the angular position of nearby stars will shift compared to those of more distant stars in the sky. This parallax, combined with the knowledge of the semi major axis of earth's orbit, allows one to measure distances to nearby stars. (In fact, the unit parsec is so defined that a star which has a parallax of one second will be at a distance of one parsec.) This method works for distances up to about 30 parsec or so.

The procedure adopted for objects in the next layer of distances uses a technique based on stellar evolution. To do this successfully we need some objects for which both the methods can be applied. The nearest star cluster (called Hyades) is used for this purpose. The distance to the Hyades cluster can be determined by geometrical methods to be about (46 ± 2) pc. (This star cluster is at a distance which makes the parallax measurements unreliable. The distance is actually determined by a variant known as statistical parallax. This cluster is moving away from us in such a way that there is a measurable decrease $(\Delta\theta)$ in the angular size (θ) with time (t). It is easy to show that the distance is $r = (vt)(\theta/\Delta\theta)$, where the velocity v could be obtained from the Doppler shift.) The known distance to Hyades is used to calibrate a class of stars called 'zero-age-main-sequence' stars. From the theory of stellar evolution, one can obtain a relation between the luminosity and surface temperature for stars that have just begun the conversion of hydrogen into helium. This relation is parametrized by the stellar mass (with more massive stars being brighter and hotter) and also depends on the abundance of elements heavier than helium (usually called the 'metallicity'). These variables can be measured from the study of the spectra of the stars without knowing the distance to the stars while the observed flux has an inverse square dependence on the distance. Therefore, comparing the zero-age-main-sequence stars in the Hyades with those in a more distant cluster will give the ratio of the distances to the two clusters. The measured distance to Hyades can be used to determine the distance of the second cluster. This method is used to cover the distances in the range 50 pc to 10 kpc.

To measure still larger distances, one can make use of a class of variable stars known as Cepheids. These are large and highly luminous stars with regular and periodic modulation of their luminosity and colour. There exists a very tight relation between their period of variation and the intrinsic luminosity (calculated either at maximum luminosity or based on an average over one period) with brighter Cepheids having longer periods. So, by comparing the period of a Cepheid with observable properties which are distance independent, one can infer its intrinsic brightness. Comparison with the observed flux will give the distance.

To do this successfully, one requires many Cepheids located at a common distance. Further, we need to know the absolute distances to a few Cepheids to calibrate the relation. The Large and Small Magellanic clouds, two dwarf satellite galaxies of the Milky Way, are usually used to provide the sample of Cepheids. This is the best procedure currently available but suffers from several intrinsic errors. The distance to the Andromeda galaxy, (710 ± 135) kpc, is determined in this manner; note that the uncertainty is already about 20 per cent. These procedures can be used up to a distance of about (1–3) Mpc.

There are several objects in the universe with well defined characteristic luminosities which are much brighter than the Cepheids. They can be used to measure the *ratios* of distances without much difficulty. However, these *ratios* can be converted to absolute distance only if the distance scale can be calibrated using some of the earlier methods. This has been the main difficulty in using these distance indicators, which are traditionally called 'secondary indicators'.

Planetary nebulae provide such a convenient secondary indicator. These are shells of gas which are formed when a star finishes its nuclear cycle and ejects the outer layers leaving behind a white dwarf. The UV-radiation from the white dwarf photoionizes the shell and causes secondary emission. These characteristic emission lines can be detected in galaxies at distances up to 20 Mpc. Observations show that the luminosities of these lines have a sharp upper bound. Comparing this upper bound in two different galaxies, the distance ratio can be determined. What is more, this procedure can be calibrated using the planetary nebula in the spheroidal bulge of Andromeda. Assuming Andromeda is at 710 kpc, the average distance to several elliptical galaxies in the core of the Virgo cluster has been determined to be (14.7 ± 1.0) Mpc.

The same principle has been used previously with globular clusters and red supergiants in a galaxy. The globular cluster population shows a well defined characteristic luminosity. Similarly, the red supergiant population has a sharp upper bound so that the luminosity of the brightest red star may be taken to be constant in all galaxies. These methods, however, have more systematic errors than the one using planetary nebulae.

The procedures outlined above were all based on properties of stellar systems. Much larger distances can be probed by using galaxies themselves as secondary indicators. We have seen before that the disc components of spiral galaxies are supported against gravity by centrifugal force. From radio and optical measurements, one can determine the quantity $v \sin i$ where v is the rotational speed and i is the angle between the axis of rotation of the galaxy and the line-of-sight. The speed v is related to the luminosity L by (1.27), viz, $L \propto v^4$. This relation allows one to determine L with about 40 per cent uncertainty. The main source of scatter being the determination of $\sin i$, which has to be inferred from the shape of the galaxy. Given L, the distance can be determined using the observed luminosity. An important feature of this method is that the measured distances have a constant *fractional* error; therefore, the absolute error increases linearly with distance. To decrease the fractional error and make sensible predictions at large distances it would be convenient if we have a large number N of galaxies in some region of space. In such a case, the statistical error can be reduced by the usual $\sqrt{N}$ factor.

A corresponding method exists for elliptical galaxies. Ellipticals are characterized by a velocity dispersion σ which seems to be well correlated with the characteristic size D by the relation

$$D \propto \sigma^{1.2}. \tag{1.64}$$

(The size D may be thought of as the effective diameter of the galaxy; it is defined in a rather complicated manner so as to reduce observational uncertainties; see chapter 7.) Once the intrinsic diameter D is found, the observed diameter can be used to estimate the distance. In practice, the scatter in the data leads to about 20 per cent error (see figure 1.4).

The last two methods are potentially very powerful and can be used to very large distances (say, 10^2 Mpc). At present, however, they suffer from statistical bias and lack of good calibrators. In addition to the methods outlined above, there have been attempts to measure distances using physical properties of supernovae, ionized gas in clusters and gravitational lens systems. These methods are, however, yet to reach the same level of acceptability as more traditional methods.

Given the distance to an extragalactic object which is far away and its line-of-sight velocity (which can be obtained from the redshift), one can

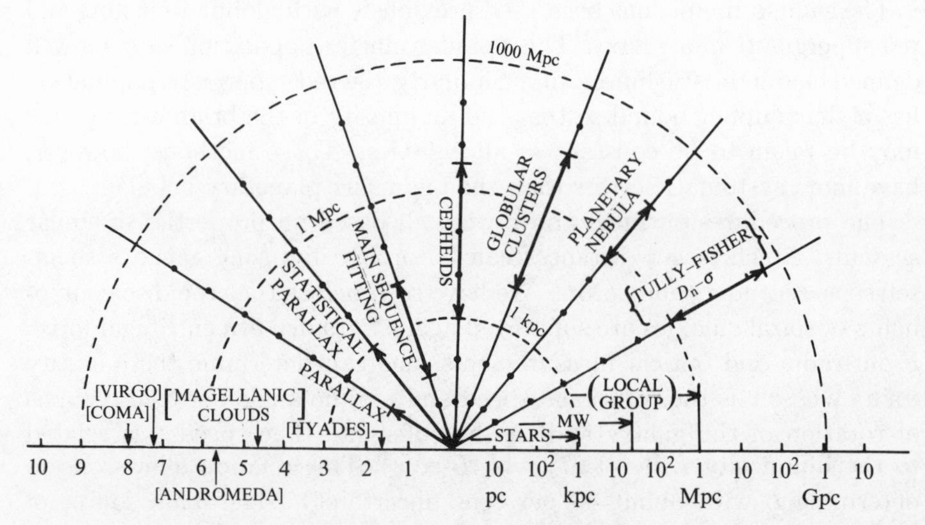

Fig. 1.4. The various distance measurement techniques are summarized. The radial lines are marked by thick, double-arrowed, segments indicating the range over which each of the method is used. The horizontal line on the right indicates the actual distance scale and the typical sizes of various structures discussed in the text. The locations of several astronomical objects are shown in the horizontal axis on the left.

attempt to determine the Hubble constant $H \simeq (v/d)$. Systematic (and statistical) errors in the measurement of d translate into inaccuracies in the determination of H leading to an uncertainty of factor two (also see exercises 1.12, 1.13).

1.11 Age of various structures

If redshifts of galaxies are interpreted as due to the expansion of the universe, then we obtain – in general – an evolving model of the universe. The age of the various structures in the universe[20] (earth, stars, galaxies ...) can then be used to impose powerful constraints on the models describing the universe.

Our galaxy contains several globular clusters with homogeneous chemical properties. They provide a simple way of determining the age of our galaxy. A typical star spends most of its lifetime in the main sequence with the time varying inversely as some power of its mass; e.g., a star of one solar mass spends about 10^{10} yr while a star of 9 solar masses spends only 2×10^7 yr. Thus, if we can determine the mass of the most massive star which is still remaining on the main sequence in a globular cluster, we can determine the age of the cluster. The mass, in turn, can be estimated from the luminosity and the colour; the main uncertainty arises from determining which stars are *just* leaving the main sequence.

To use this method effectively one needs good theoretical models for the main sequence evolution of stars of different masses. These models depend on a number of factors like the mass of the star, the relative abundance of helium to hydrogen, the abundance ratio of certain heavy elements like oxygen and iron and the details of approximations used in treating the energy transport in the star due to convection. Uncertainties in the model building translate into uncertainties in the final age. These calculations have led to ages varying as much as from 11.5 Gyr to 16 Gyr. It is probably safe to quote (14–17) Gyr as the best estimate for the oldest clusters.

Globular clusters are found prominently in the halo of our galaxy and hence, the age estimated above should be thought of as the age of the halo. The disc of our galaxy seems to be relatively younger and its age can be determined by an independent method based on the cooling time of white dwarfs. As we saw before, a white dwarf is the end stage of stellar evolution for stars with masses of the order of a solar mass. In the initial stages of its formation, a white dwarf is quite hot and reasonably luminous. It cools steadily becoming fainter; the evolution of its temperature and luminosity can be predicted from the theory. Since the cooling time increases rapidly with declining luminosity, we would expect the abundance of white dwarfs to increase rapidly as the luminosity decreases. This is indeed observed; however, there is a sudden drop in the abundance at a luminosity of about $3 \times 10^{-5}\,\mathrm{L}_\odot$. This can be interpreted as due to the fact that no white dwarf in our galaxy is old enough to have cooled to lower values of luminosity. By matching the cut-off value in the luminosity with theoretical predictions, one can estimate the time since the formation of the oldest white dwarf. This procedure gives a value of (9.3 ± 1.8) Gyr which is considerably lower than the age of the globular clusters.

Another method used for determining the age is the radioactive decay of elements with very long lifetime. This procedure works quite well in determining the age of the solar system. Consider, for example, the decay of ^{87}Rb to ^{87}Sr which has a half life of $\tau = 4.99 \times 10^{10}$ yr. Let $x(t)$ and $y(t)$ be the number of these nuclei present in some meteorite in the solar system at time t with the quantity $[x(t) + y(t)]$ being a constant. Let t_0 be the epoch at which the solar system is formed. It is easy to show that a plot of $x(t)$ against $y(t)$ will be a straight line with a slope equal to $[\exp(\lambda t_0) - 1]$ where $\lambda = \tau^{-1} \ln 2$. The abundance ratios of these elements can be measured from various meteorites; such estimates give the age of the solar system to be about 4.5×10^9 yr. Similar methods give the age of lunar rocks in the range of (4.5–4.6) Gyr.

To proceed from the age of the solar system to the age of the galaxy, one has to estimate the time which has elapsed between the formation of the galaxy and the formation of the solar system. The formation of heavy elements is supposed to have occurred in the stars. If the timescale is set to zero at the birth of the galaxies, then processes involving the evolution of stars and supernova explosions will lead to the production of the heavy elements during some interval $0 < t < T$. The time $t = T$ is the epoch at which nucleosynthesis of heavy elements declines significantly. There could be a small gap between the time $t = T$ and the time $t = T + \Delta$ at which the solar system has been formed. It is possible to estimate both T and Δ from the abundance of very long lived radioactive isotopes and from the theoretical estimates of the initial ratios of these isotopes. The procedure is again plagued by several systematic uncertainties and gives ages in the range of (6–18) Gyr.

Exercises

1.1 Classically, two positively charged nuclei of charges q_1 and q_2 cannot approach each other closer than the distance $r = (2q_1q_2/mv^2)$ where v is the relative velocity at infinite separation and m is the reduced mass. (a) Quantum mechanics allows the particles to tunnel through the Coulomb barrier to reach the zone of strong interaction. Show that the tunnelling probability depends on v through the factor

$$Q \cong \exp(-2\pi q_1 q_2/\hbar v).$$

(b) If the nuclei are in thermal equilibrium at temperature T, then the probability for the nuclei to have a large velocity v (with $mv^2 \gg T$) is determined by the factor

$$F \simeq \exp(-mv^2/2T).$$

Show that FQ is maximized for $v \simeq (2\pi q_1 q_2 T/\hbar m)^{1/3}$ and that the maximum probability is

$$P_{\text{max}} \propto \exp\left[-(T_c/T)^{1/3}\right]; \quad T_c = \left(\frac{3}{2}\right)^3 \left(\frac{2\pi q_1 q_2}{\hbar}\right)^2 m.$$

Sketch P_{max} as a function of T for the p–p reaction in the temperature range 10^5 K to 10^9 K. Based on this result what can one say about the critical temperature at which nuclear reactions can be triggered?

1.2 A star will become unstable if the radiation energy content (E_{rad}) of the star dominates over the thermal energy of matter (E_{thermal}). Show that $(E_{\text{rad}}/E_{\text{thermal}})$ is a function of M alone, if $T \propto (M/R)$. Compute the critical mass at which instability occurs.

1.3 The purpose of this exercise is to study some of the radiative processes which are important in different contexts.
(a) An electron will radiate at the rate $P = (2/3)(e^2a^2/c^3) = (2/3)(e^2a^2)$ if its acceleration is a. Consider an electromagnetic wave, with flux $F = (E^2/4\pi)$ incident on an electron. Show that the ratio between

emitted power and incident flux is

$$\sigma_T = \left(\frac{8\pi}{3}\right)\left(\frac{e^2}{mc^2}\right)^2 \simeq 6.7\times 10^{-25}\ \mathrm{cm}^2.$$

This is the Thomson scattering cross section.

(b) Atomic transitions can be understood 'classically' by treating an electron as a harmonic oscillator with natural frequency ω_0 and a damping due to radiation reaction. In this model, we assume that the electron obeys the equation:

$$\frac{d^2x}{dt^2} = -\omega_0^2 x + \frac{2}{3}\frac{e^2}{mc^3}\frac{d^3x}{dt^3}.$$

Show that this will lead to a 'natural' width γ for a spectral line with $\gamma = (2/3)(e^2\omega_0^2/mc^3)$ which is usually much smaller than ω_0. Also show that the line profile is Lorentzian and decreases as $|\omega-\omega_0|^{-2}$ for large $|\omega-\omega_0|$. The motion of a system which is emitting radiation will also lead to a spread in the spectral line width because of the Doppler effect. What will be the line profile in this case?

(c) Consider an electron moving past a proton at impact parameter b; let the relative velocity at large separation be v. Show that the total radiation emitted by the electron is

$$\delta E \simeq \frac{2}{3}\frac{e^2}{c^3}\left(\frac{e^2}{mb^2}\right)^2\left(\frac{2b}{v}\right).$$

Argue that there will be very little power at frequencies above $\omega_{max} = (v/2b)$. Assuming that the power emitted is constant for $\omega < \omega_{max}$, show that $(dE/d\omega) = (8e^6/3\pi c^3 m^2 v^2 b^2)$.

(d) For a plasma with equal number density n for protons and electrons, show that

$$\frac{dE}{d\omega dt} \simeq n^2 v \int_{b_1}^{b_2}\left(\frac{dE}{d\omega}\right) 2\pi b db \simeq \frac{16e^6}{3c^3m^2v} n^2 \ln\left(\frac{b_2}{b_1}\right).$$

What should be the values for b_2 and b_1? If the plasma is in equilibrium at temperature T, show that: (i) the specific emissivity ($\mathrm{erg\,cm^{-3}\,s^{-1}\,Hz^{-1}}$) is $J_\nu \propto n^2T^{-1/2}$, (ii) the volume emissivity ($\mathrm{erg\,cm^{-3}\,s^{-1}}$) is $\epsilon \propto n^2T^{1/2}$ and (iii) the cooling time is $t_{cool} \propto T^{1/2}n^{-1}$. (iv) In this analysis we have not included the radiation emitted during the electron–electron scatterings; is it justified?

(e) The absorption coefficient κ_ν for a radiative process is defined to be $\kappa_\nu = n\sigma_\nu$ where n is the density of absorbers and σ_ν is the cross section for absorption. If the plasma is in equilibrium with radiation, the bremstrahlung absorption and emission must match. Show that this condition requires

$$\kappa_B(\nu) \propto n^2\nu^{-2}T^{-3/2}; \quad \sigma_B(\nu) \propto n\nu^{-2}T^{-3/2}$$

at low frequencies ($h\nu \ll kT$).

(f) 'Recombination radiation' is emitted when a free electron gets bound to a nucleus. Argue that $\sigma_{\rm rec} \simeq \sigma_B$ and that the recombination rate $\alpha_{\rm rec} \propto v^{-1} \propto T^{-1/2}$. Combining with the results in previous parts, show that the cooling rate ($\mathrm{erg\,cm^{-3}\,s^{-1}}$) of a plasma can be written in the form $\epsilon \simeq \Lambda(T)n^2$ where $\Lambda(T) = aT^{-1/2} + bT^{1/2}$ with a, b being numerical constants.

1.4 (Contributed by D. Narasimha) The collapse of a gaseous cloud ('protostar') is halted and the main sequence life of a star begins when the nuclear energy generated by the star is balanced by the energy radiated away. Since the energy generated is a sensitive function of the temperature, we can assume that the central temperature of the 'zero age main sequence' (ZAMS) stars is constant at about 10^7 K. A reasonable approximation is $T = 1.5 \times 10^7\,\mathrm{K}\,(1 + \log M/\mathrm{M}_\odot)$.
(a) Show that the central density of the star decreases as M^{-2} except for a logarithmic dependence.
(b) Consider a massive star in the mass range of (5–40) $\mathrm{M}_\odot$. Assume that during the collapse of the protostar, the energy generation due to gravitational contraction is $[dL(r)/dM(r)] = \epsilon$, a constant independent of radius r at a given time t. The opacity is contributed, essentially, by the electron scattering. Show that the luminosity of this protostar during the pre-main sequence contraction remains constant.
(c) Assume that ZAMS is the stage in the evolution of the protostar at which the temperature reaches the value of 1.5×10^7 K and the luminosity reaches the value obtained in (b). For the mass range (5–40) $\mathrm{M}_\odot$ sketch the theoretical ZAMS curve in the H–R diagram.

1.5 The purpose of this exercise is to derive the behaviour of systems supported by the degeneracy pressure.
(a) As the material of the star is compressed, the individual atoms lose their identity and – if the temperature is not too high – the electrons become a degenerate Fermi gas. Show that such a gas becomes more and more ideal as the density increases and satisfies the condition:

$$\rho \gg (m_e e^2/\hbar)^3 Z^2 \left(\frac{A}{Z}\right) m_p \simeq 20Z^2\,\mathrm{g\,cm^{-3}}.$$

Estimate the degeneracy temperature of this gas and thus obtain the criterion for degeneracy.
(b) Show that the equation of state of an ideal, degenerate Fermi gas is

$$P_{NR} = \frac{(3\pi^2)^{2/3}}{5}\frac{\hbar^2}{m_e}\left(\frac{Z}{A}\right)^{5/3}\left(\frac{\rho}{m_p}\right)^{5/3} \equiv \lambda_{NR}\rho^{5/3}$$

in the non-relativistic limit, and

$$P_R = \frac{1}{4}(3\pi^2)^{1/3}\hbar c\left(\frac{Z}{A}\right)^{5/3}\left(\frac{\rho}{m_p}\right)^{4/3} \equiv \lambda_R\rho^{4/3}$$

in the relativistic limit. Further show that, for a spherically symmetric system in which pressure balances gravity, the density ρ satisfies the equation

$$\frac{1}{r^2}\frac{d}{dr}\left(\frac{r^2}{\rho^{1/3}}\frac{d\rho}{dr}\right) = -\frac{12\pi G}{5\lambda_{NR}}\rho$$

in the non-relativistic case, and

$$\frac{1}{r^2}\frac{d}{dr}\left(\frac{r^2}{\rho^{2/3}}\frac{d\rho}{dr}\right) = -\frac{3\pi G}{\lambda_R}\rho$$

in the relativistic case.

(c) Consider the non-relativistic case first. Argue from dimensional considerations that the solution must have the form $\rho(r) \propto R^{-6} f(r/R)$ where R is the radius of the star. Hence show that for such systems $R \propto M^{-1/3}$ and $\overline{\rho} \propto M^2$. With a suitable transformation of variables, reduce the equation and the boundary conditions to the form

$$\frac{1}{x^2}\frac{d}{dx}\left(x^2\frac{dy}{dx}\right) = -y^{3/2}; \quad y'(0) = 0, \quad y(1) = 0.$$

The numerical integration shows that $y(0) \cong 178$ and $y'(1) \cong -132$. From these values show that

$$MR^3 \cong (92\hbar^6/G^3 m_e^3 m_p^5)\left(\frac{Z}{A}\right)^5.$$

(d) In the relativistic case, argue that the solution must have the form $\rho(r) \propto R^{-3}\ F(r/R)$. Hence show that $M = M_0$, some fixed constant. Convert the equation into the form

$$\frac{1}{x^2}\frac{d}{dx}\left(x^2\frac{dy}{dx}\right) = -y^3; \quad y'(0) = 0, \quad y(1) = 0.$$

Numerical integration gives $y(0) \cong 6.9$ and $y'(1) \simeq -2.0$. Using these values show that

$$M_0 \cong \frac{3.1}{m_p^2}\left(\frac{Z}{A}\right)^2\left(\frac{\hbar c}{G}\right)^{3/2} \simeq 5.8\left(\frac{Z}{A}\right)^2 M_\odot$$

if the gas is made of neutrons. (For $A \simeq 2Z$, $M_0 \simeq 1.45\,\mathrm{M}_\odot$.) Using the results of (c) and the above conclusion, describe the evolution of a contracting star in its late stages.

(e) Argue why it is appropriate (and necessary) to consider a degenerate *neutron* gas in studying the relativistic case.

1.6 One intuitively associates highly flattened systems with systems having large angular momentum. Consider a disc galaxy in which all stars rotate in the same direction and are supported against gravity by centrifugal force. Is it possible to construct now another *disc* galaxy with the same spatial distribution of stars but with negligible angular momentum?

1.7 (a) Consider a gravitational encounter between two stars, each of mass m, with a relative velocity of approach v. Show that the transverse velocity induced by the encounter is $\delta v_1 \simeq (2Gm/bv)$. The surface density of stars in a galaxy with characteristic size R and total number of stars N will be $(N/\pi R^2)$. Assume that the encounters are random. Argue that the net contribution to $(\delta v_\perp)^2$ from encounters with impact parameter in the range of $(b, b+db)$ is

$$(\delta v_\perp)^2 \simeq \left(\frac{2Gm}{bv}\right)^2 \frac{2N}{R^2} b\, db.$$

Integrating this expression between $b_{\rm min}$ and $b_{\rm max}$, obtain

$$(\Delta v)^2 \simeq 8N \left(\frac{Gm}{Rv}\right)^2 \ln\left(\frac{b_{\rm max}}{b_{\rm min}}\right).$$

Argue that $b_{\rm max} \simeq R$ and $b_{\rm min} \simeq (Gm/v^2)$. If the system is in virial equilibrium, $v^2 \simeq (GNm/R)$. Hence show that the relaxation time for the system is $t_{\rm relax} \simeq n_{\rm relax}(R/v)$ where

$$n_{\rm relax} \simeq \frac{N}{8\ln\Lambda}; \quad \Lambda = \left(\frac{b_{\rm max}}{b_{\rm min}}\right) \simeq N.$$

Also show that the relaxation time $t_{\rm relax}$ is much larger than the 'age of the universe' $t_0 \approx 10^{10}$ yr, for a typical galaxy.
(b) Consider a finite D-dimensional system of N 'stars' interacting by an r^{-2} force. Show that $t_{\rm relax} \simeq N^{D-2}\, t_{\rm cross}$ for $D \leq 3$ while $t_{\rm relax} \simeq N t_{\rm cross}$ for $D \geq 3$. Why does this happen?

1.8 (a) Show that the differential equation describing the isothermal sphere can be expressed as a first order equation

$$\frac{u\, dv}{v\, du} = -\frac{u-1}{u+v-3}$$

where $u = (nx^3/m)$, $v = (m/x)$ with $n = (\rho/\rho_c)$, $m = (M(r)/4\pi\rho_c L_0^3)$, $x = (r/L_0)$ and $L_0 = 3r_0$. Can a similar transformation be effected for a general polytrope?
(b) Show that the boundary conditions in this problem are equivalent to the condition $v = 0$ at $u = 3$ with $(dv/du) = -5/3$ at $u = 3, v = 0$. Sketch the solution in the u–v plane.
(c) Show that the solution asymptotically approaches the form $\rho(r) \propto r^{-2}$, $M(r) \propto r$ in an oscillatory manner.
(d) Consider an isothermal sphere which is truncated at $r = R_{\rm max}$ by, say, confining the material by a rigid sphere of this radius. Let (u_0, v_0) be the values of u and v at the surface $r = R_{\rm max}$. Show that the isothermal solution exists only if the parameter $\lambda = (R_{\rm max}E/GM^2)$, where E and M are the energy and mass of the system, is bounded from below at some λ_c; that is, $\lambda \geq \lambda_c$. Estimate λ_c from the sketch of the solution in the u–v plane and show that $\lambda_c \simeq -0.25$.

1.9 This exercise supplies the missing details in the study of galaxy models.
(a) Verify the various results quoted in the text, as regards the polytropes, isothermal sphere, King's model, Mestel's disc and Kalnajs disc.
(b) Show that galaxy models with distribution function of the form $f(E)$ have isotropic velocity dispersion; that is $< v_r^2 > = < v_\theta^2 > = < v_\phi^2 >$. Also show that, if $f = f(E, J^2)$, then $< v_\theta^2 > = < v_\phi^2 > \neq < v_r^2 >$.

(c) Show that, for the isothermal sphere $< v^2 >= 3\sigma^2$.
(d) Show that, for Mestel's disc $< v_r^2 >= \sigma^2$ and $< v_\phi >= \sqrt{2}\Gamma(n/2+1)$ $[\Gamma(n/2 + 1/2)]^{-1}\sigma$. Hence show that, as n increases to large values, $[< v_\phi > /\sigma]$ increases as n.
(e) For the Kalnajs disc, show that $< v_\phi > /r = \Omega$ is independent of r; thus the disc rotates like a rigid body. Also show that, for $\Omega \ll \Omega_0$ the system is 'hot' with the support against gravitational force arising from random velocities while for $\Omega \approx \Omega_0$, the system is 'cold' and centrifugally supported.
(f) Consider a distribution function $f = f(E, L_z)$ which leads to a density $\rho(r, z)$. Let $f_\pm(E, L_z) = (1/2)[f(E, L_z) \pm f(E, -L_z)]$. Show that ρ is determined by f_+ and $< v_\phi >$ is determined by f_-.

1.10 It is usually said that if the intensity of radiation emitted by a source varies in a timescale Δt then the source must have a size L bounded by $c\Delta t$; i.e. $L \lesssim c\Delta t$. When does this result hold? Show that, when relativistic effects are important, it is possible to violate this bound.

1.11 This exercise explores several aspects of synchrotron radiation. Consider a particle with mass m and charge q moving in a magnetic field of strength B. Let $\gamma = (1 - v^2)^{-1/2}$ where $\mathbf{v}$ is the velocity of the particle.
(a) The power radiated by a charged particle can be expressed as $P = (2/3)q^2(a_k a^k)$ where $a^k = (d^2x^k/d\tau^2)$ is the four-acceleration of the particle. Show that the radiation emitted in the case of motion in magnetic field is

$$P = \frac{2}{3}\left(\frac{q^2}{m}\right)^2 v^2\gamma^2 \sin^2\alpha$$

where α is the angle between the magnetic field and the velocity of the particle.
(b) Argue that, in the frame of the distant observer, most of the radiation will be confined to a cone at vertical angle $\Delta\theta \simeq 2\gamma^{-1}$ about the direction of motion. (Assume $\gamma \gg 1$.) Hence show that the observer will see a pulse of radiation for a duration $\Delta t \simeq (\gamma^3\omega_B \sin\alpha)^{-1}$ in each cycle, where $\omega_B = (qB/\gamma m)$. From this, show that the electric field at the observer's location must have the form $E(t) \propto f(\omega_c t)$ with $\omega_c \equiv (3/2)\omega_B\gamma^3 \sin\alpha$.
(c) Show that the spectrum due to a given, ultra relativistic, charged particle will be of the form

$$P(\omega) \simeq \frac{\sqrt{3}}{2\pi}\left(\frac{q^3 B \sin\alpha}{m}\right) F(\omega/\omega_c),$$

where $F(x)$ is some function. Suppose there exists a distribution of electrons with energy spectrum $N(E)\,dE = CE^{-p}\,dE$ in the energy range $E_1 < E < E_2$. Show that the spectrum of radiation emitted by the whole system can be approximated as

$$P_{\text{total}} \simeq (\text{constant})\omega^{-(p-1)/2},$$

provided the range (E_1, E_2) is sufficiently broad. Hence a power law distribution of electrons leads to a power-law dependence of radiation.

1.12 Careful monitoring of a supernova will allow us to determine its flux, spectrum and the expansion velocity of the shell as functions of time. Show that these data can serve as a distance indicator. What will be the limitations of this method?

1.13 (Contributed by D. Narasimha). Let $\overline{p}(m)\, dm$ be the probability that a randomly selected member of a group (like a star in a star cluster or galaxy in a group of galaxies) has the bolometric magnitude in the interval $(m, m + dm)$; let $p(m) = N\overline{p}(m)$ where N is the total number of members in that group, with $N \gg 1$. Let $Q_r(m)$ be the probability that the rth brightest member of the group has a magnitude between m and $(m + dm)$.

(a) Derive an expression for $Q_1(m)$ in terms of $p(m)$. Derive an expression for $Q_r(m)$ in terms of $p(m)$ and $Q_{r-1}(m)$. Also show that

$$\frac{d}{dm}\left(\frac{Q_r(m)}{p(m)}\right) = Q_r(m) - Q_{r-1}(m).$$

(b) A 'standard candle' used for distance estimates should have the following properties: (i) It should be clearly identifiable in an observed image. (ii) It should have a well-defined (identifiable) characteristic with small variance in its absolute bolometric magnitude. Give arguments as to why the absolute magnitude of third or fourth brightest member will have lower variance than the brightest member if $p(< m_3 >)$ or $p(< m_4 >)$ is far greater than unity. Here

$$< m_i >= \left[\int_{-\infty}^{+\infty} mQ_i(m)\, dm\right]\left[\int_{-\infty}^{+\infty} Q_i(m)\, dm\right]^{-1}.$$

(c) Based on the known properties of star clusters, galaxies etc. decide whether one should choose the brightest, 2nd brightest, ... 10th brightest, ... 100th brightest member as a standard candle.

1.14 Most of our information about galactic and extragalactic sources is based on the study of electromagnetic radiation emitted by them. To obtain a complete picture, one has to observe the region in as many bands as possible: Radio, IR, visible, UV, X-rays, γ-rays etc. Listed below are the physical conditions which prevail in some regions of our galaxy. Describe how these objects will look in different bands:

(a) A cloud containing a fair amount of dust and complex molecules is collapsing, forming stars inside it. The newly formed stars, in turn, heat up the cloud.

(b) A star ejects its outer layers as a planetary nebula. The inner core becomes a compact white dwarf with a temperature of 10^5 K.

(c) A supernova explosion occurs leaving behind a rapidly rotating neutron star. The neutron star has a magnetic dipole axis which is misaligned with the rotation axis.

2
The Friedmann model

2.1 Introduction

The discussion in the last chapter highlights three important features of our universe: (1) It contains structures like galaxies, clusters, superclusters etc. which span a wide range of length scales. (2) These structures seem to be distributed in a statistically homogeneous manner. (3) Gravity is the dominant force which governs the large scale dynamics of the universe. To understand the nature and evolution of these large scale structures, it is usual to proceed in the following manner: We first study the overall dynamics of the universe by treating the matter distribution in it to be completely homogeneous. The inhomogeneities which we observe are then treated as deviations from this smooth universe. In chapters 2 and 3, we will study the physical properties of the smooth universe.

Since the large scale gravitational field in the universe has to be described by Einstein's theory of gravity, the model for the smooth universe should be based on general relativity. This chapter introduces the properties of such a model, called the Friedmann universe. Section 2.2 derives the form of the Friedmann metric from symmetry considerations. The kinematical properties of this metric are developed in section 2.3 and the dynamical evolution is studied in sections 2.4 and 2.5. The last section introduces the concept of Hubble radius which will be of importance in the discussions in the later chapters. The signature of the metric is $(+ - - -)$ and we will use the units in which $c = 1, \hbar = 1$ and $k_B = 1$.

Some elementary aspects of general relativity are needed in this chapter; the necessary concepts are described in appendix A. The Latin indices take the values 1, 2 and 3 while the Greek indices range over 0, 1, 2 and 3.

2.2 The Friedmann model

To construct the simplest model of the universe, we will begin by assuming that the geometrical properties of the 3-dimensional space are the same at all spatial locations and that these geometrical properties do not single out any special direction in space. Such a 3-dimensional space is called 'homogeneous' and 'isotropic'.

The geometrical properties of the space are determined by the distribution of matter through Einstein's equations. It follows, therefore, that the matter distribution should also be homogeneous and isotropic. This is certainly not true in the observed universe, in which there exists a significant degree of inhomogeneity in the form of galaxies, clusters, etc. We assume that these inhomogeneities can be ignored and the matter distribution may be described by a smoothed out average density in studying the large scale dynamics of the universe. There may be some justification in treating the matter distribution to be homogeneous at scales larger than about 100 Mpc.

The assumption of homogeneity and isotropy of the 3-space singles out a preferred class of observers, viz., those observers for whom the universe appears homogeneous and isotropic. Another observer, who is moving with a uniform velocity with respect to this fundamental class of observers, will find the universe to be anisotropic. The description of physics will be simplest if we use the coordinate system appropriate to this fundamental class of observers.

Let us determine the form of the spacetime metric in such a coordinate system (t, x^i) in which homogeneity and isotropy are self-evident. The general spacetime interval (ds^2) can be separated out as

$$\begin{aligned} ds^2 = g_{\alpha\beta}\, dx^\alpha\, dx^\beta &= g_{00}\, dt^2 + 2g_{0i}\, dt\, dx^i + g_{ij}\, dx^i\, dx^j \\ &\equiv g_{00}\, dt^2 + 2g_{0i}\, dt\, dx^i - \sigma_{ij}\, dx^i\, dx^j, \end{aligned} \tag{2.1}$$

where σ_{ij} is a positive definite spatial metric (It is assumed that any index which is repeated in a given term is summed over its range of values; see appendix A.) Isotropy of space implies that g_{0i} terms must vanish; otherwise, they identify a particular direction in space related to the 3-vector v_i with components g_{0i}. Further, in the coordinate system determined by fundamental observers, we may use the proper time of clocks carried by these observers to label the spacelike surfaces. This choice for the time coordinate t implies that $g_{00} = 1$, bringing the spacetime interval to the form:

$$ds^2 = dt^2 - \sigma_{ij}\, dx^i\, dx^j \equiv dt^2 - dl^2. \tag{2.2}$$

The problem now reduces to determining the 3-metric σ_{ij} of a 3-space which, at any instant of time, is homogeneous and isotropic.

The curvature of the 3-space is determined by the 3-dimensional Riemann tensor P_{ijkl}. For a space which is homogeneous and isotropic, P_{ijkl} cannot depend on the derivatives of the metric σ_{ij} and must be constructed out of the components σ_{ij} alone. The symmetry properties of P_{ijkl} uniquely determine this form to be

$$P_{ijkl} = q(\sigma_{ik}\sigma_{jl} - \sigma_{il}\sigma_{kj}) \tag{2.3}$$

where q is related to the scalar 3-curvature $P = \sigma^{ik}\sigma^{il}P_{ijkl} = 6q$. Since homogeneity implies the constancy of P, it follows that q is a constant.

The most general 3-space for which the property (2.3) holds is described by the line element

$$dl^2 = a^2(t)\left[\frac{dr^2}{1-kr^2} + r^2(d\theta^2 + \sin^2\theta\, d\phi^2)\right] \tag{2.4}$$

where $k = 0, +1$ or -1 and a is a constant (i.e. independent of spatial coordinates) related to the curvature by $P = (6k/a^2)$. Notice that the assumption of isotropy – which implies spherical symmetry – can be used to write the line interval in the form

$$dl^2 = a^2\left[\lambda^2(r)\, dr^2 + r^2(d\theta^2 + \sin^2\theta\, d\phi^2)\right]. \tag{2.5}$$

Computing the scalar curvature P for this space and equating it to a constant determines the form of λ to be $\lambda(r) = (1-kr^2)^{-1/2}$ (see exercise 2.1).

The prefactor a determines the overall scale of the spatial metric and, in general, can be a function of time: $a = a(t)$. Combining (2.4) and (2.2), we arrive at the full spacetime metric:

$$ds^2 = dt^2 - a^2(t)\left[\frac{dr^2}{1-kr^2} + r^2(d\theta^2 + \sin^2\theta\, d\phi^2)\right]. \tag{2.6}$$

This metric, called the Friedmann metric, describes [1] a universe which is spatially homogeneous and isotropic at each instant of time. The value of k and the form of the function $a(t)$ (called the 'expansion factor') can be determined using Einstein's equations once the matter distribution is specified.

The coordinates in (2.6) are, of course, chosen in such a way as to make the symmetries of the spacetime self-evident. This coordinate system is called the 'comoving' coordinate system. It is easy to show that world lines with x^i = constant are geodesics. To see this, consider a free material particle which is at rest at the origin of the comoving frame at some instant. No velocity can be induced on this particle by the gravitational field since no direction can be considered as special. Therefore the particle will continue to remain at the origin. Since spatial homogeneity allows us to choose any location as origin, it follows that the world lines x^i = constant are geodesics. Observers following these world lines are called fundamental (or 'comoving') observers.

The spatial hypersurfaces of the Friedmann universe have positive, zero and negative curvatures for $k = +1, 0$ and -1 respectively; the magnitude of the curvature is $(6/a^2)$ when k is non-zero. To study the geometrical properties of these spaces it is convenient to introduce a coordinate χ, defined as

$$\chi = \int \frac{dr}{\sqrt{1-kr^2}} = \begin{cases} \sin^{-1} r & (\text{for } k=1) \\ r & (\text{for } k=0) \\ \sinh^{-1} r & (\text{for } k=-1) \end{cases} . \tag{2.7}$$

In terms of (χ, θ, ϕ) the metric becomes

$$dl^2 = a^2 \left[d\chi^2 + f^2(\chi)(d\theta^2 + \sin^2\theta \, d\phi^2) \right] \tag{2.8}$$

where

$$f(\chi) = \begin{cases} \sin\chi & (\text{for } k=+1) \\ \chi & (\text{for } k=0) \\ \sinh\chi & (\text{for } k=-1) \end{cases} . \tag{2.9}$$

For $k=0$, the space is the familiar, flat, Euclidian 3-space; the homogeneity and isotropy of this space is obvious. For $k=1$, (2.8) represents a 3-sphere of radius a embedded in an abstract flat 4-dimensional Euclidian space. Such a 3-sphere is defined by the relation

$$x_1^2 + x_2^2 + x_3^2 + x_4^2 = a^2, \tag{2.10}$$

where (x_1, x_2, x_3, x_4) are the Cartesian coordinates of some abstract 4-dimensional space. We can introduce angular coordinates (χ, θ, ϕ) on the 3-sphere by the relations

$$\begin{aligned} x_1 &= a \cos\chi \sin\theta \sin\phi; \quad x_2 = a\cos\chi \sin\theta \cos\phi; \\ x_3 &= a \cos\chi \cos\theta; \quad x_4 = a \sin\chi. \end{aligned} \tag{2.11}$$

The metric on the 3-sphere can be determined by expressing dx_i in terms of $d\chi, d\theta$ and $d\phi$ and substituting in the line element

$$dL^2 = dx_1^2 + dx_2^2 + dx_3^2 + dx_4^2. \tag{2.12}$$

This leads to the metric

$$dL^2_{(\text{3-sphere})} = a^2[d\chi^2 + \sin^2\chi(d\theta^2 + \sin^2\theta \, d\phi^2)], \tag{2.13}$$

which is the same as (2.8) for $k=1$.

The entire 3-space of the $k=1$ model is covered by the range of angles $[0 \le \chi \le \pi; 0 \le \theta \le \pi; 0 \le \phi < 2\pi]$ and has a finite volume:

$$V = \int_0^{2\pi} d\phi \int_0^{\pi} d\theta \int_0^{\pi} d\chi \sqrt{g} = a^3 \int_0^{2\pi} d\phi \int_0^{\pi} \sin\theta \, d\theta \int_0^{\pi} \sin^2\chi \, d\chi = 2\pi^2 a^3. \tag{2.14}$$

The surface area of a 2-sphere, defined by χ=constant, is $S = 4\pi a^2 \sin^2\chi$. As χ increases, S increases at first, reaches a maximum value of $4\pi a^2$ at

$\chi = \pi/2$ and *decreases* thereafter. These are the properties of a 3-space which is closed but has no boundaries.

In the case of $k = -1$, (2.8) represents the geometry of a hyperboloid embedded in an *abstract* 4-dimensional Lorentzian space. (This space should not be confused with the physical spacetime.) Such a space is described by the line element

$$dL^2 = dx_1^2 + dx_2^2 + dx_3^2 - dx_4^2. \tag{2.15}$$

A 3-dimensional hyperboloid, embedded in this space, is defined by the relation

$$x_4^2 - x_1^2 - x_2^2 - x_3^2 = a^2. \tag{2.16}$$

This 3-space can be parametrized by the coordinates (χ, θ, ϕ) with

$$\begin{aligned} x_1 &= a \sinh\chi \sin\theta \sin\phi; \quad x_2 = a \sinh\chi \sin\theta \cos\phi; \\ x_3 &= a \sinh\chi \cos\theta; \quad x_4 = a \cosh\chi. \end{aligned} \tag{2.17}$$

Expressing dx_i in terms of $d\chi$, $d\theta$ and $d\phi$ and substituting into (2.15), the metric on the hyperboloid can be found to be

$$dL^2_{(\text{hyperboloid})} = a^2 \left[d\chi^2 + \sinh^2\chi (d\theta^2 + \sin^2\theta \, d\phi^2) \right], \tag{2.18}$$

which is the same as (2.8) for $k = -1$.

To cover this 3-space, we need the range of coordinates to be $[0 \le \chi \le \infty;\ 0 \le \theta \le \pi;\ 0 \le \phi < 2\pi]$. This space has infinite volume, just like the ordinary flat 3-space. The surface area of a 2-sphere, defined by $\chi =$ constant, is $S = 4\pi a^2 \sinh^2\chi$. This expression increases monotonically with χ.

The notion of homogeneity (and isotropy) of these spaces can be expressed in the following manner: On the surface of a 3-sphere, any point can be mapped on to any other point by a suitable rotation. Similarly, any point on the hyperboloid can be mapped on to any other point by a Lorentz-like transformation, which is a rotation by an imaginary angle. These mappings clearly leave the metric – and geometrical properties – of the 3-space invariant. This shows that all points on these surfaces (and all directions) are physically equivalent.

The full Friedmann metric in (2.6) can be expressed in either (t, r, θ, ϕ) coordinates or in (t, χ, θ, ϕ) coordinates. Sometimes it is convenient to use a different time coordinate η related to t by $d\eta = a^{-1}(t)\, dt$. In the $(\eta, \chi, \theta, \phi)$ coordinates the Friedmann metric becomes

$$ds^2 = a^2(\eta) \left[d\eta^2 - d\chi^2 - f^2(\chi)(d\theta^2 + \sin^2\theta \, d\phi^2) \right]. \tag{2.19}$$

In this form, all the time dependence is isolated into an overall multiplicative factor. (Some other coordinate systems are explored in exercises 2.3, 2.4.)

Friedmann universes with $k = -1, 0$ and 1 are called 'open', 'flat' and 'closed' respectively. These terms refer to the topological nature of the 3-space. The following point, however, should be noted: Our symmetry considerations (and Einstein's equations which will be discussed later) can only determine the local geometry of the spacetime and not the global topology. Consider, for example, the $k = 0$ model which has infinite volume and spatial topology of $\mathcal{R}^3$, if we allow the coordinates to take the full possible range of values: $-\infty < (x, y, z) < +\infty$. We could, however, identify the points with coordinates (x, y, z) and $(x + L, y + L, z + L)$ thereby changing the topology of this space to $\mathcal{S}^3$. Thus, our considerations do not uniquely specify the topology of the spacetime. The choices we have made for the three cases $k = 0, \pm 1$ discussed above should only be considered as the most natural choices for these models.

2.3 Kinematic properties of the Friedmann universe

The Friedmann metric in (2.6) is nonstatic because of the time dependence of $a(t)$. Since $a(t)$ multiplies the spatial coordinates, any proper distances $l(t)$ will change with time in proportion to $a(t)$:

$$l(t) = l_0 a(t) \propto a(t). \tag{2.20}$$

In particular, the proper separation between any two observers, located at constant comoving coordinates, will change which time. Let the coordinate separation between two such observers be δr, so that the proper separation is $\delta l = a(t)\delta r$. Each of the two observers will attribute to the other a velocity

$$\delta v = \frac{d}{dt}\delta l = \dot{a}\delta r = \left(\frac{\dot{a}}{a}\right)\delta l. \tag{2.21}$$

This leads to several important physical consequences of rather general nature.

To begin with, consider a narrow pencil of electromagnetic radiation which crosses any two comoving observers separated by proper distance δl. The time for transit will be $\delta t = \delta l$. Let the frequency of the radiation measured by the first observer be ω. Since the first observer sees the second one to be *receding* with velocity δv, he will expect the second observer to measure a Doppler shifted frequency $(\omega + \delta\omega)$ where

$$\frac{\delta\omega}{\omega} = -\frac{\delta v}{c} = -\delta v = -\frac{\dot{a}}{a}\delta l = -\frac{\dot{a}}{a}\delta t = -\frac{\delta a}{a}. \tag{2.22}$$

(Since the observers are separated by an infinitesimal distance of first order, δl, we can introduce a locally inertial frame encompassing both the observers. The laws of special relativity can be applied in this frame.) This equation can be integrated to give

$$\omega(t)a(t) = \text{constant}. \tag{2.23}$$

In other words, the frequency of electromagnetic radiation changes due to expansion of the universe according to the law $\omega \propto a^{-1}$. We have made implicit use of the homogeneity of the spacetime in extending the local result to a global context.

The same result can be obtained by solving Maxwell's equations for wave propagation in the Friedmann universe (see exercise 2.5). Notice that, if $\dot{a} > 0$, the frequency will be redshifted irrespective of whether the wave is travelling from a location A to location B or from B to A. Suppose that an observer receives, at time t_0, radiation from a distant source which was emitted at some earlier time $t_1 (< t_0)$. The observed and emitted frequencies are related by

$$\frac{\omega_e}{\omega_0} = \frac{a(t_0)}{a(t_e)} \equiv 1 + z \tag{2.24}$$

where the last equation defines the 'redshift' z. It is usual to say that the particular source has the redshift z. Thus, the expansion of the Friedmann model provides a natural explanation for the redshift of light emitted by distant sources. (This aspect was discussed briefly in the last chapter.)

This relation is often used to characterize a particular time (especially in the early universe) at which an event took place; that is, we *define* a function $z(t)$ by $(1 + z(t)) \equiv (a_0/a(t))$, where $a_0 = a(t_0)$, and use it instead of the time coordinate t or the value of expansion factor $a(t)$ at that time.

The expansion of the universe also affects the motion of material particles. Consider again two comoving observers separated by proper distance δl. Let a material particle pass the first observer with velocity v. When it has crossed the proper distance δl (in a time interval δt), it passes the second observer whose velocity (relative to the first one) is

$$\delta u = \frac{\dot{a}}{a}\delta l = \frac{\dot{a}}{a} v\, dt = v\frac{\delta a}{a}. \tag{2.25}$$

The second observer will attribute to the particle the velocity

$$v' = \frac{v - \delta u}{1 - v\delta u} = v - (1 - v^2)\delta u + \mathcal{O}[(\delta u)^2] = v - (1 - v^2)v\frac{\delta a}{a}. \tag{2.26}$$

This follows from the special relativistic formula for addition of velocities which is valid in an infinitesimal region around the first observer. Rewriting this equation as

$$\delta v = -v(1 - v^2)\frac{\delta a}{a} \tag{2.27}$$

and integrating, we get

$$p = \frac{v}{\sqrt{1 - v^2}} = \frac{\text{constant}}{a}. \tag{2.28}$$

In other words, the magnitude of the 3-momentum decreases as a^{-1} due to the expansion. If the particle is non-relativistic, then $v \propto p$ and velocity itself decays as a^{-1}. This result can also be derived from the study of the geodesics in the Friedmann universe (see exercise 2.7).

One conclusion which can be reached immediately from (2.28) is the following: Consider a stream of particles propagating freely in the space-time. At some time t, a comoving observer finds dN particles in a proper volume $d\mathcal{V}$, all having momentum in the range $(\mathbf{p}, \mathbf{p} + d^3\mathbf{p})$. The phase space distribution function $f(\mathbf{x}, \mathbf{p}, t)$ for the particles is defined by the relation $dN = f\, d\mathcal{V}\, d^3\mathbf{p}$. At a later instant $(t + \delta t)$ the proper volume occupied by these particles would have increased by a factor $[a(t+\delta t)/a(t)]^3$ while the volume in the momentum space will be redshifted by $[a(t)/a(t+\delta t)]^3$, showing that the phase volume occupied by the particles does not change during the free propagation. Since the number of particles dN is also conserved, it follows that f is conserved along the streamline.

The above considerations have been purely classical. It is, however, sometimes convenient to think of electromagnetic radiation as consisting of photons with zero rest mass, for which $E = \hbar\omega = p$. The decay law $p \propto a^{-1}$ then implies the redshift for radiation. A few other properties of radiation can also be derived easily by using the photon concept. If $f(\mathbf{x}, \mathbf{p}, t)$ is the phase space density of photons, then the energy density du of photons in the frequency band $(\omega, \omega + d\omega)$ flowing into a particular solid angle $d\Omega$ is proportional to $f(\hbar\omega)(p^2 dp\, d\Omega)$. The quantity 'spectral energy density' or the 'specific intensity', $I(\omega)$, of the radiation (which is the energy density per unit bandwidth) is defined by the relation

$$du \equiv I(\omega)\, d\Omega\, d\omega = \hbar\omega\, f\, p^2\, dp\, d\Omega. \tag{2.29}$$

Since $p^2 dp \propto \omega^2 d\omega$ for photons, we find that $I(\omega) \propto f\omega^3$; and further, since f is conserved during the propagation, it follows that the quantity $I(\omega)/\omega^3$ is conserved during the propagation of radiation.

This quantity $I(\omega)$ will have units (for example) $\mathrm{erg\,cm^{-3}\,Hz^{-1}}$(steradian)$^{-1}$. The 'energy flux' of radiation is sometimes defined as $F(\omega) = (c/4\pi)I(\omega)$ which will have the units $(\mathrm{erg\,s^{-1}})(\mathrm{cm})^{-2}(\mathrm{Hz})^{-1}(\mathrm{steradian})^{-1}$. It is clear from the invariance of I/ω^3 that I and F vary as a^{-3} in the expanding universe. More precisely,

$$F[\omega(1+z); z] = F[\omega; 0](1+z)^3. \tag{2.30}$$

The total flux of radiation, obtained by integrating over all frequencies, varies as $(1+z)^4$:

$$F_{\text{total}} = \int_0^\infty F d\omega \propto (1+z)^4. \tag{2.31}$$

Radiation which has intensity distribution of the form $I(\omega) = \omega^3 G(\omega/T)$ will retain the spectral shape during the expansion, with the parameter T varying with expansion as $T \propto a^{-1}$. The Planck spectrum has this form in which T corresponds to the temperature; it follows that the temperature of the radiation, which has a Planck spectrum, decreases with expansion as $T(t) \propto a(t)^{-1}$.

Cosmological observations are mostly based on electromagnetic radiation which is received from far away sources. Let an observer, located at $r = 0$, receive at time $t = t_0$ radiation from a source located at $r = r_1$. This radiation must have been emitted at some earlier time t_1 such that the events (t_1, r_1) and $(t_0, 0)$ are connected by a null geodesic. Taking the propagation of the ray to be along $\theta =$ constant, $\phi =$ constant, we can write the equation for the null geodesic to be

$$0 = ds^2 = dt^2 - a^2(t)\frac{dr^2}{1 - kr^2}. \tag{2.32}$$

Integrating this, we can find the relation between r_1 and t_1:

$$\int_{t_1}^{t_0} \frac{dt}{a(t)} = \int_0^{r_1} \frac{dr}{(1 - kr^2)^{1/2}}. \tag{2.33}$$

Suppose that the source emits two photons at t_1 and $t_1 + dt_1$, which are received by the observer at times t_0 and $t_0 + dt_0$. The right hand side of (2.33) is the same for both these photons implying

$$\int_{t_1}^{t_0} \frac{dt}{a(t)} = \int_{t_1+dt_1}^{t_0+dt_0} \frac{dt}{a(t)}. \tag{2.34}$$

This equation can be written as

$$\int_{t_1}^{t_1+dt_1} \frac{dt}{a(t)} = \int_{t_0}^{t_0+dt_0} \frac{dt}{a(t)}, \tag{2.35}$$

giving

$$\left(\frac{dt_0}{dt_1}\right) = \frac{a(t_0)}{a(t_1)}. \tag{2.36}$$

If the source has an intrinsic luminosity L, then it will emit an energy $L\,dt_1$ in a time interval dt_1. This energy, which will be received by the observer in a time interval $dt_0 = dt_1\,[a(t_0)/a(t_1)]$, would have undergone a redshift by factor $[a(t_1)/a(t_0)]$ and will be distributed over a sphere of radius $4\pi a^2(t_0) r_1^2$. The observed flux, therefore, will be

$$l = L\left(\frac{dt_1}{dt_0}\right)\frac{a(t_1)}{a(t_0)}\frac{1}{4\pi a^2(t_0) r_1^2} = \frac{L}{4\pi a_0^2 r_1^2}\left(\frac{a(t_1)}{a_0}\right)^2 = \frac{L}{4\pi a_0^2 r_1^2 (1+z)^2}. \tag{2.37}$$

It is usual to define a 'luminosity distance' $d_L(z)$ to the source at redshift z through the relation $l \equiv (L/4\pi d_L^2)$. We get

$$d_L(z) = a_0 r_1(t_1)(1+z). \tag{2.38}$$

Another observable parameter for distant sources is the angular diameter. If D is the physical size of the object which subtends an angle δ to the observer, then, for small δ, we have $D = r_1 a(t_1)\delta$. The 'angular diameter distance' $d_A(z)$ for the source is defined via the relation $\delta = (D/d_A)$; so we find that

$$d_A(z) = r_1 a(t_1) = a_0 r_1(t_1)(1+z)^{-1}. \tag{2.39}$$

Quite clearly $d_L = (1+z)^2 d_A$.

In these formulas, r_1 is related to t_1 via (2.33) and hence can be expressed as a function of z. To do this exactly, we need to know the functional form of $a(t)$. If, however, we are only interested in small z (i.e. in a small time interval $(t_0 - t_1)$) then we can Taylor expand $a(t)$ around t_0 and parametrize it by the coefficients of the Taylor expansion. It is conventional to write this expansion in the following form:

$$\begin{aligned} a(t) &= a(t_0)\left[1 + \left(\frac{\dot{a}}{a}\right)_0 (t-t_0) + \frac{1}{2}\left(\frac{\ddot{a}}{a}\right)_0 (t-t_0)^2 + \cdots\right] \\ &= a(t_0)\left[1 + H_0(t-t_0) - \frac{1}{2}q_0 H_0^2 (t-t_0)^2 + \cdots\right] \end{aligned} \tag{2.40}$$

where

$$H_0 \equiv \left(\frac{\dot{a}}{a}\right)_{t=t_0}; \qquad q_0 \equiv -\left(\frac{\ddot{a}a}{\dot{a}^2}\right)_{t=t_0}. \tag{2.41}$$

Substituting this expansion in (2.33), expanding up to quadratic order in $(t-t_0)$ and r_1 and integrating, we get the relation between r_1 and t_1:

$$r_1 = \frac{1}{a_0}\left[(t_0 - t_1) + \frac{1}{2}H_0(t_1 - t_0)^2 + \cdots\right]. \tag{2.42}$$

Inverting (2.40), we can express $(t-t_0)$ in terms of $(1+z) = (a_0/a)$:

$$(t - t_0) = -H_0^{-1}\left[z - (1 + \frac{q_0}{2})z^2 + \cdots\right]. \tag{2.43}$$

Finally, substituting (2.43) into (2.42) we can express r_1 in terms of z:

$$a_0 r_1 = H_0^{-1}\left[z - \frac{1}{2}(1+q_0)z^2 + \cdots\right]. \tag{2.44}$$

This shows that, to first order in z, the 'redshift velocity' $v \equiv cz$ is proportional to the proper distance $a_0 r_1$. We can now use this relation to express d_L (and d_A) in terms of z; we find, for example,

$$d_L(z) = a_0 r_1 (1+z) = H_0^{-1}\left[z + \frac{1}{2}(1-q_0)z^2 + \cdots\right]. \tag{2.45}$$

Notice that the null geodesics define a unique function $r_1 = r_1(z)$ in any specified cosmological model. This function allows one to use redshift to specify the radial coordinate of any object.

The quantity d_L can be determined by measuring the flux l for a class of objects for which the intrinsic luminosity L is known. If we can also measure the redshift z for these objects, then a plot of d_L against z will allow us to determine the parameters H_0 and q_0. (This procedure was described briefly in the last chapter.) In particular, observations suggest[2] that

$$H_0 = 100h \text{ km s}^{-1} \text{Mpc}^{-1}; \qquad 0.5 \lesssim h \lesssim 1 \tag{2.46}$$

The quantity H_0, called the 'Hubble constant', is of primary significance in cosmology. Its value determines the rate at which the universe is expanding today. From H_0 we can construct the time scale, $t_{\text{univ}} \equiv H_0^{-1} = 9.8 \times 10^9 h^{-1}$ yr and the length scale $l_{\text{univ}} \equiv ct_{\text{univ}} \cong 3000h^{-1}$ Mpc. These are the characteristic scales over which global effects of cosmological expansion will be important today.

2.4 The dynamics of the Friedmann model

The Friedmann metric contains a constant k and a function $a(t)$, both of which can be determined via Einstein's equations

$$G^\alpha_\beta = R^\alpha_\beta - \frac{1}{2}\delta^\alpha_\beta R = 8\pi G T^\alpha_\beta \tag{2.47}$$

if the stress-tensor for the source is specified. The assumption of homogeneity and isotropy implies that T^i_0 must be zero and that the spatial components T^j_k must have a diagonal form with $T^1_1 = T^2_2 = T^3_3$. It is conventional to write such a stress-tensor as

$$T^\alpha_\beta = \text{dia}[\rho(t), -p(t), -p(t), -p(t)]. \tag{2.48}$$

This notation is suggested by the following consideration: If the source was an ideal fluid with pressure p and density ρ then the stress-tensor for an ideal fluid, $T^\alpha_\beta = (p+\rho)u^\alpha u_\beta - p\delta^\alpha_\beta$, will have the above form in the rest frame of the fluid. The nature of the source is completely specified once the relation between ρ and p is given in the form of an equation of state $p = p(\rho)$.

The constituents of the *present day* universe are essentially radiation (in the form of zero mass particles like photons, massless neutrinos etc.) and non-relativistic matter. Radiation has the equation of state $p = (1/3)\rho$ while non-relativistic matter is well approximated as pressureless 'dust', $p \simeq 0$. This follows from the fact that, for non-relativistic particles of mass m and number density n, $\rho \simeq nmc^2$ while $p \simeq nmv^2$ with $v \ll c$. The nature – and hence the equation of state – of the constituents will change with time in an evolving universe. The equation of state at any given epoch has to be determined by studying the physical processes at that particular epoch. For most of the evolution of the universe, it is reasonable to treat the source as made up of a mixture of radiation (with $p = (1/3)\rho$) and matter (with $p \cong 0$).

The tensor G^α_β can be computed from the Friedmann metric in a straightforward manner. As is to be expected, we find that G^i_0 vanish while G^i_j is proportional to the unit matrix. The two nontrivial components are:

$$G^0_0 = \frac{3}{a^2}(\dot{a}^2 + k), \quad G^i_j = \frac{1}{a^2}(2a\ddot{a} + \dot{a}^2 + k)\delta^i_j. \tag{2.49}$$

Thus (2.47) gives two independent equations

$$\frac{\dot{a}^2 + k}{a^2} = \frac{8\pi G}{3}\rho, \tag{2.50}$$

$$\frac{2\ddot{a}}{a} + \frac{\dot{a}^2 + k}{a^2} = -8\pi G p. \tag{2.51}$$

These two equations, combined with the equation of state $p = p(\rho)$, completely determine the three functions $a(t)$, $\rho(t)$ and $p(t)$.

The value of k, which determines the curvature of spatial sections of the universe, can be determined from (2.50). This equation implies that, at the present epoch ($t = t_0$),

$$\frac{k}{a_0^2} = \frac{8\pi G}{3}\rho_0 - \left(\frac{\dot{a}}{a}\right)^2_0 = \frac{8\pi G}{3}\rho_0 - H_0^2 \equiv H_0^2(\Omega - 1) \tag{2.52}$$

where we have introduced two important definitions

$$\rho_c = \text{critical density} \equiv \frac{3H_0^2}{8\pi G} \simeq \begin{cases} 1.88 \times 10^{-29} h^2 & \text{g cm}^{-3} \\ 1.06 \times 10^4 h^2 & \text{eV cm}^{-3} \end{cases} \tag{2.53}$$

and $\Omega \equiv (\rho_0/\rho_c)$. It follows from (2.52) that the universe will be closed, flat or open depending on whether Ω is greater than, equal to or less than unity.

Equations (2.50) and (2.51) can be combined to give

$$\frac{\ddot{a}}{a} = -\frac{4\pi G}{3}(\rho + 3p). \tag{2.54}$$

For the usual kind of matter, $(\rho + 3p) > 0$ implying that $\ddot{a} < 0$. The $a(t)$ curve (which has positive $\dot{a}$ at the present epoch t_0) must be convex; in other words, a will be smaller in the past and will become zero at sometime (in the past), say, at $t = t_{\rm sing}$. It is also clear $(t_0 - t_{\rm sing})$ must be less than value of the intercept $(\dot{a}/a)_0^{-1} = H_0^{-1}$. For convenience, we will choose the time coordinate such that $t_{\rm sing} = 0$, i.e. we take $a = 0$

at $t = 0$. In that case, the present 'age' of the universe t_0 satisfies the inequality $t_0 < t_{\rm univ}$ where

$$t_{\rm univ} \equiv H_0^{-1} = 3.1 \times 10^{17} h^{-1}\,{\rm s} = 9.8 \times 10^9 h^{-1}\,{\rm yr}. \tag{2.55}$$

As a becomes smaller the components of the curvature tensor $R^{\alpha}_{\beta\gamma\delta}$ become larger and when $a = 0$ these components diverge. Such a divergence (called 'singularity') is an artifact of our theory. When the radius of curvature of the spacetime becomes comparable to the fundamental length $(G\hbar/c^3)^{1/2} \simeq 10^{-33}$ cm constructed out of $G, \hbar$ and c, quantum effects of gravity will become important, rendering the classical Einstein equations invalid. So, in reality, t_0 is the time which has elapsed from the moment at which the classical equations became valid.

The quantities ρ and p are *defined* in (2.48) as the T^0_0 and T^1_1 (say) components of the stress-tensor. The interpretation of p as 'pressure' depends on treating the source as an ideal fluid. The source for a Friedmann model should *always* have the form in (2.48); but, if the source is not an ideal fluid then it is not possible to interpret the spatial components of T^{α}_{β} as pressure. (We will see in chapter 10 that such situations can arise in the dynamics of the early universe.) It is, therefore, quite possible that the equation of state for matter at high energies does not obey the condition $(\rho+3p) > 0$. The violation of this condition may occur much before (i.e. at larger value of a) the quantum gravitational effects become important. If this happens, then the 'age of the universe' refers to the time interval since the breakdown of the condition $(\rho + 3p) > 0$. An important example in which the condition $(\rho + 3p) > 0$ is violated is discussed in exercise 2.14.

From equation (2.50), we see that $\rho a^3 = (3/8\pi G)a(\dot{a}^2 + k)$; differentiating this expression and using equation (2.51) we get

$$\frac{d}{dt}(\rho a^3) = -3a^2\dot{a}p = -p\frac{da^3}{dt}. \tag{2.56}$$

Or, equivalently,

$$\frac{d}{da}(\rho a^3) = -3a^2 p. \tag{2.57}$$

Given the equation of state $p = p(\rho)$, we can integrate (2.57) to obtain $\rho = \rho(a)$. Substituting this relation into (2.50) we can determine $a(t)$.

For an equation of state of the form $p = w\rho$, (2.57) gives $\rho \propto a^{-3(1+w)}$; in particular, for non-relativistic matter ($w = 0$) and radiation ($w = 1/3$) we find $\rho_{\rm NR} \propto a^{-3}$ and $\rho_R \propto a^{-4}$. Thus

$$\rho_{\rm NR}(t) = \rho_{\rm NR}(t_0)\left(\frac{a_0}{a}\right)^3 = \rho_c \Omega_{\rm NR}(1+z)^3 \tag{2.58}$$

and

$$\rho_R(t) = \rho_R(t_0)\left(\frac{a_0}{a}\right)^4 = \rho_c\Omega_R(1+z)^4 \tag{2.59}$$

where $\Omega_{\rm NR} = (\rho_{\rm NR}/\rho_c)$ etc. (The notation which we will use while referring to Ω is as follows: The value of (ρ/ρ_c) at the present epoch will be called Ω or Ω_0; the latter form will be used only when we need to distinguish between the value of (ρ/ρ_c) at the present epoch from the value of a similar quantity defined at sometime in the past. Thus, unless otherwise specified explicitly, $\Omega \equiv \Omega_0 \equiv [\rho(t_0)/\rho_c]$. The fraction of the critical density contributed at the present epoch by some species labelled by x will be denoted by Ω_x; that is $\Omega_x \equiv [\rho_x(t_0)/\rho_c]$). Observations suggest that, at the present epoch,

$$\Omega_{\rm total} \equiv \Omega \simeq \Omega_{\rm NR} \geq 0.2; \quad \Omega_R h^2 \simeq 2.56 \times 10^{-5}.$$

(Observations which determine $\Omega_{\rm NR}$ will be discussed in detail in chapter 11; the properties of MBR were mentioned in the last chapter and will be studied in detail in chapter 6.) Thus matter dominates over radiation at present. But from (2.58) and (2.59) it follows that the radiation density grows faster than the matter density as we go to earlier phases of the universe (i.e. to higher redshifts). At some time $t = t_{\rm eq}$ in the past (corresponding to a value $a = a_{\rm eq}$ and redshift $z = z_{\rm eq}$) the radiation and matter will have equal energy densities. From (2.59) and (2.58) we get

$$(1+z_{\rm eq}) = \frac{a_0}{a_{\rm eq}} = \frac{\Omega_{\rm NR}}{\Omega_R} \simeq 3.9 \times 10^4(\Omega h^2). \tag{2.60}$$

Since the temperature of the radiation grows as a^{-1}, the temperature of the universe at this epoch will be

$$T_{\rm eq} = T_{\rm now}(1+z_{\rm eq}) = 9.24(\Omega h^2)\,{\rm eV}. \tag{2.61}$$

For $t \ll t_{\rm eq}$ the energy density in the universe is dominated by radiation (with $p = (1/3)\rho$) while for $t \gg t_{\rm eq}$, the energy density is dominated by matter (with $p \simeq 0$).

In deriving the above expressions for $T_{\rm eq}$ and $z_{\rm eq}$, we have assumed that the energy density of relativistic particles today, $\rho_R(t_0)$, arises essentially from MBR photons. Among all electromagnetic radiation present in the universe, MBR definitely makes the dominant contribution. However, it is quite possible that there exist other relativistic particles – say, massless neutrinos – in the universe today contributing to ρ_R. In that case the numerical values for $\Omega_R, T_{\rm eq}$ and $z_{\rm eq}$ will change. This issue will be discussed in detail in the next chapter.

Let us now solve (2.50) to determine $a(t)$. Though it can be done for the general case, some of the facts noted above allow us to simplify the analysis considerably. At any redshift z, the curvature term ka^{-2} in the left hand side of (2.50) has the magnitude

$$\frac{1}{a^2} = \frac{1}{a_0^2}(1+z)^2 = H_0^2|(1-\Omega)|(1+z)^2 \tag{2.62}$$

while the radiation density on the right hand side of (2.50) contributes

$$\frac{8\pi G}{3}\rho = \frac{8\pi G}{3}\rho_c\Omega_R(1+z)^4 = H_0^2\Omega_R(1+z)^4. \tag{2.63}$$

Comparing these two expressions we see that the curvature term is completely negligible (compared to the radiation energy density) at redshifts $z \gg z_c$ where

$$(1+z_c) = [\Omega_R^{-1}(1-\Omega)]^{1/2} \simeq 200|(1-\Omega)|^{1/2}h. \tag{2.64}$$

Since both the factors in $|(1-\Omega)|^{1/2}h$ are less than unity, z_c is less than 200. This redshift is quite small compared to $z_{\rm eq}$ in (2.60). So we can ignore the curvature term in (2.50) for $t \lesssim t_{\rm eq}$ without incurring any significant error.

This does not necessarily mean that the curvature term dominates for $z < z_c$. It is, of course, true that curvature dominates over *radiation* for $z < z_c$. But, since $z_c \ll z_{\rm eq}$, the universe is *matter dominated* near $z \simeq z_c$ and we should really compare the *matter density* with the curvature to decide which term is dominant.

During the matter dominated phase, the energy density varies as $\rho = \rho_c\Omega(1+z)^3$ while the curvature term grows as $a^{-2} = H_0^2\ |(\Omega-1)|\ (1+z)^2$. If $\Omega < 1$, then the two terms on the right hand side of Einstein's equation become comparable at some $z = z_{\rm flat} = z_f$ where z_f is determined by equating the two terms on the right hand side:

$$a_0^{-2}(1+z_f)^2 = H_0^2(1-\Omega)(1+z_f)^2 = \frac{8\pi G}{3}\rho = H_0^2\Omega(1+z_f)^3. \tag{2.65}$$

This gives $z_f = (\Omega^{-1}-2)$. It is possible for the curvature term to dominate over matter density at sufficiently small z; say, for $z \ll z_f$ provided $\Omega < 0.5$; then the curvature term can dominate over matter for $0 < z \ll z_f$. During this phase, the effect of ρ is ignorable and the expansion factor $a(t)$ grows as t. For $z \gg z_f$ we can again ignore the curvature term. Since observations suggest that $\Omega \gtrsim 0.2$, we must have $z_f < 3$. So, except at small redshifts (say, in the range, $z \simeq$ (1–10)) the curvature term is completely ignorable in Einstein's equations.

Equation (2.50), therefore, can be written as

$$\frac{\dot{a}^2}{a^2} + \frac{k}{a^2} \cong \frac{\dot{a}^2}{a^2} = \frac{8\pi G}{3}\rho = \frac{8\pi G}{3}\rho_{\rm eq}\left[\left(\frac{a_{\rm eq}}{a}\right)^4 + \left(\frac{a_{\rm eq}}{a}\right)^3\right] \tag{2.66}$$

for $z \gg z_c$. In terms of the dimensionless variables $x = (a/a_{\rm eq})$ and $y = t(8\pi G\rho_{\rm eq}/3)^{1/2} \equiv (tH_{\rm eq}/\sqrt{2})$, this equation becomes

$$\left(\frac{dx}{dy}\right)^2 = \frac{1}{x^2} + \frac{1}{x} = \frac{1}{x^2}(1+x). \tag{2.67}$$

Integrating, and using the initial condition $x = 0$ at $y = 0$, we find the solution to be

$$H_{\rm eq}t = \frac{2\sqrt{2}}{3}\left[\left(\frac{a}{a_{\rm eq}} - 2\right)\left(\frac{a}{a_{\rm eq}} + 1\right)^{1/2} + 2\right]. \tag{2.68}$$

This gives the expansion factor of the universe $a(t)$ in terms of the two (known) parameters:

$$\begin{aligned} H_{\rm eq}^2 &= \frac{16\pi G}{3}\rho_{\rm eq} = \frac{16\pi G}{3}\rho_c\Omega_R(1+z_{\rm eq})^4 = 2H_0^2\Omega_R(1+z_{\rm eq})^4 \\ &= 2H_0^2\Omega(1+z_{\rm eq})^3, \end{aligned} \tag{2.69}$$

$$a_{\rm eq} = a_0(1+z_{\rm eq})^{-1} = H_0^{-1}|(\Omega-1)|^{-1/2}(\Omega_R/\Omega). \tag{2.70}$$

From (2.68) we can find the value of $t_{\rm eq}$; setting $a = a_{\rm eq}$ gives $H_{\rm eq}t_{\rm eq} \simeq 0.552$, or

$$\begin{aligned} t_{\rm eq} &= \frac{2\sqrt{2}}{3}H_{\rm eq}^{-1}(2-\sqrt{2}) \simeq 0.39H_0^{-1}\Omega^{-1/2}(1+z_{\rm eq})^{-3/2} \\ &= 1.57\times 10^{10}(\Omega h^2)^{-2}\,{\rm s}. \end{aligned} \tag{2.71}$$

From (2.68), we can also find two limiting forms for $a(t)$ valid for $t \gg t_{\rm eq}$ and $t \ll t_{\rm eq}$:

$$\left(\frac{a}{a_{\rm eq}}\right) = \begin{cases} (3/2\sqrt{2})^{2/3}(H_{\rm eq}t)^{2/3} \\ (3/\sqrt{2})^{1/2}(H_{\rm eq}t)^{1/2} \end{cases}. \tag{2.72}$$

Thus $a \propto t^{2/3}$ in the matter-dominated phase (when the curvature is negligible) and $a \propto t^{1/2}$ in the radiation-dominated phase.

If we use the $a(t)$ relation for the matter-dominated phase and define $t_{\rm eq}$ as the time at which $a = a_{\rm eq}$, then we will get

$$t_{\rm eq} \simeq \frac{2}{3}H_0^{-1}\Omega^{-1/2}(1+z_{\rm eq})^{-3/2} = 1.71(t_{\rm eq})_{\rm exact} = 2.7\times 10^{10}(\Omega h^2)^{-2}\,{\rm s}. \tag{2.73}$$

The values of $t_{\rm eq}$ in (2.73) and (2.71) differ slightly. For practical calculations we will use the expression in (2.73) rather than (2.71).

The formulas derived above are valid for $z \gtrsim 10$. A different approximation is useful for $z \ll z_{\rm eq}$ when we can ignore the contribution due to radiation and treat the universe as matter-dominated. For lower redshifts, however, we need to take into account the curvature term (k/a^2). From (2.50) we get,

$$\dot{a}^2 + k = \frac{8\pi G}{3}\rho a^2 = H_0^2 \Omega a_0^2 \left(\frac{a_0}{a}\right). \tag{2.74}$$

Or, using the dimensionless variables $x = (a/a_0)$ and $y = H_0 t$,

$$\left(\frac{dx}{dy}\right) = (1 - \Omega + \Omega x^{-1})^{1/2}. \tag{2.75}$$

Integrating and using the fact that $(a/a_0) = x = (1+z)^{-1}$, we get

$$H_0 t = \int_0^{(1+z)^{-1}} \frac{dx}{[1 - \Omega + \Omega x^{-1}]^{1/2}} \tag{2.76}$$

To be precise, we should set the lower limit of integration to be $(1+z_{\text{eq}})^{-1}$, when matter domination begins; but we will be using (2.76) for $z \ll z_{\text{eq}}$. Since $z_{\text{eq}}^{-1} \ll z^{-1}$, the lower limit can be taken to be zero with sufficient accuracy. The value of the integral depends on Ω; we get, for $\Omega > 1$:

$$H_0 t = \frac{\Omega}{2(\Omega - 1)^{3/2}} \left[\cos^{-1}\left(\frac{\Omega z - \Omega + 2}{\Omega z + \Omega}\right) - \frac{2(\Omega - 1)^{1/2}(\Omega z + 1)^{1/2}}{\Omega(1+z)}\right] \tag{2.77}$$

and for $\Omega < 1$:

$$H_0 t = \frac{\Omega}{2(1 - \Omega)^{3/2}} \left[\frac{2(1 - \Omega)^{1/2}(\Omega z + 1)^{1/2}}{\Omega(1+z)} - \cosh^{-1}\left(\frac{\Omega z - \Omega + 2}{\Omega z + \Omega}\right)\right]. \tag{2.78}$$

These expressions, together with the relation $a = a_0(1+z)^{-1} = H_0^{-1}|(1-\Omega)|^{-1/2}\,(1+z)^{-1}$, completely determine $a(t)$. (Some properties of the matter dominated model are explored in exercises 2.11, 2.12.)

Setting z to zero in the expressions derived above will give us the *present* age of the universe for a given value of Ω:

$$\begin{aligned} t_0 &= H_0^{-1} \frac{\Omega}{2(\Omega - 1)^{3/2}} \left[\cos^{-1}(2\Omega^{-1} - 1) - 2\Omega^{-1}(\Omega - 1)^{1/2}\right] \quad (\text{for } \Omega > 1) \\ &= H_0^{-1} \frac{\Omega}{2(1 - \Omega)^{3/2}} \left[2\Omega^{-1}(1 - \Omega)^{1/2} - \cosh^{-1}(2\Omega^{-1} - 1)\right] (\text{for } \Omega < 1) \end{aligned} \tag{2.79}$$

and $t_0 = (2/3)H_0^{-1}$ for $\Omega = 1$. We can use this expression, written in the form,

$$t_0 = H_0^{-1} A(\Omega) = 9.78 h^{-1} A(\Omega) \text{ Gyr} \tag{2.80}$$

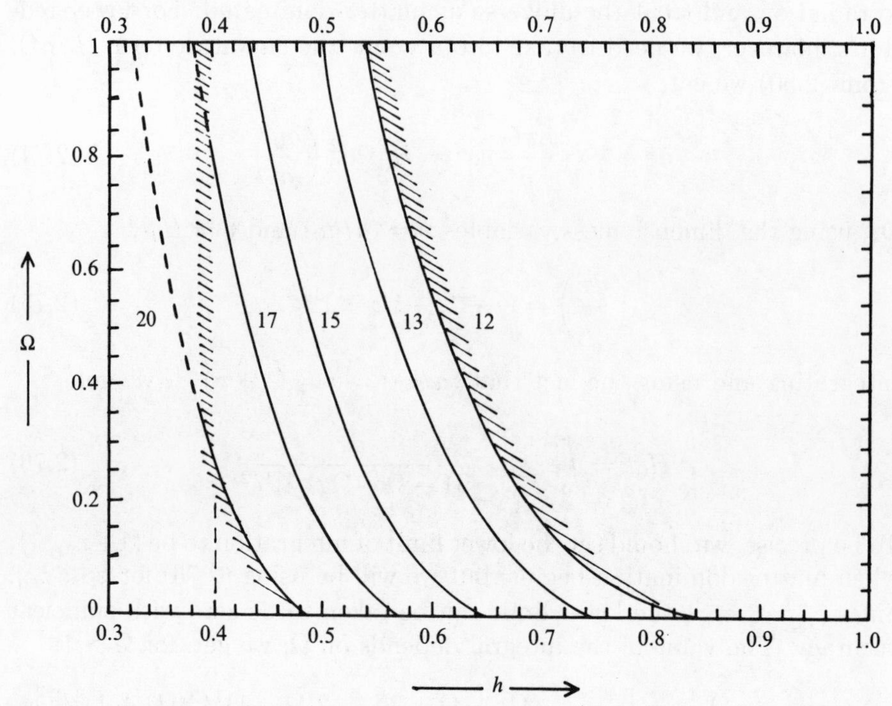

Fig. 2.1. Constraints on Ω and h arising from the age of the universe are shown. The curves are parametrized by the age of the universe in Gyr. If the age is between 12 Gyr and 20 Gyr and $0.4 < h < 1$, then the parameters are confined to the region between the shaded curves.

(where $A(\Omega)$ is the dimensionless 'age function') to constrain the parameters Ω and h. The function $A(\Omega)$ decreases monotonically with Ω; thus, for a fixed h, models with larger Ω will have smaller age.

The observations discussed in chapter 1 suggest that the age of the universe is in the range (10–20) Gyr, and that $h \geq 0.5$. Taking $t_0 \geq 10\,\text{Gyr}$ we get the bound $\Omega < 3.1$; if $t_0 \geq 12\,\text{Gyr}$ and $h > 0.5$ then $\Omega \leq 1.5$. For $\Omega = 1$, we get $t_0 = 6.52 \times 10^9 h^{-1}\,\text{yr}$ which is consistent with the condition $t_0 > 10\,\text{Gyr}$ if $h < 0.65$; if $\Omega = 1$ and $t_0 > 12\,\text{Gyr}$ we must have $h < 0.54$ (see figure 2.1).

The mass density contributed by a particular species-x today is $\rho_x = \Omega_x \rho_c = 1.88 \times 10^{-29}\ (\Omega_x h^2)\ \text{g}\,\text{cm}^{-3}$, in which the combination $(\Omega_x h^2)$ occurs. It is, therefore, convenient to have a bound on (Ωh^2) from observations. From (2.80), it follows that

$$B(\Omega) \equiv \Omega A^2(\Omega) = (\Omega h^2)\left(\frac{t_0}{9.78\,\text{Gyr}}\right)^2. \tag{2.81}$$

The function $B(\Omega)$ increases monotonically and reaches the value $(\pi^2/4) \simeq 2.47$ as $\Omega \to \infty$; thus $B(\Omega) \leq 2.47$, implying $(\Omega h^2) < 2.47$

$(t_0/9.78\,\mathrm{Gyr})^{-2}$. Since $t_0 \geq 10\,\mathrm{Gyr}$, this gives the bound $(\Omega h^2) \leq 2.4$ irrespective of the value of h. If we further use the constraint $h \geq 0.5$, then can improve this bound and get $\Omega h^2 < 0.8$. This result can be used to put powerful constraints on ρ_x contributed by various species.

These relations also give the age of the universe (t) at any given redshift for arbitrary Ω. Note that the limit $\Omega \to 1$ can be taken in these expressions without experiencing any discontinuity. For $\Omega = 1$, we will get

$$t = \frac{2}{3} H_0^{-1} (1+z)^{-3/2}. \tag{2.82}$$

From the general solution, we can also determine the explicit form of the function $r_1(t_1)$ which was needed in the last section to specify the angular diameter distance and luminosity distance. By definition of $r_1(t_1)$, we have,

$$\int_0^{r_1} \frac{dr}{(1-kr^2)^{1/2}} = \int_{t_1}^{t_0} \frac{dt}{a(t)}. \tag{2.83}$$

We write this relation as

$$\int_0^{r_1} \frac{dr}{(1-kr^2)^{1/2}} = \int_{t_1}^{t_0} \frac{dt}{a(t)} = \int_{a_1}^{a_0} \frac{da}{a\dot{a}}. \tag{2.84}$$

Expressing $a = a_0(1+z)^{-1}, a_1 = a_0(1+z_1)^{-1}$ and substituting for $\dot{a}$ using Einstein's equations we can easily obtain:

$$r_1 = \frac{|(1-\Omega)|^{1/2}}{\Omega^2(1+z_1)} [2\Omega z_1 + (2\Omega - 4)(\sqrt{1+\Omega z} - 1)]. \tag{2.85}$$

The angular diameter distance to a source at redshift z is $d_A = a_0 r_1(z)(1+z)^{-1} = H_0^{-1}|(1-\Omega)|^{1/2} r_1(z)(1+z)^{-1}$. Substituting for r_1, we get

$$d_A(z) = 2H_0^{-1}\Omega^{-2}[\Omega z + (\Omega - 2)(\sqrt{1+\Omega z} - 1)](1+z)^{-2}. \tag{2.86}$$

For small z this will reduce to the form discussed in the last section; for $z \gg 1$, this expression becomes

$$d_A(z) \cong 2(H_0\Omega)^{-1} z^{-1}. \tag{2.87}$$

We will require this result in chapter 6.

The following point is worth emphasizing regarding the Friedmann model: In general relativity, the choice of coordinates used in describing a spacetime is completely arbitrary. We may, if we want, choose $a(t)$ to be the *time* coordinate of the Friedmann universe (during the expanding phase), rather than the conventional t-coordinate. In this case, the metric can be written as

$$\begin{aligned} ds^2 &= dt^2 - a^2(t)[d\chi^2 + f^2(\chi)(d\theta^2 + \sin^2\theta\, d\phi^2)] \\ &= N(a)da^2 - a^2[d\chi^2 + f^2(\chi)(d\theta^2 + \sin^2\theta\, d\phi^2)] \end{aligned} \tag{2.88}$$

where $N(a) = \dot{a}^{-2}$ can be expressed in terms of a using the equation

$$\frac{\dot{a}^2}{a^2} = \frac{8\pi G}{3}\rho(a) - \frac{k}{a^2}. \tag{2.89}$$

In other words,

$$N(a) = \left[\frac{8\pi G}{3}\rho(a)a^2 - k\right]^{-1}. \tag{2.90}$$

Once the form of $\rho(a)$ is known – by integrating $d(\rho a^3) = -p\, d(a^3)$, say – the full metric can be written as

$$ds^2 = \left[\frac{8\pi G}{3}\rho(a)a^2 - k\right]^{-1} da^2 - a^2[d\chi^2 + f^2(\chi)(d\theta^2 + \sin^2\theta\, d\phi^2)]. \tag{2.91}$$

Further, if we use $a = a_0(1+z)^{-1}$, the spacetime can be expressed in the coordinates (z, χ, θ, ϕ). This form of the metric is often simpler than the more conventional one; it also illustrates how redshift works as a valid spacetime coordinate.

The solution $a(t)$ for the matter dominated universe can be expressed and interpreted in a different manner which is often quite useful. We have seen that all proper distances in the universe scale with the expansion factor. Consider, for example, a fiducial region in the universe with the coordinate radius r. Due to the overall expansion of the universe, the *proper* radius of this region will keep increasing:

$$\text{proper radius} = l(t) = a(t)r = \left(\frac{a(t)}{a_0}\right) l_0 \propto a(t). \tag{2.92}$$

The differential equation satisfied by $l(t)$ follows from (2.50). We get, in the matter dominated phase,

$$\dot{l}^2 + k\frac{l_0^2}{a_0^2} = \frac{8\pi G}{3}(\rho_0 l_0^3)\,\frac{1}{l} \equiv \frac{2GM}{l}, \tag{2.93}$$

where $M = (4\pi\rho_0 l_0^3/3) = (4\pi\rho(t)l(t)^3/3)$ is the mass contained in this region (which has the proper radius l_0 today). Notice that M does not change during the expansion because $\rho \propto a^{-3}$ and $l^3 \propto a^3$. Using

$$\frac{k}{a_0^2} = H_0^2(\Omega - 1) = \frac{8\pi G}{3}\rho_c(\Omega - 1) = \frac{8\pi G}{3}\rho_0\frac{(\Omega - 1)}{\Omega} \tag{2.94}$$

we can write (2.93) as

$$\frac{1}{2}\dot{l}^2 - \frac{GM}{l} = -\frac{GM}{l_0}\frac{(\Omega - 1)}{\Omega} \equiv E. \tag{2.95}$$

This equation has an obvious (Newtonian) interpretation: It represents the motion of a particle with energy E in the gravitational field of mass M. The energy is negative for $\Omega > 1$; therefore, $l(t)$ increases to a maximum value of $l_0\Omega(\Omega-1)^{-1}$ and then begins to decrease. If $\Omega < 1$, then $E > 0$ and $l(t)$ increases without bound. The solution to (2.95) can be expressed in a parametric form as

$$\frac{l}{l_0} = \frac{\Omega}{2(\Omega-1)}(1-\cos\theta); \quad t = \left(\frac{2GM}{l_0^3}\right)^{-1/2} \frac{1}{2}\frac{\Omega^{3/2}}{(\Omega-1)^{3/2}}(\theta - \sin\theta) \tag{2.96}$$

for $\Omega > 1$, and as

$$\frac{l}{l_0} = \frac{\Omega}{2(1-\Omega)}(1-\cosh\theta); \quad t = \left(\frac{2GM}{l_0^3}\right)^{-1/2} \frac{1}{2}\frac{\Omega^{3/2}}{(1-\Omega)^{3/2}}(\sinh\theta - \theta) \tag{2.97}$$

for $\Omega < 1$. These expressions are, of course, identical to the $[t(z), a(z)]$ relations derived earlier. But the present form reveals the dynamical evolution of the matter dominated model in an explicit form.

The above analysis also reveals an interesting way of parametrizing the size of a region in the universe. It is usually convenient to specify the size of a region of the universe by some quantity which does not change with expansion. The amount of non-relativistic mass $M(\lambda)$, contained inside a sphere of proper radius $(\lambda/2)$,

$$M(\lambda) \equiv \frac{4\pi}{3}\rho_{\rm NR}\left(\frac{\lambda}{2}\right)^3 = 1.45 \times 10^{11}\,{\rm M}_\odot(\Omega h^2)\left(\frac{\lambda_0}{1\,{\rm Mpc}}\right)^3 \tag{2.98}$$

is such a quantity. As the universe expands, $\lambda \propto a(t)$ while $\rho_{\rm NR} \propto a^{-3}$ keeping $M(\lambda)$ constant. Thus we can specify the proper size λ of a region by specifying the amount of mass associated with it ; since this quantity is conserved during the expansion we do not have to specify when this quantity is measured.

Notice that the quantity $M(\lambda)$ is computed using the smoothed-out density of the homogeneous universe. According to (2.98), a typical galactic mass ($\sim 10^{11}\,{\rm M}_\odot$) will correspond to a proper size of about 1 Mpc; actual galaxies are much smaller because they ceased to expand with the cosmic medium sometime in the past, and are now dominated by self-gravity. This fact, of course, is irrelevant to the scaling arguments given above which deal with a (hypothetical) smooth universe.

The Friedmann model allows an even more remarkable generalization. It is possible to remove the mass $M = (4\pi/3)\rho R^3$ contained in a sphere of radius R of a Friedmann model and replace it by a more condensed, spherically symmetric, configuration of matter with the same mass M (and smaller radius R_1, say) inside the cavity. (The latter could even be a point mass with $R_1 = 0$). It turns out that Einstein's equations can still be satisfied. Further, this exercise can be repeated at several locations in a Friedmann universe simultaneously, producing what is known as the 'Swiss-cheese model'. We will discuss this model in chapter 8.

2.5 Radiative processes in expanding universe

We have seen in the last chapter that many sources of radiation like galaxies, quasars etc. exist in the redshift band of, say, $z = 0.5$ to 5. This region falls in the matter dominated region in which curvature term of Einstein's equation cannot be ignored and we have to use the full form of $a(t)$ derived above. Very often, however, we will only need the differential form of these equations and one can express quantities of physical interest in a compact and general form. We shall now discuss some of these situations.

Consider some emission process which takes place at a redshift of z for which the emissivity (which has units $\mathrm{erg\,cm^{-3}s^{-1}Hz^{-1}}$) is $J_\omega(z)$. In an interval $(t, t+dt)$ this process will produce a spectral density of radiation $dI_\omega(z) = J_\omega(z)\,dt$. At the present epoch we will observe this radiation density as

$$dI_{\omega(1+z)^{-1}}(z=0) = \frac{J_\omega(z)\,dt}{(1+z)^3} \tag{2.99}$$

where we have used the results $\omega \propto (1+z)$ and $I \propto (1+z)^3$, derived earlier. If the sources are distributed in a redshift interval (z_1, z_2), then the total spectral intensity of radiation observed at the present epoch will be

$$I[\omega_0; z=0] = \int_{z_1}^{z_2} \frac{J[\omega_0(1+z); z]}{(1+z)^3}\left|\frac{dt}{dz}\right| dz. \tag{2.100}$$

The modulus sign on (dt/dz) takes into account the fact that t decreases as z increases. The Ω dependence of the result arises only through this factor. From equation (2.76) we find that

$$\frac{dt}{dz} = -\frac{1}{H_0(1+z)^2(1+\Omega z)^{1/2}}. \tag{2.101}$$

Hence

$$I(\omega_0; z=0) = \int_{z_1}^{z_2} \frac{J[\omega_0(1+z); z]}{H_0(1+z)^5(1+\Omega z)^{1/2}} dz. \tag{2.102}$$

The flux of radiation $F(\omega) = (c/4\pi)I(\omega)$ is

$$F(\omega_0; z=0) = \frac{c}{4\pi H_0}\int_{z_1}^{z_2} \frac{J[\omega_0(1+z); z]dz}{(1+z)^5(1+\Omega z)^{1/2}}. \tag{2.103}$$

As examples in using this formula, let us consider two important emission processes in cosmology: (i) 21-cm line emission by neutral hydrogen and (ii) bremsstrahlung by a hot ionized plasma.

The $n = 1$ state of the hydrogen atom, in which the electron and the proton have parallel spins, has a slightly higher energy than the state with antiparallel spins. The radiation emitted when the atom makes the transition between these two states has a wavelength of 21 cm (corresponding to a frequency of $\nu_H = c/\lambda = 1420$ MHz). Ignoring the natural and Doppler widths, the emissivity can be written as

$$J(\nu) = \frac{3}{4} A n_H h\nu \delta(\nu - \nu_H) \tag{2.104}$$

where $A = 2.85 \times 10^{-15}\,\mathrm{s}^{-1}$ is the rate of spontaneous transitions (which is rather small because this is a magnetic dipole transition) and n_H is the number density of hydrogen atoms. Suppose the observed frequency is ν_0; then $\nu = \nu_0(1+z)$. Further, if $x(z)$ is the fraction of hydrogen gas which is in the ionized form at the redshift of z, then

$$n_H = n_0(1+z)^3[1 - x(z)] \tag{2.105}$$

where n_0 is the present number density of hydrogen atoms *and* ions. Substituting in (2.103) we find, for $\nu_0 < \nu_H$:

$$\begin{aligned} F(\nu_0; z=0) &= \frac{3c(h\nu_0)}{16\pi H_0}(An_0)\int_0^\infty [1 - x(z)]\frac{\delta[\nu_0(1+z) - \nu_H]}{(1+z)(1+\Omega z)^{1/2}}dz \\ &= \frac{3ch}{16\pi H_0}(An_0)\frac{[1 - x(z_c)]}{(1+z_c)(1+\Omega z_c)^{1/2}} \end{aligned} \tag{2.106}$$

with $(1 + z_c) \equiv (\nu_H/\nu_0)$. We will, therefore, observe the flux at all frequencies $\nu_0 < \nu_H$ and, of course, no flux at $\nu_0 > \nu_H$. The discontinuity at ν_H will be by the amount

$$\Delta F = \frac{3ch}{16\pi H_0}(An_0)[1 - x(z=0)] = \frac{3ch}{16\pi H_0}An_H(0) \tag{2.107}$$

where $n_H(0)$ is the density of neutral atoms today. The observed value of ΔF can be used to estimate $n_H(0)$; when unobserved, bounds on ΔF can be converted into bounds on $n_H(0)$.

As a second example, consider the bremsstrahlung emission by the ionized gas in the universe. If the ionized gas has a temperature higher than 10^6 K, the most dominant radiation loss is through bremsstrahlung. The emissivity for this process (in units $\mathrm{erg\,cm^{-3}\,s^{-1}\,Hz^{-1}}$) is

$$J[\nu] = AT_e^{-1/2}n_e^2\left[\exp\left(-\frac{h\nu}{kT_e}\right)\right]g(\nu, T_e) \tag{2.108}$$

where T_e is the electron temperature, n_e is the number density of electrons and g is a correction factor which is nearly constant for $h\nu \ll kT_e$ and is about $(h\nu/kT_e)^{1/2}$ for $h\nu \gg kT_e$. Substituting into (2.103), we get

$$\begin{aligned} F(\nu_0; z=0) = \frac{cA}{4\pi H_0}n_e^2(0)\int_0^{z_{\max}} dz\, g(z)\frac{(1+z)}{(1+\Omega z)^{1/2}}T_e^{-1/2}(z) \\ \left[\exp\left(-\frac{h\nu_0(1+z)}{kT_e(z)}\right)\right] \end{aligned} \tag{2.109}$$

where $g(z) = g[\nu_0(1+z), T_e(z)]$. This integral can be worked out if the temperature history of the gas $T_e(z)$ is known. This process will be discussed in detail in chapter 6.

Another important radiation process in cosmology is the absorption (and scattering) of radiation by matter which is in the line of sight between the source and the observer. This process is best described by the parameter called optical depth. If the density of absorbers (scatterers) is n and the cross section for the process is σ, then the optical depth is $\tau \simeq n\sigma l$ for a path length of l if we ignore cosmological effects. To incorporate the effects of expansion, we will proceed as follows: A photon received with frequency ν_0 will have the frequency $\nu = \nu_0(1+z)$ at the redshift z. During the interval $(t, t+dt)$ corresponding to the redshift range $(z, z+dz)$ the optical depth increases by $d\tau_\nu = \sigma(\nu)n(z)c\,dt = c\sigma(\nu)n(z)|(dt/dz)|dz$. Integrating this expression, we find the total optical depth due to matter in the redshift range $(0, z)$ to be

$$\tau[\nu_0; z] = \int_0^z \sigma(\nu)n(z)c\left|\frac{dt}{dz}\right| dz = \frac{c}{H_0}\int_0^z \frac{\sigma[\nu_0(1+z)]n(z)}{(1+z)^2(1+\Omega z)^{1/2}}dz. \tag{2.110}$$

As an example in the use of this formula, consider the absorption of radiation by neutral hydrogen at wavelength 21 cm. The absorption cross section for this process is

$$\sigma(\nu) = \frac{A}{4\pi}\left(\frac{3}{4}\right)\left(\frac{h\nu}{2kT_{\rm sp}}\right)\left(\frac{c}{\nu}\right)^2 \delta(\nu - \nu_H) \tag{2.111}$$

where $T_{\rm sp}$ is the so called 'spin-temperature.' It is defined by the relation

$$\frac{n_{\rm up}}{n_{\rm down}} = \exp\left(-\frac{h\nu_H}{kT_{\rm sp}}\right) \tag{2.112}$$

where $n_{\rm up}$ and $n_{\rm down}$ denote the number of atoms in the upper and lower energy levels. Substituting $\sigma(\nu)$ in (2.110)we get

$$\tau[\nu_0; z] = \frac{3A}{32\pi}\left(\frac{h\nu_H}{kT_{\rm sp}}\right)\left(\frac{c}{\nu_H}\right)^3 \frac{n_H(0)}{H_0}\frac{(1+z_c)^2}{(1+\Omega z_c)^{1/2}}[1-x(z_c)] \;\; (\text{for } z > z_c) \tag{2.113}$$

where $(1+z_c) = (\nu_H/\nu_0)$. Suppose we are receiving the radiation emitted by a class of sources at typical redshift of $z \simeq z_{\rm source}$. Absorption can take place anywhere in the redshift range of $(0, z_{\rm source})$; hence the intensity of the light received by us will be diminished at all frequencies in the band $[\nu_H(1+z_{\rm source})^{-1}, \nu_H]$. The discontinuity in τ at ν_H is

$$\Delta\tau = \frac{3A}{32\pi}\left(\frac{h\nu_H}{kT_{\rm sp}}\right)\left(\frac{c}{\nu_H}\right)^3 \frac{n_H(0)}{H_0}[1-x(0)] \tag{2.114}$$

which allows us to determine (or put bounds on) the quantity $n_H(0)[1 - x(0)]$.

Another important example of this type is the Lyman-α absorption by neutral hydrogen. This process can be studied in a similar manner and will be discussed in detail in chapter 9.

2.6 The Hubble radius

The time scale over which the physical quantities will change in an expanding universe is clearly $(\dot{a}/a)^{-1}$. This time scale corresponds to a length scale which we will call the Hubble radius:

$$d_H(t) \equiv \left(\frac{\dot{a}}{a}\right)^{-1} \equiv H(t)^{-1}. \tag{2.115}$$

This length $d_H(t)$ is typically the size over which physical processes operate coherently. This is also the length-scale at which general relativistic effects become important. For $L \ll d_H$, Newtonian gravity is often adequate.

During most phases of the expansion of the universe, we can approximate the expansion factor by $a(t) \propto t^n$ with a suitable n which is *less than* unity. Then the Hubble radius will be proportional to t; it follows that the Hubble radius grows at a *faster* rate than the proper length. Given the cosmological evolution of the model – that is, the function $a(t)$ – we can determine $d_H(t)$ at any given time. Consider, for example, the Hubble radius at $t_{\rm eq}$. From (2.73), it follows that

$$d_H(t_{\rm eq}) \equiv H_{\rm eq}^{-1} = \left(\frac{t_{\rm eq}}{0.552}\right) \cong 0.85 \times 10^{21} (\Omega h^2)^{-2}\,{\rm cm}. \tag{2.116}$$

A region with proper radius $d_H(t_{\rm eq})$ at $t = t_{\rm eq}$ would have expanded by the factor $[a_0/a(t_{\rm eq})] = (1 + z_{\rm eq})$ from $t = t_{\rm eq}$ till today. Therefore, the proper radius today for that region – which was as big as the Hubble radius $d_H(t_{\rm eq})$ at $t = t_{\rm eq}$ – will be

$$l_{\rm eq}({\rm today}) = d_H(t_{\rm eq})(1 + z_{\rm eq}) \cong 11\,{\rm Mpc}\,(\Omega h^2)^{-1} \tag{2.117}$$

This size, of course, is much smaller than the Hubble radius today: d_H (today) $\simeq 3000 h^{-1}$ Mpc.

Reversing the above argument, we can draw an important conclusion: Consider a region of proper size λ today (with $\lambda < d_H$ (today)). As we go back in time, the proper radius of this region will shrink as $a(t) \propto t^n$ with $n < 1$; but the Hubble radius of the universe decreases *faster*, as t. Therefore, there will be some time $t = t_{\rm enter}$ (λ) in the past, when the proper radius of this region will equal the Hubble radius of the universe. For $t < t_{\rm enter}(\lambda)$, the proper radius will be bigger than the Hubble radius.

It is usual to say that 'the length scale λ enters the Hubble radius' at $t = t_{\text{enter}}(\lambda)$ (see figure 2.2).

It follows from the earlier analysis that a length of $\lambda_{\text{eq}} \equiv 11\,\text{Mpc}(\Omega h^2)^{-1}$ enters the Hubble radius at $t = t_{\text{eq}}$, the time at which the universe made the transition from the radiation dominated phase to the matter dominated phase. Smaller regions ($\lambda < \lambda_{\text{eq}}$) would have entered the Hubble radius earlier, in the radiation dominated phase while larger regions ($\lambda > \lambda_{\text{eq}}$) will enter later, in the matter dominated phase. Given the explicit form of $a(t)$, we can easily compute the time $t_{\text{enter}}(\lambda)$ by solving the equation

$$\left(\frac{\dot{a}}{a}\right)^{-1}_{t=t_{\text{enter}}} = \lambda \left(\frac{a(t)}{a_0}\right)_{t=t_{\text{enter}}} \tag{2.118}$$

We mentioned earlier that the numerical values for $z_{\text{eq}}, t_{\text{eq}}$ etc. depend on the value of Ω_R. The expressions derived so far are based on the assumption that only MBR photons contribute to Ω_R. If there exist other massless particles in the universe today Ω_R will be higher, z_{eq} will be lower and t_{eq} will be higher. In the later chapters, we will need the expressions for $t_{\text{eq}}, d_H(t_{\text{eq}}), t_{\text{enter}}$ etc. for the case in which Ω_R is contributed by photons as well as 3 massless neutrinos. We will see in

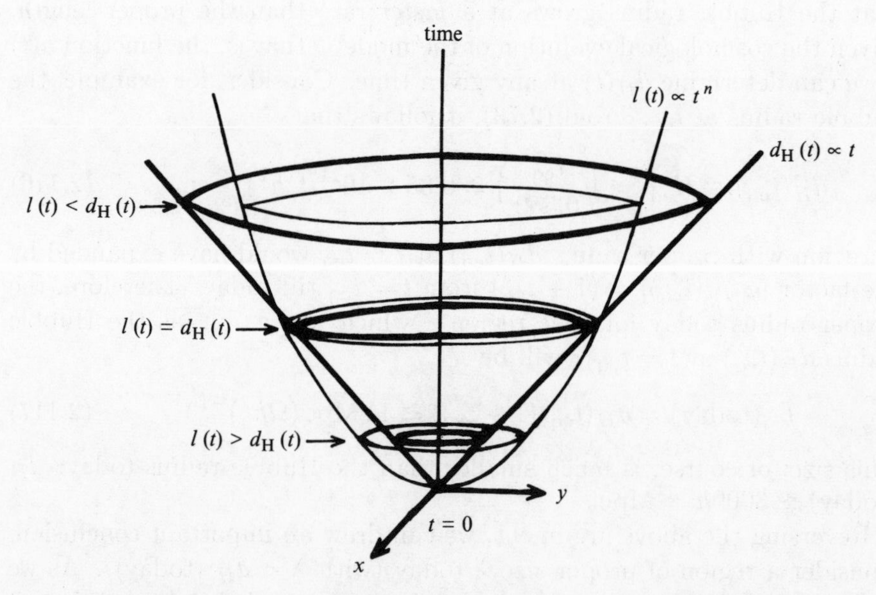

Fig. 2.2. The growth of the Hubble radius (thick line) and proper length (thin line) are shown in a spacetime diagram. For small t the proper length is bigger than the Hubble radius; it enters the Hubble radius at some instant of time and remains smaller than the Hubble radius thereafter. It is assumed that $a(t) \propto t^n$ with $n < 1$.

chapter 3 that, in this case,

$$d_H(t_{\rm eq}) \simeq 1.7 \times 10^{21} (\Omega h^2)^{-2}\,{\rm cm}, \qquad l_{\rm eq}(t_0) \simeq 13\,{\rm Mpc}\,(\Omega h^2)^{-1}. \tag{2.119}$$

Evaluating $t_{\rm enter}(\lambda)$ we find:

$$t_{\rm enter}(\lambda) = \begin{cases} 2.6 \times 10^7\,{\rm s}\,(\Omega h^2)\,(\lambda/1\,{\rm Mpc})^3\,; & (\lambda > \lambda_{\rm eq}) \\ 6.1 \times 10^8\,{\rm s}\,(\lambda/1\,{\rm Mpc})^2\,; & (\lambda < \lambda_{\rm eq}) \end{cases} \tag{2.120}$$

Given $t_{\rm enter}$, one can compute the temperature of the universe at that epoch using the relation $T(z) = T_0(1+z)$. We get:

$$T_{\rm enter}(\lambda) = \begin{cases} 948\,{\rm eV}\,(\Omega h^2)^{-1}(\lambda/1\,{\rm Mpc})^{-2}; & (\lambda > \lambda_{\rm eq}) \\ 63\,{\rm eV}\,(\lambda/1\,{\rm Mpc})^{-1}; & (\lambda < \lambda_{\rm eq}) \end{cases} \tag{2.121}$$

It should be noted that, in the above formulas, λ refers to proper distance *today*. These formulas for $t_{\rm enter}$ and $T_{\rm enter}$ can be more conveniently reexpressed in terms of the mass M contained in the region rather than λ. A region containing a mass

$$M_{\rm eq} \equiv M(\lambda_{\rm eq}) = 3.2 \times 10^{14}\,{\rm M}_\odot\,(\Omega h^2)^{-2} \tag{2.122}$$

will come into the Hubble radius at $t = t_{\rm eq}$. Smaller regions will enter earlier and larger masses later. The relation

$$(1 + z_{\rm enter}) = \begin{cases} 1.41 \times 10^5 (\Omega h^2)^{1/3} (M/10^{12}\,{\rm M}_\odot)^{-1/2}; & M < M_{\rm eq} \\ 1.1 \times 10^6 (\Omega h^2)^{-1/3} (M/10^{12}\,{\rm M}_\odot)^{-2/3}; & M > M_{\rm eq} \end{cases} \tag{2.123}$$

gives the redshift at which a region containing mass M enters the Hubble radius. In deriving these expressions, we have ignored the curvature term. This approximation is justified because we will only be interested in those wavelengths for which $z_{\rm enter} \gg z_{\rm flat}$.

We conclude with a comment on another important length scale in cosmology, viz. the horizon size. Suppose for a moment that the universe is described by the expansion factor $a(t) = a_0 t^n$ with $n < 1$ for *all* $t \geq 0$. Then, during the time interval $(0, t)$, a photon can travel a maximum coordinate distance of

$$\xi(t) = \int_0^t \frac{dx}{a(x)} = \frac{1}{a_0} \frac{t^{1-n}}{(1-n)} \tag{2.124}$$

which corresponds to the proper distance:

$$h(t) = a(t)\xi(t) = (1-n)^{-1} t. \tag{2.125}$$

This quantity differs from the Hubble radius $(\dot{a}/a)^{-1} = n^{-1} t$ only by a constant factor of order unity. To avoid any possible confusion between these two quantities, we would like to emphasize the following fact: Notice that $d_H(t)$ is a local quantity and its value at t is essentially decided by the behaviour of $a(t)$ near t; in contrast, the value of $h(t)$ depends on the entire past history of the universe. In fact, $h(t)$ depends very sensitively on the behaviour of $a(t)$ near $t = 0$ – something about which we know nothing. If, for example, $a(t) \propto t^m$ with $m \geq 1$ near $t = 0$, then $h(t)$ is infinite for all $t \geq 0$. Thus there can be several physical situations in which $h(t)$ and $d_H(t)$ differ widely; in such cases, one should examine each case carefully and decide which quantity is physically relevant.

Exercises

2.1 (a) Argue that a homogeneous, isotropic spatial hypersurface must have the line element

$$d\sigma^2 = a^2 \left[\lambda^2(r)\, dr^2 + r^2 d\Omega^2\right]; \quad a = \text{constant}$$

because of spherical symmetry.
(b) Compute the value of scalar curvature P for this line element and show that

$$P = \frac{3}{2a^2 r^3} \frac{d}{dr}\left[r^2\left(1 - \frac{1}{\lambda^2}\right)\right].$$

(c) Homogeneity implies that P is a constant. Equate P to a constant and integrate the resulting equation to obtain

$$r^2\left(1 - \frac{1}{\lambda^2}\right) = Ar^4 + B.$$

(d) Give arguments as to why B should be zero, thereby obtaining $\lambda^2(r) = (1 - Ar^2)^{-1}$. Reduce this expression to the standard form by rescaling r.

2.2 (a) Though the 3-sphere has a formal similarity with the 2-sphere, there are certain significant differences. Show that the spatial sections of the $k = 1$ Friedmann model possess a translation symmetry which leaves *no* points fixed. (This will not be true in 2 dimensions; it is not possible to 'comb' the 2-sphere smoothly.)
(b) Consider a 3-dimensional space of velocities (v_x, v_y, v_z). Define a distance between any two nearby points A and B in this space to be the relative velocity of the observers having the velocities A and B. (The relative velocity, of course, has to be found using special relativity.) Show that the metric in this space has the form of a ($k = -1$) Friedmann model

$$dV^2 = dv_r^2 + \sinh^2 v_r (dv_\theta^2 + \sin^2\theta\, dv_\phi^2).$$

2.3 The metric of the $k = 0$ Friedmann universe can be expressed in the form $g_{\alpha\beta} = \Omega^2 \eta_{\alpha\beta}$ where $\eta_{\alpha\beta}$ is the flat (Lorentzian) metric and $\Omega = \Omega(T)$. Even for the $k = \pm 1$ models, the metric can be reduced to the form $g_{\alpha\beta} = \Omega^2 \eta_{\alpha\beta}$ where $\Omega = \Omega(t, \mathbf{x})$ now depends on spatial coordinates as well. Construct this coordinate system.

2.4 (a) Show that the coordinate transformation

$$R = ra(t), \quad T = F(q), \quad q \equiv q(r,t) = \int^r \frac{x\,dx}{1-kx^2} + \int^t \frac{dy}{a(y)\dot{a}(y)}$$

where $F(q)$ is an arbitrary function of its argument, reduces the Friedmann metric to the form

$$ds^2 = e^{\nu} dT^2 - e^{\lambda} dR^2 - R^2(d\theta^2 + \sin^2\theta\, d\phi^2).$$

Determine the form of ν and λ.
(b) As a special case, consider the transformation

$$R = ra(t), \quad T = t - t_0 + \frac{1}{2} a\dot{a}r^2 + \mathcal{O}(r^4)$$

in which only terms up to quadratic order are retained. Show that, up to this order, the metric becomes

$$\begin{aligned} ds^2 &\cong (1 - \frac{\ddot{a}}{a}R^2)dT^2 - \left(1 + \frac{k}{a^2}R^2 + \frac{\dot{a}^2}{a^2}R^2\right) dR^2 \\ &\quad - R^2(d\theta^2 + \sin^2\theta\, d\phi^2) \\ &= \left[1 + \frac{4\pi G}{3}(\rho + 3p)R^2\right] dT^2 - \left[1 + \frac{8\pi G}{3}\rho R^2\right] dR^2 \\ &\quad - R^2(d\theta^2 + \sin^2\theta\, d\phi^2). \end{aligned}$$

The time derivatives in $\dot{a}, \ddot{a}$ etc are evaluated at $t = t_0$; so ρ and p refer to values at this instant.
(c) For $p = 0$, this metric gives an effective (Newtonian) gravitational potential of $\phi_N = -(2\pi G\rho R^2/3)$. Compute the redshift z of a photon emitted at a point R and received at the origin, taking into account both gravitational and Doppler redshifts. Determine the function $R(z)$ correct up to order z^2. (Why should the accuracy be limited to this order?) In the Newtonian theory, R should be interpreted as angular diameter distance. (Why?) Show that this Newtonian analysis reproduces the general relativistic result (discussed in the text) up to order z^2 correctly. Is this to be expected?

2.5 The dynamics of the electromagnetic field in curved spacetime is described by the action

$$A_{\text{elec}} = \frac{1}{16\pi} \int F_{\alpha\beta} F^{\alpha\beta} \sqrt{-g}\, d^4x; \quad F_{\alpha\beta} = \frac{\partial A_\beta}{\partial x^\alpha} - \frac{\partial A_\alpha}{\partial x^\beta}.$$

(a) Show that this action is invariant under the transformation

$$A_\alpha \to A_\alpha; \quad x^\alpha \to x^\alpha; \quad g_{\alpha\beta} \to \Omega^2 g_{\alpha\beta}; \quad g^{\alpha\beta} \to \Omega^{-2} g^{\alpha\beta}$$

Note that $(A^\alpha)_{\text{new}} = g^{\alpha\beta}_{\text{new}}\,(A_\beta)_{\text{new}} = \Omega^{-2} A^\alpha_{\text{old}}$.
(b) Using the coordinate system $(\eta, \chi, \theta, \phi)$, show that the electromagnetic waves in the Friedmann metric have the time dependence

$$A_\alpha \propto \exp(-ik\eta) = \exp\left[-ik \int \frac{dt}{a(t)}\right]$$

(c) Since the time derivative of the phase of the wave defines the (instantaneous) frequency, conclude that

$$\omega(t)a(t) = \text{constant}.$$

2.6 A material particle is released with some initial velocity in a Friedmann model.
(a) Show that, in the $k = -1$ model as $t \to \infty$, the velocity of the particle approaches that of some fundamental observer but the position of the particle is at constant proper distance from this observer.
(b) What happens in the $k = 0$ and $k = +1$ models?

2.7 Derive the result $p(t) \propto a(t)^{-1}$ by studying the geodesic equation in the Friedmann universe.
(a) Consider a particle travelling along θ = constant, ϕ = constant. Show that the zeroth component of the geodesic equation reads as

$$\frac{d^2t}{ds^2} + \frac{a\dot{a}}{1 - kr^2}\left(\frac{dr}{ds}\right)^2 = 0$$

(b) Eliminate (dr/ds) between the above equation and the first integral

$$\left(\frac{dt}{ds}\right)^2 - \frac{a^2}{1 - kr^2}\left(\frac{dr}{ds}\right)^2 = 1$$

to obtain

$$\frac{d^2t}{ds^2} + \frac{\dot{a}}{a}\left[\left(\frac{dt}{ds}\right)^2 - 1\right] = 0$$

(c) Integrate this to obtain $a[(dt/ds)^2 - 1]$=constant. If $p^\alpha = (dx^\alpha/ds)$ is the four-momentum of the particle, then the condition $p^\alpha p_\alpha = 1$ reduces to $[(dt/ds)^2 - \sigma_{\alpha\beta}\, p^\alpha p^\beta] = 1$. Show that $\sigma_{\alpha\beta}\, p^\alpha p^\beta \equiv |\mathbf{p}|^2$=(constant$/a^2$).
(d) The geodesic for a particle, moving in a spacetime with metric $g_{\alpha\beta}$, can be obtained most efficiently from the Hamilton–Jacobi equation

$$g^{\alpha\beta}\frac{\partial A}{\partial x^\alpha}\frac{\partial A}{\partial x^\beta} = m^2$$

where A is the action. Write down this equation in the Friedmann metric. Reduce the problem of determining the radial geodesics to quadrature.

2.8 In deriving the relation $p(t) \propto a(t)^{-1}$ in the text, we did not seem to account for the gravitational attraction of cosmic matter on the particle. In exercise 2.7 above, the gravitational effects are included (through the $\Gamma^\alpha_{\beta\mu}\, \dot{x}^\beta \dot{x}^\mu$ term of the geodesic equation). Why do both approaches give the same result?

2.9 Let $n(t_0)$ be the proper number density of a class of sources which are distributed uniformly in the universe. We assume that the number of sources in unit comoving volume is a constant; i.e. $n(t)a^3(t)$=constant. Let $N(z)$ be the number of such sources with redshifts less than z (as observed from the earth today).

(a) Show that

$$N(z) = 4\pi n(t_0)a_0^3 \int_{t_z}^{t_0} \frac{r^2(t_1)}{a(t_1)} dt_1$$

where t_z is the time corresponding to redshift z and $r(t_1)$ is the null geodesic connecting (t_1, r) and $(t_0, 0)$.
(b) Show that, for small-z, this expression can be approximated as

$$N(z) = \frac{4\pi}{3} \frac{n(t_0)}{H_0^3} z^3 \left[1 - \frac{3}{2} z(1 + q_0) + \mathcal{O}(z^2) \right].$$

(c) Suppose the sources have the same intrinsic luminosity L. Show that the number of sources with fluxes (energy per unit time per unit area) greater than S, as observed from the earth today, is

$$N(S) = \int_{t_s}^{t_0} 4\pi a^2(t_1) r^2(t_1) n(t_1)\, dt_1$$

where t_s is determined through

$$\frac{r^2(t_s)}{a^2(t_s)} = \frac{L}{4\pi S a_0^4}.$$

(d) Show that, for small $(t_0 - t_s)$, we get the approximate expression

$$N(S) = \frac{4\pi}{3} n(t_0) \left(\frac{L}{4\pi S} \right)^{3/2} \left[1 - 3H_0 \left(\frac{L}{4\pi S} \right)^{1/2} + \cdots \right].$$

(e) Compute $N(z)$ exactly for the $k = 0$, matter dominated universe. Estimate numerically the number of galaxies which may be expected to intervene between us and a quasar of redshift $z = 3$.

2.10 The surface brightness S of a source is defined to be $(dF/d\Omega)$ where dF is the flux received by us from the part of the source which subtends a solid angle $d\Omega$. Consider a class of sources with the same intrinsic properties. Show that $(dF/d\Omega) \propto (1 + z)^{-4}$.

2.11 (a) For a matter dominated universe discussed in the text, show that

$$a_0 r_1 = H_0^{-1} \Omega^{-2} (1 + z)^{-1} \left[2\Omega z + 2(\Omega - 2)(\sqrt{\Omega z + 1} - 1) \right].$$

(b) From this, derive the expressions

$$d_L = H_0^{-1} q_0^{-2} \left[z q_0 + (q_0 - 1)(\sqrt{2 q_0 z + 1} - 1) \right],$$

$$d_A = H_0^{-1} q_0^{-2} (1 + z)^{-2} \left[z q_0 + (q_0 - 1)(\sqrt{2 q_0 z + 1} - 1) \right].$$

(c) Are these expressions monotonic functions of z?

2.12 The maximum proper distance a photon can travel in the interval $(0, t)$ is given by the horizon size

$$h(t) = a(t) \int_0^t \frac{dx}{a(x)}.$$

Show that, for a matter dominated universe

$$h(z) = H_0^{-1}(1+z)^{-1}(\Omega-1)^{-1/2}\cos^{-1}\left[1-\frac{2(\Omega-1)}{\Omega(1+z)}\right] \quad (\text{for } \Omega > 1)$$
$$= 2H_0^{-1}(1+z)^{-3/2} \quad (\text{for } \Omega = 1)$$
$$= H_0^{-1}(1+z)^{-1}(1-\Omega)^{-1/2}\cosh^{-1}\left[1+\frac{2(1-\Omega)}{\Omega(1+z)}\right] \quad (\text{for } \Omega < 1).$$

Show also that $d_H \simeq 3H_0^{-1}\,\Omega_0^{-1/2}\,(1+z)^{-3/2} = 3t$ for $(1+z) \gg \Omega^{-1}$.

2.13 (a) Show that the line element

$$ds^2 = dt^2 - t^2\left[\frac{dr^2}{1+r^2} + r^2(d\theta^2 + \sin^2\theta\, d\phi^2)\right]$$

is a solution to Friedmann equations with $p = \rho = 0$. Also show that there exists a coordinate transformation which will transform the metric into the Lorentzian form:

$$ds^2 = dT^2 - dR^2 - R^2(d\theta^2 + \sin^2\theta\, d\phi^2).$$

(b) The metric in (t, r, θ, ϕ) coordinates is called the Milne universe. Interpret the origin of the relation $p \propto a^{-1} \propto t^{-1}$ in this metric. Calculate d_A, d_L and h and interpret these results physically.

2.14 (a) Solve Einstein's equations for $k = 1$ and $k = 0$, when the equation of state for matter is $p = -\rho$. (It is believed that matter in our universe is described by such an equation of state, during the very early stages of its evolution, for a short period of time.) Show that the line elements have the form

$$ds^2 = dt^2 - e^{2Ht}\left[dr^2 + r^2(d\theta^2 + \sin^2\theta\, d\phi^2)\right],$$
$$ds^2 = dT^2 - \cosh^2 HT\left[\frac{dR^2}{1-R^2} + R^2(d\theta^2 + \sin^2\theta\, d\phi^2)\right].$$

(b) Since the source is the same, we expect the two line elements above to represent the same spacetime. Prove that this is indeed the case by finding the coordinate transformation between (t, r) and (T, R).

(c) Consider a $k = 0$ Friedmann model for which the source is made up of ordinary matter (with $p = 0, \rho_{NR} \propto a^{-3}$) and the 'vacuum energy' (with $p_v = -\rho_v$; $\rho_v > 0$). Thus $T_0^0 = (\rho_{NR} + \rho_v)$ and $T_1^1 = \rho_v$ etc. Show that the equation $d(\rho a^3) = p\, d(a^3)$ is satisfied by the choice, $\rho_{NR} \propto a^{-3}$, $\rho_v =$ constant. Integrate the Einstein equations to find $a(t)$. Show that the 'age' of the universe is given by

$$t_0 = \frac{2}{3}H_0^{-1}\Omega_v^{-1/2}\ln\left[\frac{1+\Omega_v^{1/2}}{(1-\Omega_v)^{1/2}}\right]$$

where $\Omega_v = (\rho_v/\rho_c)$. Note that $t_0 > H_0^{-1}$ for $\Omega_v \gtrsim 0.74$. This is an example in which the bound $t_0 < H_0^{-1}$ derived in the text does not apply. Explain why.

2.15 The last two problems illustrate situations in which the specification of the source does not uniquely specify the value of k. (In exercise 2.13, the trivial source $p = 0, \rho = 0$ leads to spacetimes with $k = 0$ or $k = -1$; in exercise 2.14, the source with equation of state $p = -\rho$ leads to spacetimes with $k = 0$ or $k = 1$.)
(a) Investigate the geometric origin of this non-uniqueness.
(b) Show that (fortunately!) there does not exist any other spacetime which can be represented as Friedmann universes with different values of k.

2.16 Consider a closed Friedmann model which is matter dominated. The total proper volume of the universe at present ($t = t_0$) is $2\pi^2 a_0^3$.
(a) What is the total proper volume that we see, looking out into the sky?
(b) What is the total proper volume *now* occupied by matter which we see, looking out into the sky?

3
Thermal history of the universe

3.1 Introduction

We saw in the last chapter that the physical conditions in an expanding universe change with time. To understand the features of the universe today, it is necessary to grasp the past history of the universe. We shall now tackle this question.

The physical processes which occur in the early universe are described in this chapter. Section 3.2 develops the basic thermodynamics needed to understand these processes. In sections 3.3 and 3.4, we consider the possible existence of a relic background of massless or massive fermions (like the neutrinos) in our universe today. Section 3.5 discusses the primordial nucleosynthesis and its observational relevance. Finally, the decoupling of matter from radiation is studied in section 3.6.

3.2 Distribution functions in the early universe

The analysis in chapter 2 showed that the universe was dominated by radiation at redshifts higher than $z_{\text{eq}} \simeq 3.9 \times 10^4(\Omega h^2)$. In the radiation dominated phase, the temperature of the radiation will be higher than $T_{\text{eq}} \simeq 9.2\ (\Omega h^2)\,\text{eV} \simeq 1.07 \times 10^5(\Omega h^2)\,\text{K}$ and will be increasing as $T \propto (1+z)$.

The contents of the universe, at these early epochs, will be in a form very different from that in the present day universe. Atomic and nuclear structures have binding energies of the order of a few tens of eV and MeV respectively. When the temperature of the universe was higher than these values, such systems could not have existed as bound objects. Further, when the temperature T of the universe becomes higher than the rest mass m of a charged particle (say, electron or muon) then the photon energy will be large enough to produce these particles and their antiparticles in large numbers. For example, when $T \gg m_{\text{elec}} \simeq 0.5\,\text{MeV} \simeq 5.8 \times 10^9\,\text{K}$

there will be a large number of positrons in the universe. The typical energy of these particles will be T, making them ultra-relativistic.

Thus, depending on the temperature T, the early universe would be populated by different kinds of elementary particles at different times. To work out the physical processes at some time t, we need to know the distribution function $f_A(\mathbf{x}, \mathbf{p}, t) \equiv f_A(\mathbf{p}, t)$ of these particles. Here $A = 1, 2 \cdots$ labels different species of particles, like electron, muon etc; the dependence of f_A on the space coordinates is ruled out because of the homogeneity of the universe.

To determine the form of $f_A(\mathbf{p}, t)$ we may reason as follows: The different species of particles will be interacting constantly through various forces, scattering off each other and exchanging energy and momentum. If the rate of these reactions, $\Gamma(t)$, is much higher than the rate of expansion of the universe, $H(t) = (\dot{a}/a)^{-1}$, then these interactions can produce (and maintain) thermodynamic equilibrium among the interacting particles with some temperature $T(t)$. All these interactions which occur between the particles have a short range. (The Coulomb force between charged particles has a long range; but, in a plasma, the process of Debye shielding reduces this range, making it effectively a short-range force.) Therefore we may assume that the role of these interactions is limited to providing a mechanism for thermalization and ignore their effects in deciding the *form* of the distribution function (see exercise 3.1). In that case, the particles may be treated as an *ideal* Bose or Fermi gas with the distribution function:

$$f_A(\mathbf{p}, t)\, d^3\mathbf{p} = \frac{g_A}{(2\pi)^3} \left\{\exp[(E_\mathbf{p} - \mu_A)/T_A(t)] \pm 1\right\}^{-1} d^3\mathbf{p}, \tag{3.1}$$

where g_A is the spin degeneracy factor of the species, $\mu_A(T)$ is the chemical potential, $E(\mathbf{p}) = (\mathbf{p}^2 + m^2)^{1/2}$ and $T_A(t)$ is the temperature characterizing this species at time t. The upper sign $(+1)$ corresponds to fermions and the lower sign (-1) is for bosons.

At any instant in time, the universe will also contain a black-body distribution of photons with some characteristic temperature $T_\gamma(t)$. If a particular species couples to the photon directly or indirectly, and if the rate of these A–γ interactions is high enough (i.e. $\Gamma_{A\gamma} \gg H$), then these particles will have the same temperature as photons : $T_A = T_\gamma$. Since this is usually the case, one often refers to the photon temperature by the term 'temperature of the universe'. Of course, any set of particle species $A, B, C \cdots$ which are interacting among themselves at a high enough rate will also have the same temperature $T_A = T_B = T_C \cdots$.

As the universe evolves, the temperature $T(t)$ changes due to expansion in a timescale of the order of $H^{-1}(t) \equiv (\dot{a}/a)^{-1}$; the *rate* at which temperature is changing is given by $H(t)$. The rate of interaction (per particle) can be expressed as $\Gamma \equiv n < \sigma v >$ where n is the number density of target particles, v is the relative velocity and σ is the interaction cross section. Since σ is usually a function of energy, $< \sigma v >$ denotes an average value for this combination. As long as $\Gamma \gg H$, the interactions can maintain equilibrium. In that case, f will evolve adiabatically, maintaining the form of the equilibrium distribution given by (3.1), with the temperature corresponding to the instantaneous value.

The assumption of spatial homogeneity has played a crucial role in the above description. This can be seen as follows: In the characteristic timescale $H^{-1}(t)$, over which the parameters in the universe change, the particles can travel only a maximum distance $cH^{-1}(t)$. So, if two regions in the universe had different temperatures at some time t_1, particle interactions may not be able to bring them to the same temperature at a later time $t > t_1$. By imposing strict homogeneity for the entire universe *at all times*, we have bypassed this problem.

It could happen that, at some instant, the *total* interaction rate $\Gamma_A(t)$ of a species A (taking into account interactions among themselves as well as with all other species) falls below the expansion rate $H(t)$: $\Gamma_A(t) \lesssim H(t)$; but the interaction rate among all other species Γ_{other} could still be much higher than the expansion rate: $\Gamma_{\text{other}} \gg H$. In such a situation, the distribution function of all species other than A will be still given by (3.1) with a common temperature T. The particle species A, however, would have completely decoupled; its distribution function will not in general be given by (3.1). Its form, however, can be ascertained by the following argument.

Once the species A is completely decoupled, each of the A particles will be travelling along a geodesic in the spacetime. We have seen in chapter 2 that f is conserved during such free propagation. This allows one to obtain the function f_{dec} after the species has decoupled, from the known form of f_{equi} before decoupling. For simplicity, let us assume that the decoupling occurs instantaneously at some time $t = t_D$ when the temperature was T_D and the expansion factor was a_D. For $t < t_D$, the distribution function is given by (3.1). At some later time, $t > t_D$, let the distribution function be $f_{\text{dec}}(p, t)$. Because of the redshift in momentum, all particles with momentum p at time t must have had momentum $p[a(t)/a(t_D)]$ at $t = t_D$. Therefore,

$$f_{\text{dec}}(p, t) = f_{\text{equi}}\left(p\frac{a(t)}{a(t_D)}, t_D\right) \qquad (\text{for } t > t_D) \qquad (3.2)$$

where f_{equi} is the equilibrium distribution function of (3.1). Thus, as long as the species A was in equilibrium at *some* time, we can determine its distribution function at all later times.

From the distribution function (3.1), we can calculate the number density n, energy density ρ and pressure p. Suppressing the time dependence and the subscript A for simplicity, we have:

$$n = \int f(\mathbf{k})\, d^3\mathbf{k} = \frac{g}{2\pi^2}\int_m^\infty \frac{(E^2 - m^2)^{1/2} E\, dE}{\exp[(E-\mu)/T] \pm 1}, \tag{3.3}$$

$$\rho = \int E f(\mathbf{k})\, d^3\mathbf{k} = \frac{g}{2\pi^2}\int_m^\infty \frac{(E^2 - m^2)^{1/2} E^2\, dE}{\exp[(E-\mu)/T] \pm 1}, \tag{3.4}$$

$$p = \int \frac{1}{3}\frac{|k|^2}{E} f(\mathbf{k})\, d^3\mathbf{k} = \frac{g}{6\pi^2}\int_m^\infty \frac{(E^2 - m^2)^{3/2}\, dE}{\exp[(E-\mu)/T] \pm 1}. \tag{3.5}$$

We will use the symbol k to denote the momentum when the pressure is denoted by letter p. For a collection of relativistic particles, the pressure p corresponding to velocity v is $p = m(v^2/3\sqrt{1-v^2})$, which is the same as $p = (k^2/3E)$.

A useful identity can be derived from these expressions. Differentiating (3.5) with respect to T, and treating μ as some specified function of T, we get

$$\frac{dp}{dT} = \frac{4\pi}{3}\int_0^\infty \frac{k^4\, dk}{E} f^2 \left[\exp\frac{(E-\mu)}{T}\right]\left[\frac{E}{T^2} + \frac{d}{dT}\left(\frac{\mu}{T}\right)\right]. \tag{3.6}$$

Using the fact that

$$\frac{df}{dk} = -\frac{k}{ET} f^2 \exp\frac{(E-\mu)}{T}, \tag{3.7}$$

we can rewrite (dp/dT) as

$$\frac{dp}{dT} = -\frac{4\pi}{3}\int_0^\infty dk\,(k^3 T)\left(\frac{df}{dk}\right)\left[\frac{E}{T^2} + \frac{d}{dT}\left(\frac{\mu}{T}\right)\right]. \tag{3.8}$$

Integrating by parts and using the definitions of ρ and p we find

$$\frac{dp}{dT} = \frac{1}{T}(\rho + p) + nT\frac{d}{dT}\left(\frac{\mu}{T}\right). \tag{3.9}$$

From Einstein's equations we have the relation $d(\rho a^3) = -p\, d(a^3)$, which can be written as

$$\frac{d}{dT}[(\rho + p)a^3] = a^3\frac{dp}{dT}. \tag{3.10}$$

Substituting for (dp/dT) from (3.9) and rearranging the terms, we get:

$$d(sa^3) \equiv d\left\{\frac{a^3}{T}(\rho + p - n\mu)\right\} = \left(\frac{\mu}{T}\right) d(na^3). \tag{3.11}$$

In most cases of interest to us either $\mu \ll T$ *or* (na^3) will be approximately constant. The above relation shows that, in either case, the quantity (sa^3) will be conserved.

When $\mu \ll T$, the expression for s reduces to $T^{-1}(\rho+p)$. Expanding the quantity $T\, d[a^3 T^{-1}(\rho + p)]$ and using the relation $(dp/dT) \simeq T^{-1}(p+\rho)$ derived earlier, we get

$$\begin{aligned} T\, d(sa^3) = T\, d\left[\frac{(\rho+p)a^3}{T}\right] &= d\left[(\rho+p)a^3\right] - (\rho+p)a^3\frac{dT}{T} \\ &\simeq d[(\rho+p)a^3] - a^3\, dp = d(\rho a^3) + p\, d(a^3). \end{aligned} \tag{3.12}$$

Comparing with the familiar thermodynamic relation $T\, dS = dE + p\, dV$, we see that $s = T^{-1}\ (\rho + p)$ may be interpreted as the entropy density. Then (3.11) shows that entropy density $s \propto a^{-3}$ during expansion, provided $\mu \ll T$. Note that s is an additive quantity.

The expressions for n, ρ and p simplify considerably in some limiting cases. When the particles are highly relativistic ($T \gg m$) and nondegenerate ($T \gg \mu$), we get

$$\rho \cong \frac{g}{2\pi^2}\int_0^\infty \frac{E^3\, dE}{\mathrm{e}^{E/T} \pm 1} = \begin{cases} g_B(\pi^2/30)T^4 & \text{(bosons)} \\ \frac{7}{8}g_F(\pi^2/30)T^4 & \text{(fermions)} \end{cases}. \tag{3.13}$$

Thus the total energy density contributed by all the relativistic species together can be expressed as

$$\rho_{\text{total}} = \sum_{i=\text{boson}} g_i\left(\frac{\pi^2}{30}\right)T_i^4 + \sum_{i=\text{fermion}} \frac{7}{8}g_i\left(\frac{\pi^2}{30}\right)T_i^4 = g_{\text{total}}\left(\frac{\pi^2}{30}\right)T^4 \tag{3.14}$$

where

$$g_{\text{total}} \equiv \sum_{\text{boson}} g_B\left(\frac{T_B}{T}\right)^4 + \sum_{\text{fermion}} \frac{7}{8}g_F\left(\frac{T_F}{T}\right)^4. \tag{3.15}$$

In writing g_{total}, we have explicitly taken into account the possibility that all the species may have a thermal distribution but may not have the same temperature. If all species have the same temperature, then $g = g_{\text{boson}} + (7/8)g_{\text{fermion}}$.

The pressure due to relativistic species is $p \simeq (\rho/3) = g(\pi^2/90)T^4$; so the entropy density of the relativistic species of particles will be

$$s \cong \frac{1}{T}(\rho + p) = \frac{2\pi^2}{45}qT^3 \tag{3.16}$$

with

$$q \equiv q_{\text{total}} = \sum_{\text{boson}} g_B\left(\frac{T_B}{T}\right)^3 + \frac{7}{8}\sum_{\text{fermion}} g_F\left(\frac{T_F}{T}\right)^3. \tag{3.17}$$

Clearly, $q_{\text{total}} = g_{\text{total}}$ if all the particles have the same temperature. Our previous analysis shows that the quantity $S = qT^3a^3$ is conserved during the expansion.

The number density of relativistic particles can be computed in the same way:

$$n \cong \frac{g}{2\pi^2}\int_0^\infty \frac{E^2\,dE}{e^{E/T}\pm 1} = \begin{cases} (\zeta(3)/\pi^2)g_B T^3 & \text{(boson)} \\ \frac{3}{4}(\zeta(3)/\pi^2)g_F T^3 & \text{(fermion)} \end{cases} . \tag{3.18}$$

where $\zeta(3) \simeq 1.202$ is the Riemann zeta function of order 3. Combining this with (3.13), we find that the mean energy of the particles $< E >\equiv (\rho/n)$ is about $2.7T$ for bosons and $3.15T$ for fermions. Note that s is proportional to the number density of relativistic particles, if all species have the same temperature; in fact, $s \simeq 1.8qn_\gamma$ where n_γ is the photon number density.

In the opposite limit of $T \ll m$, the exponential in (3.1) is large compared to unity. Then we get, for *both* bosons and fermions, the expression:

$$\begin{aligned} n &\cong \frac{g}{2\pi^2}\int_0^\infty p^2\,dp\,\exp\left[-\frac{(m-\mu)}{T}\right]\exp\left(-\frac{p^2}{2mT}\right) \\ &= g\left(\frac{mT}{2\pi}\right)^{3/2}\exp\left[-\frac{1}{T}(m-\mu)\right]. \end{aligned} \tag{3.19}$$

In this limit $\rho \simeq nm$ and $p = nT \ll \rho$.

A comparison of (3.18) and (3.19) shows that the number (and energy) density of non-relativistic particles are exponentially damped by the factor $\exp -(m/T)$ with respect to that of the relativistic particles. So, for $t < t_{\rm eq}$, in the radiation dominated phase, we may ignore the contribution of non-relativistic particles to ρ. We have seen in chapter 2 that, during the radiation dominated phase, $a(t) \propto t^{1/2}$; therefore

$$\left(\frac{\dot a}{a}\right)^2 = H^2(t) = \frac{1}{4t^2} = \frac{8\pi G}{3}\rho = \frac{8\pi G}{3}g\left(\frac{\pi^2}{30}\right)T^4. \tag{3.20}$$

It is convenient to express these results in terms of the Planck energy $m_{\rm Pl} = G^{-1/2} = 1.22\times 10^{19}$ GeV:

$$H(T) \cong 1.66 g^{1/2}\left(\frac{T^2}{m_{\rm Pl}}\right), \tag{3.21}$$

$$t \cong 0.3 g^{-1/2}\left(\frac{m_{\rm Pl}}{T^2}\right) \simeq 1\,{\rm s}\left(\frac{T}{1\,{\rm MeV}}\right)^{-2} g^{-1/2}. \tag{3.22}$$

The factor g in these expressions counts the degrees of freedom of those particles which are *still* relativistic at this temperature T. As the temperature decreases, more and more particles will become non-relativistic and g and q will decrease; thus $g = g(T)$ and $q = q(T)$ are slowly varying, decreasing functions of T. In the currently popular models for particle interactions,[1] $g \simeq 10^2$ at $T \gtrsim 300\,\text{GeV}$; $g \simeq 10$ for $T \simeq (100\text{–}1)\,\text{MeV}$ and $g \simeq 3$ for $T < 1\,\text{MeV}$.

The slow variation of $q(T)$ has one important consequence. The expression for the conserved entropy in the radiation dominated phase is $s \propto q(T)\, T^3 a^3$ (see (3.16)). So the temperature T decreases as a^{-1} only as long as q is constant. If the number of degrees of freedom changes, then T will decrease slightly more slowly than a^{-1}; the correct relation is $q^{1/3}(T) T \propto a^{-1}$.

Finally, let us consider the distribution function for a species which has already decoupled. The expression in (3.2) simplifies considerably if the decoupling occurs either when the species is ultra-relativistic ($T_D \gg m$) or when it is non-relativistic ($T_D \ll m$). In the first case,

$$f_{\text{dec}}(p) = f_{\text{equi}}\left(p\frac{a(t)}{a(t_D)}, T_D\right) \cong \frac{g}{(2\pi)^3}\left[\exp\frac{1}{T_D}\left(p\frac{a(t)}{a(t_D)}\right) \pm 1\right]^{-1}. \tag{3.23}$$

This has the same form as the f_{equi} for a relativistic species with the temperature

$$T(t) = T_D[a(t_D)/a(t)], \tag{3.24}$$

even though this species is not in thermodynamic equilibrium any longer. The 'temperature' in this distribution function falls *strictly* as a^{-1}; the entropy of these particles, $S_A = (s_A a^3)$ is separately conserved. Note that for the species which are still in thermal equilibrium, $T \propto q^{-1/3} a^{-1}$ falls more slowly.

The number density of these decoupled particles will be given by (3.18):

$$n = g_{\text{eff}}\left(\frac{\zeta(3)}{\pi^2}\right) T_D^3 \left(\frac{a_D}{a}\right)^3 \tag{3.25}$$

where $g_{\text{eff}} = (3g/4)$ for fermion and $g_{\text{eff}} = g$ for boson. (Here g refers to the spin degeneracy factor of the particular species which has decoupled.) This number density will be comparable to the number density of photons at any given time. In particular, any such decoupled species will continue to exist in our universe today as a relic background, with number densities comparable to the number density of photons.

The following point should be noted: Suppose that a species with mass m decouples at the temperature T_D with $T_D \gg m$. At the time of decoupling, most of these particles will be ultra-relativistic and their (mean) momentum $p(t_D)$ and energy $E(t_D) = (p^2(t_D)+m^2)^{1/2} \simeq p(t_D)$ will be of order T_D. Their distribution function at $t = t_D$ is well-approximated by the f_{equi} of zero-mass particles. Decoupling 'freezes' the distribution function in this form. At a later time ($t > t_D$), the mean momentum of the particles will get redshifted to a value $p(t) = p(t_D)\,(a_D/a) \simeq T_D(a_D/a)$. For $t \gg t_D$, most of the particles will have momentum $p(t)$ which is much smaller than m. Thus the individual particles would have become non-relativistic when the universe has expanded sufficiently, which will happen when the temperature of the universe drops below $T_{\rm nr} \simeq m$: that is, when $(a/a_D) \gtrsim (T_D/m)$. The energy of each of these particles will now be $E(t) = (p^2(t) + m^2)^{1/2} \simeq m$. *But the distribution function (and the number density) of the particles will still be given by the (frozen-in) form which corresponds to relativistic particles.* Thus, for $t \gg t_D$, the number density of these particles will be similar to those of relativistic species but the energy density will be $\rho_{\rm dec} \simeq nm$.

Consider next the other extreme case, that of a species which decouples when most of the particles are already non-relativistic: $(T_D \ll m)$. In this case,

$$
\begin{aligned}
f_{\rm dec}(p) &= f_{\rm equi}(p\frac{a}{a_D}, T_D) \\
&\cong \frac{g}{(2\pi)^3}\exp\left[-\frac{(m-\mu)}{T_D}\right]\exp\left[-\frac{p^2}{2m}\frac{1}{T_D}\left(\frac{a}{a_D}\right)^2\right] \\
&\cong \frac{g}{(2\pi)^3}e^{-m/T_D}\exp\left[-\frac{p^2}{2mT_D}\left(\frac{a}{a_D}\right)^2\right]
\end{aligned}
\tag{3.26}
$$

where we have further assumed that $\mu \ll T_D$. This distribution function has the same form as that of a non-relativistic Maxwell–Boltzmann gas with a 'temperature' $T(t) \equiv T_D(a_D/a)^2$ which decreases as the *square* of the expansion factor. The corresponding number density is given by (3.19):

$$
\begin{aligned}
n &= g\left(\frac{mT_D}{2\pi}\right)^{3/2}\left(\frac{a_D}{a}\right)^3\exp\left(-\frac{1}{T_D}(m-\mu)\right) \\
&\cong g\left(\frac{mT_D}{2\pi}\right)^{3/2}\left(\frac{a_D}{a}\right)^3\exp\left(-\frac{m}{T_D}\right) \quad \text{(for } \mu \ll T_D\text{)}.
\end{aligned}
\tag{3.27}
$$

As is to be expected, $n \propto a^{-3}$. The energy density of these particles will be $\rho \simeq nm$.

To any species of particle which is not being created or destroyed ($n \propto a^{-3}$), we can assign a conserved number $N \propto na^3$; since $a^3 \propto s^{-1}$, we can conveniently define this number to be $N \equiv (n/s)$. From our expressions (3.16), (3.18) and (3.19) (valid for $\mu \ll T$) it follows that

$$
N = \begin{cases}
\left[45\zeta(3)/2\pi^4\right]\left[g_{\rm eff}/q\right] \cong 0.28\,(g_{\rm eff}/q) \\
\left[45/2\pi^4\right](\pi/8)^{1/2}(g/q)\,(m/T)^{3/2}\,e^{-m/T} \\
\qquad\qquad \cong 0.15\,(g/q)\,(m/T)^{3/2}\,e^{-m/T}
\end{cases}
\tag{3.28}
$$

where $g_{\rm eff} = g$ for bosons and $g_{\rm eff} = (3g/4)$ for fermions.

The assumption that decoupled particles travel along geodesics is equivalent to ignoring 'gravitational collisions' between these particles. In other words, the gravitational force on each relic particle is assumed to be entirely due to the gravitational field produced by the smooth distribution of matter. We have seen in chapter 1 that the timescale for gravitational collisions is N times the dynamical timescale. Since the number of relic elementary particles inside a Hubble radius is enormous, the neglect of gravitational collisions is perfectly justified.

3.3 Relic background of relativistic particles

The discussion so far was based on general principles. Given the mass spectrum and the interactions of the elementary particles, the formalism developed above can be used to make concrete predictions.

To understand the processes which occur in the early universe when the temperature was T, we need to know the physics of particle interactions at energies $E \simeq T$. Based on our current knowledge of the latter, the study of early universe can be divided into three different phases:

(i) Our understanding of particle interactions is reasonably complete for energies below 1 GeV. Correspondingly, we should be able to follow the evolution of the universe from the temperature of $T \simeq 1\,\text{GeV} \simeq 1.2 \times 10^{13}\,\text{K}$ downwards with reasonable accuracy.

(ii) There are several theoretical models which attempt to describe the particle interactions in the energy range 1 GeV to 10^{16} GeV. These models are comparatively more speculative, with the uncertainties increasing with the energy. (The range (1–100) GeV is somewhat better understood because it is accessible in particle accelerators.) Given a specific particle physics model, the evolution of the universe in this temperature range $10^{16}\,\text{GeV} \simeq 1.2 \times 10^{29}\,\text{K}$ to $1\,\text{GeV} \simeq 1.2 \times 10^{13}\,\text{K}$ can be worked out. Since the models are not unique, we cannot obtain unique *predictions.* However, it is often possible to work out *some* consequences of these models which can be tested by cosmological observations.

(iii) The physics at energies above 10^{16} GeV is very uncertain. Quantum gravitational effects, about which we know very little, will be significant at energies $E \gtrsim m_{\text{Pl}} \simeq 1.22 \times 10^{19}\,\text{GeV}$. The very basis for our discussion, classical general relativity, breaks down at these energies.

The uncertainty in our knowledge of particle interactions at high energies prevents us from *predicting* a unique material composition or evolutionary history for our universe. To make any progress, it is necessary to make reasonable assumptions about the material content of the universe at some moment in time and work out the consequences. Since the physical processes are relatively well understood at $T \lesssim 10^{12}\,\text{K}$, we will discuss this part of the evolution first.

Let us begin by ascertaining the composition of the universe at $T \lesssim 10^{12}\,\mathrm{K}$. Since the rest mass of the electron $m_e \simeq 0.5\,\mathrm{MeV} = 6\times 10^9\,\mathrm{K}$, we expect a significant number density (i.e. number density comparable to that of photons) of ultra-relativistic electrons (e) and positrons ($\overline{e}$) at $T > 6\times 10^9\,\mathrm{K}$. The only other particle species which could be relativistic at the temperature $10^{12}\,\mathrm{K}$ is the neutrino (ν). It is known that there exist 3 kinds ('flavours') of neutrinos, viz. electron-neutrino (ν_e), muon-neutrino (ν_μ) and tau-neutrino (ν_τ). The masses of these particles are uncertain; the experimental *upper bounds* on the masses are 13 eV, 0.25 MeV and 35 MeV for ν_e, ν_μ and ν_τ respectively. We shall assume, for the time being, that $m_\nu = 0$ for all the three species. There could be interesting physical consequences if any of the neutrinos has non-zero mass; this possibility needs to be discussed separately and we will take it up in section 3.4.

The neutrons (n) and protons (p) contained in the present day universe must have existed at $T \simeq 10^{12}\,\mathrm{K}$ as well, since these particles could not have been produced at $T < 10^{12}\,\mathrm{K}$. The ratio between the number density of baryons (n_B) and the number density of photons (n_γ) remains approximately constant from $T \simeq 10^{12}\,\mathrm{K}$ till today. This number in the present day universe is $(n_B/n_\gamma)_0 \simeq (\rho_c\Omega_B\,/m_B n_\gamma)_0 \simeq 10^{-8}$ to 10^{-10}. The smallness of this (conserved) number shows that we may ignore the effect of n_B on the overall dynamics of the radiation-dominated universe (see exercise 3.3). There could also be a tiny fraction of muon–antimuon pairs; we will ignore them for simplicity.

Since photons are not conserved, the chemical potential for photons is identically zero. The reaction $e\overline{e} \leftrightarrow \gamma\gamma$ maintains the equilibrium between $e, \overline{e}$ and γ at this temperature. The conservation of chemical potential in this reaction implies that $(\mu_e + \overline{\mu}_e) = 0$; that is $\overline{\mu}_e = -\mu_e$. (We will denote a particle–antiparticle pair by $A, \overline{A}$ and the corresponding chemical potentials by $\mu_A, \overline{\mu}_A$.) The excess of electrons over positrons will then be

$$
\begin{aligned}
n - \overline{n} &= \frac{g}{2\pi^2}\int_m^\infty E(E^2 - m^2)^{1/2} \\
&\qquad \times \left[\frac{1}{\exp[(1/T)(E-\mu)]+1} - \frac{1}{\exp[(1/T)(E+\mu)]+1}\right] dE \\
&\cong \frac{gT^3}{6\pi^2}\left[\pi^2\left(\frac{\mu}{T}\right) + \left(\frac{\mu}{T}\right)^3\right] \qquad (\text{for } T \gg m).
\end{aligned}
\tag{3.29}
$$

As the universe cools to temperature $T \ll m_e$, electrons and positrons will annihilate in pairs and only this small excess will survive (see exercise 3.4).

The only other *charged* particle which will be present in the universe is the proton. Since our universe appears to be electrically neutral (the bound on net number density of free charges being $(n_Q/s) \lesssim 10^{-27}$; see exercise 3.5), the electron excess $(n - \overline{n})$ should be equal to the number density of protons n_p. Since $(n_p/n_\gamma) \simeq 10^{-8}$ it follows that $[(n - \overline{n})/n_\gamma] \simeq 10^{-8}$. Using (3.18) and (3.29), we can write

$$\frac{n - \overline{n}}{n_\gamma} \cong \left(\frac{g_e}{g_\gamma}\right) \frac{\pi^2}{6\zeta(3)} \left[\frac{\mu}{T} + \frac{1}{\pi^2}\left(\frac{\mu}{T}\right)^3\right] \cong 1.33\left(\frac{\mu}{T}\right) \simeq 10^{-8}. \tag{3.30}$$

Clearly $(\mu/T) \ll 1$ and we can set $\mu \cong 0$ for both electrons and positrons.

Similarly, from the reaction $\nu\overline{\nu} \leftrightarrow e\overline{e}$, it follows that $\mu_\nu + \overline{\mu}_\nu = \mu_e + \overline{\mu}_e = 0$. The excess of neutrinos over antineutrinos will again be given by an expression similar to (3.29). Unfortunately, the value of $[(n_\nu - \overline{n}_\nu)/n_\gamma]$ for our universe is not known. If this number is large, then our universe will have a large lepton number (L), which is far in excess of the baryon number B. Since our universe does not seem to have large values for any quantum number, a large value for L will require special choice of initial conditions. This suggests that L should be small. In that case, $\mu_\nu = -\overline{\mu}_\nu \cong 0$. We will make this assumption in what follows.

The above arguments show that, at $T \simeq 10^{12}$ K, the energy density of the universe is essentially contributed by $e, \overline{e}, \nu, \overline{\nu}$ and photons. Since the interactions among them maintain the equilibrium, they all have the same temperature. Taking $g_B = g_\gamma = 2$, $g_e = \overline{g}_e = 2$, $g_\nu = \overline{g}_\nu = 1$ and including 3 flavours of neutrinos, we find:

$$g_{\rm total} = g_B + \frac{7}{8} g_F = 2 + \frac{7}{8}[2 + 2 + 2 \times 3] = \frac{43}{4} = 10.75\,. \tag{3.31}$$

The g-values for electrons and positrons represent the two possible spin states for massive, spin-1/2 fermions. Though photons have spin-1, they have only two accessible states (corresponding to two states of polarization) giving $g_\gamma = 2$. Massless spin-1/2 fermions, like neutrinos, exist only in left-handed or right-handed states, making $g_\nu = 1$. From (3.21) and (3.22), we can find the precise time–temperature relationship for this phase of the evolution:

$$H(T) \cong 5.44\left(\frac{T^2}{m_{\rm Pl}}\right); \quad t \cong 0.09\left(\frac{m_{\rm Pl}}{T^2}\right). \tag{3.32}$$

Since neutrinos have no electric charge, they have no direct coupling with photons. Their interaction with baryons can be ignored because of the low density of baryons. So they are kept in equilibrium essentially through reactions like $\nu\overline{\nu} \leftrightarrow e\overline{e}$, $\nu e \leftrightarrow \nu e$ etc. The cross section, $\sigma(E)$, for these weak interaction processes is[2] of the order of $(\alpha^2 E^2/m_x^4)$ where $\alpha \simeq 2.8 \times 10^{-2}$ is related to the gauge coupling constant $g_s \simeq 0.6$ by $\alpha = (g^2/4\pi)$ and $m_x \simeq 50\,{\rm GeV}$ is the mass of the gauge vector boson mediating the weak interaction (see exercise 3.6). Defining the 'Fermi coupling constant' $G_F = (\alpha/m_x^2) \simeq 1.17 \times 10^{-5}(\mathrm{GeV})^{-2} = (293\,\mathrm{GeV})^{-2}$ and using the fact that $E \simeq T$, we can write $\sigma \simeq G_F^2 E^2 \simeq G_F^2 T^2$. Since the number density of interacting particles is $n \cong (\zeta(3)g/\pi^2)T^3 \cong 1.3T^3$ and $< v > \simeq c = 1$, the rate of interactions is given by

$$\Gamma = n\sigma|v| \simeq 1.3 G_F^2 T^5. \tag{3.33}$$

The rate of expansion, from (3.32) is $H \cong 5.4\ (T^2/m_{\mathrm{Pl}})$. So

$$\frac{\Gamma}{H} \simeq 0.24 T^3 \left(\frac{m_{\mathrm{Pl}}}{G_F^{-2}}\right) \simeq \left(\frac{T}{1.4\,\mathrm{MeV}}\right)^3 = \left(\frac{T}{1.6 \times 10^{10}\,\mathrm{K}}\right)^3 \tag{3.34}$$

The interaction rate of neutrinos becomes lower than the expansion rate when the temperature drops below $T_D \simeq 1\,\mathrm{MeV}$. At lower temperatures, the neutrinos are completely decoupled from the rest of the matter.

Since the neutrinos are taken to be massless, they are clearly relativistic at the time of decoupling. (This conclusion will be true even if neutrinos have a mass m_ν with $m_\nu \ll T_D \simeq 1\,\mathrm{MeV}$.) Their distribution function at later times is given by (3.23) with $T_\nu \propto a^{-1}$. The present day universe should contain a relic background of these neutrinos.

At the time of decoupling, the photons, neutrinos and the rest of the matter had the same temperature. As long as the photon temperature decreases as a^{-1}, neutrinos and photons will continue to have the same temperature even though the neutrinos have decoupled. However, the photon temperature will decrease at a slightly lower rate if the g-factor is changing. In that case, T_γ will become higher than T_ν as the universe cools. Such a change in the value of g occurs when the temperature of the universe falls below $T \simeq m_e$. The electron rest mass $m_e \simeq 0.5\,\mathrm{MeV}$ corresponds to a temperature of $5 \times 10^9\,\mathrm{K}$. When the temperature of the universe becomes lower than this value the mean energy of the photons will fall below the energy required to create $e\bar{e}$ pairs. Thus the backward reaction in $e\bar{e} \leftrightarrow \gamma\gamma$ will be severely suppressed. The forward reaction will continue to occur resulting in the disappearance of the $e\bar{e}$ pairs.

This process clearly changes the value of g. At $T_D > T \gtrsim m_e$, the neutrinos have decoupled and their entropy is separately conserved; but the photons ($g = 2$) are in equilibrium with electrons ($g = 2$) and positrons ($g = 2$). This gives $g(\gamma, e, \bar{e}) = 2 + (7/8) \times 4 = (11/2)$. For $T \ll m_e$, when the $e\bar{e}$ annihilation is complete, the only relativistic species left in this set is the photon ($g = 2$). The conservation of $S = q(Ta)^3$, applied to particles which are in equilibrium with radiation, shows that the quantity $q(T_\gamma a)^3 = g(aT_\gamma)^3$ remains constant during expansion. (Since γ, e and $\bar{e}$ all have the same temperature, $q = g$.) Because g decreases during the $e\bar{e}$ annihilation, the value of $(aT_\gamma)^3$ after the $e\bar{e}$ annihilation will be higher than its value before:

$$\left[(aT_\gamma)^3_{\mathrm{after}}/(aT_\gamma)^3_{\mathrm{before}}\right] = [g_{\mathrm{before}}/g_{\mathrm{after}}] = \frac{11}{4}. \tag{3.35}$$

The neutrinos, since they are decoupled, do not participate in this process. They are characterized by a temperature $T_\nu(t)$ which falls *strictly* as a^{-1} and their entropy $(s_\nu a^3)$ is separately conserved. Let $T_\nu = Ka^{-1}$; originally, before $e\bar{e}$ annihilations began, the photons and neutrinos had the same temperature: $(aT_\nu)_{\rm before} = (aT_\gamma)_{\rm before} = K$. It follows that:

$$\begin{aligned}(aT_\gamma)_{\rm after} &= \left(\frac{11}{4}\right)^{1/3}(aT_\gamma)_{\rm before} = \left(\frac{11}{4}\right)^{1/3}(aT_\nu)_{\rm before} \\ &= \left(\frac{11}{4}\right)^{1/3}(aT_\nu)_{\rm after} \simeq 1.4(aT_\nu)_{\rm after}.\end{aligned} \tag{3.36}$$

The first equality follows from (3.35), the second from the fact that $T_\gamma = T_\nu$ at $T \gtrsim m_e$ and the third from strict constancy of (aT_ν). Thus the $e\bar{e}$ annihilations increase the temperature of photons compared to that of neutrinos by a factor $(11/4)^{1/3} \simeq 1.4$.

It can be easily verified that the photons released by this process ($e\bar{e} \to 2\gamma$), get thermalized rapidly due to the scattering with charged particles. (We will study these scattering processes in detail later in section 3.6). The analysis above, of course, is based on this tacit assumption.

After the $e\bar{e}$ annihilations, the g factor does not change. Both T_γ and T_ν fall as a^{-1} and the ratio $T_\nu = (4/11)^{1/3}T_\gamma \cong 0.71 T_\gamma$ is maintained till today. The relic ν background today should have the distribution given by (3.23) with $(T_\nu)_{\rm now} \cong 0.71 \times 2.7\,{\rm K} = 1.9\,{\rm K}$ (see exercise 3.7).

Thus the species of particles which remain relativistic today will be photons ($g_\gamma = 2$) with a temperature $T_\gamma \cong 2.7\,{\rm K}$ and 3 flavours of massless neutrinos and antineutrinos ($g_F = 3 + 3 = 6$) with a temperature $T_\nu = (4/11)^{1/3}T_\gamma$. From (3.15) and (3.17) we find

$$\begin{aligned}g({\rm now}) &= 2 + \frac{7}{8} \times 6 \times \left(\frac{4}{11}\right)^{4/3} \cong 3.36, \\ q({\rm now}) &= 2 + \frac{7}{8} \times 6 \times \left(\frac{4}{11}\right) \cong 3.91.\end{aligned} \tag{3.37}$$

The energy and entropy densities of these relativistic particles in the present day universe are

$$\begin{aligned}\rho_R &= \frac{\pi^2}{30} gT^4 = 8.09 \times 10^{-34}\,{\rm g\,cm^{-3}} \\ s &= \frac{2\pi^2}{45} qT^3 \cong 2.97 \times 10^3\,{\rm cm^{-3}}.\end{aligned} \tag{3.38}$$

This ρ_R corresponds to the $\Omega_R = 4.3 \times 10^{-5}h^{-2}$. Note that $\rho_R = (g_{\rm total}/g_\gamma)\,\rho_\gamma \simeq 1.68\rho_\gamma$; similarly, $\Omega_R = 1.68\Omega_\gamma$.

The matter density today is $\rho_{NR} = 1.88 \times 10^{-29}\ \Omega h^2\,{\rm g\,cm^{-3}}$. The redshift $z_{\rm eq}$ at which matter and radiation had equal energy densities is determined by the relation $(1+z_{\rm eq}) = (\rho_{NR}/\rho_R)$. This quantity $z_{\rm eq}$ was calculated in chapter 2 assuming that Ω_R is contributed by photons alone. The correct value, if there are three massless neutrino species, is

$$(1+z_{\rm eq}) = \left(\frac{\Omega_{NR}}{\Omega_R}\right) = 2.3 \times 10^4 (\Omega h^2). \tag{3.39}$$

This corresponds to the temperature

$$T_{\rm eq} = T_0(1+z_{\rm eq}) = 5.5(\Omega h^2)\,{\rm eV} \tag{3.40}$$

and time

$$t_{\rm eq} \cong \frac{2}{3} H_0^{-1} \Omega^{-1/2} (1+z_{\rm eq})^{-3/2} = 5.84 \times 10^{10} (\Omega h^2)^{-2}\,{\rm s}. \tag{3.41}$$

As explained in chapter 2, we have defined $t_{\rm eq}$ using the expression for $a(t)$ valid in the matter dominated phase (using the more precise form of $a(t)$ will give $t_{\rm eq}({\rm exact}) = 0.585 t_{\rm eq} = 3.41 \times 10^{10}(\Omega h^2)^{-2}\,{\rm s}$). The size of the Hubble radius at $t_{\rm eq}$ is about $d_H(t_{\rm eq}) \simeq ct_{\rm eq} \simeq 1.75 \times 10^{21}(\Omega h^2)^{-2}\,{\rm s}$. This corresponds to the length scale $\lambda_{\rm eq} = d_H(t_{\rm eq})(1+z_{\rm eq}) \simeq 13\,{\rm Mpc}$ today.

The entropy density s was used earlier to define the conserved quantities $N_A = (n_A/s)$ for those species of particles with $n_A \propto a^{-3}$. Knowing the present value of s, we can explicitly compute this number. For example, the present baryon number density $n_B = 1.13 \times 10^{-5}\ (\Omega_B h^2)\,{\rm cm}^{-3}$ corresponds to the ratio:

$$\left(\frac{n_B}{s}\right) = \left(\frac{\rho_c \Omega_B}{s m_B}\right) = 3.81 \times 10^{-9}(\Omega_B h^2). \tag{3.42}$$

Often, the number density of photons, n_γ, is used rather than s to define a 'baryon-to-photon' ratio, $\eta \equiv (n_B/n_\gamma)$, for our universe. Since

$$n_\gamma = \left(\frac{2\zeta(3)}{\pi^2}\right) T_0^3 = \left(\frac{45\zeta(3)}{\pi^4 q}\right) s \simeq 0.142 s \simeq 422\,{\rm cm}^{-3}, \tag{3.43}$$

this ratio is

$$\eta = \left(\frac{n_B}{n_\gamma}\right) = \left(\frac{n_B}{s}\right)\left(\frac{s}{n_\gamma}\right) \simeq 7\left(\frac{n_B}{s}\right) = 2.67 \times 10^{-8}(\Omega_B h^2). \tag{3.44}$$

This value is extremely low; the corresponding number in stellar interiors, for example, is about 10^2.

As we shall see later, the dominant matter density in the universe is *not* contributed by the baryons. It seems very likely that non-baryonic dark matter contributes a fraction $\Omega_{\rm DM} \simeq 0.2$–$1$ while baryons contribute only $\Omega_B \simeq 0.01$ or so. Hence $\Omega = \Omega_B + \Omega_{\rm DM} \approx \Omega_{\rm DM}$. When the universe becomes matter dominated at $t \simeq t_{\rm eq}$, it will be essentially dominated by the non-baryonic dark matter and *not* by the baryons. The baryonic energy density will dominate over radiation when $\rho_B = \Omega_B \rho_c (1+z)^3$

becomes larger than $\rho_R = \Omega_R \rho_c (1+z)^4$. This occurs at the redshift $z_{\rm RB}$ where

$$(1 + z_{\rm RB}) = \frac{\Omega_B}{\Omega_R} = 2.3 \times 10^4 (\Omega_B h^2). \tag{3.45}$$

For $\Omega_B \simeq 0.01, h = 0.5$, this is about $z_{\rm RB} \simeq 60$.

In this section we have started with a particular composition for the universe at $T \simeq 10^{12}$ K and worked out the consequences. This initial composition contained a small fraction of protons and neutrons but no antiprotons or antineutrons. Similarly, there was a small excess of electrons over positrons, so that after the $e\bar{e}$ annihilations an excess of electrons survived. At temperatures higher than a few GeV, the universe would have contained large numbers of antiprotons, antineutrons etc. as well, but there should have been a slight excess of protons over antiprotons. As the universe cools through the temperature $T \simeq m_{\rm proton} \simeq 1\,{\rm GeV}$, the protons and antiprotons will annihilate each other leaving a small excess of protons. It is, therefore, clear that we have put in by hand an excess of baryons over antibaryons in the initial conditions, so as to reproduce correctly the present day universe. A truly fundamental theory should explain how this baryon excess arises in the earlier phases of the universe. Such an explanation is indeed possible in some of the particle physics models. These models contain interactions which can change the 'baryon number' and thus produce an excess of baryons, starting from a baryon–antibaryon symmetric state.

3.4 Relic background of wimps

The discussion in section 3.3 assumed that the neutrinos are massless. If they have non-zero rest mass, then the physical consequences will be quite different. These consequences can be predicted[3] in a rather general manner for any weakly interacting massive particle (usually called a 'wimp') which could be a neutrino or some other particle.

Let us first consider the case of a wimp which decouples while it is still relativistic; i.e. $T_D \gg m$ where m is the mass of the particle and T_D is the decoupling temperature. Such a particle is characterized by the conserved quantity

$$N = 0.28 \left(\frac{g_{\rm eff}}{q}\right)_{T=T_D} = 0.21 \left(\frac{g}{q(T_D)}\right) \tag{3.46}$$

where we have assumed the particle to be a fermion and set $g_{\rm eff} = (3g/4)$ in (3.28). They will have a number density

$$n_0 = N s_0 = 2.97 \times 10^3 N\,{\rm cm}^{-3} \cong 619 g [q(T_D)]^{-1}\,{\rm cm}^{-3} \tag{3.47}$$

in the present universe.

These particles would have become non-relativistic at some temperature $T_{\rm nr} \simeq m$ in the past as long as $m > T_0$; i.e. $m \gtrsim 1.7 \times 10^{-4}$ eV. (If this is not the case, the particles will be relativistic even today and will behave just like the massless neutrinos discussed in the last section.) The energy contributed by each particle today is $E \simeq m$; so the energy density of these particles today will be

$$\rho = n_0 m = 6.19 \times 10^3 g[q(T_D)]^{-1} \left(\frac{m}{10\,\text{eV}}\right) \text{eV}\,\text{cm}^{-3} \tag{3.48}$$

which corresponds to

$$(\Omega h^2)_{\rm wimp} = 0.59 \left(\frac{m}{10\,\text{eV}}\right) \left[\frac{g}{q(T_D)}\right]. \tag{3.49}$$

In chapter 2, we obtained the constraint $\Omega h^2 \lesssim 1$ from the observations related to the age of the universe. Combining this constraint with (3.49), we get a bound on the mass m:

$$m \lesssim 17 q(T_D) g^{-1}\,\text{eV}. \tag{3.50}$$

This bound is valid for any wimp which decouples while being still relativistic. The precise value of the right hand side depends on the value of q at the time of decoupling.

Neutrinos or other wimps with masses less than about 1 MeV decouple at $T_D \simeq$ (1–3) MeV; at this temperature $q = 10.75$. So, for a single massive species with $g = 2$, we get:

$$\Omega_\nu h^2 = \left(\frac{m_\nu}{91.5\,\text{eV}}\right); \quad m_\nu \lesssim 91.5\,\text{eV}; \quad (T_D \simeq 1\,\text{MeV}) \tag{3.51}$$

The value of q will be higher at higher T_D. At temperatures above 300 GeV, the standard model predicts that 8 gluons, $W, \overline{W}, Z$, one Higgs doublet and 3 generations of quarks and leptons will all be relativistic, making $q \cong 106.5$. If a wimp decouples at $T_D \gtrsim 300$ GeV then the corresponding bound will be

$$(\Omega h^2)_{\rm wimp} = \left(\frac{m}{910\,\text{eV}}\right); \quad m \lesssim 910\,\text{eV}; \quad (T_D \gtrsim 300\,\text{GeV}) \tag{3.52}$$

However, it should be noted that decoupling at such high energies is possible only if the particles have non-standard interactions.

Consider next the wimps which decouple when they are non-relativistic ($m \gtrsim 3T_D$) (see exercises 3.8, 3.9 and 3.10). The value of N for these particles is given by

$$N = 0.145 \left[\frac{g_A}{q(T_D)}\right] \left(\frac{m}{T_D}\right)^{3/2} e^{-m/T_D}. \tag{3.53}$$

Unlike the previous case, N now depends on m quite strongly. To obtain a numerical estimate, we have to determine T_D by the criterion $\Gamma = H$.

The reactions which are capable of changing the number of the wimps of type A are of the form $A\overline{A} \leftrightarrow X\overline{X}$, where X is some generic species of particle. (It will be assumed that the Xs are in thermal equilibrium.) The average value of σv for such annihilation processes can be expressed in the form

$$< \sigma v > \equiv \sigma_0 \left(\frac{T}{m}\right)^k . \tag{3.54}$$

The value of k depends on the details of the dominant annihilation process (s-wave, p-wave$\cdots$ etc.); it is usually of order unity. The value of σ_0 depends on m and has a simple form in the two extreme cases of $m \ll m_Z$ and $m \gg m_Z$, where $m_Z \simeq 10^2$ GeV is the mass of the Z boson mediating weak interaction. Let us consider the $m \ll m_Z$ case first. For wimps with $m < m_Z$, the cross section σ_0 can be expressed as

$$\sigma_0 \cong \frac{c}{2\pi} G_F^2 m^2 . \tag{3.55}$$

The value of the constant c depends on the type of the 'type' of the fermion. Fermions with spin-1/2 are classified as 'Dirac' type or 'Majorana' type. A Dirac-type fermion will be distinct from its antiparticle while a Majorana-type fermion will be its own antiparticle (see appendix B). For Dirac-type fermions, which we will consider, $c \simeq 5$. The reaction rate, therefore, is

$$\begin{aligned}\Gamma = n < \sigma v > &\simeq g_A \left(\frac{mT}{2\pi}\right)^{3/2} \mathrm{e}^{-m/T} \sigma_0 \left(\frac{T}{m}\right)^k \\ &= \frac{\sigma_0 g_A}{(2\pi)^{3/2}} T^3 \left(\frac{m}{T}\right)^{3/2-k} \mathrm{e}^{-m/T} .\end{aligned} \tag{3.56}$$

The expansion rate is $H = 1.66 g^{1/2}\ (T^2/m_{\rm Pl})$; thus the condition $(\Gamma/H) = 1$ becomes

$$\frac{\Gamma}{H} = 3.825 \times 10^{-2} \left(\frac{g_A}{q^{1/2}}\right) \left(\frac{m}{T}\right)^{1/2-k} \mathrm{e}^{-m/T} (\sigma_0 m m_{\rm Pl}) = 1. \tag{3.57}$$

Solving this equation for $\exp[-(m/T)]$ and substituting in (3.53) we find

$$\begin{aligned}N &\cong \frac{3.79}{q^{1/2}} \left(\frac{m}{T_D}\right)^{k+1} (\sigma_0 m m_{\rm Pl})^{-1} \\ &\cong 2.87 \times 10^{-9} q^{-1/2} \left(\frac{m}{T_D}\right)^{k+1} \left(\frac{m}{1\,\mathrm{GeV}}\right)^{-3} .\end{aligned} \tag{3.58}$$

This corresponds to a number density

$$n_0 = N s_0 = 8.523 \times 10^{-6} q^{-1/2} \left(\frac{m}{T_D}\right)^{k+1} \left(\frac{m}{1\,\mathrm{GeV}}\right)^{-3} \tag{3.59}$$

and the density parameter:

$$(\Omega h^2)_{\text{wimp}} = 0.81 q^{-1/2} \left(\frac{m}{T_D}\right)^{k+1} \left(\frac{m}{1\,\text{GeV}}\right)^{-2}. \quad (3.60)$$

To make numerical estimates, (3.57) has to be solved for T_D. Taking logarithms, (3.57) can be written as

$$\frac{m}{T_D} = 17.966 + \ln\left(\frac{g_A}{q^{1/2}}\right) + \left(\frac{1}{2} - k\right) \ln\left(\frac{m}{T_D}\right) + 3 \ln\left(\frac{m}{1\,\text{GeV}}\right). \quad (3.61)$$

This condition will determine the value of T_D in terms of m. Since g is a slowly varying function of T, it is best to solve this equation iteratively.

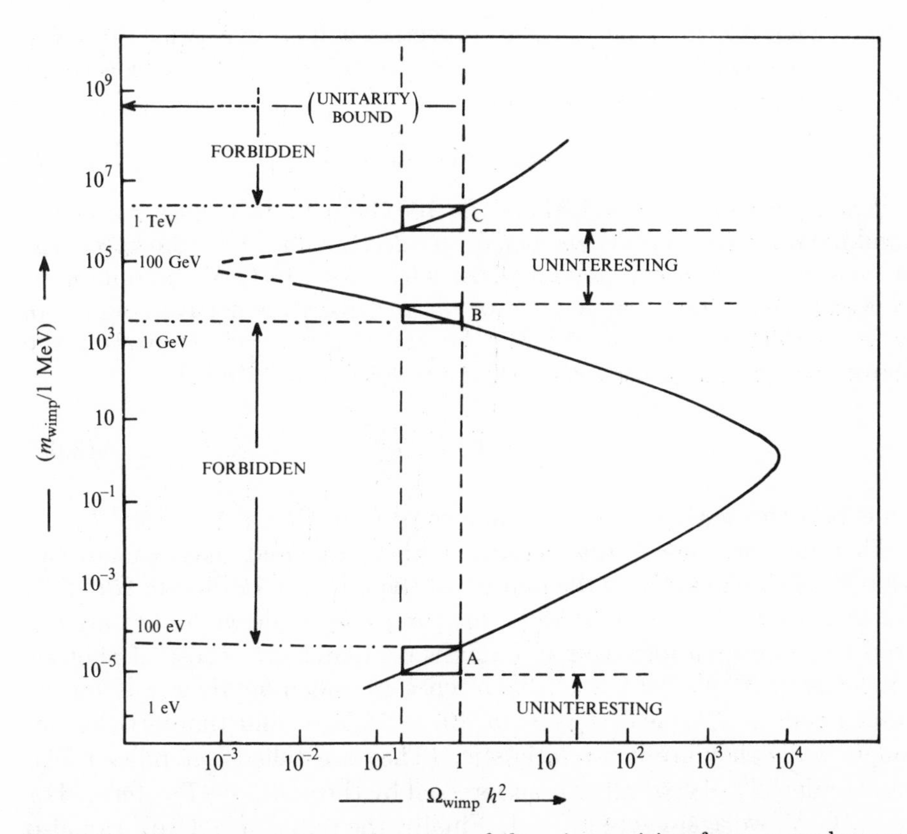

Fig. 3.1. The constraints on the mass of the wimp arising from cosmology are shown. The two vertical, broken, lines indicate the allowed range of values for the combination Ωh^2. If the mass of the wimp is far less than 100 eV or is in the 10 GeV to 0.1 TeV range, the wimps do not contribute significantly to the energy density. The mass ranges from 100 eV to a few GeV and above a few TeV are cosmologically forbidden. The allowed regions of interest are indicated by rectangular boxes. Also shown is the unitarity bound on the wimp mass which will be discussed in Chapter 11.

Consider, as a typical example, a wimp with $m \gtrsim 1\,\mathrm{GeV}$. To leading order $(m/T_D) \simeq 17.966$; assuming, for simplicity, $k = 0$, the $\ln(m/T_D)$ term corrects this to $(m/T_D) \simeq 19.41$ giving $T_D \simeq 52\,\mathrm{MeV}\,(m/1\,\mathrm{GeV})$; at this temperature $g \simeq 10^2$. This gives $\ln(q^{1/2}/g_A) \simeq \ln 5 \simeq 1.61$ correcting (m/T_D) further to 17.8. Thus, to this order of iteration

$$\left(\frac{m}{T}\right) \simeq 17.8 + 3 \ln \left(\frac{m}{1\,\mathrm{GeV}}\right) \tag{3.62}$$

On substitution into (3.58) and (3.60) we find

$$N \simeq 5.11 \times 10^{-9} \left(\frac{m}{1\,\mathrm{GeV}}\right)^{-3} \quad ; \quad \Omega_A h^2 \simeq 1.44 \left(\frac{m}{1\,\mathrm{GeV}}\right)^{-2} . \tag{3.63}$$

The fermion A and its antiparticle $\overline{A}$ together will provide twice this value to Ω, corresponding to $\Omega_{A\overline{A}} h^2 \simeq 2.88\ (m/1\,\mathrm{GeV})^{-2}$. The constraint $\Omega h^2 \lesssim 1$ then gives the mass bound

$$m \gtrsim 2\,\mathrm{GeV}. \tag{3.64}$$

Finally, consider relics with $m \gg 100\,\mathrm{GeV}$. In this mass range the annihilation cross section σ_0 begins to *decrease* as m^2. Since σ_0 was increasing as m^2 in the previous case with $m < 100\,\mathrm{GeV}$, we only have to change the value of N by a factor m^4 to obtain the correct result. So, for $m > 100\,\mathrm{GeV}$, we will get $N \propto m$ and $\Omega_A h^2 \propto m^2$. Repeating the above analysis and substituting the numbers, we will find that

$$\Omega_A h^2 \simeq \left(\frac{m}{1\,\mathrm{TeV}}\right)^2 . \tag{3.65}$$

Thus particles with $m \simeq 1\,\mathrm{TeV}$ can also provide $\Omega h^2 \simeq 1$.

This analysis reveals that wimps in three different mass ranges can contribute significantly to the density of the universe leading to $\Omega h^2 \lesssim 1$. For $m \lesssim 10^2\,\mathrm{eV}$, $T_D \simeq (1\text{–}3)\,\mathrm{MeV}$, the wimps decouple while they are relativistic. Their number density today is comparable to those of photons and for $m \simeq 10^2\,\mathrm{eV}$, $\Omega h^2 \simeq 1$. (Such relics are called *hot* relics.) If, on the other hand, $m \gtrsim 1\,\mathrm{GeV}$, $T_D \simeq (m/19) \simeq 52\,\mathrm{MeV}$ and the particles decouple while they are non-relativistic. (They are called *cold* relics.) The number density of cold relics is suppressed by the $\exp(-m/T_D)$ term. For $m \approx 2\,\mathrm{GeV}$, we again get $\Omega h^2 \simeq 1$. Finally, the range $m \simeq 1\,\mathrm{TeV}$ can also lead to $\Omega h^2 \simeq 1$; such a particle will also be, of course, a cold relic. It is interesting to note that purely cosmological considerations (viz. $\Omega h^2 \lesssim 1$) rule out the possible existence of stable, weakly interacting fermions in the mass range $100\,\mathrm{eV} < m < 2\,\mathrm{GeV}$.

It was pointed out in chapter 1 that the universe contains a significant amount of dark matter. Wimps, if they exist, can provide much of the mass in the form of dark matter. For most of the discussion in the future chapters, we shall assume that this is indeed the case. Depending on the nature of the wimp, the dark matter will also be called 'hot' or 'cold'.

3.5 Synthesis of light nuclei

The binding energies of the first four light nuclei, ^{2}H, ^{3}H, ^{3}He and ^{4}He are 2.22 MeV, 6.92 MeV, 7.72 MeV and 28.3 MeV respectively. As the universe cools below these temperatures, one expects these bound structures to form. The abundance of light elements, which are synthesized in the early universe, can be used to obtain important constraints on the cosmological parameters.

Though the energy considerations suggest that these nuclei could be formed when the temperature of the universe is in the range (1–30) MeV, the actual synthesis takes place only at a much lower temperature, $T_{\rm nuc} = T_n \simeq 0.1\,\text{MeV}$. The main reason for this delay is the 'high entropy' of our universe, i.e., the high value for the photon-to-baryon ratio, η^{-1}. This fact can be understood as follows.

Let us assume, for a moment, that the nuclear (and other) reactions are fast enough to maintain thermal equilibrium between various species of particles and nuclei. In thermal equilibrium, the number density of a nuclear species ^AN_z with atomic mass A and charge Z will be

$$n_A = g_A \left(\frac{m_A T}{2\pi}\right)^{3/2} \exp\left[-\left(\frac{m_A - \mu_A}{T}\right)\right]. \tag{3.66}$$

In particular, the equilibrium number densities of protons and neutrons will be

$$\begin{aligned} n_p &= 2\left(\frac{m_p T}{2\pi}\right)^{3/2} \exp\left[-\frac{1}{T}(m_p - \mu_p)\right] \\ &\cong 2\left(\frac{m_B T}{2\pi}\right)^{3/2} \exp\left[-\frac{1}{T}(m_p - \mu_p)\right], \end{aligned} \tag{3.67}$$

$$\begin{aligned} n_n &= 2\left(\frac{m_n T}{2\pi}\right)^{3/2} \exp\left[-\frac{1}{T}(m_n - \mu_n)\right] \\ &\cong 2\left(\frac{m_B T}{2\pi}\right)^{3/2} \exp\left[-\frac{1}{T}(m_n - \mu_n)\right]. \end{aligned} \tag{3.68}$$

The mass difference between the proton and the neutron $Q \equiv m_n - m_p = 1.293\,\text{MeV}$ has to be retained in the exponent but can be ignored in the prefactor $m_A^{3/2}$; we have set, in the prefactor, $m_n \cong m_p \cong m_B$, an average value.

Since the chemical potential is conserved in the reactions producing ^AN_z out of Z protons and $(A - Z)$ neutrons, μ_A for any species can be expressed in terms of μ_p and μ_n:

$$\mu_A = Z\mu_p + (A-Z)\mu_n. \tag{3.69}$$

Writing

$$\begin{aligned}\exp\left[\frac{1}{T}(\mu_A - m_A)\right] &= \exp\left[\frac{1}{T}[Z\mu_p + (A-Z)\mu_n]\right]\exp(-m_A/T)\\ &= [\exp(\mu_p/T)]^Z[\exp(\mu_n/T)]^{(A-Z)}\exp(-m_A/T)\end{aligned} \tag{3.70}$$

and substituting for $\exp(\mu_p/T)$ and $\exp(\mu_n/T)$ from (3.67) and (3.68), we get

$$\begin{aligned}\exp\left[\frac{1}{T}(\mu_A - m_A)\right] &= 2^{-A}n_p^Z n_n^{(A-Z)}\left(\frac{2\pi}{m_B T}\right)^{3A/2}\\ &\quad\times\exp\left[\frac{1}{T}(Zm_p + (A-Z)m_n - m_A)\right]\\ &= 2^{-A}n_p^Z n_n^{(A-Z)}\left(\frac{2\pi}{m_B T}\right)^{3A/2}\exp(B_A/T)\end{aligned} \tag{3.71}$$

where $B_A \equiv Zm_p + (A-Z)m_n - m_Z$ is the binding energy of the nucleus. Therefore, the number density in (3.66) becomes

$$n_A = g_A 2^{-A}A^{3/2}\left(\frac{2\pi}{m_B T}\right)^{3(A-1)/2} n_p^Z n_n^{A-Z}\exp(B_A/T). \tag{3.72}$$

We define the 'mass fraction' of the nucleus A by $X_A = (An_A/n_B)$, where n_B is the number density of baryons in the universe. Substituting for n_A, n_p and n_n in (3.72) by $n_A = n_B\ (A^{-1}X_A) = \eta n_\gamma(A^{-1}X_A)$, $n_p = n_B X_p = \eta n_\gamma X_p$ and $n_n = n_B X_n = \eta n_\gamma X_n$, where $\eta = 2.68\times 10^{-8}$ $(\Omega_B h^2)$ is the baryon-to-photon ratio and $n_\gamma = [2\zeta(3)/\pi^2]\ T_0^3$ is the number density of photons, we get

$$X_A = F(A)(T/m_B)^{3(A-1)/2}\eta^{A-1}X_p^Z X_n^{A-Z}\exp(B_A/T) \tag{3.73}$$

where

$$F(A) = g_A A^{5/2}[\zeta(3)^{A-1}\pi^{(1-A)/2}2^{(3A-5)/2}]. \tag{3.74}$$

Equation (3.73) shows why the high entropy of the universe, i.e. small value of η, hinders the formation of nuclei. To get $X_A \simeq 1$, it is *not* enough that the universe cools to the temperature $T \lesssim B_A$; it is necessary that it cools *still further* so as to offset the small value of the η^{A-1} factor. The temperature T_A at which the mass fraction of a particular species A will be of order unity ($X_A \simeq 1$) is given by

$$T_A \simeq \frac{B_A/(A-1)}{\ln(\eta^{-1}) + 1.5\ln(m_B/T)}. \tag{3.75}$$

This temperature will be much lower than B_A; for ^{2}H, ^{3}He and ^{4}He the value of T_A is 0.07 MeV, 0.11 MeV and 0.28 MeV respectively. Comparison with the binding energy of these nuclei shows that these values are lower than B_A by a factor of about 10, at least.

Thus, even when the thermal equilibrium is maintained, significant synthesis of nuclei can occur only at $T \lesssim 0.3$ MeV and not at higher temperatures. If such is the case, then we would expect significant production ($X_A \lesssim 1$) of nuclear species A at temperatures $T \lesssim T_A$. It turns out, however, that the rate of nuclear reactions is *not* high enough to maintain thermal equilibrium between various species. We have to determine the temperatures up to which thermal equilibrium can be maintained and redo the calculations to find non-equilibrium mass fractions.

In particular, we used the equilibrium densities for n_p and n_n in the above analysis. In thermal equilibrium, the inter-conversion between n and p is possible through the weak interaction processes $(\nu + n \leftrightarrow p + e)$, $(\overline{e} + n \leftrightarrow p + \overline{\nu})$ and the 'decay' $(n \leftrightarrow p + e + \overline{\nu})$. From the conservation of the chemical potential in these reactions, we find that $\mu_n + \mu_\nu = \mu_p + \mu_e$ giving $(\mu_n - \mu_p) = (\mu_e - \mu_\nu) \simeq 0$, since $\mu_e \simeq 0$, $\mu_\nu \simeq 0$. The equilibrium (n/p) ratio will, therefore, be

$$\left(\frac{n_n}{n_p}\right) = \frac{X_n}{X_p} = \exp(-Q/T), \tag{3.76}$$

where $Q = m_n - m_p$. This ratio will be maintained only as long as the n–p reactions are rapid enough. But when this reaction rate Γ falls below the expansion rate $H \simeq 5.5\ (T^2/m_{\rm Pl})$ at some temperature T_D, say, the (n/p) ratio will get 'frozen' at the value $\exp(-Q/T_D)$. The only process which can continue to change this ratio thereafter will be the beta decay, $n \to p + e + \overline{\nu}$, of the free neutron. The neutron decay will continue to decrease this ratio until all the neutrons are used up in forming bound nuclei.

To determine T_D we have to estimate Γ and use the condition $\Gamma = H$. The rate Γ of the n–p reactions can be calculated from the theory of weak interactions[4]. The total reaction rate $\Gamma(n \to p)$ can be written as the sum of the rates for the reactions $n + \overline{e} \to p + \overline{\nu}$, $n + \nu \to p + e$ and $n \to p + e + \nu$; similarly, the total reaction rate for $\Gamma(p \to n)$ can be written as the sum of the rates for $p + e \to n + \nu$, $p + \nu \to n + \overline{e}$ and $p + e + \overline{\nu} \to n$. All these rates can be expressed as functions of temperature T by evaluating them numerically. For example, the interaction $(pe \to \nu n)$ has the rate

$$\Gamma(pe \to \nu n) = \frac{G_F^2}{2\pi^3}(1 + 3g_A^2)m_e^5 \int_q^\infty dx \frac{x(x-q)(x^2-1)^{1/2}}{[1 + \exp ax][1 + \exp((q-x)b)]} \tag{3.77}$$

where $q = (Q/m_e)$, $a = (m_e/T_\gamma)$, $b = (m_e/T_\nu)$ and $g_A \simeq 1.26$ is the axial-vector coupling constant. It is convenient to express these rates in terms of the decay rate of the free neutron. The neutron lifetime τ_n (which has[5] an observed value of about 915.4 s), is defined as $\tau_n = \Gamma^{-1}$ $(n \to pe\bar{\nu})$ where

$$\begin{aligned}\Gamma(n \to pe\bar{\nu}) &= \frac{G_F^2}{2\pi^3}(1+3g_A^2)m_e^5 \int_0^q dx\, x(x-q)^2(x^2-1)^{1/2} \\ &\simeq 1.636 \frac{G_F^2}{(2\pi)^3}(1+3g_A^2)m_e^5.\end{aligned} \tag{3.78}$$

On expressing $G_F^2(1+3g_A^2)$ in terms of τ_n, we can write

$$\tau_n \Gamma(pe \to \nu n) = 0.61 \int_q^\infty dx \frac{x(x-q)^2(x^2-1)^{1/2}}{[1+\exp ax][1+\exp((q-x)b)]}. \tag{3.79}$$

Clearly $\tau_n \Gamma$ is a function of temperature alone. Similar calculation can be performed for all other reactions allowing us to compute $\tau_n \Gamma(n \to p)$ and $\tau_n \Gamma(p \to n)$ as functions of T. At high temperatures, $T \gg Q \simeq 1.3\,\mathrm{MeV}$, both the rates vary as T^5; in the range of $T \simeq (0.1\text{–}1)\,\mathrm{MeV}$, the rate $\Gamma(n \to p) \propto T^{4.42}$ while $\Gamma(p \to n)$ decreases faster; for $T < 0.1\,\mathrm{MeV}$, $\Gamma(n \to p) \simeq \tau_n^{-1}$ is essentially dominated by the neutron decay while $\Gamma(p \to n)$ drops exponentially. These functions are plotted in figure 3.2.

Also plotted is the Hubble constant $H(T) \simeq 5.5(T^2/m_{\mathrm{Pl}}) \simeq 0.179(T/m_e)^2\,\mathrm{s}^{-1}$. The reaction rates are equal to $t^{-1} = 2H(T)$ around $T_D \simeq (0.8\text{–}0.7)\,\mathrm{MeV}$. Thus thermal equilibrium between neutrons and protons exists only for $T \gtrsim T_D \simeq 0.8\,\mathrm{MeV}$. At $T_D \simeq 0.7\,\mathrm{MeV}$, when the assumption of thermal equilibrium becomes invalid, the (n/p) ratio will be

$$\left(\frac{n}{p}\right) = \exp\left(-\frac{Q}{T_D}\right) \simeq \frac{1}{6} \tag{3.80}$$

giving $X_n \simeq (1/7)$, $X_p \simeq (6/7)$. Since T_A calculated in (3.75) is lower than T_D for all the light nuclei, none of the light nuclei will exist in significant quantities at this temperature T_D. For example, ^{2}H contributes only a mass fraction $X_2 \simeq 10^{-12}$ at this temperature.

As the temperature falls further to $T = T_{\mathrm{He}} \simeq 0.28\,\mathrm{MeV}$, a significant amount of He could have been produced if the nuclear reaction rates were high enough. These reactions (D(D,n) ^{3}He(D,p) ^{4}He, D(D,p) ^{3}H(D,n) ^{4}He, D(D,γ) ^{4}He) are all based on D, ^{3}He and ^{3}H and do not occur rapidly enough because the mass fractions X_A of D, ^{3}He and ^{3}H are still quite small (10^{-12}, 10^{-19} and 5×10^{-19} respectively) at $T \simeq 0.3\,\mathrm{MeV}$.

These abundances become nearly unity only at $T \lesssim 0.1\,\mathrm{MeV}$; therefore, only at $T \lesssim 0.1\,\mathrm{MeV}$ can these reactions be fast enough to produce an equilibrium abundance of ^{4}He. When the temperature becomes $T \lesssim$

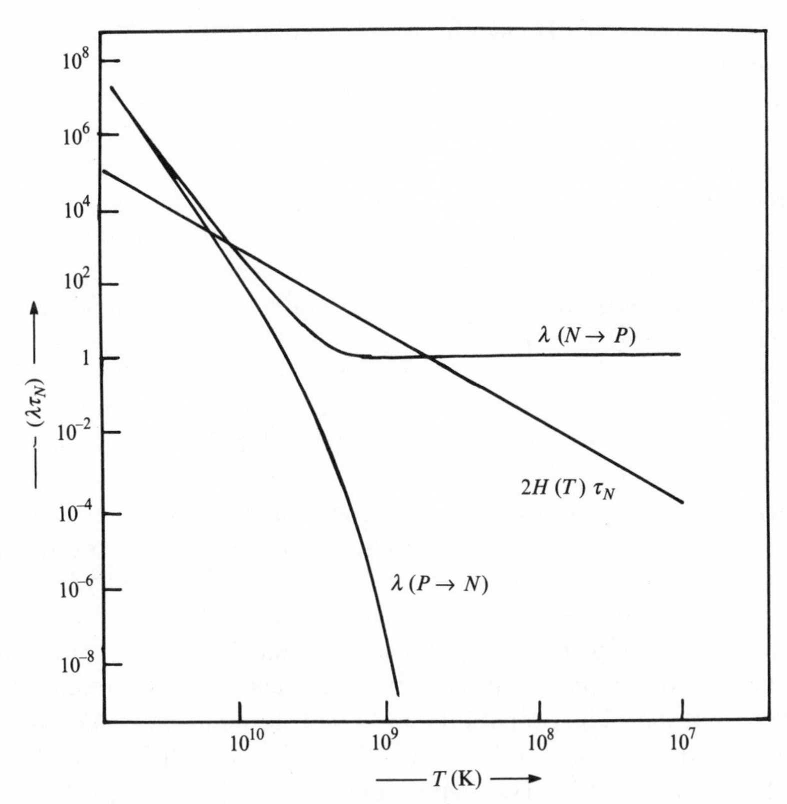

Fig. 3.2. The reaction rates $\lambda(N \to P)$ and $\lambda(P \to N)$ are plotted as a function of the temperature. The reaction rates are parametrized by the decay rate τ_N of the free neutron. Also shown is the expansion rate as a function of the temperature.

0.1 MeV, the abundance of D and ^{3}H builds up and these elements further react to form ^{4}He. A good fraction of D and ^{3}H is converted to ^{4}He. The resultant abundance of ^{4}He can be easily calculated by assuming that almost all neutrons end up in ^{4}He. Since each ^{4}He nucleus has two neutrons, $(n_n/2)$ helium nuclei can be formed (per unit volume) if the number density of neutrons is n_n. Thus the mass fraction of ^{4}He will be

$$Y = \frac{4(n_n/2)}{n_n + n_p} = \frac{2(n/p)}{1 + (n/p)}. \tag{3.81}$$

The (n/p) ratio at the time of 'freeze-out' (t_D) was $(1/6)$; from t_D, till the time of nucleosynthesis (t_N) a certain fraction of neutrons would have decayed, lowering this ratio. Since the freeze-out occurred at $T_D \simeq 1\,\mathrm{MeV}$, $t_D \simeq 1\,\mathrm{s}$ and ^{4}He synthesis occurred at $T_N \simeq 0.1\,\mathrm{MeV}$, $t_N \simeq 3$ min, the decay factor will be $\exp(-t_N/\tau_n) \simeq 0.8$. Therefore, the value of (n/p) at the time of ^{4}He synthesis will be about $0.8 \times (1/6) \simeq (1/7)$. This gives $Y \simeq 0.25$.

As the reactions converting D and ^{3}H to ^{4}He proceed, the number density of D and ^{3}H is depleted and the reaction rates – which are proportional to $\Gamma \propto X_A(\eta n_\gamma) < \sigma v >$ – become small. These reactions soon freeze-out leaving a residual fraction of D and ^{3}H (a fraction of about 10^{-5} to 10^{-4}). Since $\Gamma \propto \eta$ it is clear that the fraction of (D, ^{3}H) left unreacted will decrease with η. In contrast, ^{4}He synthesis – which is not limited by any reaction rate – is fairly independent of η and depends only on the (n/p) ratio at $T \simeq 0.1\,\text{MeV}$.

The production of still heavier elements – even those like ^{16}C, ^{16}O which have higher binding energies than ^{4}He – is highly suppressed in the early universe. Two factors are responsible for this suppression: (1) Direct reactions between two He nuclei or between H and He will lead to nuclei with atomic masses 8 or 5. Since there are no tightly bound isotopes with masses 8 or 5, these reactions do not lead to any further synthesis. (The three-body interaction, $^4\text{He} + {}^4\text{He} + {}^4\text{He} \to {}^{12}\text{C}$, is suppressed because of the low number density of ^{4}He nuclei; it is this 'triple-α' reaction which helps further synthesis in stellar interiors.) (2) For nuclear reactions to proceed, the participating nuclei must overcome their Coulomb repulsion. The probability to tunnel through the Coulomb barrier is governed by the factor $F = \exp[-2A^{1/3}\,(Z_1Z_2)^{2/3}(T/1\,\text{MeV})^{-1/3}]$ where $A^{-1} = A_1^{-1} + A_2^{-1}$. For heavier nuclei (with larger Z), this factor suppresses the reaction rate.

Small amounts (about 10^{-10} to 10^{-9} by mass) of ^{7}Li are produced by $^4\text{He}(^3\text{H}, n)^7\text{Li}$ or by $^4\text{He}(^3\text{He}, \gamma)^7\text{Be}$ followed by a decay of ^{7}Be to ^{7}Li. The first process dominates if $\eta \lesssim 3 \times 10^{-10}$ and the second process, for $\eta \gtrsim 3 \times 10^{-10}$. In the second case, a small amount (10^{-11}) of ^{7}Be is left as a residue.

Given the various nuclear reaction rates, the primordial abundances of all the light elements can be computed by numerical integration of the relevant equations. The most uncertain input in these calculations is the neutron decay rate; the half-life of the neutron $\tau = 10.5 \pm 0.2$ min is known to only 2 per cent accuracy. All the weak reaction rates $\Gamma \propto G_F^2$ $(1 + 3g_A^2)T^5$ are proportional to $T^5\tau^{-1}$. An increase in τ decreases all Γs and makes the freeze out (determined by $H(T_D) = \Gamma(T_D)$) occur at a higher temperature. Since $H \propto T^2$, $\Gamma \propto \tau^{-1}T^5$ we find that $T_D \propto \tau^{1/3}$. When T_D increases, so will the (n/p) value at freeze out, resulting in a higher value for ^{4}He. The changes in the other nuclear abundances are not significant because they are within the present observational errors.

The two main *cosmological* parameters on which the results depend are the number of degrees of freedom, g (at $T \simeq 1\,\text{MeV}$) and η. An increase in g increases $H(T) \propto g^{1/2}T^2$ and leads to higher freeze out temperature

$T_D \propto g^{1/6}$, and higher value of ^{4}He abundance. The dependence on η is more complicated. Since the mass fractions $X_A \propto \eta^{(A-1)}$, a larger value of η will allow the D, ^{3}He and ^{3}H abundances to build up earlier and thus lead to an earlier formation of ^{4}He. Since the (n/p) ratio is higher at earlier times, this will result in more ^{4}He. But the (n/p) ratio is only varying slowly with time at $T \simeq 0.1\,\text{MeV}$, and hence this dependence is rather mild. The amount of residual D and ^{3}H, however, depends more strongly on η; it decreases as η^{-k} with $k \simeq 1.3$. The dominant channels for ^{7}Li production are different for $\eta \lesssim 3 \times 10^{-10}$ and for $\eta \gtrsim 3 \times 10^{-10}$. This fact leads to a trough in the abundance of ^{7}Li around $\eta \simeq 3 \times 10^{-10}$ where neither process is very efficient.

The calculated and observed values for several primordial abundances are shown in figure 3.3. The observations indicate[6], with reasonable certainty that: (i) (D/H) $\gtrsim 1 \times 10^{-5}$. (ii) [(D + ^{3}He)/H] $\simeq (1\text{–}8) \times 10^{-5}$ (iii) (^{7}Li/H) $\sim 10^{-10}$ and (iv) $0.236 <$ (^{4}He/H) < 0.254. These observations are consistent with the predictions if $10.3\,\text{min} \lesssim \tau \lesssim 10.7\,\text{min}$, and

$$\eta = (3 - 10) \times 10^{-10}. \tag{3.82}$$

Since $\eta = 2.68 \times 10^{-8} \Omega_B h^2$, this leads to the important conclusion:

$$0.011 \leq \Omega_B h^2 \leq 0.037. \tag{3.83}$$

When combined with the known bounds on h, $0.4 \leq h \leq 1$, we can constrain the baryonic density of the universe to be:

$$0.011 \leq \Omega_B \leq 0.23. \tag{3.84}$$

These are the best bounds on Ω_B available today. It shows that, if Ω_{total} has no contribution other than Ω_B, then the universe must be open.

The abundance of elements depends, as we have seen, on the (n/p) ratio at the time of freeze out. If some physical mechanism can be devised which will distribute neutrons and protons *differently* in space, then the results of the nucleosynthesis will vary from location to location. Such models involving inhomogeneous nucleosynthesis will change the Ω_B bounds derived above. It turns out, however, that it is still not possible[6] to produce viable models in which $\Omega_B = 1$.

Since ^{4}He production depends on g, the observed value of ^{4}He restricts the number (N_ν) of light neutrinos (that is, neutrinos with $m_\nu \lesssim 1\,\text{MeV}$ which would have been relativistic at $T \simeq 1\,\text{MeV}$). The observed abundance is best explained by $N_\nu = 3$, is barely consistent with $N_\nu = 4$ and rules out $N_\nu > 4$. The laboratory bounds[7] on the total number of particles including neutrinos which are less massive than $(m_z/2) \simeq 46\,\text{GeV}$ and couple to the Z^0 boson is $N_\nu = 2.79 \pm 0.63$. This is determined by measuring the decay width of the Z^0; each particle with mass less than

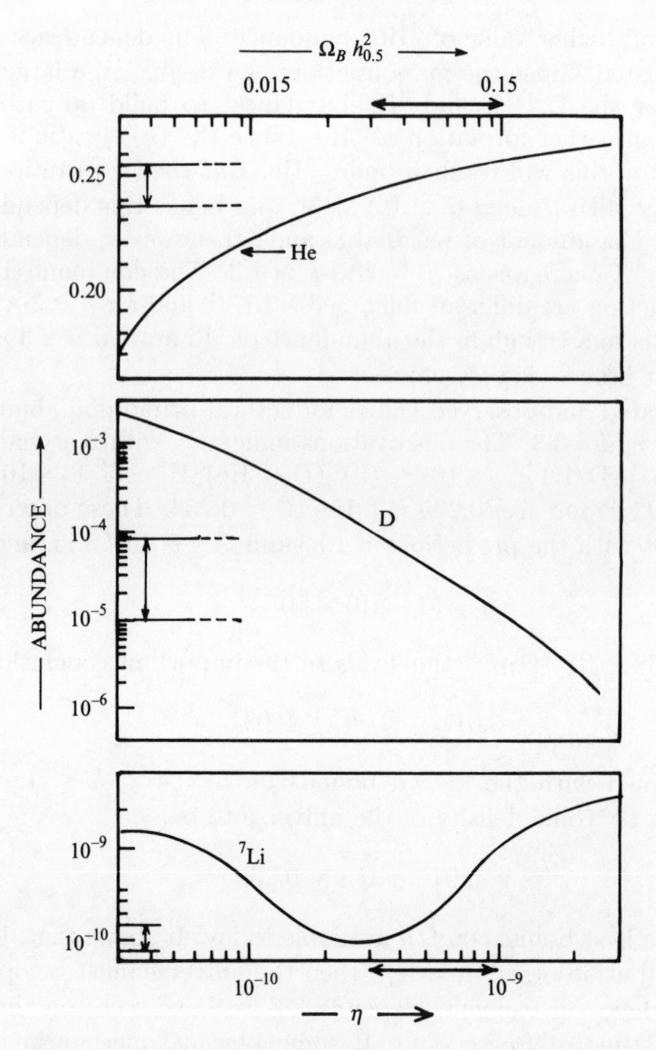

Fig. 3.3. The theoretical calculations lead to abundances of light nuclei like the ones shown here. (The precise values depend on several input parameters but the overall trend is as shown.) The observed values of the light elements (marked in the vertical axis) permit only a narrow range for $\Omega_B h^2$ (marked in the horizontal axis) and exclude the value $\Omega_B = 1$.

$(m_z/2)$ contributes about 180 MeV to this decay width. This bound is consistent with the cosmological observations.

3.6 Decoupling of matter and radiation

At temperatures below 0.1 MeV, the main constituents of the universe will be the hydrogen nucleus (i.e., proton), helium-4 nucleus, electrons, photons and decoupled neutrinos. Since $m_e \simeq 0.5\,\text{MeV}$, the ions and electrons may be considered non-relativistic. These constituents interact among themselves and with the photons through various electromagnetic processes, like bremsstrahlung, Compton (and Thomson) scattering, recombination ($p + e \leftrightarrow H + \gamma$) and Coulomb scattering between charged particles. To decide whether these processes can maintain thermal equilibrium among the constituents, we should compare the various interaction rates with the expansion rate[8].

To begin with, the mean-free-time for Coulomb scatterings among electrons is given by

$$t_{ee} \simeq (n_e \sigma v)^{-1} \propto n_e^{-1} \left(\frac{m}{T}\right)^{1/2} \left(\frac{e^2}{T}\right)^{-2} = n_e^{-1} \left(\frac{m}{T}\right)^{1/2} \left(\frac{e^2}{T}\right)^{-2} \left(\frac{1}{\pi \ln \Lambda}\right) \tag{3.85}$$

where $\ln \Lambda$ originates[9] from distant collisions and is about 30. The n_e in this expression refers to the number density of *free* charged particles; since recombination changes this number, we will write $n_e = x_e n_B$, where x_e is the fraction of charged particles which have *not* combined to form atoms. We will see later that x_e is nearly unity for $T \gtrsim 1\,\text{eV}$ and drops rapidly to a low value of about 10^{-5} by $T \approx 0.1\,\text{eV}$. Inserting the numerical values, we find

$$t_{ee} = 1.35\,\text{s}\,(T/1\,\text{eV})^{-3/2}(x_e \Omega_B h^2)^{-1} \tag{3.86}$$

which is far smaller than the expansion timescale

$$H^{-1}(T) = \begin{cases} 1.46 \times 10^{12}\,\text{s}\,(T/1\,\text{eV})^{-2} & (\text{for } t < t_{\text{eq}}) \\ 1.13 \times 10^{12}\,\text{s}\,(\Omega h^2)^{-1/2}(T/1\,\text{eV})^{-3/2} & (\text{for } t > t_{\text{eq}}) \end{cases} \tag{3.87}$$

at all times. (Even after x_e has fallen to a low final value of 10^{-5}, $t_{ee} \ll H^{-1}$.) A similar calculation will show that the ions and the electrons also interact at a sufficiently rapid rate. Thus these scatterings can easily maintain the thermal distribution for the matter.

The equality of temperature *between* matter and radiation has to be maintained by the interaction between photons and electrons. The simplest form of encounter between photons and electrons is the non-relativistic, low-energy (Thomson) scattering with the mean-free-time of

$$t_{Th} = \frac{1}{n_e \sigma_T c} = 6.14 \times 10^7\,\text{s}\,(T/1\,\text{eV})^{-3}(x_e \Omega_B h^2)^{-1} \tag{3.88}$$

where $\sigma_T = (8\pi/3)(e^2/m)^2$ is the Thomson scattering cross section. This process, however, cannot help in thermalization because there is no energy exchange between the photon and electron in the Thomson scattering limit. Scattering with energy exchange is usually called Compton scattering. In addition to Compton scattering, bremsstrahlung and its inverse process (free–free absorption) can also help in maintaining thermal equilibrium between photons and matter. These are the processes which need to be considered. As we shall see, Compton scattering is the dominant mechanism at high temperatures.

The timescale for Compton scattering of photons can be estimated as follows: The mean-free-path for a photon between two collisions is $\lambda_\gamma = (\sigma_T n_e)^{-1}$ where σ_T is the Thomson scattering cross section. Thus in travelling a distance l the photon, performing a random walk, will undergo N collisions where $N^{1/2}\lambda_\gamma = l$. The fractional frequency change in each Compton scattering is $(\delta\nu/\nu) \simeq (v/c)^2 \simeq (T/m)$. Treating this as another random walk in the frequency space, we can determine the net change in the frequency after N collisions to be $(\delta\nu/\nu) \simeq N^{1/2}(T/m) = (l/\lambda_\gamma)(T/m)$. Energy exchange by Compton scattering can be considered effective when $(\delta\nu/\nu) \simeq 1$, which gives the timescale

$$t_c \simeq \lambda_\gamma \left(\frac{m}{T}\right) = \frac{1}{n_e \sigma_T}\left(\frac{m}{T}\right). \tag{3.89}$$

Notice that t_c is larger than the Thomson scattering timescale by the factor (m/T). This factor is much larger than unity in the temperature range we are interested in. Substituting the numbers, we get,

$$t_c = 3.0 \times 10^{13}\,\mathrm{s}\,(\Omega_B h^2 x_e)^{-1}(T/1\,\mathrm{eV})^{-4}. \tag{3.90}$$

Comparing with H^{-1} we find that

$$\frac{t_c}{H^{-1}} = \left[\frac{T}{4.54 x_e^{-1/2}(\Omega_B h^2)^{-1/2}\,\mathrm{eV}}\right]^{-2}. \tag{3.91}$$

Thus this process is of importance in maintaining thermal equilibrium between matter and radiation for $T \gtrsim 5\,\mathrm{eV}$, corresponding to $z > 1.7 \times 10^4$. The photons are strongly influenced by charged particles up to this redshift.

For the sake of comparison, let us consider the timescale for free–free absorption (see exercise 3.13). For a thermal plasma at temperature T, the timescale for free–free absorption $t_{\mathrm{ff}}(\omega)$, at frequency ω is given by

$$t_{\mathrm{ff}} \cong \frac{3\sqrt{6\pi}}{4}\frac{(mT)^{1/2}m}{e^6 n_e^2}\frac{\omega^3}{(2\pi)^3}(1 - \mathrm{e}^{-\omega/T})^{-1}. \tag{3.92}$$

For photons with frequency $\omega \simeq T$, we get

$$t_{\mathrm{ff}} = 2 \times 10^{14}\,\mathrm{s}\,(\Omega_B h^2 x_e)^{-2}(T/1\,\mathrm{eV})^{-5/2} \tag{3.93}$$

which gives

$$\frac{t_{\mathrm{ff}}}{H^{-1}} \cong (T/1.9 \times 10^4\,\mathrm{eV})^{-1/2}(\Omega_B h^2)^{-2} \quad (\text{for } x_e \simeq 1). \tag{3.94}$$

Thus $t_{\rm ff}$ ceases to be effective at a very high temperature $T \gtrsim 10^4\,$eV. In fact, even when $T \gtrsim 10^4\,$eV, free–free absorption is subdominant because Compton scattering is far more effective. The ratio between these two timescales will be

$$\frac{t_{\rm ff}}{t_c} = \frac{1}{(\Omega_B h^2 x_e)}\left(\frac{T}{0.3\,{\rm eV}}\right)^{3/2}. \tag{3.95}$$

Thus for all the way down to $T \simeq 0.3\,$eV, we have $t_c < t_{\rm ff}$ and hence Compton scattering is more effective.

The timescale for photon production by free–free emission (bremstrahlung) is comparable to $t_{\rm ff}$ for frequencies $\omega \simeq T$ (see exercise 3.13). So similar conclusions apply for this process as well. However, notice that the effectiveness of these processes is strongly frequency dependent. A more careful calculation is needed if we are interested in the $\omega \ll T$ (or $\omega \gg T$) region of the spectrum; we will have occasion to comment about these effects in chapter 6. Lastly, one should note that there exists another photon production mechanism ($e+\gamma \to e+2\gamma$) called a 'double-Compton process' with the timescale of $t_{2C} \simeq 10^{20}\,{\rm s}\,(T/1\,{\rm eV})^{-5}(\Omega_B h^2)^{-1}$ which dominates over the free–free processes at $T \gtrsim 1\,$keV.

The above calculation, however, is somewhat oversimplified because the expression in (3.93) does not take into account the effect of Thomson scattering on the photons. If l_{ff} is the mean distance a typical photon travels before getting absorbed, it would have experienced $N = (l_{ff}/\lambda_\gamma)$ scatterings in this trip. Hence, it would have travelled a root-mean-square distance of $\lambda_{\rm rms} = N^{1/2}\lambda_\gamma = (l_{ff}\lambda_\gamma)^{1/2}$ before getting absorbed. This correction takes into account the fact that Thomson scatterings increase the effective path length for photon absorption. The corresponding values are

$$\begin{aligned} \bar{t} \equiv \frac{\lambda_{\rm rms}}{c} &= 1.1\times 10^{11}\,{\rm s}\,(T/1\,{\rm eV})^{-11/4}(\Omega_B h^2 x_e)^{-3/2}, \\ \frac{\lambda_{\rm rms}}{cH^{-1}} = \frac{\bar{t}}{H^{-1}} &= (\Omega_B h^2 x_e)^{-3/2}(T/0.03\,{\rm eV})^{-3/4} \end{aligned} \tag{3.96}$$

$$\frac{\bar{t}}{t_c} = (\Omega_B h^2 x_e)^{-1/2}(T/88\,{\rm eV})^{5/4}. \tag{3.97}$$

The timescale $\bar{t}$ is still longer than t_c for $T \gtrsim 90\,$eV where Compton scattering dominates. At lower temperatures (90 eV–1 eV), the (effective) free–free absorption can dominate over Compton scattering.

Another difference between Compton scattering and free–free transitions is worth mentioning at this stage. Compton scattering does not change the number of photons; thus this process alone can never lead to a Planck spectrum *if the system originally had the incorrect number of photons for a given total energy.* Since free–free absorption and emission can change the photon number they can lead to true thermalization. In this particular context, however, the system *does* have the correct number of photons because of the initial conditions.

The above analysis shows that there are several processes which can keep the radiation and matter tightly coupled until a temperature of about a few electron volts provided that a sufficient number of free charged particles exist (i.e. $x_e \approx 1$).

A depletion of free charged particles occurs through the process of 're'combination, in which electrons and ions combine to form neutral, atomic systems. For the sake of simplicity, we will consider only the formation of neutral hydrogen. The binding energy of atomic hydrogen is 13.6 eV. As the universe cools to temperatures below this value, the formation of bound atomic structure is *energetically* favoured. However, just as in the case of nucleosynthesis, the high entropy of the universe delays this process until the temperature drops to a much lower value, $T_{\text{atom}} \simeq 0.29\,\text{eV}$. Below this temperature, most of the protons and electrons combine together to form neutral atoms leaving behind only a small fraction ($\sim 10^{-5}$) of free electrons (and protons). The disappearance of free charged particles reduces the scattering cross section between photons and charged particles, thereby increasing the mean-free-path (λ_γ) of the photons. When $\lambda_\gamma \gtrsim H^{-1}$, the photons decouple from the rest of the matter; this happens at $T = T_{\text{dec}} \simeq 0.26\,\text{eV}$.

The temperature, T_{atom}, at which hydrogen atoms are formed can be computed by methods similar to those used in the last section if we make two simplifying assumptions: (i) The system is in thermodynamic equilibrium and (ii) the recombination proceeds through the electron and proton combining to form a hydrogen atom in the ground state. (Neither of these assumptions is quite correct; we will discuss the modifications which are required in a more realistic scenario in the end). In thermodynamic equilibrium, the number densities of protons (n_p), electrons (n_e) and hydrogen atoms (n_{H}) are given by the usual formula

$$n_i = g_i \left(\frac{m_i T}{2\pi}\right)^{3/2} \exp\left[\frac{1}{T}(\mu_i - m_i)\right] \tag{3.98}$$

where $i = e$, p or H. The equilibrium is maintained by the reaction $p + e \leftrightarrow n_{\text{H}} + \gamma$. Balance of chemical potential in this reaction implies that $\mu_p + \mu_e = \mu_{\text{H}}$. Using this relation and expressing μ_p and μ_e in terms of n_p and n_e, n_{H} becomes

$$n_{\text{H}} = \left(\frac{g_{\text{H}}}{g_p g_e}\right) n_p n_e \left(\frac{m_e T}{2\pi}\right)^{-3/2} \exp(B/T) \tag{3.99}$$

where $B = m_p + m_e - m_{\text{H}} = 13.6\,\text{eV}$ is the binding energy and we have set $m_p \approx m_{\text{H}}$ in the prefactor, $m^{-3/2}$. Introducing the 'fractional ionization', x_i, for each of the particle species and using the facts $n_p = n_e$ and $n_p + n_{\rm H} = n_B$, it follows that $x_p = x_e$ and $x_{\rm H} = (n_{\rm H}/n_B) = 1 - x_e$. Equation (3.99) can now be written in terms of x_e alone:

$$\frac{1-x_e}{x_e^2} = \frac{4\sqrt{2}\zeta(3)}{\sqrt{\pi}}\eta\left(\frac{T}{m_e}\right)^{3/2}\exp(B/T) \cong 3.84\eta(T/m_e)^{3/2}\exp(B/T) \tag{3.100}$$

where $\eta = 2.68 \times 10^{-8}(\Omega_B h^2)$ is the baryon-to-photon ratio. We may define $T_{\rm atom}$ as the temperature at which 90 per cent of the electrons have combined with protons: i.e. when $x_e = 0.1$. This leads to the condition:

$$(\Omega_B h^2)^{-1}\tau^{-3/2}\exp(-13.6\tau^{-1}) = 3.13 \times 10^{-18} \tag{3.101}$$

where $\tau = (T/1\,{\rm eV})$. For a given value of $(\Omega_B h^2)$, this equation can be easily solved by iteration. Taking logarithms and iterating once we find

$$\tau^{-1} \cong 3.084 - 0.0735\ln(\Omega_B h^2) \tag{3.102}$$

with the corresponding redshift $(1+z) = (T/T_0)$ given by

$$(1+z) = 1367[1 - 0.024\ln(\Omega_B h^2)]^{-1}. \tag{3.103}$$

For $\Omega_B h^2 = 1, 0.1, 0.01$ we get $T_{\rm atom} \cong 0.324\,{\rm eV}$, $0.307\,{\rm eV}$, $0.292\,{\rm eV}$ respectively. These values correspond to the redshifts of 1367, 1296 and 1232. For the sake of definiteness, the mean values

$$(1+z)_{\rm atom} \equiv 1300; \quad T_{\rm atom} = T_0(1+z)_{\rm atom} = 0.308\,{\rm eV} \tag{3.104}$$

will be adopted in future discussion as the epoch of formation of atoms.

Since the above analysis was based on equilibrium densities, it is important to check that the rate of the reactions $p + e \leftrightarrow {\rm H} + \gamma$ is fast enough to maintain equilibrium. The thermally averaged cross section for the process of recombination is given by

$$< \sigma v > \cong 4.7 \times 10^{-24}\left(\frac{T}{1\,{\rm eV}}\right)^{-1/2} {\rm cm}^2. \tag{3.105}$$

The cross section for interaction is proportional to the square of the de Broglie wavelength, $\lambda^2 \propto v^{-2}$; hence σv has $v^{-1} \propto T^{-1/2}$ dependence (see exercise 3.14). The reaction rate, therefore, will be

$$\begin{aligned}\Gamma &= n_p < \sigma v > = (x_e \eta n_\gamma) < \sigma v > \\ &= 2.374 \times 10^{-10}\,{\rm cm}^{-1}\,\tau^{7/4}{\rm e}^{-(6.8/\tau)}(\Omega_B h^2)^{1/2}\end{aligned} \tag{3.106}$$

In arriving at this expression, we have approximated equation (3.100) by

$$x_e \approx \left[\frac{\pi}{4\sqrt{2}\zeta(3)}\right]^{1/2}\eta^{-1/2}\left(\frac{T}{m_e}\right)^{-3/4}\exp\left(-\frac{6.8}{\tau}\right) \tag{3.107}$$

which is valid for $x_e \ll 1$. This Γ has to be compared with the expansion rate H of the universe. Taking the universe to be matter-dominated at this temperature, we have

$$H = 2.945 \times 10^{-23}\,\mathrm{cm}^{-1}\,(\Omega h^2)^{1/2}\tau^{3/2}. \tag{3.108}$$

Equating the two, we get

$$\tau^{-1/4}\exp(6.8/\tau) = 8.06 \times 10^{12}(\Omega_B/\Omega)^{1/2}. \tag{3.109}$$

This equation can also be solved by taking logarithms and iterating the solution. We find that

$$\tau^{-1} \cong 4.316 - 0.074\ln\left(\frac{\Omega}{\Omega_B}\right); \quad (1+z) = 977\left[1 - 0.017\ln\left(\frac{\Omega}{\Omega_B}\right)\right]^{-1}. \tag{3.110}$$

For $\Omega \simeq 10\Omega_B$, this gives $T_D \simeq 0.24\,\mathrm{eV}$. The fact that $T_D < T_{\mathrm{atom}}$ justifies the assumption of thermal equilibrium used in the earlier calculation. There are, however, some other difficulties with this assumption which will be discussed later.

When the reaction rate falls below the expansion rate, the formation of neutral atoms ceases. The remaining electrons and protons have negligible probability for combining with each other. The residual fraction can be estimated as the fraction present at $T = T_D$, i.e., $x_e(T_D)$. Combining (3.107) and (3.109) , we find

$$x_e(T_D) \cong 7.382 \times 10^{-6}\left(\frac{T_D}{1\,\mathrm{eV}}\right)^{-1}\left(\frac{\Omega^{1/2}}{\Omega_B h}\right). \tag{3.111}$$

For $T_D \simeq 0.24\,\mathrm{eV}$, this gives

$$x_e(T_D) \simeq 3 \times 10^{-5}\left(\frac{\Omega^{1/2}}{\Omega_B h}\right). \tag{3.112}$$

Thus a small fraction ($\sim 10^{-5}$) of electrons and protons will remain free in the universe.

The formation of atoms affects the photons, which were in thermal equilibrium with the rest of the matter through various scattering processes described above. It is easy to verify that the timescales for Compton scattering and free–free absorption become much larger than the expansion timescale when x_e drops to its residual value. The only scattering which is still somewhat operational is the Thomson scattering; this process merely changes the direction of the photon without any energy exchange. Its only effect is to make any given photon perform a random walk. When the number density of charged particles decreases, even this interaction rate Γ of the photons drops and, eventually, at some $T = T_{\mathrm{dec}}$, becomes lower than the expansion rate. For $T < T_{\mathrm{dec}}$, the photons are decoupled from the rest of the matter. The rate of Thomson scattering is given by

$$\begin{aligned}\Gamma &= \sigma n_e = \sigma x_e n_B = \sigma x_e \eta n_\gamma \\ &= 3.36 \times 10^{-11} (\Omega_B h^2)^{1/2} \tau^{9/4} \exp(-6.8/\tau)\,\mathrm{cm}^{-1}.\end{aligned} \tag{3.113}$$

Comparing this with the expansion rate

$$H = 2.945 \times 10^{-23} (\Omega h^2)^{1/2} \tau^{3/2}\,\mathrm{cm}^{-1}, \tag{3.114}$$

we get the condition

$$\tau^{-3/4} \mathrm{e}^{6.8/\tau} = 1.14 \times 10^{12} (\Omega_B/\Omega)^{1/2}. \tag{3.115}$$

where $\tau = (T_{\mathrm{dec}}/1\,\mathrm{eV})$. Solving this with one iteration we get

$$\tau^{-1} \cong 3.927 + 0.0735 \ln(\Omega_B/\Omega), \tag{3.116}$$

which corresponds to the parameters

$$T_{\mathrm{dec}} \cong 0.26\,\mathrm{eV}; \quad (1 + z_{\mathrm{dec}}) \cong 1100. \tag{3.117}$$

For $T \lesssim 0.2\,\mathrm{eV}$, the neutral matter and photons evolve as uncoupled systems. The parameter T characterizing the Planck spectrum continues to fall as a^{-1} because of the redshift of photons. The neutral matter behaves as a gaseous mixture of hydrogen and helium.

It should be stressed that three distinct events take place in the universe around $T \simeq (0.3$–$0.2)\,\mathrm{eV}$: (i) Most of the protons and electrons combine to form H atoms (ii) The process of recombination stops, leaving a small fraction of free electrons and protons, when the interaction rate for $pe \leftrightarrow \mathrm{H}\gamma$ drops below the expansion rate. (iii) The photon mean-free-path becomes larger than H^{-1}, decoupling radiation from matter. These events occur at almost the same epoch because $\eta \simeq 10^{-8}$ and $\Omega h^2 \simeq 1$, $\Omega_B \lesssim 1$. For a different set of values for these parameters, these events could occur at different epochs.

After decoupling, the temperature of the neutral atoms fall faster than that of radiation. The decrease of matter temperature is governed by the equation

$$\frac{dT_m}{dt} + 2\frac{\dot{a}}{a} T_m = \frac{4\pi^2}{45} \sigma_T x_e \left(\frac{T^4}{m}\right) (T - T_m) \tag{3.118}$$

where T is the radiation temperature (see exercise 3.15). The term $2(\dot{a}/a)$ T_m describes the cooling due to expansion while the term on the right hand side accounts for the energy transfer from radiation to matter. This process is now governed by the relaxation time for *matter*, which is given by:

$$t_{\mathrm{matter}} \simeq \frac{1}{\sigma_T x_e n_\gamma} \left(\frac{m}{T}\right). \tag{3.119}$$

At high temperatures ($x_e \simeq 1$), $t_{\text{matter}} \simeq t_c\,(n_e/n_\gamma) \ll t_c$; then $T_m \approx T$ to a high degree of accuracy. As x_e becomes smaller, t_{matter} increases and the energy transfer from the radiation to matter becomes less and less effective. The adiabatic cooling term makes the matter temperature fall faster than the radiation temperature. The relaxation time t_{matter} becomes of the order of expansion time at about $T \simeq 0.27\,\text{eV}$ ($z \simeq 1150$). At lower temperatures the matter temperature falls slightly slower than $T_m \propto a^{-2}$.

The small fraction of *ionized* matter ($n_e \simeq n_p \simeq 10^{-5} n_B$), however, continues to be affected by the photons. The *electron* mean-free-path, $\lambda_e = (\sigma_T n_\gamma)^{-1}$ which governs this process is much smaller than the photon mean-free-path because $n_\gamma \gg n_e$. The corresponding timescale $t_{\text{elec}} = (n_e/n_\gamma) t_c$ will be

$$t_{\text{elec}} = 2.15 \times 10^6\,\text{s}\,(T/1\,\text{eV})^{-4} \tag{3.120}$$

which leads to the ratio

$$\frac{t_{\text{elec}}}{H^{-1}} = (\Omega_B h^2)^{1/2} (T/60\,\text{K})^{-5/2}. \tag{3.121}$$

Thus the free electrons are tied to the radiation till a redshift of 20 or so. (In other words, the small number of electrons have many collisions with a *small* number of photons, though *most* of the photons are unaffected.) This process is, therefore, capable of maintaining the temperature of free electrons at the same value as T in the photon spectrum right up to $z \approx 20$. Of course, this interaction has very little effect on the photons because of the small number of charged particles present.

We shall now discuss the various approximations which have been made in the above calculation.[10] To begin with, notice that we assumed a recombination process which directly produces a H atom in the *ground* state. This will release a photon with energy of 13.6 eV in each recombination. If $n_\gamma(B)$ is the number density of photons in the background radiation with energy $B = 13.6\,\text{eV}$, then,

$$\frac{n_\gamma(B)}{n} \simeq \frac{16\pi}{n} T^3 \exp\left(-\frac{B}{T}\right) \simeq \frac{3 \times 10^7}{(\Omega_B h^2)} \exp\left(-\frac{13.6}{T}\right). \tag{3.122}$$

This ratio is unity at about 0.8 eV (i.e. at a redshift of $z \approx 3300$) and decreases rapidly at lower temperatures. Thus, at lower temperatures, the addition of 13.6 eV photons due to recombination significantly enhances the availability of ionizing photons. These energetic photons have a high probability of ionizing neutral atoms formed a little earlier. (That is, the 'backward' reaction $H + \gamma \to p + e$ is enhanced.) Hence this process is not very effective in producing a *net* number of neutral atoms.

The dominant process which actually operates is the one in which recombination proceeds through an excited state: ($e + p \to \text{H}^\star + \gamma$;

$H^\star \to H+\gamma_2$). This will produce two photons each of which has *lesser* energy than the ionization potential of the hydrogen atom. The $2P$ and $2S$ levels provide the most rapid route for recombination; the decay from $2P$ state produces a single photon while the decay from $2S$ state is through two photons. Since the reverse process does not occur at the same rate, this is non-equilibrium recombination.

Because of the above complication, the recombination proceeds at slower rate compared to that predicted by Saha's equation. The actual fractional ionization is higher than the value predicted by Saha's equation at temperatures below about 1300. For example, at $z = 1300$, these values differ by a factor 3; at $z \simeq 900$, they differ by a factor of 200. The values of $T_{\rm atom}$, $T_{\rm dec}$ etc., however, do not change significantly.

As an example, consider the value of $T_{\rm dec}$. In the redshift range of $800 < z < 1200$, the fractional ionization is given (approximately) by the formula,

$$x_e = 2.4 \times 10^{-3} \frac{(\Omega h^2)^{1/2}}{(\Omega_B h^2)} \left(\frac{z}{1000}\right)^{12.75} . \tag{3.123}$$

(This is obtained by fitting a curve to the numerical solution[11].) Using this expression, we can compute the optical depth for photons to be

$$\tau = \int_0^t n(t) x_e(t) \sigma_T \, dt = \int_o^z n(z) x_e(z) \sigma_T \left(\frac{dt}{dz}\right) dz \simeq 0.37 \left(\frac{z}{1000}\right)^{14.25} \tag{3.124}$$

where we have used the relation $H_0 \, dt \cong -\Omega^{-1/2} \, dz$ which is valid for $z \gg 1$. This optical depth is unity at $z_{\rm dec} = 1072$. Our approximate calculation earlier gave a value of 1100 which is quite close to the exact value.

From the optical depth, we can also compute the probability that the photon was last scattered in the interval $(z, z + dz)$. This is given by $(\exp -\tau)$ $(d\tau/dz)$ which can be expressed as

$$P(z) = e^{-\tau} \frac{d\tau}{dz} = 5.26 \times 10^{-3} \left(\frac{z}{1000}\right)^{13.25} \exp\left[-0.37 \left(\frac{z}{1000}\right)^{14.25}\right] . \tag{3.125}$$

This $P(z)$ has[11] a sharp maximum at $z \simeq 1067$ and a width of about $\Delta z \cong 80$. It is therefore reasonable to assume that decoupling occurred at $z \simeq 1070$ in an interval of about $\Delta z \simeq 80$. We shall see in chapter 6 that the finite thickness of the surface of last scattering has important observational consequences.

The photons emitted during recombination cannot be thermalized effectively, and thus distort the Planck spectrum. The distortions are on the high frequency part ($\nu/T > 30$) of the spectrum which corresponds to $\nu \gtrsim 1.5 \times 10^{12}$ Hz ($\lambda \lesssim 0.02$ cm) today. (The Planck spectrum has only 10^{-10} of all photons in the range $\nu/T > 30$.) Unfortunately, galactic nuclei and dust emit very strongly in the region $\lambda < 3 \times 10^{-2}$ cm, completely swamping this primordial signal.

Exercises

3.1 Consider a plasma with electron density n_e and temperature T. Under what circumstance can one treat the electrons as an *ideal* gas? Are these conditions satisfied in the early universe?

3.2 (a) Show that

$$(i) \int_0^\infty \frac{z^{x-1}\,dz}{e^z + 1} = \left(1 - \frac{1}{2^{x-1}}\right) \int_0^\infty \frac{z^{x-1}\,dz}{e^z - 1}$$

$$(ii) \int_0^\infty \frac{z^{x-1}\,dz}{e^z - 1} = \Gamma(x)\zeta(x) \quad (\text{for} \quad x > 1).$$

(b) Derive the expressions for n and ρ, discussed in the text, for massless particles with $\mu = 0$.

(c) In the case of relativistic ($T \gg m$), degenerate ($\mu \gg T$) fermions, show that

$$\rho \cong (1/8\pi^2)g\mu^4; \quad n = (1/6\pi^2)g\mu^3; \quad p = (1/24\pi^2)g\mu^4.$$

(d) For relativistic particles (bosons or fermions) with $\mu < 0$ and $|\mu| < T$, show that

$$\rho = (3g/\pi^2)T^4 \exp(\mu/T); \quad n = (g/\pi^2)T^3 \exp(\mu/T);$$
$$p = (g/\pi^2)T^4 \exp(\mu/T).$$

3.3 (a) Compute the redshift at which the density of the universe is comparable to the nuclear density. What is the temperature at this epoch? Should the ideal gas approximation be modified here?

(b) The discussion in the text did not take into account muons. Is this justified in the temperature range which was studied?

3.4 Consider an electron gas which has a density $n_0 = (N/V)$ in the absence of pair production. Let n_+ and $(n_- - n_0)$ denote the number densities of the positron–electron pairs present in equilibrium; $n_- - n_0 = n_+$. Show that, for $T \ll m$,

$$n_+ n_- = 4\left(\frac{mT}{2\pi}\right)^3 \exp(-2m/T)$$

and

$$n_+ = -\frac{1}{2}n_0 + \left\{\frac{1}{4}n_0^2 + [(mT)^3/2\pi^3]\exp(-2m/T)\right\}^{1/2}.$$

3.5 (a) Can a $k = +1$ Friedmann universe contain net electric charge? What about net baryon number or lepton number?

(b) Consider an open Friedmann model with net charge density n_Q (distributed homogeneously). Show that the constraint $\Omega h^2 \lesssim 1$ implies $(n_Q/s) \lesssim 10^{-27}$.
(c) A massless, degenerate ($|\mu_\nu| \gg T$) neutrino will contribute an energy density $\rho_\nu \cong (\mu_\nu^4/8\pi^2)$. Show that the constraint $\Omega h^2 \lesssim 1$ implies the bounds

$$(\mu_\nu/T_\nu) \lesssim 53; \quad \frac{(n_\nu - n_{\overline{\nu}})}{n_\gamma} \lesssim 3.7 \times 10^3.$$

3.6 (a) Assume that weak interactions are mediated by a vector boson with mass m_x. Let $\sigma(E)$ be a typical cross section for a $2 \leftrightarrow 2$ process. Argue that for $E \lesssim m_x$, $\sigma(E) \simeq (\alpha^2 E^2/m_x^4)$. How does $\sigma(E)$ behave for $E \gg m_x$?
(b) Assume that the rate Γ for some reaction is proportional to T^n with $n > 3$. Let t_D be defined to be the time of decoupling, at which $\Gamma(t_D) = H(t_D)$ with $t_D \ll t_{eq}$. Show that the number of *further* interactions which take place is less than unity.

3.7 (a) Show that, during the $e\overline{e}$ annihilation, the relation aT = constant gets modified to the form $aTf(m/T)$=constant, where

$$f^3(x) = 1 + \frac{15}{2\pi^4}\int_0^\infty \frac{y^2\,dy}{\sqrt{x^2+y^2}}\,\frac{3x^2+4y^2}{[\exp\sqrt{x^2+y^2}+1]}.$$

(b) Show that the energy density during $e\overline{e}$ annihilation is given by

$$\epsilon(x) = 1 + \left(\frac{21}{8}\right)\left(\frac{4}{11}\right)^{4/3} f^4 + \frac{30}{\pi^4}\int_0^\infty \frac{y^2\sqrt{x^2+y^2}\,dy}{[\exp\sqrt{x^2+y^2}+1]}.$$

(c) Using these, estimate how long it takes for the ratio (T_γ/T_ν) to become, say, 1.39.

3.8 (a) The non-relativistic Boltzmann equation has the form:

$$\hat{D}_{\rm NR} f = \left(\frac{\partial}{\partial t} + \mathbf{v}\cdot\nabla + \mathbf{F}\cdot\nabla_{\mathbf{p}}\right) f(\mathbf{x},\mathbf{p},t) = \hat{C}[f]$$

where $\hat{C}[f]$ is the collision term. Show that a proper, relativistic generalization of this equation will be

$$\hat{D}_R f \equiv \left(p^\mu \frac{\partial}{\partial x^\mu} - \Gamma^\mu_{\alpha\beta}p^\alpha p^\beta \frac{\partial}{\partial p^\mu}\right) f = \hat{C}[f].$$

Also show that, in the context of Friedmann universes, with $f(\mathbf{x},\mathbf{p},t)$ $= f(|\mathbf{p}|,t) = f(E,t)$, the operator on the left hand side will be

$$\hat{D}_R = E\frac{\partial}{\partial t} - \left(\frac{\dot{a}}{a}\right)|\mathbf{p}|^2\frac{\partial}{\partial E}.$$

(b) Integrate this equation over the momentum to obtain

$$\frac{dn}{dt} + 3\frac{\dot{a}}{a}n = \frac{g}{(2\pi)^3}\int \hat{C}[f]\frac{d^3p}{E}$$

where $n(t) = (g/8\pi^3) \int d^3p\, f(E,t)$. Interpret the various terms in this equation.

(c) Consider a process $A\overline{A} \leftrightarrow X\overline{X}$. Assume that: (i) All the distribution functions satisfy the condition $f \ll 1$. (ii) $X, \overline{X}$ are in thermal equilibrium with zero chemical potential. Show that, in this case,

$$\dot{n}_A + 3\left(\frac{\dot{a}}{a}\right) n_A = - < \sigma v > [n_A^2 - f_A^2]$$

where f_A is the equilibrium number density and

$$< \sigma v > f_A^2 = \int d\Lambda \, |M|^2 \exp\left[-\frac{1}{T}(E_X + E_{\overline{X}})\right]$$

where

$$d\Lambda = d\Gamma_A \, d\Gamma_{\overline{A}} \, d\Gamma_X \, d\Gamma_{\overline{X}} \, (2\pi)^4 \delta(p_A + p_{\overline{A}} - p_X - p_{\overline{X}})$$

with $d\Gamma = (g/8\pi^3)(d^3p/2E)$ and $|M|^2$ is the square of the matrix element for the process $A\overline{A} \leftrightarrow X\overline{X}$. Interpret the various terms in this equation.

(d) The $3(\dot{a}/a)n$ term can be eliminated by changing the variable to $N \propto (na^3) \equiv (n/s)$ defined in the text. Further, let $\tau \equiv (T/m)$ where m is mass of the particle -A. Show that

$$\frac{\tau}{N}\left(\frac{dN}{d\tau}\right) = \left[\frac{\Gamma_A}{H(\tau)}\right]\left[\left(\frac{N}{N_{eq}}\right)^2 - 1\right]$$

where $\Gamma_A = N_{eq} < \sigma v >$, $H(\tau) = (\dot{a}/a)_\tau$ and N_{eq} is the equilibrium value for $N = (n/s)$ discussed in the text. Interpret this equation.

(e) Let $< \sigma v >= \sigma_0 \tau^k$, and τ_D be the 'freeze out' point: $\tau_D = (T_D/m)$. For $\tau \gg \tau_D$, $(N - N_{eq})$ is small and slowly varying. Show that, in this limit $(N - N_{eq}) \simeq a\tau^{-(k+2)}$ where $a = 1.89(g^{1/2}/q)\,(m_{pl} m \sigma_0)^{-1}$. In the second limiting case with $\tau \ll \tau_D$ we have $N \gg N_{eq}$. Using this fact, show that the final value of N is $N_\infty = N(\tau = \infty) = 2a(k+1)\tau_D^{-(k+1)}$.

3.9 Repeat the analysis in the text for a cold relic, keeping the value of σ_0 arbitrary and taking $k = 0$. Show that

$$(\Omega_A h^2) \simeq \mathcal{O}(1).\left(\frac{\sigma_0}{10^{-37}\,\text{cm}^2}\right).$$

This shows why the annihilation cross section should be that of a weak interaction process if $\Omega_A h^2 \simeq 1$.

3.10 In the text, we only discussed stable relics. Suppose there exists a wimp – say, a heavy neutrino, ν_H – with mass m_H which decays into a light neutrino ν_L of mass m_L and a very light boson. ($\nu_H \to \nu_L b$). Let the lifetime be τ_H and assume, for simplicity, that the decay is instantaneous. Compute the contribution to Ω today from ν_L and b if (i) both ν_L and b are relativistic today or (ii) b is still relativistic but ν_L is non-relativistic. What are the constraints on m_H and τ_H in this case?

3.11 When the nucleons form ^{4}He, a certain amount of energy will be released. Assuming this energy is emitted as photons, which subsequently get thermalized, estimate the increase in the radiation temperature.

3.12 If the universe contained many more ν than $\overline{\nu}$ would it have produced more ^{4}He or less? What about D?

3.13 (a) The opacity due to free–free absorption can be understood by the following simple model. If the number of photons at some frequency ω, $n(\omega)$, is large ($n(\omega) \gg 1$)) and if $\hbar\omega \ll kT$, then the photons can be treated as producing a classical electromagnetic field with $\mathbf{E} = \mathbf{E}_0 \cos \omega t$. Show that the velocity induced by this wave on a free electron is

$$\mathbf{v} = \mathbf{v}_0 + \frac{e\mathbf{E}_0}{m\omega} \sin \omega t.$$

(b) Show that the mean energy of the electron in the radiation field can be taken to be

$$< \epsilon >= \frac{3}{2}T + \frac{e^2 E_0^2}{4m\omega^2}.$$

(c) Assume that the electromagnetic part of the energy gets thermalized in the scatterings. Argue that the rate of collisions is $\Gamma = \sigma n_e v$ with $\sigma \simeq (e^4/T^2)$.

(d) The energy dissipated per unit volume per unit time is $n_e\Gamma(e^2E_0^2/4m\omega^2)$. By definition, this is equal to the product of opacity (κ) and energy flux ($E_0^2/8\pi$). Equating the two, show that

$$\kappa \cong \frac{e^6 n_e^2}{(mT)^{3/2}\omega^2}.$$

A more careful analysis gives

$$\kappa_{\text{exact}} = \frac{4}{3}\frac{e^6 n_e^2}{m\sqrt{6\pi m T}}\frac{(2\pi)^3}{\omega^3}(1 - \mathrm{e}^{-\omega/T}).$$

Evaluate the free–free absorption timescale from κ_{exact} for $\omega = T$.

(e) Using the result for free–free absorption, show that the rate of free–free emission (bremstrahlung) will be

$$I(\omega) = \frac{8}{3}(2\pi)^3 \frac{e^6 n_e^2}{m\sqrt{6\pi m T}} \mathrm{e}^{-\omega/T}.$$

(f) The number of photons produced per unit volume per unit time through free–free emission will be

$$\frac{dN_\gamma}{dV\,dt} = \int_0^\infty \frac{4\pi I(\omega)}{\omega}\,d\omega.$$

This integral diverges in the lower limit. Argue that this integral should be cut off at $\omega = \omega_c$ where ω_c is the frequency at which the free–free absorption timescale is the same as the Compton timescale; that is,

$$\kappa(\omega_c) = \sigma_T n_e \left(\frac{T}{m}\right)$$

compute $(dN_\gamma/dV\,dt)$ with this lower cutoff.

(g) Define the timescale for photon production to be t_γ where

$$t_\gamma^{-1} = \frac{1}{n_\gamma}\left(\frac{dN_\gamma}{dV\,dt}\right).$$

Show that, near $(\omega \simeq T)$, $t_\gamma \simeq t_{ff}$.

3.14 Show that the rate of recombination of electrons and protons in a thermal distribution (forming hydrogen atoms in the ground state) is given by

$$< \sigma v > \simeq \frac{4\pi^2 e^2}{m_e^2}\frac{B}{(3mT)^{1/2}}$$

where B is the binding energy of the hydrogen atom.

3.15 (a) Consider an electron which is moving with a velocity v through a radiation bath of temperature T. Show that the electron will feel a 'drag' force

$$F = -\frac{4}{3}\sigma_T\left(\frac{\pi^2}{15}T^4\right)v = -\frac{4\pi^2}{45}\sigma_T T^4 v.$$

Derive the result classically but provide a quantum mechanical interpretation as well.

(b) Show that the net rate of transfer of energy from radiation to matter per electron is

$$\frac{dQ}{dt} = \frac{4\pi^2}{15}\sigma_T\left(\frac{T^4}{m}\right)(T - T_e)$$

where T_e is the matter temperature and T is the radiation temperature. Explain the origin of the first term in the right hand side.

(c) If the matter is fully ionized, then the kinetic energy per electron is $3T_e$ so that

$$\dot{T}_e = \frac{4\pi^2}{45}\sigma_T T^4 (T - T_e).$$

(d) How does this equation get modified in an expanding universe, with only a fraction of the matter being ionized?

3.16 (a) Consider a plasma in which photons and electrons are at temperatures T_r and T_m respectively. Show that the distribution function $n(\nu, t)$ of photons changes due to free–free processes as

$$\frac{\partial n}{\partial t} = \kappa\frac{\mathrm{e}^{-x}}{x^3}\left[1 - n(\mathrm{e}^x - 1)\right],$$

with

$$\kappa = \frac{32\pi^2}{3}\frac{e^6 n_e^2}{(T_r)^3 m\sqrt{6\pi m T}}; \quad x = \frac{h\nu}{T_m}.$$

(b) Estimate the relaxation time for the equilibrium to be achieved.

Part two

The clumpy universe

4
Linear theory of perturbations

4.1 Introduction

The discussion in the last two chapters was based on the assumption that the matter in our universe is distributed homogeneously. The real universe, of course, contains inhomogeneous structures like galaxies, clusters etc. In this chapter, we shall discuss a possible (and popular) mechanism for the formation of these inhomogeneities.

In section 4.2, we discuss the general formalism of linear perturbation theory and analyze, in a qualitative way, how a small inhomogeneity can grow due to gravitational instability. The mathematics of linear perturbation theory, both in the fully relativistic case and in Newtonian approximation, is developed in section 4.3. (This section uses fairly advanced concepts from general relativity; readers who are not familiar with general relativity may skip the latter half of this section.) The solutions to these equations are studied in sections 4.4 and 4.5. In section 4.6, we consider the effects of various dissipative processes on the evolution of inhomogeneities. Finally, all these results are combined together in section 4.7 to derive the form of the power spectrum of perturbations.

4.2 Suppression and growth of inhomogeneities

The formalism developed in the last two chapters needs to be modified to take into account the inhomogeneities present in our universe. We shall try to reconstruct the observed universe in the following manner: We assume that, at some time in the past, there were small deviations from homogeneity in our universe. These deviations can grow due to gravitational instability over a period of time and, eventually, form galaxies, clusters etc. As long as these inhomogeneities are small, their growth can be studied by the linear perturbation theory which will be developed in this chapter[1]. Once the deviations from the smooth universe become large, linear theory fails and we have to use other techniques to understand the nonlinear evolution. The nonlinear evolution will be discussed in chapter 8.

It should be noted that this approach *assumes* the existence of small inhomogeneities at some initial time. To be considered complete, the cosmological model should also *produce* these initial inhomogeneities by some viable physical mechanism. Some attempts in this direction will be discussed in chapter 10.

Consider a perturbation of the metric $g_{\alpha\beta}(x)$ and the source $T_{\alpha\beta}$ into the form $(g_{\alpha\beta} + \delta g_{\alpha\beta})$ and $(T_{\alpha\beta} + \delta T_{\alpha\beta})$, where the set $(g_{\alpha\beta}, T_{\alpha\beta})$ corresponds to the smooth background universe, while the set $(\delta g_{\alpha\beta}, \delta T_{\alpha\beta})$ denotes the perturbation. Assuming the latter to be 'small', we can linearize Einstein's equations to obtain a second-order differential equation of the form

$$\hat{\mathcal{L}}(g_{\alpha\beta})\delta g_{\alpha\beta} = \delta T_{\alpha\beta} \tag{4.1}$$

where $\hat{\mathcal{L}}$ is a linear differential operator depending on the background space-time. Since this is a linear equation, it is convenient to expand the solution in terms of some suitably chosen mode functions. For the sake of simplicity, let us consider the spatially flat $(\Omega = 1)$ universe. The mode functions will then be plane waves and by Fourier transforming the variables we can obtain a set of separate equations $\hat{\mathcal{L}}_{(k)}\delta g_{(k)} = \delta T_{(k)}$, for each mode, labelled by a wave vector $\mathbf{k}$. Solving this set of equations, we can determine the evolution of each mode separately. (A similar procedure, of course, works for the case with $\Omega \neq 1$. In this case, the mode functions will be more complicated than the plane waves; but, with a suitable choice of orthonormal functions, we can obtain a similar set of equations.) There is, however, one major conceptual difficulty in carrying out this programme. In general relativity, the form (and numerical value) of the metric coefficients $g_{\alpha\beta}$ (or the stress-tensor components $T_{\alpha\beta}$) can be changed by a relabelling of coordinates $x^\alpha \to x^{\alpha'}$. By such a trivial change we can make a small $\delta T_{\alpha\beta}$ large or even generate a component which was originally absent. Thus the perturbations may grow at different rates – or even decay – when we relabel coordinates. It is necessary to tackle this ambiguity before we can meaningfully talk about the growth of inhomogeneities.

There is a simple way of handling this problem for modes which have proper wavelengths much smaller than the Hubble radius. The general relativistic effects due to the curvature of space-time will be negligible at length scales far smaller than the Hubble radius. In such regions, we shall see that there exists a natural choice of coordinates in which Newtonian gravity is applicable. All physical quantities can be unambiguously defined in this context.

Though such a Newtonian analysis provides valuable insight into the behaviour of inhomogeneities, it suffers from the following difficulty: The proper wavelength of any mode will be bigger than the Hubble radius at sufficiently early epochs. We saw in chapter 2 that any proper length λ (as measured today) with $\lambda \ll H^{-1}$ (today) would have entered the Hubble radius at some time $t_{\rm enter}(\lambda)$ in the past. So Newtonian analysis can be used to study a mode labelled by λ *only* for times $t \gg t_{\rm enter}(\lambda)$, when the mode is well within the Hubble radius. Thus, the early evolution of *any* mode needs to be tackled by general relativity and the coordinate ambiguities have to be settled.

There are two different ways of handling such difficulties in general relativity. The first method is to resolve the problem by force: We may choose a particular coordinate system and compute everything in that coordinate system. If the coordinate system is physically well motivated, then the quantities computed in that system can be interpreted easily; for example, we will treat δT^0_0 to be the perturbed mass (energy) density even though it is coordinate dependent. The difficulty with this method is that one cannot fix the gauge *completely* by simple physical arguments; the residual gauge ambiguities do create some problems.

The second approach is to construct quantities – linear combinations of various perturbed physical variables – which are scalars under coordinate transformations. Einstein's equations are then rewritten as equations for these gauge invariant quantities. This approach, of course, is manifestly gauge invariant from start to finish[2]. However, it is more complicated than the first one; besides, the gauge invariant objects do not, in general, possess any simple interpretation. We shall be mainly concerned with the first approach.

In principle, therefore, the perturbation theory can proceed in two steps: (i) Given a mode λ, we know $t_{\rm enter}(\lambda)$. For $t < t_{\rm enter}(\lambda), \lambda > d_H$ and we should use some form of a general relativistic perturbation theory to evolve $\delta\rho_\lambda(t)$ from some initial time $t = t_i$ to $t = t_{\rm enter}(\lambda)$. (ii) For $t > t_{\rm enter}(\lambda), \lambda < d_H$ and the evolution of $\delta\rho_\lambda$ can be studied using Newtonian theory.

In spite of the complications mentioned above, the main results regarding the evolution of the perturbations can be understood by fairly elementary reasoning. Since this approach highlights the various physical effects which are present in the problem, we shall adopt this approach in this section. The rigorous mathematical theory of linear perturbations will be developed in the next section.

The material content of the smooth universe has three main components – baryons (ρ_B), dark matter ($\rho_{\rm DM}$) and relativistic matter like photons (ρ_R). For the sake of definiteness, we shall assume that the dark matter is made of wimps of mass m, which could be either hot or cold. The wimps would have decoupled at some temperature $T = T_D$ and would have become non-relativistic at $T = T_{\rm nr} \simeq m$. In all the scenarios we are interested in, $m > 5.5\,{\rm eV}$; hence $T_{\rm nr} > T_{\rm eq}$ and $t_{\rm nr} < t_{\rm eq}$. Since baryons are directly coupled to photons till $t \approx t_{\rm dec}$, while wimps are not, the evolution of these two components is quite different. It is therefore best to treat the growth of inhomogeneities in each component separately.

Each of these components can be specified by the equation of state which connects the pressure p_x of the component x with the density ρ_x. For non-relativistic matter, treated as dust, $p_x \approx 0$, while for radiation $p_R = (1/3)\rho_R$. For most of our purposes, it is enough to consider a simple equation of state of the form $p_x = w_x \rho_x$ where w_x is a constant. In such a medium, a density perturbation $\delta\rho_x$ will generate a corresponding pressure perturbation $\delta p_x = w_x \delta\rho_x$. It is conventional to define the velocity dispersion of a particular component by the relation $v_x^2 = (\dot{p}_x/\dot{\rho}_x)$ so that $\delta p_x = v_x^2 \delta\rho_x$. In this simple case, in which $w_x = (p_x/\rho_x)$ is a constant, $v_x^2 = w_x$. The situation becomes somewhat more complex when no single component dominates the dynamics. This will be discussed in the next section.

Having characterized the fluctuations in each component x by a velocity dispersion v_x^2, we can study their evolution. Consider first the situation in which the wavelength λ of the perturbation is much larger than the Hubble radius d_H. Since processes like pressure, viscosity etc. act at scales much smaller than d_H, they do not affect the evolution of the modes with $\lambda > d_H$. Even though a rigorous study of such a mode is complicated, the final result can be obtained by the following simple argument.

Consider a spherical region of radius $\lambda (> d_H)$ containing matter with a mean density ρ_1, embedded in a $k = 0$ Friedmann universe of density ρ_0 (with $\rho_1 = \rho_0 + \delta\rho$; $\delta\rho$ small and positive). It follows from spherical symmetry (see chapter 2) that the inner region is not affected by the matter outside; hence the inner region evolves as a $k = +1$ Friedmann universe. Therefore, we can write, for the two regions:

$$H_1^2 + \frac{1}{a_1^2} = \frac{8\pi G}{3}\rho_1; \quad H_0^2 = \frac{8\pi G}{3}\rho_0, \tag{4.2}$$

where $H_0 = (\dot{a}_0/a_0)$ and $H_1 = (\dot{a}_1/a_1)$. We will compare the perturbed universe with the background universe *when their expansion rates are equal*; i.e. we compare their densities at a time t when $H_1 = H_0$. We then get

$$\frac{8\pi G}{3}(\rho_1 - \rho_0) = \frac{1}{a_1^2}, \tag{4.3}$$

or

$$\left(\frac{\rho_1 - \rho_0}{\rho_0}\right) = \frac{\delta\rho}{\rho_0} = \frac{3}{8\pi G(\rho_0 a_1^2)}. \tag{4.4}$$

In general, if $H_0 = H_1$ at some time, then $a_0 \neq a_1$ at that time. But, if $(\delta\rho/\rho_0)$ is small, then a_1 and a_0 will differ by only a small quantity and we can set $a_1 \approx a_0$ in the right hand side of (4.4). This allows one to find how $(\delta\rho/\rho_0)$ scales with a. Since $\rho_0 \propto a^{-4}$ in the radiation dominated phase $t < t_{\rm eq}$ and $\rho_0 \propto a^{-3}$ in the matter dominated phase $t > t_{\rm eq}$ we get

$$\left(\frac{\delta\rho}{\rho}\right) \propto \begin{cases} a^2 & (\text{for}\, t < t_{\rm eq}) \\ a & (\text{for}\, t > t_{\rm eq}) \end{cases}. \tag{4.5}$$

Thus, the amplitude of the mode with $\lambda > d_H$ always grows; as a^2 in the radiation dominated phase and as a in the matter dominated phase. We will prove this result more rigorously in the next section. (Also see exercises 4.5 and 4.8).

Let us next consider what happens to this mode after it enters the Hubble radius when $\lambda < d_H$. There are two processes which can prevent the amplitude from growing.

The first one is the familiar pressure support. If the pressure distribution of matter can re-adjust itself fast enough – i.e., if sufficient pressure can build up before gravity crushes the perturbed region by its weight – then the pressure will prevent gravitational enhancement of the density contrast. The condition for this is

$$\begin{aligned}&\{\text{timescale for pressure readjustment}\}\\ &\quad < \{\text{timescale for gravitational collapse}\}.\end{aligned} \tag{4.6}$$

That is

$$t_{\rm pressure} \simeq \frac{\text{wavelength}}{\text{vel. dispersion}} = \frac{\lambda}{v} < \frac{1}{(G\rho)^{1/2}} = \text{free fall time} \simeq t_{\rm grav}. \tag{4.7}$$

This condition for stability implies that growth is suppressed in modes with wavelengths λ less than a critical wavelength $\lambda_J \sim v(G\rho)^{-1/2}$. It is conventional to define this 'Jeans length' with an extra $\sqrt{\pi}$ factor:

$$\lambda_J \equiv \sqrt{\pi}\frac{v}{(G\rho)^{1/2}}. \tag{4.8}$$

If the universe contains only one species of particle, then both v and ρ will correspond to that species. In a multi-component medium, v will be the velocity dispersion of the perturbed component (it is the perturbed component that provides the pressure support), but ρ will be the density of the component which is most dominant gravitationally (it is this component which is making the perturbation collapse). In general, of course, these two components will not be the same.

The pressure in a baryonic gas is essentially provided by collisions. But in the dark matter component, the collisions are quite ignorable. The 'pressure' support in a collisionless system arises from the readjustment of orbits. In both the cases, however, the timescale $t_{\rm pressure}$ is set by the velocity dispersion, v (see exercise 4.7).

There is a second process which can prevent the growth of perturbations. This occurs when (i) the perturbed species is *not* the dominant species (which governs the expansion rate) *and* (ii) the dominant species is smoothly distributed. Suppose that $t_{\rm grav}$ for the perturbed species (say, dark matter) is indeed less than $t_{\rm pressure}$; thus the condition $\lambda > \lambda_J$ is satisfied and pressure cannot prevent the collapse. But suppose that we are in a radiation dominated phase, when the expansion timescale $t_{\rm exp} \sim (G\rho_{\rm dominant})^{-1/2} \sim (G\rho_R)^{-1/2}$ is smaller than $t_{\rm grav}$. Then the universe will be expanding too fast for the collapsing region to condense out. We have here a situation with $t_{\rm exp} < t_{\rm grav} < t_{\rm pressure}$; that is,

$$\frac{1}{(G\rho_R)^{1/2}} < \frac{1}{(G\rho_{\rm DM})^{1/2}} < \frac{\lambda}{v}. \tag{4.9}$$

It is the rapid background expansion rather than pressure support which prevents the growth in this case.

If neither of these processes is operational, then the amplitude of the perturbation will grow. It is clear that the second process will prevent growth in all modes with $\lambda < d_H$ in the radiation dominated phase. Thus, in the radiation dominated phase, only modes with $\lambda > d_H$ will grow and they grow as a^2 (see (4.5)). In the matter dominated phase, for all $\lambda \gg \lambda_J$, we can ignore pressure effects; thus the analysis leading to (4.5) is valid even for modes with $\lambda \gg \lambda_J$ (i.e for modes with $\lambda > d_H$ *as well as* for modes with $d_H > \lambda \gg \lambda_J$). So we conclude that, for all $\lambda \gg \lambda_J$ in the matter dominated phase, the amplitude grows as a. Modes with $\lambda \gtrsim \lambda_J$ also grow, but at a slower rate, because of the effects of the pressure.

We can now put all the pieces together and study the evolution of a perturbation with wavelength λ. The relevant scalings are shown in the left half of figure 4.1. Suppose that this mode (with proper wavelength $\lambda \propto a$) enters the Hubble radius in the radiation dominated phase at some $a = a_{\rm enter}$. Let us consider the perturbations in the dark matter component at different epochs first. For dark matter, the velocity dispersion $v \simeq 1$ when the particles are relativistic ($a < a_{\rm nr}$) and decays as $v \propto a^{-1}$ (due to the redshifting of the momentum $p \propto mv \propto a^{-1}$ when the particles are non-relativistic, $a > a_{\rm nr}$). Since $\rho_{\rm dom} = \rho_R$ for

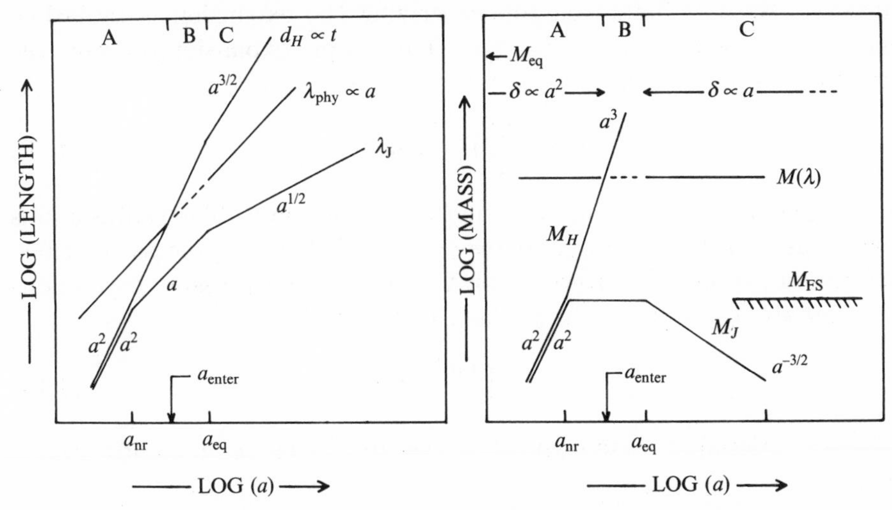

Fig. 4.1. The Jeans length for dark matter component is shown on the left side of the picture (in a log–log scale). The three stages A, B and C discussed in the text are marked. Also plotted are the Hubble radius and the physical wavelength of the perturbation. The part of the line shown in dots corresponds to the epoch during which the growth is suppressed and is only logarithmic. The corresponding Jeans mass is plotted against the expansion factor in the right side. Also shown are the mass contained inside the Hubble radius and the mass contained in a perturbation with a wavelength λ, for which $a_{\rm nr} < a_{\rm enter} < a_{\rm eq}$. The damping scale due to free-streaming of the dark matter is also marked in the diagram for the Jeans mass.

$a < a_{\rm eq}$ and $\rho_{\rm dom} = \rho_{\rm DM}$ for $a > a_{\rm eq}$, the quantity $\rho_{\rm dom}^{-1/2}$ will scale as $\rho_{\rm dom}^{-1/2} = \rho_R^{-1/2} \propto a^2$ for $a > a_{\rm eq}$ and as $\rho_{\rm dom}^{-1/2} \propto \rho_{\rm DM}^{-1/2} \propto a^{3/2}$ for $a < a_{\rm eq}$. Combining, we find that, for dark matter,

$$\lambda_J \propto \frac{v}{\rho_{\rm dom}^{1/2}} \propto \begin{cases} a^2 & (a < a_{\rm nr}) \\ a & (a_{\rm nr} < a < a_{\rm eq}) \\ a^{1/2} & (a_{\rm eq} < a) \end{cases} . \tag{4.10}$$

There are essentially three stages in the evolution of a mode which enters the Hubble radius between $a_{\rm nr}$ and $a_{\rm eq}$ (as we shall see later, these are the modes most relevant to astrophysics): (a) Stage A ($a < a_{\rm enter}$): The wavelength of the perturbation is bigger than the Hubble radius; from our earlier discussion it follows that

$$\delta \equiv \left(\frac{\delta\rho}{\rho}\right) \propto a^2. \tag{4.11}$$

(b) Stage B ($a_{\rm enter} < a < a_{\rm eq}$): The wavelength is inside the Hubble radius and is bigger than λ_J; so, pressure support cannot stop the collapse.

However, the dominant component driving the expansion is radiation, and since $\rho_R > \rho_{\rm DM}$, $t_{\rm exp} < t_{\rm collapse}$. Thus, rapid expansion prevents the growth of perturbations in this stage:

$$\delta = \left(\frac{\delta\rho}{\rho}\right) = \text{constant}. \tag{4.12}$$

(c) Stage C ($a_{\rm eq} < a$): The wavelength is inside the Hubble radius and is bigger than λ_J; further $\rho_{\rm dom}$ now *is* $\rho_{\rm DM}$ itself. Neither process described before can prevent the growth. For $\lambda \gg \lambda_J$ (so that pressure corrections are ignorable), the growth is as described by (4.5):

$$\delta = \left(\frac{\delta\rho}{\rho}\right) \propto a. \tag{4.13}$$

It is conventional to present the above results in terms of a quantity called Jeans mass:

$$M_J \equiv \frac{4\pi}{3}\rho\left(\frac{\lambda_J}{2}\right)^3 \tag{4.14}$$

where ρ is the component under discussion. Since $\rho_{\rm dom} \propto \rho_R \propto a^{-4}$ in the radiation dominated phase and $\rho_{\rm dom} \propto \rho_{\rm DM} \propto a^{-3}$ in the matter dominated phase, we see that

$$M_J \propto \rho_{\rm DM}\lambda_J^3 \propto \begin{cases} a^2 & (a < a_{\rm nr}) \\ \text{constant} & (a_{\rm nr} < a < a_{\rm eq}) \\ a^{-3/2} & (a_{\rm eq} < a) \end{cases} . \tag{4.15}$$

Notice that the ρ which appears in the definition of λ_J is ρ_R for $a < a_{\rm eq}$ and $\rho_{\rm DM}$ for $a > a_{\rm eq}$. But the scaling density in M_J is always the density of the component under discussion – which is $\rho_{\rm DM}$ here. The mass inside the Hubble radius, M_H, is similarly defined to be

$$M_H = \frac{4\pi}{3}\rho_{\rm DM}\left(\frac{d_H}{2}\right)^3 \propto \begin{cases} a^2 & (a < a_{\rm nr}) \\ a^3 & (a_{\rm nr} < a < a_{\rm eq}) \\ a^{3/2} & (a_{\rm eq} < a) \end{cases} . \tag{4.16}$$

The perturbation, of course, has a constant mass $M(\lambda)$. All the previous three stages can, of course, be re-expressed in terms of these masses, as shown in the right half of figure 4.1.

From the scalings given above, and the known value of $M_{\rm eq}$, the Jeans mass can be estimated numerically. For example, $M_J = 3.2 \times 10^{14}$ $M_\odot(\Omega h^2)^{-2}(a/a_{\rm eq})^{-3/2}$ for $a > a_{\rm eq}$. Also note that $M_J \simeq M_H$ for $a < a_{\rm nr}$. In this case, M_J is not of much physical relevance.

Let us next consider the perturbations in the baryonic component. Since baryons and photons are tightly coupled for $a < a_{\rm dec}$, the pressure and density of the coupled baryon–photon system is well correlated. Let $a_{B\gamma}$ be the epoch at which $\rho_\gamma = \rho_B$. This $a_{B\gamma}$ corresponds to the redshift $(1 + z_{B\gamma}) = (\Omega_B/\Omega_\gamma) = 3.9 \times 10^4(\Omega_B h^2)$. Since $z_{\rm dec} \simeq 1100$, $z_{B\gamma} > z_{\rm dec}$ for $\Omega_B h^2 > 0.026$. In order to describe the most general situation, we will assume that this is indeed the case. For $a < a_{B\gamma}$, $p_R \gg p_B$ and $\rho_R \gg \rho_B$; so

$$v^2 = \left(\frac{\partial p}{\partial \rho}\right) \simeq \frac{p}{\rho} = \frac{p_R + p_B}{\rho_R + \rho_B} \approx \frac{p_R}{\rho_R} = \frac{1}{3} \qquad (a < a_{B\gamma}) \tag{4.17}$$

For $a_{B\gamma} < a < a_{\rm dec}$, baryons are still coupled to photons, maintaining pressure equilibrium ($p = p_R + p_B \approx p_R$), but the dominant density is $\rho = \rho_B + \rho_R \approx \rho_B$. So

$$v^2 \simeq \frac{p}{\rho} \approx \frac{p_R}{\rho_B} \propto a^{-1} \qquad (a_{B\gamma} < a < a_{\rm dec}). \tag{4.18}$$

For $a_{\rm dec} < a$, baryons are decoupled from photons and there is no pressure equilibrium. The $v^2 \propto (p_B/\rho_B)$ now is just the velocity dispersion of a gaseous mixture of hydrogen and helium. So

$$v^2 \propto a^{-2} \qquad (a_{\rm dec} < a). \tag{4.19}$$

Notice that, at decoupling, v^2 drops from (p_R/ρ_B) to (p_B/ρ_B). Since $p_R \propto n_R T$ while $p_B \propto n_B T$ with $(n_R/n_B) \simeq 10^8 \gg 1$, this is a large drop in v^2 and consequently in λ_J. More precisely, $v^2 = (\partial p/\partial \rho)_{\rm adia}$ drops from the value $(1/3)(\rho_\gamma/\rho_B) = (1/3)(\Omega_\gamma/\Omega_B)(1 + z_{\rm dec})$ to the value $(5/3)(T_{\rm dec}/m_B) = (5/3)(T_0/m_B)(1 + z_{\rm dec})$, which is a reduction by a factor $F_1 = 6.63 \times 10^{-8}(\Omega_B h^2)$. This changes the Jeans mass by $F_1^{3/2} \cong 1.7 \times 10^{-11}(\Omega_B h^2)^{3/2}$.

If $\Omega_B h^2 < 0.026$, which is a more likely situation, $z_{B\gamma} < z_{\rm dec}$; in that case, the intermediate situation does not arise and v^2 drops from $(1/3)$ to $(5/3)(T_0/m_B)(1 + z_{\rm dec})$ by a factor $F_2 = 1.9 \times 10^{-9}$, changing the Jeans mass by $F_2^{3/2} \cong 8.3 \times 10^{-14}$.

The rest of the analysis is similar to that of dark matter and the behaviour of λ_J and M_J are shown in figure 4.2.

The scalings for λ_J and $M_{\rm JB}$ are:

$$\lambda_J \propto \begin{cases} a^2 & (a < a_{\rm eq}) \\ a^{3/2} & (a_{\rm eq} < a < a_{B\gamma}) \\ a & (a_{B\gamma} < a < a_{\rm dec}) \\ a^{1/2} & (a_{\rm dec} < a) \end{cases} \tag{4.20}$$

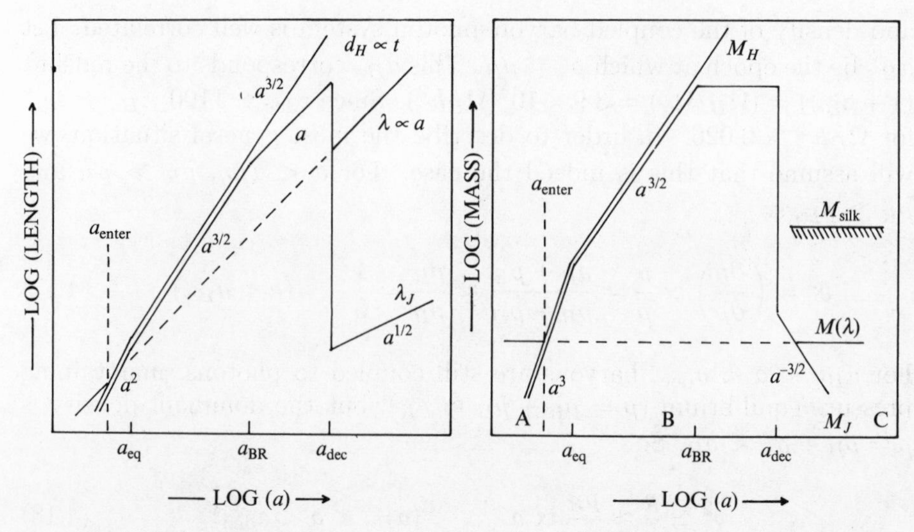

Fig. 4.2. The Jeans length for the baryons is plotted against the expansion factor in a logarithmic scale on the left hand side. The most general case, in which $a_{B\gamma} < a_{\rm dec}$, is shown. If $a_{B\gamma} > a_{\rm dec}$ a part of the curve will be absent. The corresponding Jeans mass for the baryons is plotted against the expansion factor in the right hand side. Also shown are the mass inside the Hubble radius and the mass contained in the perturbation. The three stages A, B and C discussed in the text are marked. The Silk-damping scale for the baryons is shown on the Jeans mass diagram.

and

$$M_{\rm JB} \propto \begin{cases} a^3 & (a < a_{\rm eq}) \\ a^{3/2} & (a_{\rm eq} < a < a_{B\gamma}) \\ \text{constant} & a_{B\gamma} < a < a_{\rm dec}) \\ a^{-3/2} & (a_{\rm dec} < a) \end{cases} \tag{4.21}$$

and the actual values are

$$\begin{aligned} M_{\rm JB} &= 3.2 \times 10^{14}\,{\rm M}_\odot \left(\frac{\Omega_B}{\Omega}\right)(\Omega h^2)^{-2}\left(\frac{a}{a_{\rm eq}}\right)^3 (a < a_{\rm eq}) \\ &= 3.2 \times 10^{14}\,{\rm M}_\odot \left(\frac{\Omega_B}{\Omega}\right)(\Omega h^2)^{-2}\left(\frac{a}{a_{\rm eq}}\right)^{3/2} \quad (a_{\rm eq} < a < a_{B\gamma}) \\ &= 3.2 \times 10^{14}\,{\rm M}_\odot \left(\frac{\Omega_B}{\Omega}\right)(\Omega h^2)^{-2}\left(\frac{a_{B\gamma}}{a_{\rm eq}}\right)^{3/2} \quad (a_{B\gamma} < a < a_{\rm dec}) \\ &\cong 2.5 \times 10^3\,{\rm M}_\odot(\Omega_B/\Omega)(\Omega h^2)^{-1/2}\left(\frac{a}{a_{\rm dec}}\right)^{-3/2} \quad (a_{\rm dec} < a) \end{aligned} \tag{4.22}$$

where we have used the fact $(a_{B\gamma}/a_{\rm eq}) \cong 0.595(\Omega/\Omega_B)$. The Jeans mass just before decoupling, in the interval $a_{B\gamma} < a < a_{\rm dec}$ is

$$M_{\rm JB}(t \lesssim t_{\rm dec}) = 1.47 \times 10^{14}\,{\rm M}_\odot \left(\frac{\Omega}{\Omega_B}\right)^{1/2} (\Omega h^2)^{-2} \tag{4.23}$$

while just after decoupling it is

$$M_{\rm JB}(t \gtrsim t_{\rm dec}) = 2.5 \times 10^3\,{\rm M}_\odot(\Omega_B/\Omega)(\Omega h^2)^{-1/2}. \tag{4.24}$$

This drop is quite sudden and significant. If there is no intermediate phase (i.e., if $\Omega_B h^2 \leq 0.026$), then the $M_J(t < t_{\rm dec})$ is higher:

$$\begin{aligned} M_J(t < t_{\rm dec}) &\cong 3.2 \times 10^{14}\,{\rm M}_\odot \left(\frac{\Omega_B}{\Omega}\right)(\Omega h^2)^{-2}\left(\frac{a_{\rm dec}}{a_{\rm eq}}\right)^{3/2} \\ &= 3.1 \times 10^{16}\,{\rm M}_\odot \left(\frac{\Omega_B}{\Omega}\right)(\Omega h^2)^{-1/2} \end{aligned} \tag{4.25}$$

where we have used the fact that $(a_{\rm dec}/a_{eq}) = (T_{\rm eq}/T_{\rm dec}) \cong [5.5(\Omega h^2)\,{\rm eV}/0.26\,{\rm eV}] \simeq 21\Omega h^2$. In this case, the reduction is also by a larger factor of $F_2^{3/2} \cong 8.3 \times 10^{-4}$ leading to the same M_J after decoupling. (Of course, $M_J(t > t_{\rm dec})$ cannot depend on the past history.)

The time evolution of perturbations during the various phases can be determined just as in the case of dark matter. We find:

$$\delta_B \equiv \left(\frac{\delta\rho}{\rho}\right)_B \propto \begin{cases} a^2 & (a < a_{\rm enter}) \\ \text{constant} & (a_{\rm enter} < a < a_{\rm dec}) \\ a & (a_{\rm dec} < a) \end{cases} . \tag{4.26}$$

The last part deserves special attention. Notice that perturbations in dark matter can grow from $a_{\rm eq}$ onwards, while perturbations in baryons grow only from $a_{\rm dec}$. During the time from $a_{\rm eq}$ to $a_{\rm dec}$, the perturbations in dark matter would have grown by a factor $(a_{\rm dec}/a_{\rm eq}) = (T_{\rm eq}/T_{\rm dec}) \cong 21\Omega h^2$. When the baryons decouple, their perturbation will feel the *perturbed* gravitational potential of dark matter and will be driven by it. (We may say that the baryons 'fall into' the potential wells created by the dark matter). This implies that δ_B will grow *rapidly* for a short time after $a_{\rm dec}$ and will equalize with the value of $\delta_{\rm DM}$; after that, both δ_B and $\delta_{\rm DM}$ will grow as a.

Throughout the above discussion, we have assumed that the background universe is a high density universe with Ω close to unity. If, instead, Ω is small (say $\Omega = 0.1$) then there could be another additional complication viz. that the universe may become 'curvature dominated' for some $z < z_c$. (Such a transition to curvature dominated universe could have occurred in the past ($z_c > 0$) only if $\Omega < 0.5$; for example, if $\Omega = 0.2$, $z_c \simeq 3$; see chapter 2.) If this transition occurs, then the growth of perturbations stops around z_c. This suppression occurs for the same reason as the suppression in the radiation dominated phase: the expansion, dominated by a *smooth* a^{-2} term, is too rapid for the growth of perturbations: $t_{\rm exp} < t_{\rm collapse}$. Thus, in a low density ($\Omega < 0.5$) universe there is no growth of perturbations after $z \approx z_c$.

4.3 The linear perturbation theory

To establish the results of the last section rigorously, we have to derive the differential equations governing the growth of inhomogeneities. We shall now develop the mathematical background for studying the evolution of perturbations[3].

In the linear regime each mode evolves independently. At any given time t, the modes can be separated into two sets, those for which the proper wavelength $\lambda(t)$ is much smaller than the Hubble radius, $d_H(t)$, and those for which $\lambda(t) \gtrsim d_H(t)$. The former set of modes can be analyzed in Newtonian approximation while the study of the latter set requires general relativity. We will derive both the Newtonian perturbation theory and the general relativistic version, in this section. The properties of solutions to these equations will be studied in the next section.

To avoid any misunderstanding, we emphasize the following fact: It is not possible to study cosmology using Newtonian gravity without invoking procedures which are mathematically dubious. However, if we are only interested in regions much smaller than the characteristic length scale set by the curvature of space-time, then one can introduce a valid Newtonian *approximation* to general relativity. It is possible to study growth of perturbations using this approximation. Even in this context, we can analyze density perturbations only in the non-relativistic component. Perturbations in the relativistic component, at all scales, can be correctly handled only by the full theory.

There is a systematic procedure (involving expansion in powers of c^{-1}) which will allow one to determine the Newtonian limit of a given metric. For our purpose, this can be most easily done by transforming the $k = 0$ Friedmann metric

$$ds^2 = c^2\,dt^2 - a^2(t)\left[dR^2 + R^2(d\theta^2 + \sin^2\theta\,d\phi^2)\right] \tag{4.27}$$

to a coordinate system which is locally inertial at the origin. It can be easily shown that (see exercise 2.14), the metric in such a coordinate system is given by

$$ds^2 \cong c^2\left(1 + \frac{2\phi_b(t,\mathbf{x})}{c^2}\right)dt^2 - (dx^2 + dy^2 + dz^2) \tag{4.28}$$

where

$$\phi_b(t,\mathbf{x}) = -\frac{1}{2}\left(\frac{\ddot{a}}{a}\right)|\mathbf{x}|^2; \quad |\mathbf{x}| = a(t)|\mathbf{R}|. \tag{4.29}$$

To the lowest non-trivial order, for $|\mathbf{x}| \ll d_H$, we may treat ϕ_b as an equivalent Newtonian potential due to the uniform, homogeneous background.

Consider now the Newtonian limit of the equations of motion for matter ($T^{\alpha}_{\beta;\alpha} = 0$). For the ideal fluid, in the non-relativistic approximation, they reduce to:

$$\dot{\rho} \equiv \frac{d\rho}{dt} \equiv \frac{\partial \rho}{\partial t} + (v^i \partial_i)\rho = -\rho(\nabla \cdot \mathbf{v}) \tag{4.30}$$

$$\dot{v}^i = -\partial^i \phi - \rho^{-1} \partial^i p. \tag{4.31}$$

(These are the equations of mass conservation and pressure support.) They can be satisfied, for the potential in (4.28), by the following ansatz: $p_b = 0$, $\rho_b(\mathbf{x}, t) = \rho_b(t)$, $\mathbf{v}_b(t, \mathbf{x}) = f(t)\mathbf{x}$. We then get:

$$\frac{\partial \rho_b}{\partial t} + 3\rho_b f(t) = 0 \tag{4.32}$$

and

$$\dot{f} + f^2(t) = \left(\frac{\ddot{a}}{a}\right). \tag{4.33}$$

The last equation integrates to give $f(t) = (\dot{a}/a) = H_b(t)$; substituting this result in the first equation we discover that $\rho_b \propto a^{-3}$. This set determines the Newtonian limit of the Friedmann universe.

We now perturb this solution. As long as the perturbations are linear and have a scale length much smaller than d_H (so that the entire perturbed region can be covered by the region in which (4.28) is valid), we can simply add the perturbed potential $\delta\phi$ (due to the perturbed density $\delta\rho$) to the background potential $\phi_b(t, \mathbf{x})$. Let $\phi, \mathbf{v}, p$ and ρ denote variables containing perturbed parts as well. By writing down the linearized versions of (4.30) and (4.31) one can easily obtain the perturbation equations for variables like $\delta\rho$. We will, however, proceed in a (more complicated) manner which has the advantage that it can be easily adapted to a fully relativistic situation.

We begin by decomposing the gradient of the velocity field $(\partial_i v_j)$ into an antisymmetric part, symmetric traceless part and the trace by writing

$$\partial_j v_i = \omega_{ij} + \sigma_{ij} + H\delta_{ij} \tag{4.34}$$

where $2\omega_{ij} = (\partial_j v_i - \partial_i v_j)$, $2\sigma_{ij} = \partial_j v_i + \partial_i v_j - 2H\delta_{ij}$ and $H(t, \mathbf{x}) = [\partial_i v^i(\mathbf{x}, t)/3]$ is the trace. (In the absence of perturbations, $H(t, \mathbf{x})$ reduces to $H_b(t)$, the Hubble constant of the background universe; in this sense we may consider $\nabla \cdot \mathbf{v}$ to be proportional to the 'perturbed Hubble constant' $H(t, \mathbf{x}) = H_b(t) + \delta H(t, \mathbf{x})$.) It is easy to verify that

$$(\partial_j v_i)(\partial^i v^j) = 3H^2 + 2(\sigma^2 - \omega^2) \tag{4.35}$$

where $2\sigma^2 = \sigma_{ij}\sigma^{ij}$ and $2\omega^2 = \omega_{ij}\omega^{ij}$. Taking the divergence of the Euler equation

$$\frac{\partial v^i}{\partial t} + (v^j \partial_j) v^i + \partial^i \phi = -\rho^{-1} \partial^i p \tag{4.36}$$

and using $3H(t, \mathbf{x}) = \partial_i v^i$ and (4.35) we get

$$\frac{\partial}{\partial t}(\nabla \cdot \mathbf{v}) + (v^j \partial_j)(\nabla \cdot \mathbf{v}) + (\partial_i v^j)(\partial_j v^i) + \nabla^2 \phi = -\partial_i(\rho^{-1} \partial^i p) \tag{4.37}$$

or

$$3\dot{H} + 3H^2 + 2(\sigma^2 - \omega^2) + \nabla^2 \phi = -\partial_i \left(\frac{\partial^i p}{\rho} \right). \tag{4.38}$$

This equation is exact in the Newtonian limit; we now linearize it retaining only first order corrections to the background variables. We can ignore σ^2 and ω^2 since they are of second order (note that $\sigma^2 = \omega^2 = 0$ for the unperturbed case); we can also replace ρ by ρ_b in the right hand side because p is actually δp (since $p_b = 0$). We thus get

$$\delta\dot{H} = -2H_b \delta H - \frac{1}{3} \nabla^2 \delta\phi - \frac{1}{3} \frac{\nabla^2 \delta p}{\rho_b} = -2H_b \delta H - \frac{4\pi G}{3} \delta\rho - \frac{1}{3} v^2 \frac{\nabla^2 \delta\rho}{\rho_b} \tag{4.39}$$

where we have set $\nabla^2 \delta\phi = 4\pi G \delta\rho$ and $\delta p = v^2 \delta\rho$. The continuity equation

$$\dot{\rho} = -3H(t, \mathbf{x})\rho \tag{4.40}$$

can similarly be linearized to give

$$\delta\dot{\rho} = -3H_b \delta\rho - 3\rho_b \delta H. \tag{4.41}$$

The density contrast δ is defined by the relation $\delta \equiv (\delta\rho/\rho_b)$. Using this definition we find that:

$$\delta H = -H_b \left(\frac{\delta\rho}{\rho_b} \right) - \frac{1}{3} \frac{\delta\dot{\rho}}{\rho_b} = -H_b \delta - \frac{1}{3} \left(\frac{\dot{\rho}_b}{\rho_b} \delta + \dot{\delta} \right) = -\frac{1}{3} \dot{\delta}. \tag{4.42}$$

Substituting this into (4.39), we obtain

$$\ddot{\delta} + 2H_b \dot{\delta} - 4\pi G \rho_b \delta - v^2 \nabla^2 \delta = 0. \tag{4.43}$$

Notice that the 'overdots' in this equation stand for the operation $[(\partial/\partial t) + v^i \partial_i]$ and not just $(\partial/\partial t)$. However, when operating on δ, we only need to retain the zeroth order part of v^i which is just $Hx^i = (\dot{a}/a)x^i$. That is

$$\dot{\delta} \equiv \left(\frac{\partial \delta}{\partial t} \right)_x + v^i \partial_i \delta \simeq \left(\frac{\partial \delta}{\partial t} \right)_x + \left(\frac{\dot{a}}{a} \right) x^i \partial_i \delta. \tag{4.44}$$

We will now reintroduce the Friedmann coordinates (X, Y, Z) related to (x, y, z) by $x = a(t)X$ etc. Clearly, for any function $f(t, x)$

$$\begin{aligned} df &= \left(\frac{\partial f}{\partial t}\right)_x dt + \left(\frac{\partial f}{\partial x}\right)_t dx = \left(\frac{\partial f}{\partial t}\right)_x dt + \left(\frac{\partial f}{\partial x}\right)_t [\dot{a}X\, dt + a\, dX] \\ &= \left[\left(\frac{\partial f}{\partial t}\right)_x + \left(\frac{\partial f}{\partial x}\right)_t Hx\right] dt + \left(\frac{\partial f}{\partial x}\right)_t a\, dX \end{aligned} \tag{4.45}$$

showing

$$\left(\frac{\partial}{\partial t}\right)_x + Hx^i \left(\frac{\partial}{\partial x^i}\right)_t = \left(\frac{\partial}{\partial t}\right)_X. \tag{4.46}$$

Thus, in the (t, X) coordinate system, the overdot merely means partial derivative with respect to t; however, $\nabla_x^2 = a^{-2}\nabla_X^2$. Thus we get the final equation for the perturbed density contrast:

$$\ddot{\delta} + 2H_b\dot{\delta} - v^2 a^{-2}\nabla^2\delta = 4\pi G\rho_b\delta. \tag{4.47}$$

This equation, expressed in the original Friedmann coordinates, describes the growth of perturbations in the Newtonian limit.

The discussion can be easily generalized to the situation with several components; let the density contrasts be $\delta_A = (\delta\rho_A/\rho_A)$ with $A = 1, 2, 3 \cdots$. In this case, we get

$$\ddot{\delta}_A + 2H\dot{\delta}_A - v_A^2 a^{-2}\nabla^2\delta_A = 4\pi G\sum_K \rho_K\delta_K \tag{4.48}$$

where the sum on the right hand side is over all components (including, of course, the component A). Notice that any smoothly distributed component (with $\delta = 0$) will not make any contribution to the right hand side; however, the smooth energy density of this component does contribute to background expansion and affects $H(t)$ in the $2H\dot{\delta}$ term. Since the curvature term ka^{-2} in the Friedmann equation can also be thought of as a smoothly distributed source, with $\rho \propto a^{-2}$, it follows that our perturbation equation is valid for $k \neq 0$ as well. The solutions to this equation will be discussed in the next section.

The behaviour of *velocity* perturbations in a matter dominated universe with $p = 0$ is of considerable significance. Writing $v^i = Hx^i + u^i$ where u^i is the (linear) perturbation, and expanding the equation (4.36), the linear term can be written as

$$\frac{\partial u^i}{\partial t} + (H_b x^i \partial_j)u^i + (u^j\partial_j)(H_b x^i) + \partial^i\delta\phi = 0 \tag{4.49}$$

Changing to Friedmann coordinates, $X^i = a^{-1}x^i$, this becomes

$$\left(\frac{\partial \mathbf{u}}{\partial t}\right)_X + \left(\frac{\dot{a}}{a}\right)\mathbf{u} = -\frac{1}{a}\nabla_X\delta\phi \equiv \mathbf{g}. \tag{4.50}$$

Similarly, linearizing the equation $3H = \partial_i v^i$ and using the result $3\delta H = -\dot{\delta}$, we find

$$a^{-1}\nabla_X \cdot \mathbf{u} + \dot{\delta} = 0. \tag{4.51}$$

This equation can be solved in the following manner: Since $\nabla_x^2 \delta\phi = 4\pi G\rho_b\delta$, it follows that $\nabla_x \cdot \mathbf{g} = -4\pi G\rho_b\delta$ or $\nabla_X \cdot \mathbf{g} = -4\pi G\rho_b a\delta$. So (4.51) can be written as

$$\frac{1}{a}\nabla \cdot \mathbf{u} = \frac{\partial}{\partial t}\left(\frac{\nabla \cdot \mathbf{g}}{4\pi G\rho_b a}\right). \tag{4.52}$$

Or, equivalently,

$$\mathbf{u} = a\frac{\partial}{\partial t}\left(\frac{\mathbf{g}}{4\pi G\rho_b a}\right) \tag{4.53}$$

where we have ignored any solution of the homogeneous equation (see exercise 4.4). The equation (4.50) now becomes

$$\frac{\partial}{\partial t}\left[a^2\frac{\partial}{\partial t}\left(\frac{\mathbf{g}}{4\pi G\rho_b a}\right)\right] = \mathbf{g}a. \tag{4.54}$$

In Fourier space, with $\mathbf{g}_k = -a^{-1}\,(i\mathbf{k})\delta\phi_k = -i\mathbf{k}\pi G\rho_b a\delta_{\mathbf{k}}$, this equation reads as

$$\frac{1}{a^2}\frac{d}{dt}\left[a^2\frac{d\delta_k}{dt}\right] = 4\pi G\rho_b\delta_k \tag{4.55}$$

which is identical to the perturbation equation. Therefore, the correct solution to our equations is

$$\mathbf{u} = a\frac{\partial}{\partial t}\left(\frac{\mathbf{g}}{4\pi G\rho_b a}\right); \quad \nabla \cdot \mathbf{g} = -4\pi G\rho_b a\delta. \tag{4.56}$$

This can be rewritten in a more useful form. Since $\mathbf{g}_k \propto \rho_b a\delta_k$, it follows that

$$\mathbf{u}_k = \frac{a}{4\pi G}\left(\frac{\mathbf{g}_k}{\rho_b a\delta_k}\right)\frac{\partial\delta_k}{\partial t} = \frac{\mathbf{g}_k}{4\pi G\rho_b}\left(\frac{1}{\delta_k}\frac{d\delta_k}{dt}\right). \tag{4.57}$$

In other words, the peculiar velocity is proportional to the peculiar acceleration, mode by mode, with a proportionality constant which depends on time:

$$\mathbf{u}_k = \frac{\mathbf{g}_k}{4\pi G\rho_b}\frac{\dot{a}}{\delta_k}\frac{d\delta_k}{da} = \frac{H_b}{4\pi G\rho_b}\frac{a}{\delta_k}\frac{d\delta_k}{da}\mathbf{g}_k = \frac{2f}{3H_b\Omega}\mathbf{g}_k \tag{4.58}$$

with $f_k(a) = (a/\delta_k)\,(d\delta_k/da)$. For growing modes, we can ignore the pressure corrections; in such a case, $f_k(a)$ will be independent of k and will depend only on t. We can, therefore, switch back to the $\mathbf{x}$ space and write

$$\mathbf{u} = \frac{2f}{3H_b\Omega}\mathbf{g} = -\frac{2f}{3H_b\Omega}\frac{1}{a}\nabla\delta\phi \tag{4.59}$$

where $\delta\phi$ is the 'Newtonian' potential generated by the excess density $\rho_b\delta$.

The value of f at the present epoch depends only on H_0 and Ω. Hence, by measuring $\mathbf{g}$ and $\mathbf{u}$ in the present universe and using this relation, we can obtain information about Ω. This aspect will be discussed in more detail in chapter 7.

We shall now derive the corresponding perturbation equation in general relativity. It turns out that there is a formal similarity between the exact, general relativistic equation (written in a gauge called the 'comoving' gauge) and its Newtonian counterpart. The two basic equations of the perturbation theory in the Newtonian approximation were for the perturbed density and perturbed Hubble constant:

$$\begin{aligned} \delta\dot{\rho} &= -3H_b\delta\rho - 3\rho_b\delta H \\ \delta\dot{H} &= -2H_b\delta H - \frac{4\pi G}{3}\delta\rho - \frac{1}{3}\frac{\nabla^2\delta p}{\rho_b}. \end{aligned} \tag{4.60}$$

The correct, general relativistic, perturbation equations are identical to the above set except for the change of ρ_b to $(\rho_b + p_b)$. (Note that in the Newtonian limit $p_b \ll \rho_b$.) Thus the correct equations are

$$\delta\dot{\rho} = -3H_b\delta\rho - 3(\rho_b + p_b)\delta H \tag{4.61}$$

$$\delta\dot{H} = -2H_b\delta H - \frac{4\pi G}{3}\delta\rho - \frac{1}{3}\frac{\nabla^2\delta p}{(\rho_b + p_b)}. \tag{4.62}$$

For a single component model, we can set $\delta p = v^2\delta\rho$, eliminate δH and obtain a second order equation for $\delta\rho$. It is usual to use a Fourier transform with proper wavelength so that ∇^2 is equivalent to $a^{-2}k^2$.

We will now derive (4.61) and (4.62). The rest of the material in this section uses fairly advanced concepts from general relativity[4] and may be omitted in the first reading.

We will be using these general relativistic equations to study either the modes which are bigger than the Hubble radius or the perturbations in the relativistic components in the radiation dominated phase at $z > 10^4$. The modes which are astrophysically relevant enter the Hubble radius at $z > 10^3$. Thus their wavelengths are bigger than the Hubble radius only at $z \gtrsim 10^3$. We have seen earlier that, for these redshifts, the curvature term in the Friedmann equation can be ignored. Thus, we may take the background universe to be one with $k = 0$. The same conclusion is applicable to the study of relativistic components in the radiation dominated phase as well.

At each event in spacetime we choose an orthonormal basis such that the momentum density vanishes. In this frame, the four-velocity u^α has the components $u^0 = u_0 = 1$ and $u^j = 0$. If D_α is the covariant derivative operator, then the 'overdot' will be used to denote the directional derivative along u^α, i.e. the operator $u^\alpha D_\alpha$. Surfaces orthogonal to the comoving world lines (for which u^α are the tangent vectors) will be called comoving hypersurfaces; the projection tensor onto these surfaces will be $h_{\alpha\beta} = g_{\alpha\beta} - u_\alpha u_\beta$ (in the comoving basis, only nonzero components of this tensor will be $h_{ij} = g_{ij}$). Using D_α and $h_{\alpha\beta}$ we can construct the natural derivative $h^\beta_\alpha D_\beta$ on the comoving surfaces and the Laplacian $\nabla^2 = h^\beta_\alpha D_\beta h^{\alpha\mu} D_\mu$. In the absence of perturbations, $\nabla^2 = a^{-2}\,\partial^i\partial_i$ because the comoving surfaces are flat (at any given t); in the presence of metric fluctuations, these surfaces show deviation from flatness but this deviation will only produce a second order effect when ∇^2 is applied to the perturbed quantity. Thus, as far as perturbation equations are concerned, one can continue to use $\nabla^2 = a^{-2}\partial_i\partial^i$ even when the hypersurfaces are not flat. This procedure defines our gauge.

With these preliminaries, we can start the derivation of the fluctuation equations. The relativistic analogues of the continuity and Euler equations can be written down from $T^\alpha_{\beta;\alpha} = 0$:

$$\dot\rho = -3H(\rho + p), \tag{4.63}$$

$$\dot u_\alpha = -\frac{h^\beta_\alpha D_\beta p}{(\rho + p)} \tag{4.64}$$

where all quantities refer to the background metric. We now proceed exactly in the same manner as before by starting with the relation

$$\begin{aligned} D_\alpha\,\dot u^\alpha &= D_\alpha(u^\beta D_\beta u^\alpha) = (D_\alpha u^\beta)(D_\beta u^\alpha) + u^\beta D_\alpha D_\beta u^\alpha \\ &= (D_\alpha u^\beta)(D_\beta u^\alpha) + u^\beta D_\beta D_\alpha u^\alpha + u^\beta [D_\alpha, D_\beta] u^\alpha \\ &= (D_\alpha u^\beta)(D_\beta u^\alpha) + u^\beta D_\beta (D_\alpha u^\alpha) + u^\beta R_{\alpha\beta} u^\alpha . \end{aligned} \tag{4.65}$$

Since u^α has constant norm, $u^\alpha D_\beta u_\alpha = 0$ and hence $D_\beta u^0 = 0$ in the comoving basis. Therefore $(D_\alpha u^\beta)(D_\beta u^\alpha)$ and $D_\alpha u^\alpha$ reduce to purely spatial terms $(D_i u^j)\,(D_j u^i)$ and $(D_i u^i)$. We can now separate this tensor into σ_{ij}, ω_{ij} and H exactly as before (with ∂_i replaced by D_i). The last term in (4.65) is

$$u^\alpha R_{\alpha\beta} u^\beta = R_{00} = 8\pi G\left(T_{00} - \frac{1}{2}T_{\alpha\beta}g^{\alpha\beta}\right) = 4\pi G(\rho + 3p). \tag{4.66}$$

Using this result and the identities

$$D_i u^i = 3H; \quad (D_i u^j)(D_j u^i) = 3H^2 + 2(\sigma^2 - \omega^2), \tag{4.67}$$

we get

$$D_\alpha \dot u^\alpha = 3\dot H + 3H^2 + 2(\sigma^2 - \omega^2) + 4\pi G(\rho + 3p). \tag{4.68}$$

The left hand side can be related to $\nabla^2 p$ by using the Euler equation. We have

$$\begin{aligned} D_\alpha \dot{u}^\alpha &= h^\beta_\alpha D_\beta \dot{u}^\alpha + u_\alpha u^\beta D_\beta \dot{u}^\alpha \\ &= h^\beta_\alpha D_\beta \dot{u}^\alpha + u^\beta D_\beta (u_\alpha \dot{u}^\alpha) - u^\beta \dot{u}^\alpha D_\beta u_\alpha \end{aligned} \tag{4.69}$$

in which the middle term vanishes ($u_\alpha \dot{u}^\alpha = 0$) and the last term is of second order ($\dot{u}^\alpha (u^\beta D_\beta u_\alpha) = \dot{u}^\alpha \dot{u}_\alpha$). Using (4.64), the first term can be written as

$$h^\beta_\alpha D_\beta \dot{u}^\alpha = -h^\beta_\alpha D_\beta \left[\frac{h^{\alpha\beta} D_\beta p}{(\rho + p)} \right] \approx -\frac{\nabla^2 p}{(\rho + p)} \tag{4.70}$$

which is correct to linear order. Therefore

$$D_\alpha \dot{u}^\alpha = -\frac{\nabla^2 p}{(\rho + p)}. \tag{4.71}$$

Substituting back into (4.68) and noticing that σ^2 and ω^2 are also of second order we get

$$\dot{H} = -H^2 - \frac{4\pi G}{3}(\rho + 3p) - \frac{1}{3}\frac{\nabla^2 p}{(\rho + p)}. \tag{4.72}$$

This equation, along with equation (4.63), constitute the basic set. Before we separate out the quantities as $H = H_b + \delta H$ etc., we have to take note of one additional complication. The interval $d\tau$ along a comoving world line between two adjacent comoving surfaces is position dependent on the surface and hence cannot be taken to be a valid time label for the hypersurfaces. Suppose t stands for valid ordering label for the hypersurfaces; then we can prove (we will do so later) that

$$\frac{d\tau}{dt} = 1 - \frac{\delta p}{\rho + p}. \tag{4.73}$$

Given this relation, our equations can be recast in terms of the derivatives with respect to t. The continuity equation becomes

$$\begin{aligned} \frac{d\rho}{d\tau} &= \frac{d(\rho_b + \delta\rho)}{dt\,(1 - \delta p(\rho_b + p_b)^{-1})} \cong (\dot{\rho}_b + \delta\dot{\rho}) \left[1 + \frac{\delta p}{\rho_b + p_b} \right] \\ &\simeq \dot{\rho}_b + \delta\dot{\rho} + \frac{\dot{\rho}_b}{\rho_b + p_b}\delta p \\ &= -3H_b(\rho_b + p_b) - 3\delta H(\rho_b + p_b) - 3H_b(\delta\rho + \delta p) \end{aligned} \tag{4.74}$$

where we have kept only the linear terms and used the overdot symbol to denote derivatives with respect to t. Equating the zeroth order terms we get

$$\dot{\rho}_b = -3H_b(\rho_b + p_b). \tag{4.75}$$

Using this in the first order equations, we get

$$\delta\dot{\rho}+\frac{\dot{\rho}_b}{(\rho_b+p_b)}\delta p=\delta\dot{\rho}-3H_b\delta p=-3\delta H(\rho_b+p_b)-3H_b\delta\rho-3H_b\delta p \quad (4.76)$$

or

$$\delta\dot{\rho}=-3(\rho_b+p_b)\delta H-3H_b\delta\rho \quad (4.77)$$

which is one of the perturbation equations (4.61). The $\dot{H}$ equation can be obtained along the same lines: We begin with

$$\begin{aligned}\frac{dH}{d\tau}&=(\dot{H}_b+\delta\dot{H})\left(1+\frac{\delta p}{\rho_b+p_b}\right)\\&=-H_b^2-2H_b\delta H-\frac{4\pi G}{3}(\rho_b+3p_b)-\frac{4\pi G}{3}(\delta\rho+3\delta p)-\frac{1}{3}\frac{\nabla^2\delta p}{(\rho_b+p_b)}\end{aligned} \quad (4.78)$$

where we have used the result $\nabla^2 p_b=0$. The zeroth order term is

$$\dot{H}_b=-H_b^2-\frac{4\pi G}{3}(\rho_b+3p_b) \quad (4.79)$$

while the first order term is

$$\frac{\dot{H}_b}{(\rho_b+p_b)}\delta p+\delta\dot{H}=-2H_b\delta H-\frac{4\pi G}{3}\delta\rho-4\pi G\delta p-\frac{1}{3}\frac{\nabla^2\delta p}{(\rho_b+p_b)}. \quad (4.80)$$

Substituting the value of $\dot{H}_b$ in this, we get

$$\begin{aligned}\delta\dot{H}+2H_b\delta H+\frac{4\pi G}{3}\delta\rho+\frac{1}{3}\frac{\nabla^2\delta p}{(\rho_b+p_b)}&=-\frac{\delta p}{(\rho_b+p_b)}\left[\dot{H}_b+4\pi G(\rho_b+p_b)\right]\\&=-\frac{\delta p}{(\rho_b+p_b)}\left[-H_b^2-\frac{4\pi G}{3}(\rho_b+3p_b)+4\pi G(\rho_b+p_b)\right]\\&=-\frac{\delta p}{(\rho_b+p_b)}\left[-H_b^2+\frac{8\pi G}{3}\rho_b\right]=0.\end{aligned} \quad (4.81)$$

So we get the second of the perturbation equations (4.62) :

$$\begin{aligned}\delta\dot{H}&=-2H_b\delta H-\frac{4\pi G}{3}\delta\rho-\frac{1}{3}\frac{\nabla^2\delta p}{(\rho_b+p_b)}\\&=-2H_b\delta H-\frac{4\pi G}{3}\delta\rho-\frac{v^2}{3}\frac{\nabla^2\delta\rho}{(\rho_b+p_b)}.\end{aligned} \quad (4.82)$$

In arriving at the last equation we have used the definition $\delta p=v^2\delta\rho$. The rest of the analysis proceeds as in the non-relativistic case; rewriting the continuity equation as

$$\delta H=-\frac{1}{3(\rho_b+p_b)}\left[\delta\dot{\rho}+3H_b\delta\rho\right]=-\frac{1}{3(1+w)}\left[\dot{\delta}-3H_bw\delta\right] \quad (4.83)$$

where $\delta = (\delta\rho/\rho)$ and $w = (p/\rho)$ and substituting into the $\delta\dot{H}$ equation, we get

$$\begin{aligned}
\delta\dot{H} &= \frac{\dot{w}}{3(1+w)^2}\left[\dot{\delta} - 3H_b w\delta\right] \\
&\quad - \frac{1}{3(1+w)}\left[\ddot{\delta} - 3H_b w\dot{\delta} - 3H_b\dot{w}\delta - 3\dot{H}_b w\delta\right] \\
&= -\frac{1}{3(1+w)}\ddot{\delta} + \dot{\delta}\left[\frac{\dot{w}}{3(1+w)^2} + \frac{H_b w}{(1+w)}\right] \\
&\quad - \delta\left[\frac{H_b w\dot{w}}{(1+w)^2} - \frac{H_b\dot{w}}{(1+w)} - \frac{\dot{H}_b w}{(1+w)}\right].
\end{aligned} \tag{4.84}$$

It is easy to verify that

$$\begin{aligned}
\dot{w} &= 3H_b(w - v^2)(1+w), \\
\dot{H}_b &= -H_b^2 - \frac{1}{2}H_b^2(1+3w) = -\frac{3}{2}H_b^2 - \frac{3}{2}wH_b^2 = -\frac{3}{2}H_b^2(1+w).
\end{aligned} \tag{4.85}$$

Using these we can write

$$\delta\dot{H} = -\frac{1}{3(1+w)}\left\{\ddot{\delta} - 3H_b(2w - v^2)\dot{\delta} + \frac{9}{2}H_b^2\delta(2v^2 + w^2 - w)\right\}. \tag{4.86}$$

Therefore the equation (4.82) becomes

$$\begin{aligned}
&\frac{1}{3(1+w)}\left\{\ddot{\delta} - 3H_b\dot{\delta}(2w - v^2) + \frac{9}{2}H_b^2\delta(2v^2 + w^2 - w)\right\} \\
&\qquad = -2H_b\frac{1}{3(1+w)}\left[\dot{\delta} - 3H_b w\delta\right] + \frac{1}{2}H_b^2\delta + \frac{v^2}{3}\frac{1}{(1+w)}\nabla^2\delta.
\end{aligned} \tag{4.87}$$

Equivalently,

$$\ddot{\delta} + H_b\dot{\delta}[2 - 3(2w - v^2)] - \frac{3}{2}H_b^2\delta\left[1 - 6v^2 - 3w^2 + 8w\right] = -\left(\frac{kv}{a}\right)^2\delta \tag{4.88}$$

in which we have introduced the Fourier transform such that $\nabla^2\delta = -(k/a)^2\delta$. This is the relativistic analogue of the Newtonian perturbation equation. The Newtonian limit is obtained by setting $w \approx 0$ and $v \approx 0$ in the left hand side and using $H_b^2 = (8\pi G\rho/3)$.

To complete the derivation we have to obtain equation (4.73). There are several ways of doing this, the easiest being the following:

Let (t, x_i) be a set of coordinates with t labelling the comoving hypersurfaces and let $(\lambda\mathbf{u}, \mathbf{e}_i)$ be a coordinate basis associated with this set, with $\lambda = (d\tau/dt)$. Obviously, $\mathbf{u}\cdot\mathbf{e}_i = 0$ which implies that

$$\dot{\mathbf{u}}\cdot\mathbf{e}_i = -\dot{\mathbf{e}}_i\cdot\mathbf{u} = \dot{u}_i. \tag{4.89}$$

On the other hand, since the basis vectors are coordinate induced, all Lie brackets vanish: $[\lambda \mathbf{u}, \mathbf{e}_i] = 0$. On expanding this, we get

$$(\lambda \dot{\mathbf{e}}_i)^\alpha - D_i(\lambda \mathbf{u})^\alpha = 0 \tag{4.90}$$

which is equivalent to

$$(\dot{\mathbf{e}}_i)^\alpha - D_i u^\alpha = \lambda^{-1}(\partial_i \lambda) u^\alpha. \tag{4.91}$$

Taking the dot product with u_α, the second term vanishes; the first term gives $\dot{u}_i$ because of (4.89). Thus we find

$$\dot{u}_i = \lambda^{-1}(\partial_i \lambda). \tag{4.92}$$

Combining this with (4.64) and working up to first order, we find

$$\lambda^{-1}\partial_i \lambda = -\frac{\partial_i \delta p}{\rho_b + p_b} \tag{4.93}$$

giving

$$\lambda = \exp\left[-\frac{\delta p}{(\rho_b + p_b)}\right] \simeq 1 - \frac{\delta p}{\rho_b + p_b} = \frac{d\tau}{dt}. \tag{4.94}$$

This is the result to be proved.

Lastly, let us consider the case of several uncoupled fluids. In the Newtonian limit, this poses no special problems; we only have to use the H contributed by all matter and replace the driving term $4\pi G \delta\rho$ by $4\pi G \sum \delta\rho_i$. But the situation is more complicated in the relativistic case.

In a manner very similar to that for a single component fluid, one can derive the following equations:

$$\begin{aligned} \dot{\rho}_N &= -3H_N(\rho_N + p_N), \\ \frac{1}{3} D_\alpha \dot{u}_N^\alpha &= \dot{H}_N + H_N^2 + \frac{4\pi G}{3}(\rho + 3p), \\ D_\alpha \dot{u}_N^\alpha &= -\frac{\nabla^2 p_N}{\rho_N + p_N} + \frac{3(H - H_N)\dot{p}_N}{\rho_N + p_N} \end{aligned} \tag{4.95}$$

where $N = 1, 2, 3 \cdots$ denotes the various fluids (radiation, matter etc.) and

$$\rho = \sum_N \rho_N, \qquad p = \sum_N p_N, \qquad H = \sum_N H_N \tag{4.96}$$

etc. The last term in the third equation does not occur in the case of a single component. The mean values of all the variables like H_N etc. are defined by averaging over each spacelike hypersurface. It, therefore, follows that

$$< H_N >= H \tag{4.97}$$

giving

$$H - H_N = \delta H - \delta H_N. \tag{4.98}$$

It is also easy to show that

$$(\rho + p)H = \sum (\rho_N + p_N)H_N. \tag{4.99}$$

Combining all these equations, we can derive the perturbation equations:

$$\begin{aligned}
\delta\dot{\rho}_N &= -3(\rho_N + p_N)\delta H_N - 3H\delta\rho_N - 3H(\delta p_N - \theta_N \delta p) \\
\delta\dot{H}_N &= -2H\delta H_N - \frac{4\pi G}{3}\delta\rho - \frac{1}{3}\frac{\nabla^2 \delta p_N}{\rho_N + p_N} \\
&\quad + \frac{\dot{p}_N}{\rho_N + p_N}\left[\sum_M (\theta_M \delta H_M) - \delta H_N\right]
\end{aligned} \tag{4.100}$$

where

$$\theta_N = \frac{\rho_N + p_N}{\rho + p}. \tag{4.101}$$

To each component, we can associate a velocity dispersion

$$v_N^2 = \frac{\delta p_N}{\delta \rho_N} = \frac{\dot{p}_N}{\dot{\rho}_N}. \tag{4.102}$$

However, this does not lead to a corresponding relation between the *total* δp and *total* $\delta\rho$. Because of this complication, it is best to use a different set of variables.

This feature can be illustrated clearly in the case of a two-component fluid (which could be, for example, dark matter and radiation). Let the equations of state for the fluids be $p_1 = w_1\rho_1$, $p_2 = w_2\rho_2$, with constant w_1 and w_2. Then, for individual components, $\delta p_1 = w_1\delta\rho_1 = w_1\rho_1\delta_1$, and $\delta p_2 = w_2\delta\rho_2 = w_2\rho_2\delta_2$; the velocity dispersions in these components are $v_1^2 = (\dot{p}_1/\dot{\rho}_1) = w_1$ and $v_2^2 = (\dot{p}_1/\dot{\rho}_2) = w_2$. Consider now the full system with pressure $p = p_1 + p_2$ and density $\rho = \rho_1 + \rho_2$. In analogy with the individual components, we may define

$$w \equiv \frac{p}{\rho} = \frac{p_1 + p_2}{\rho_1 + \rho_2}; \quad v^2 = \frac{\dot{p}}{\dot{\rho}} = \frac{\dot{p}_1 + \dot{p}_2}{\dot{\rho}_1 + \dot{\rho}_2}. \tag{4.103}$$

Notice that w and v^2 are not constants; hence the relation $w = v^2$ does not hold for the full system. To characterize this difference, we will define a variable N by the relation,

$$pN = \delta p - \frac{\dot{p}}{\dot{\rho}}\delta\rho = (v_1^2\rho_1\delta_1 + v_2^2\rho_2\delta_2) - \frac{(v_1^2\dot{\rho}_1 + v_2^2\dot{\rho}_2)}{(\dot{\rho}_1 + \dot{\rho}_2)}(\rho_1\delta_1 + \rho_2\delta_2). \tag{4.104}$$

Clearly, N measures the deviation between $(\delta p/\delta\rho)$ and $(\dot{p}/\dot{\rho})$. Using the fact that

$$\dot{\rho}_1 = -3(1+w_1)\rho_1\left(\frac{\dot{a}}{a}\right), \tag{4.105}$$

this expression can be simplified further. A straight-forward calculation gives

$$pN = \frac{(\rho_1+p_1)(\rho_2+p_2)}{(\rho+p)}(v_1^2 - v_2^2)\left[\frac{\delta_1}{1+w_1} - \frac{\delta_2}{1+w_2}\right]. \tag{4.106}$$

A perturbation for which $N = 0$ is usually called 'adiabatic', because of the following reason: Consider a two fluid model with radiation and dust as the constituents, for which $w_1 = 0$ and $w_2 = (1/3)$. The condition $N = 0$ implies $\delta_1 = \delta_2(3/4)$; i.e.

$$\left(\frac{\delta n}{n}\right)_{\text{matter}} = \frac{3}{4}\left(\frac{\delta\rho_R}{\rho_R}\right). \tag{4.107}$$

But since $\rho_R \propto T^4$ and $s_R \propto T^3$, where s_R is the entropy of the radiation, it follows that $(3/4)\ (\delta\rho_R/\rho_R) = (\delta s_R/s_R)$. Thus, the condition $N = 0$ is equivalent to demanding that $(\delta n/n)_{\text{matter}} = (\delta s_R/s_R)$, or,

$$\delta\left(\frac{s_R}{n}\right) = \frac{\delta s_R}{n} - \frac{s_R\delta n}{n^2} = \frac{\delta s_R}{n} - \frac{s_R}{n}\frac{\delta s_R}{s_R} = 0. \tag{4.108}$$

Thus adiabatic perturbations conserve the value of entropy-per-matter particle.

In contrast, one can also consider perturbations of the radiation–dust system for which the *total* density contrast δ is zero, while N is non-zero. In this case, $(\delta_m\rho_m + \delta_R\rho_R) = 0$; that is

$$\frac{\delta_R}{\delta_m} = -\frac{\rho_m}{\rho_R} \tag{4.109}$$

and

$$S = \frac{\delta_m}{(1+w_m)} - \frac{\delta_R}{(1+w_R)} = \delta_m - \frac{3}{4}\delta_R = \delta_m\left(1 + \frac{3}{4}\frac{\rho_m}{\rho_R}\right). \tag{4.110}$$

In the radiation dominated phase $\rho_m \ll \rho_R$ implying $\delta_R \ll \delta_m$; hence these modes are usually called 'isothermal'. A more accurate term would be 'isocurvature', since these modes – which do not change the energy density in the spacetime – cannot alter the curvature of the background spacetime.

A general perturbation, of course, need not be either isothermal or adiabatic. However, since δ_1 and δ_2 can be expressed in terms of the total density contrast δ and S, we can always consider an arbitrary perturbation to be a linear superposition of adiabatic and isothermal components.

Using the general set of equations for the multicomponent medium discussed above, we can write down the differential equations for δ and S. The procedure is straightforward but tedious. We finally arrive at the equations:

$$\ddot{\delta} + [2 - 3(2w - v^2)]H\dot{\delta} - \frac{3}{2}H^2(1 - 6v^2 + 8w - 3w^2)\delta = -\frac{k^2}{a^2}(v^2\delta + wN) \tag{4.111}$$

and

$$\ddot{S} + (2 - 3u^2)H\dot{S} = \frac{k^2}{a^2}[-u^2 S + (v_2^2 - v_1^2)(1 + w)^{-1}\delta] \tag{4.112}$$

with

$$N = \frac{(\rho_1 + p_1)(\rho_2 + p_2)}{(\rho + p)p}(v_1^2 - v_2^2)S \tag{4.113}$$

and

$$v^2 = \frac{\dot{p}}{\dot{\rho}}; \qquad v_A^2 = \frac{\dot{p}_A}{\dot{\rho}_A}; \tag{4.114}$$

$$u^2 = \frac{(\rho_1 + p_1)v_2^2 + (\rho_2 + p_2)v_1^2}{(\rho + p)}. \tag{4.115}$$

Quite clearly (4.111) reduces to the density perturbation equation derived earlier if $N = 0$ and there is only one component. When there exists more than one component, δ and S are coupled; S generates δ and vice versa. We will consider the nature of the solutions to these equations in the next section.

4.4 Gravitational instability in the relativistic case

We shall now consider the solutions to the perturbation equations in different contexts and justify rigorously the conclusions arrived at in section 4.2.

The separate cases which need to be considered are the following: (1) Growth of perturbations in the *dominant* component when the wavelength of the perturbation is bigger than the Hubble radius: The *dominant* component will be the non-relativistic matter for $t > t_{\rm eq}$, and the relativistic particles for $t < t_{\rm eq}$. This case can be studied by the relativistic perturbation equation for the single component. (2) Perturbations in the relativistic, dominant component for $t < t_{\rm eq}$, for wavelengths *smaller* than the Hubble radius. Since the component is relativistic, this case also needs to be studied using the exact equation. (3) The evolution of adiabatic and isothermal modes in a two-component fluid consisting of radiation and dust. Since the coupling between the two modes is dominantly a relativistic phenomenon, it is best to study this evolution using the relativistic, two-component equations for δ and S. (4) Perturbations in the non-relativistic matter, in the matter dominated phase, for modes with $\lambda \ll d_H$. Since $\lambda \ll d_H$, this can be analyzed in the Newtonian approximation. These modes can be further divided into the range $d_H \gg \lambda \gg \lambda_J$, for which the pressure support is ineffective, and $\lambda_J \gtrsim \lambda$, for which pressure support can prevent the growth of perturbations. (5) Perturbations in the matter component, during the radiation dominated phase, for modes with $\lambda_J \ll \lambda \ll d_H$. (We saw in section 4.2 that these modes – though they are bigger than the Jeans length – do not grow because the universe is expanding too rapidly). This can be studied in the Newtonian approximation, using the equation for the multicomponent fluid. This case is already covered in (3) above, but the Newtonian analysis is more transparent. (6) Perturbations in the baryonic component after the decoupling, when the universe is dominated by dark matter. This can also be studied using the Newtonian approximation to the multicomponent fluid.

It is clear that the first three cases require relativistic theory while the last three can be studied using the Newtonian approximation. We will consider relativistic perturbation theory in this section; the Newtonian approximation will be considered in section 4.5.

Before we take up the specific cases, it is worthwhile to examine the general structure of the perturbation equation. Consider a perturbation, labelled by a wave number k, in the linear approximation. To understand the growth (or suppression) of δ_A it is better to rewrite the perturbation equation in the following way:

$$\frac{d^2\delta_A}{dt^2} = -\alpha_A H \frac{d\delta_A}{dt} + \left(\mu_A H^2 - \frac{k^2 v_A^2}{a^2}\right)\delta_A \tag{4.116}$$

where $H = (\dot{a}/a)$ (we drop the subscript b on H_b hereafter), v_A is the velocity dispersion which decides the relation between perturbed pressure and perturbed density, and the variables μ_A, α_A are quantities which depend on the equation of state of the background matter. The specific form of these functions, of course, is crucial in deciding the behaviour of δ_A as a function of a. Notice that, if (p/ρ) is a constant, then these functions reduce to constants independent of time.

The first term on the right hand side always dampens the growth; this is purely an effect of expansion. The expansion rate $(\dot{a}/a)$, of course, is determined by *all* matter in a multicomponent medium. If the dominant, smooth component is not the species A under consideration, then the damping due to expansion will suppress the growth; this is what happens in radiation dominated or curvature dominated phases.

The second term represents the conflict between pressure support and gravity. This term will act as a 'restoring force' and prevent growth if (ignoring the numerical factor μ_A):

$$a^2H^2 < k^2v_A^2. \tag{4.117}$$

Since $H^2 = (8\pi G\rho_{\text{dom}}/3)$ and $k^2 = (2\pi/\lambda)^2$ we can write this as

$$\frac{(\lambda a)}{v} = \frac{\lambda_{\text{proper}}}{v} < \frac{1}{(G\rho)^{1/2}}; \qquad \lambda < \lambda_J \equiv \frac{v}{(G\rho)^{1/2}} \tag{4.118}$$

which is precisely the condition $t_{\text{pressure}} < t_{\text{grav}}$ discussed before. In a single component medium, we have only two terms on the right hand side. The first term (expansion) always dampens the growth; the second term (pressure-gravity) may assist the growth or suppress it depending on the values of λ and λ_J. These results agree with the conclusions reached in section 4.2.

(1) Perturbations in the dominant component; $\lambda > d_H$

Let us now consider the various special cases mentioned earlier. In the general relativistic perturbation equation

$$\ddot{\delta} + [2 - 3(2w - v^2)]H\dot{\delta} - \frac{3H^2}{2}(1 - 6v^2 + 8w - 3w^2)\delta = -\frac{k^2}{a^2}v^2\delta, \tag{4.119}$$

where $H = (\dot{a}/a)$, $w = (p/\rho)$ and $v^2 = (\dot{p}/\dot{\rho})$, we change the independent variable from t to a. Since

$$\frac{d}{dt} = \dot{a}\frac{d}{da} = H(a)a\frac{d}{da}; \quad \dot{H} = -\frac{3}{2}(1 + w)H^2, \tag{4.120}$$

the perturbation equation becomes:

$$a^2\frac{d^2\delta}{da^2} + Aa\frac{d\delta}{da} + \left(B + \frac{k^2v^2}{H^2a^2}\right)\delta = 0 \tag{4.121}$$

where

$$A = \frac{3}{2}(1 - 5w + 2v^2); \qquad B = -\frac{3}{2}(1 - 6v^2 + 8w - 3w^2). \tag{4.122}$$

In the radiation dominated case $w = v^2 = (1/3)$, giving $A = 0, B = -2$; the equation reduces to

$$a^2\frac{d^2\delta}{da^2} + \left(\frac{k^2}{3H^2a^2} - 2\right)\delta = 0. \tag{4.123}$$

For the matter dominated case we may set $w = v^2 = 0$ in the definition of A and B, obtaining $A = (3/2)$, $B = -(3/2)$; this corresponds to the approximation $(v/c) \ll 1$. However the (v^2k^2/H^2a^2) term should be retained in the equation if we want to discuss the effect of pressure support. Then the equation is

$$a^2\frac{d^2\delta}{da^2}+\frac{3}{2}a\frac{d\delta}{da}+\left(\frac{k^2v^2}{H^2a^2}-\frac{3}{2}\right)\delta=0. \tag{4.124}$$

These equations have fairly simple solutions. Consider first the modes for which the quantity (k^2v^2/H^2a^2) is far less than unity. Then the equations can be approximated as:

$$\begin{aligned} &a^2\delta''-2\delta\approx 0 \quad \text{(radiation dominated phase)},\\ &a^2\delta''+\frac{3}{2}a\delta'-\frac{3}{2}\delta\approx 0 \quad \text{(matter dominated phase)}. \end{aligned} \tag{4.125}$$

The solutions can be written down by inspection; the growing modes are

$$\delta=\begin{cases} a^2 & \text{(radiation dominated phase)}\\ a & \text{(matter dominated phase)}. \end{cases} \tag{4.126}$$

These are exactly the results which were obtained earlier in section 4.3 by a more qualitative method. Notice that the condition $(k^2v^2/H^2a^2)\ll 1$ has a different meaning in the radiation dominated phase and in the matter dominated phase. In the radiation dominated phase, this condition selects the modes which are bigger than the *Hubble radius*; while in the matter dominated phase it selects modes which are bigger than the *Jeans length*. Thus the Hubble radius plays the role of Jeans length in the radiation dominated phase.

(2) Perturbations in the relativistic component; $\lambda < d_H$; $t < t_{\rm eq}$

In this case, with $(k^2/3H^2a^2)\gg 1$ (i.e $\lambda \ll d_H$) and $t<t_{\rm eq}$ (radiation dominated phase), we are interested in the perturbations in the dominant, relativistic component, which cannot be handled by the Newtonian approximation. Ignoring the factor 2 in the second term, the equation (4.123) becomes

$$\frac{d^2\delta}{da^2}+\frac{k^2}{3H^2a^4}\delta\cong 0. \tag{4.127}$$

Notice that $H^2\propto\rho\propto a^{-4}$ in the radiation dominated phase, making the coefficient of δ a constant. The solution is therefore just a plane wave:

$$\delta=\exp\left(i\frac{k}{\sqrt{3}Ha^2}\right)a=\exp\left[i\frac{d_H(a)}{\lambda_{phy}(a)}\right] \tag{4.128}$$

which oscillates rapidly because $d_H\gg\lambda_{phy}$. Because of this oscillation, the average value of the density contrast, $<\delta>$, at any location, will be zero. The relativistic component is distributed smoothly in regions with sizes smaller than d_H.

It is clear that the solution of (4.123) behaves very differently in the two limiting cases depending on the relative values of λ and d_H. In fact, equation (4.123) can be solved exactly thereby exhibiting the transition in the behaviour of the solution. Writing $\delta = aQ$ in (4.123), the equation for Q turns out to be

$$\frac{d^2Q}{da^2} + \frac{2}{a}\frac{dQ}{da} + \left[\mu^2 - \frac{2}{a^2}\right] Q = 0 \tag{4.129}$$

where $\mu^2 = (k^2 v^2 / H_{\rm eq}^2 a_{\rm eq}^4)$ and we have used the fact $H^2 a^2 = H_{\rm eq}^2 (a_{\rm eq}^4 / a^2)$. The general solution to this equation is

$$Q = C j_1(\mu a) + D n_1(\mu a) \tag{4.130}$$

where j_1 and n_1 are spherical Bessel functions. The transition from oscillatory behaviour (for $\lambda \ll d_H$) to secular growth (for $\lambda \gg d_H$) can be clearly seen in this solution. For small μa, the Bessel functions oscillate with the amplitude decreasing as a^{-1}; hence $\delta = Qa$ behaves as an oscillatory function with constant amplitude. For large μa, the asymptotic behaviour of the Bessel functions leads to the following form of the solution: $Q \simeq c_1 a + c_2 a^{-2}$, leading to

$$\delta \cong c_1 a^2 + c_2 a^{-1}. \tag{4.131}$$

Thus the growing mode scales as a^2. These are the same asymptotic forms discovered earlier. The parameter $\mu^2 a^2 = (k^2 v^2 a^2 / H_{\rm eq}^2 a_{\rm eq}^4) = (k^2 v^2 / H^2 a^2) \simeq (d_H^2 / \lambda^2)$ is essentially the square of the ratio between the Hubble radius and proper wavelength of the perturbation. Thus the nature of the solution – whether it is damped or growing – depends on this ratio.

(3) Perturbations in a two-component system

Let us next consider the situation with more than one component, say, radiation and dark matter. Though the equations cannot be solved exactly in this case, it is possible to obtain the main results by suitable approximation.

In a system containing radiation and dark matter, the background densities are $\rho_{\rm DM} \propto a^{-3}$ and $\rho_R \propto a^{-4}$. It is convenient to use $x = (a/a_{\rm eq})$ as the independent variable and express all the quantities in terms of x. To begin with, the densities can be written as

$$\frac{\rho_{\rm DM}}{\rho_{\rm eq}} = \frac{1}{2x^3}; \quad \frac{\rho_R}{\rho_{\rm eq}} = \frac{1}{2x^4}; \quad \frac{\rho}{\rho_{\rm eq}} = \frac{\rho_{\rm DM} + \rho_R}{\rho_{\rm eq}} = \frac{1}{2x^4}(x+1). \tag{4.132}$$

Since $p_{\rm DM} = 0$, $p_R = (1/3)\rho_R$, we have $w_R = v_R^2 = (1/3)$, $w_{\rm DM} = v_{\rm DM}^2 = 0$; also note that $(p/\rho_{\rm eq}) = (p_{\rm DM} + p_R)/\rho_{\rm eq} = (1/6x^4)$. For the full system,

$$w = \frac{p_{\rm DM} + p_R}{\rho_{\rm DM} + \rho_R} = \frac{1}{3(1+x)}; \quad v^2 = \frac{\dot{p}_{\rm DM} + \dot{p}_R}{\dot{\rho}_{\rm DM} + \dot{\rho}_R} = \frac{4}{9} \frac{1}{(x+4/3)}. \tag{4.133}$$

Using these relations, we find

$$u^2 = \frac{(\rho_R + p_R)v_{\rm DM}^2 + (\rho_{\rm DM} + p_{\rm DM})v_R^2}{(\rho + p)} = \frac{1}{3} \frac{x}{(x+4/3)} \tag{4.134}$$

and

$$H^2(x) = H_{\rm eq}^2 \frac{1}{2x^4}(1+x). \tag{4.135}$$

The perturbations δ_R and $\delta_{\rm DM}$ are characterized by the wavenumber k. It is, however, more convenient to use the ratio $2\pi\omega = [d_H(t_{\rm eq})/\lambda(a_{\rm eq})]$ between the Hubble radius at $a = a_{\rm eq}$ and the wavelength of the perturbation at $a = a_{\rm eq}$ to label the perturbations. Perturbations with $\omega > 1$ will enter the Hubble radius in the radiation dominated phase, at $a < a_{\rm eq}$, while the modes with $\omega < 1$ will enter the Hubble radius in the matter dominated phase, at $a > a_{\rm eq}$. We can express the combination (k^2/H^2a^2) in terms of ω:

$$\frac{k^2}{H^2a^2} = \frac{2x^2}{(1+x)} \left[\frac{d_H}{(a/k)}\right]_{\rm eq} = \frac{2x^2}{(1+x)}\omega^2. \tag{4.136}$$

The coupled differential equations for the two-component system were given in section 4.3 in terms of the independent variables

$$\delta = \frac{\delta\rho_R + \delta\rho_{\rm DM}}{\rho_R + \rho_{\rm DM}} = \frac{\rho_R\delta_R + \rho_{\rm DM}\delta_{\rm DM}}{\rho_R + \rho_{\rm DM}} \tag{4.137}$$

and

$$S = \frac{\delta_{\rm DM}}{1 + w_{\rm DM}} - \frac{\delta_R}{1 + w_R} = \delta_{\rm DM} - \frac{3}{4}\delta_R. \tag{4.138}$$

Given δ and S, we can solve for $\delta_{\rm DM}$ and δ_R; we find that:

$$\begin{aligned} \Delta_R \equiv \frac{3}{4}\delta_R = \frac{\delta_R}{1 + w_R} &= \left[\frac{\delta}{1+w} - \frac{(\rho_{\rm DM} + p_{\rm DM})}{(\rho + p)} S\right], \\ \Delta_{\rm DM} \equiv \delta_{\rm DM} = \frac{\delta_{\rm DM}}{1 + w_{\rm DM}} &= \left[\frac{\delta}{1+w} - \frac{(\rho_R + p_R)}{(\rho + p)} S\right]. \end{aligned} \tag{4.139}$$

Since all the quantities which appear in the coupled equations are now given in terms of x, we can explicitly write down these equations. The independent variable can be changed from t to x by using the relations

$$\frac{d}{dt} = Ha\frac{d}{da} = H(x)x\frac{d}{dx} \equiv H\hat{D}$$
$$\frac{d^2}{dt^2} = H^2a\frac{d}{da}\left(a\frac{d}{da}\right) - \frac{3}{2}H^2(1+w)a\frac{d}{da} \equiv H^2\hat{D}^2 - \frac{3}{2}H^2(1+w)\hat{D} \tag{4.140}$$

where $\hat{D} = x(d/dx)$ and we have used the fact that $\dot{H} = -(3/2)\,(1+w)H^2$. With these modifications, the equations for δ and S become

$$\left[\hat{D}^2 + \left(\frac{5x}{2(1+x)} - \frac{x}{(x+4/3)} - 1\right)\hat{D} + \left(\frac{1}{2}\frac{x^2}{(1+x)^2} + \frac{3x}{4} + \frac{9x^2}{4(x+4/3)} - \frac{3x^2}{1+x} - 2\right)\right]\delta = \frac{8}{9}\frac{\omega^2x^2}{(x+1)^2(x+4/3)}[xS - (1+x)\delta] \tag{4.141}$$

and

$$\left[\hat{D}^2 + \left(\frac{x}{2(1+x)} - \frac{x}{(x+4/3)}\right)\hat{D} + \frac{2}{3}\omega^2\frac{x^3}{(1+x)(x+4/3)}\right]S = \frac{2}{3}\frac{\omega^2x^2}{(x+4/3)}\delta. \tag{4.142}$$

Using the relation between (S,δ) and $(\Delta_R, \Delta_{\rm DM})$ we can also change the independent variables from (S,δ) to $(\Delta_R, \Delta_{\rm DM})$. For Δ_R and $\Delta_{\rm DM}$ we get the equations

$$\left[\hat{D}^2 + \left(\frac{1}{2}\frac{x}{1+x} - 1\right)\hat{D} + \left(\frac{2}{3}\frac{\omega^2x^2}{(1+x)} + \frac{4}{3}\frac{1}{(x+4/3)}\left\{\frac{x}{x+4/3} - 2\right\}\right)\right]\Delta_R = \left[\frac{3}{2}\frac{x}{(1+x)} - \frac{x}{(x+4/3)}\hat{D}\right]\Delta_{\rm DM} \tag{4.143}$$

and

$$\left[\hat{D}^2 + \frac{1}{2}\frac{x}{(1+x)}\hat{D} - \frac{3}{2}\frac{x}{(1+x)}\right]\Delta_{\rm DM} = \frac{4}{3}\frac{1}{(x+4/3)}\left[\hat{D} + 2 - \frac{x}{x+4/3}\right]\Delta_R. \tag{4.144}$$

Though these equations cannot be solved exactly, all the important properties of the solution can be obtained by suitable approximations. We shall now discuss these properties.

A particular mode, labelled by the parameter ω, enters the Hubble radius in the radiation dominated phase if $\omega > 1$ and in the matter dominated phase if $\omega < 1$. The x_{enter} and ω are related by the condition $\omega^2 x_{\text{enter}}^2 = 2\pi^2(1 + x_{\text{enter}})$. (This can be easily derived from the relation $2\pi k^{-1} a_{\text{enter}} = H^{-1}(a_{\text{enter}})$.) Thus for $\omega \gg 1$, $x_{\text{enter}} \simeq \omega^{-1}$ while for $\omega \ll 1$, $x_{\text{enter}} \simeq \omega^{-2}$. It is, therefore, best to consider $\omega \ll 1$ and $\omega \gg 1$ cases separately. In each of these cases, we expect the nature of the solutions to be different for $x < x_{\text{enter}}$ and $x > x_{\text{enter}}$.

Consider first the simpler case of $\omega \ll 1$ with $x_{\text{enter}} \approx \omega^{-2}$. The two ranges of values $x \ll 1 \ll x_{\text{enter}}$ and $x > x_{\text{enter}} \gg 1$ may be dealt with separately. For $x \ll 1$, we can obtain the lowest order solutions to the (δ, S) by a series of expansions in the small parameter ω. We find that there are four *sets* of independent solutions:

$$\delta_{ga} = x^2 \left[1 - \frac{17}{16}x + \cdots\right] - \frac{\omega^2 x^4}{15}[1 + \cdots]; \; S_{ga} = \frac{\omega^2 x^4}{32}\left[1 - \frac{28}{25}x + \cdots\right],$$
$$\delta_{da} = x^{-1}\left[1 - \frac{1}{2}x + \cdots\right] + \frac{1}{2}\omega^2 x[1 + \cdots]; \; S_{da} = \frac{1}{2}\omega^2 x\left[1 - \frac{1}{4}x + \cdots\right],$$
$$\delta_{di} = \frac{\omega^2 x^3}{6}\left[1 - \frac{17}{10}x + \cdots\right]; \; S_{di} = 1 - \frac{\omega^2 x^3}{18}[1 + \cdots],$$
$$\delta_{gi} = \frac{\omega^2 x^3}{6}\left[\ln x - \frac{5}{4} + \cdots\right]; \; S_{gi} = \ln x + \frac{1}{4}x[1 + \cdots]. \tag{4.145}$$

These solutions are classified as adiabatic (for which $S \ll \delta$; labelled as 'a') or isothermal (for which $\delta \ll S$; labelled as 'i') and also as 'growing' (labelled by 'g') or 'decaying' (labelled by 'd'). This labelling is somewhat arbitrary except for δ_{ga} and δ_{da}.

Several properties of the modes can be easily understood from these solutions. To begin with, notice that the coupling between δ and S is through a term of the form $\omega^2 x^n$. For the $\omega \ll 1$, $x \ll 1$ case which we are considering (that is, when the modes are bigger than the Hubble radius), this coupling is very weak. The distinction between adiabatic and isothermal modes is well defined for these modes; the evolution preserves this distinction (and does not mix the two) as long as the modes are bigger than the Hubble radius.

This result is quite general and can be seen directly from (4.141) and (4.142). The coupling between δ and S is through terms like $\omega^2 x^n$ at *all* x; as long as this quantity is small there will be no mixing. It can also be seen that if the $\omega^2 x^3$ and $\omega^2 x^2$ terms in (4.142) can be neglected, then $S=$ constant is a valid solution. In other words, there exist solutions for which S is (approximately) conserved when the modes are bigger than the Hubble radius. This result is also valid for any other two component system with a reasonable equation of state.

Among the adiabatic modes, $\delta_{ga} \simeq x^2$ and $\delta_{da} \simeq x^{-1}$ deserve the terminology 'growing' and 'decaying'. The corresponding S are given the same labels. These constitute the first two sets of solutions.

In contrast, isothermal solutions have $\delta \ll S$. Since δ is negligible, there is no (significant) perturbation in the energy density. Consequently, the geometry of the spacetime does not change significantly. These modes correspond to a redistribution of energy densities of dust and radiation. The dominant isothermal mode is called 'growing' as a matter of convention. The last two sets give the 'growing' and 'decaying' isothermal modes.

Which of these solutions (adiabatic or isothermal) is actually realized in the universe depends on the physical origin of perturbations. We shall concentrate here on the adiabatic modes, because – as we will see in chapter 10 – they seem to be easier to generate by physical mechanisms. Between δ_{ga} and δ_{da}, the growing mode will soon dominate.

For the $\omega^2 \ll 1$ case we are considering, the modes enter the Hubble radius in the matter dominated phase. In this case, we can find the solutions for the $x^2\omega^2 \gg 1$, $x \gg 1$ range in a similar manner. For this case, the strong coupling (large $x\omega$) ties together δ and S. The dominant mode is

$$\delta_g = S_g = x; \qquad \delta_d = S_d = x^{-3/2}. \tag{4.146}$$

Since $S = \delta$, the distinction between isothermal and adiabatic modes is no longer relevant.

It should be noted that the two components in our system interact with each other only through gravity; this is because dark matter has no direct coupling with radiation. The situation can be different if non-gravitational couplings exist between the components. For example, consider baryons and photons. Because of the strong coupling which exists in the radiation dominated era, the behaviour of baryons will be completely dictated by that of the radiation. If we start with adiabatic modes then the condition $\delta_B \cong (3/4)\delta_R$ will be maintained till decoupling and the baryonic perturbations will not grow.

Let us next consider the case with $\omega^2 \gg 1$, which provides more features. Here, *three* cases need to be distinguished. When $x\omega \ll 1$, (which also implies $x \ll 1$), the mode is bigger than the Hubble radius; when $\omega^{-1} \ll x \ll 1$, the mode has entered the Hubble radius but the universe is still radiation dominated (this case was absent previously for $\omega \ll 1$); finally, when $x \gg 1$ (which also implies $\omega x \gg 1$), the mode is inside the Hubble radius in a *matter* dominated universe.

To study these cases, it is easier to work with (4.143) and (4.144), which will directly give the behaviour of the density contrast δ_R and $\delta_{\rm DM}$. When $\omega x \ll 1$ (implying $x \ll 1$ as well) these equations can be approximated as

$$(\hat{D}^2 - \hat{D} - 2)\Delta_R \cong 0; \qquad \hat{D}^2\Delta_{\rm DM} \cong (\hat{D} + 2)\Delta_R. \tag{4.147}$$

In this limit, various pressure terms are unimportant because the wavelength is bigger than the Hubble radius. Since the universe is radiation dominated, the perturbation $\Delta_{\rm DM}$ is driven by Δ_R while the 'back reaction' of $\Delta_{\rm DM}$ on Δ_R is negligible. (Notice that, for comparable Δ_R and $\Delta_{\rm DM}$, the $\delta\rho_R \gg \delta\rho_{\rm DM}$ since $\rho_R \gg \rho_{\rm DM}$.) The dominant mode is

$$\Delta_R = \Delta_{\rm DM} = x^2; \quad \delta_{\rm DM} = x^2; \quad \delta_R = \frac{4}{3}x^2. \tag{4.148}$$

Since $S = \Delta_R - \Delta_{\rm DM} \cong 0$, the mode is clearly adiabatic; both radiation and dust perturbations grow as x^2.

The nature of the equations changes when the mode has entered the Hubble radius, but the universe is still radiation dominated. Here we have $\omega x \gg 1$ and $x \ll 1$. The equations can now be approximated as

$$\left[\hat{D}^2 - \hat{D} + \frac{2}{3}\omega^2 x^2\right]\Delta_R \cong 0; \quad \hat{D}^2\Delta_{\rm DM} \cong \hat{D}\Delta_R. \tag{4.149}$$

(After this set of equations is solved, one can explicitly verify that the terms which were neglected are smaller than those which are retained.) The radiation still drives $\Delta_{\rm DM}$ and the back reaction is ignorable. But the equation governing Δ_R now has a different structure. Since $\hat{D}^2 - \hat{D} = x^2(d/dx^2)$, we find the solutions to this equation to be

$$\Delta_R \cong A\exp(\pm i\nu x); \quad \nu^2 = \frac{2}{3}\omega^2 \gg 1, \tag{4.150}$$

which is oscillating rapidly because $\omega \gg 1$. (This solution was obtained earlier by solving (4.123).) Substituting for Δ_R in the second equation, we find that

$$\Delta_{\rm DM} = \frac{A}{i\nu x}\exp(\pm i\nu x) + B\ln x + C \simeq B\ln x + C. \tag{4.151}$$

Thus the perturbations in dark matter do not grow significantly (the growth is only logarithmic) during the period $\omega^{-1} \ll x \ll 1$. Notice that the wavelength of the dark matter perturbations is bigger than the (dark matter) Jeans length. It is *not* the pressure but the rapid expansion which is preventing the growth of dark matter perturbations. This aspect was already discussed in section 4.2.

Finally, consider the range $x \gg 1$ corresponding to the matter dominated phase with the mode well within the Hubble radius. The equations

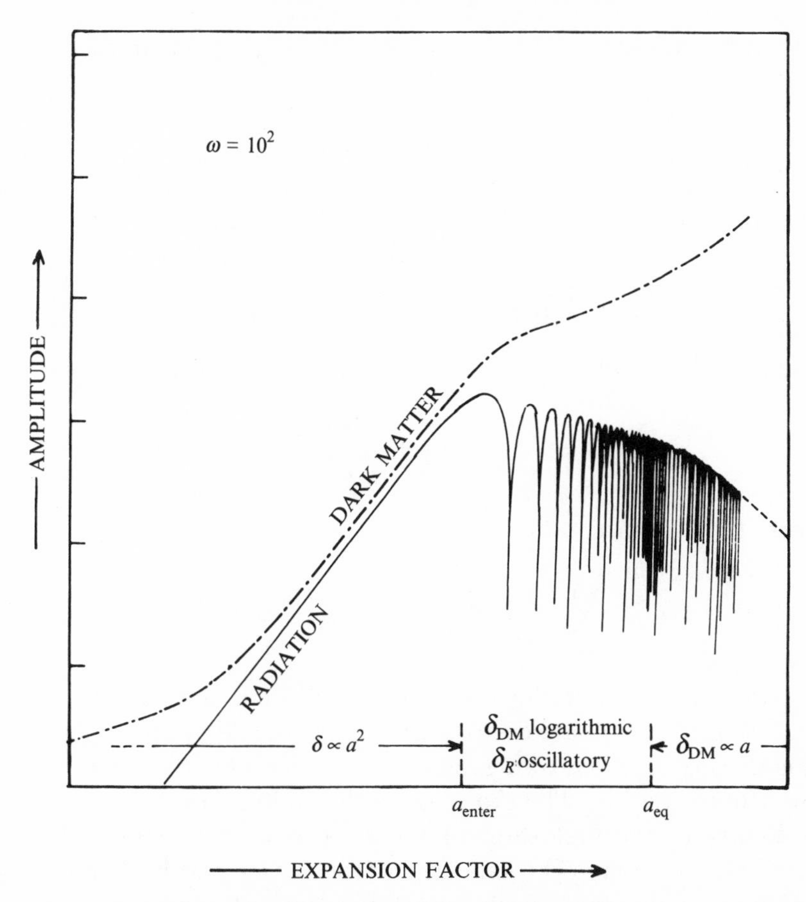

Fig. 4.3. The behaviour of the perturbation amplitudes of dark matter and radiation is shown. The solid line depicts radiation while the broken line indicates dark matter. Both grow as a^2 for $a < a_{\rm enter}$. For $a > a_{\rm enter}$ the radiation amplitude oscillates rapidly with a decreasing profile. The amplitude of dark matter perturbation grows logarithmically until $a = a_{\rm eq}$ and grows as a thereafter.

now become

$$\begin{gathered}\left[\hat{D}^2 + \frac{1}{2}\hat{D} - \frac{3}{2}\right]\Delta_{\rm DM} \cong 0, \\ \left[\hat{D}^2 + \frac{1}{2}\hat{D} + \frac{2}{3}\omega^2 x\right]\Delta_R \cong -\left[\hat{D} - \frac{3}{2}\right]\Delta_{\rm DM}.\end{gathered} \tag{4.152}$$

Now the roles have been reversed and the radiation perturbation is driven by the dark matter. Solving the first equation we find that

$$\Delta_{\rm DM} = Ax + Bx^{-3/2} \cong Ax \tag{4.153}$$

where we have only retained the growing mode. Using this in the second equation we get

$$\frac{d}{dx}\left(x\frac{d\Delta_R}{dx}\right)+\frac{1}{2}\frac{d\Delta_R}{dx}+\frac{2}{3}\omega^2\left(\Delta_R-\frac{3A}{4\omega^2}\right)=0. \tag{4.154}$$

Substituting, $\Delta_R = (3A/4\omega^2) + x^{-3/4}f(x)$, the equation for f becomes

$$\ddot{f}=-\left(\frac{3}{16}\frac{1}{x^2}+\frac{2}{3}\frac{\omega^2}{x}\right)f\cong-\frac{2}{3}\frac{\omega^2}{x}f. \tag{4.155}$$

This can solved, to sufficient accuracy, in the WKB approximation:

$$f(x)=\frac{x^{1/4}}{\sqrt{\omega}}\exp\left(\pm i\sqrt{\frac{8}{3}}\omega x^{1/2}\right). \tag{4.156}$$

Therefore, Δ_R becomes,

$$\Delta_R\cong\frac{3A}{4\omega^2}+B\frac{1}{\sqrt{\omega x}}\exp\left(\pm i\sqrt{\frac{8}{3}}\omega x^{1/2}\right). \tag{4.157}$$

In other words, the oscillations imposed on Δ_R (through the $(2/3)\omega^2 x$ term), continue to dominate over the driving by the $\Delta_{\rm DM}$ term. Thus the radiation density has no growing modes at wavelengths which are smaller than the Hubble radius. The evolution is shown in figure 4.3.

This completes the first three of the six cases mentioned in the beginning of this section. The remaining three cases can be handled by Newtonian approximation which we will take up in the next section.

4.5 Solutions to the Newtonian perturbation equation

If $\lambda \ll d_H$, then the perturbations can be analyzed by Newtonian theory. The non-relativistic, Newtonian limit of (4.119) is obtained by setting $w \approx 0, v \approx 0$ in that equation. Using further the fact that $H^2 = (8\pi G\rho/3)$ we get

$$\ddot{\delta}_A+\frac{2\dot{a}}{a}\dot{\delta}_A+\frac{k^2v_A^2}{a^2}\delta_A=4\pi G\rho_A\delta_A. \tag{4.158}$$

Since there are no gauge ambiguities, one can keep the density contrast δ as the dependent variable and t as the independent variable.

If there are more than one species of particles populating the universe, then the right hand side of (4.158) contains contributions from all the perturbed species; the equation gets modified to

$$\ddot{\delta}_A+\frac{2\dot{a}}{a}\dot{\delta}_A+\frac{k^2v_A^2}{a^2}\delta_A=\sum_{\text{all } B}4\pi G\rho_B\delta_B. \tag{4.159}$$

The structure of (4.159) is quite similar to that of the exact relativistic equation; there is one term giving dilution due to expansion $(2\dot{a}\dot{\delta}/a)$, one representing pressure support $(k^2v^2/a^2)\delta$ and the driving force due to the gravitational field of all perturbed matter $(4\pi G\sum\rho\delta)$. All the qualitative considerations mentioned in the previous two sections can be easily seen to hold in this case as well.

(1) Perturbations in non-relativistic component; $\lambda < d_H$; $t > t_{\rm eq}$

The solutions to this equation can be discussed in two limiting cases. We start with the one for which $\lambda \gg \lambda_J$.

If $\lambda \gg \lambda_J$ (i.e. if the pressure is negligible compared with gravity) then the $(k^2v^2/a^2)\delta$ term can be ignored, relative to the right hand side. However, notice that the Newtonian approximation is valid only for $\lambda \ll d_H$; therefore, we are considering the modes with $\lambda_J \ll \lambda \ll d_H$. Since $\rho \propto a^{-3} \propto t^{-2}$, it is easy to verify that the growing solution to this equation is $\delta \propto t^{2/3} \propto a$ for an $\Omega = 1$ universe. This is precisely what we found in the last section for $\lambda \gg \lambda_J$ modes.

This range with $\lambda_J \ll \lambda \ll d_H$ is of considerable importance in the study of structure formation. In fact, most of the scales which are relevant to astrophysics satisfy $\lambda \ll d_H$ for $z \lesssim 10^3$ and they grow only when $\lambda \gg \lambda_J$. Because of this importance, we will discuss the solutions to this equation in greater detail.

To begin with, notice that we now have to consider background models with $\Omega \neq 1$. (In the last section, we could approximate the background universe as $\Omega \simeq 1$, because we were essentially interested in the $z \gtrsim 10^3$ phase.) The solutions for a universe with $\Omega \neq 1$ can be obtained by the following argument.

Let $\rho(t)$ be a solution to the background Friedmann model dominated by pressureless dust. Consider now the function $\rho_1(t) \equiv \rho(t+\tau)$ where τ is some constant. Since the Friedmann equations contain t only through the derivative, $\rho_1(t)$ is also a valid solution. If we now take τ to be small, then $[\rho_1(t) - \rho(t)]/\rho(t)$ will be a small density contrast. Clearly,

$$\delta(t) = \frac{\rho_1(t) - \rho(t)}{\rho(t)} = \frac{\rho(t+\tau) - \rho(t)}{\rho(t)} \cong \tau\frac{d\ln\rho}{dt} = -3\tau H(t) \qquad (4.160)$$

where the last relation follows from the fact that $\rho \propto a^{-3}$ and $H = (\dot{a}/a)$. Since τ is a constant, it follows that $H(t)$ is a solution to be the perturbation equation. This curious fact, of course, can be verified directly: From the equations describing the Friedmann model, it follows that $\dot{H} + H^2 = (-4\pi G\rho/3)$. Differentiating this relation and using $\dot{\rho} = -3H\rho$ we immediately get $\ddot{H} + 2H\dot{H} - 4\pi G\rho H = 0$. Thus H satisfies the same equation as δ.

Since $\dot{H} = -H^2 - (4\pi G\rho/3)$, H is a decreasing function of time, and the solution $\delta = H \equiv \delta_d$ is a decaying mode. The growing solution $(\delta \equiv \delta_i)$ can be easily found by using the fact that, for any two linearly independent solutions of (4.158), the Wronskian $(\dot{\delta}_i\delta_d - \dot{\delta}_d\delta_i)$ has a value a^{-2}. This implies

$$\delta_i = \delta_d \int \frac{dt}{a^2\delta_d^2} = H(t) \int \frac{dt}{a^2H^2(t)}. \tag{4.161}$$

Thus, we have the important result that the $H(t)$ of the background spacetime allows one to completely determine the evolution of density contrast. It is more useful to express δ as a function of z rather than as a function of t. For $\Omega \neq 1$, we have the relations (see chapter 2):

$$\begin{aligned} a(z) &= a_0(1+z)^{-1}, \\ H(z) &= H_0(1+z)(1+\Omega z)^{1/2} \end{aligned} \tag{4.162}$$

and

$$H_0\, dt = -(1+z)^{-2}(1+\Omega z)^{-1/2}dz. \tag{4.163}$$

Taking $\delta_d = H(z)$, we get

$$\begin{aligned} \delta_i &= \delta_d(z) \int a^{-2}\delta_d^{-2}(z) \left(\frac{dt}{dz}\right) dz \\ &= (a_0H_0)^{-2}(1+z)(1+\Omega z)^{1/2} \int_z^\infty dx(1+x)^{-2}(1+\Omega x)^{-3/2}. \end{aligned} \tag{4.164}$$

This integral can be expressed in terms of elementary functions:

$$\begin{aligned} \delta_i = &\frac{1+2\Omega+3\Omega z}{(1-\Omega)^2} \\ &- \frac{3}{2}\frac{\Omega(1+z)(1+\Omega z)^{1/2}}{(1-\Omega)^{5/2}} \ln\left[\frac{(1+\Omega z)^{1/2}+(1-\Omega)^{1/2}}{(1+\Omega z)^{1/2}-(1-\Omega)^{1/2}}\right]. \end{aligned} \tag{4.165}$$

Thus $\delta_i(z)$ for an arbitrary Ω can be given in closed form. The solution in (4.165) is not normalized in any manner; that can be achieved by multiplying δ_i by some constant depending on the context (see exercise 4.6).

The asymptotic forms of δ_i are of interest. For large z (i.e., early times), $\delta_i \propto z^{-1}$. This is to be expected because for small t, the curvature term can be ignored and the Friedmann universe can be approximated as a $\Omega = 1$ model. (The large z expansion of the logarithm in (4.165) has to be taken up to $O(z^{-5/2})$ to get the correct result; it is much easier to obtain the asymptotic form directly from the integral.) For $\Omega \ll 1$, one can see that $\delta_i \simeq$ constant for $z \ll \Omega^{-1}$. This is the curvature dominated phase, in which the growth of perturbations is halted by rapid expansion. All these features have been obtained in section 4.2 by more elementary reasoning.

Given the form of the density contrast, one can determine the peculiar velocity field related to this density contrast. As we have seen earlier, the peculiar velocity $\mathbf{v}_k$ in the Fourier space is proportional to the peculiar acceleration $\mathbf{g}_k$ with the coefficient of proportionality $(2f/3H\Omega)$ where

$$f(a) = \frac{a}{\delta}\frac{d\delta}{da} = -\frac{(1+z)}{\delta}\frac{d\delta}{dz}. \tag{4.166}$$

The peculiar velocities and accelerations observed in our universe at the present epoch are related by the value of f at $z = 0$, which – in turn – will depend only on Ω. Though $f(\Omega, z = 0)$ can be calculated exactly from (4.165), its functional form is not convenient for further manipulations. It turns out, however, that $f(\Omega, z = 0)$ is very well approximated by the power law $f(\Omega) \approx \Omega^{0.6}$, which is often used, instead of the exact form, for estimates[5]. With this approximation,

$$\mathbf{u} = \frac{2}{3H\Omega}\Omega^{0.6}\mathbf{g}, \tag{4.167}$$

where all the quantities are evaluated at the present time and $\mathbf{g}$ is the peculiar acceleration generated by the density contrast.

Lastly, we have to consider the case with $(k^2v^2/H^2a^2) \gg 1$. Here the pressure term dominates over matter and we can ignore the $4\pi G\rho$ term. Changing the variable from t to a, the perturbation equation becomes

$$a\frac{d^2\delta}{da^2} + \frac{3}{2}\frac{d\delta}{da} + \frac{k^2v^2}{H^2a^3}\delta \approx 0. \tag{4.168}$$

(This is, of course, the same as (4.124) with the (3/2) factor ignored.) When the baryons and the radiation are tightly coupled, $v^2 \propto T_{\rm matter} = T_{\rm rad}$ and hence $v^2 \propto a^{-1}$. Let $v^2 = v_0^2(a_0/a)$ where v_0 and a_0 are some constants. Changing the independent variable to $x = (a/a_0)^{1/2}$, we get the equation:

$$\frac{d^2\delta}{dx^2} + \frac{2}{x}\frac{d\delta}{dx} + \frac{\omega^2\delta}{x^2} = 0; \quad \omega^2 \equiv \frac{4k^2v_0^2}{H^2a^3} \tag{4.169}$$

or

$$\frac{1}{x^2}\frac{d}{dx}\left(x^2\frac{d\delta}{dx}\right) + \frac{\omega^2\delta}{x^2} = 0. \tag{4.170}$$

Note that ω^2 is a constant because $H^2 \propto \rho \propto a^{-3}$. This has the solution $\delta = x^n$ with $n \cong [(-1/2) \pm i\omega]$ for $\omega \gg 1$. Therefore, using $x \propto a^{1/2} \propto t^{1/3}$, the density contrast can be written as

$$\delta = t^{-1/6}\exp\left(\pm\frac{i\omega}{3}\ln t\right) \tag{4.171}$$

which decays as $t^{-1/6}$.

This decay law has a simple interpretation. The oscillations in the baryon density can be thought of as acoustic vibrations in a medium. The frequency of these vibrations scales as $\nu = (v/\lambda) \propto a^{-1/2}a^{-1} \propto a^{-3/2} \propto t^{-1}$ while the energy of the vibrations in a volume V will vary as $E \propto \rho_b(v\delta)^2 V \propto v^2\delta^2 \propto \delta^2 a^{-1} \propto \delta^2 t^{-2/3}$. In the case of $(kv \gg Ha)$ there will be several acoustic oscillations within one expansion timescale, and hence, the expansion may be treated as (adiabatically) slow. In such a case, the quantity (E/ν), which is an adiabatic invariant, will remain constant. Since $\nu \propto t^{-1}$ and $E \propto t^{-2/3}\delta^2$, it follows that $t^{1/3}\delta^2 =$ constant or $\delta \propto t^{-1/6}$.

(2) Perturbations in dark matter component; $\lambda < d_H$; $t < t_{\rm eq}$

As further illustrations of the Newtonian approximation, we discuss the two other situations mentioned in section 4.2: (a) Suppression of growth of $\delta_{\rm DM}$ in the radiation dominated phase and (b) rapid growth of δ_B just after $t_{\rm dec}$. Of these two the first case has already been studied in the last section when we discussed the two component system; we will repeat the analysis in the Newtonian approximation to illustrate the connection between the two and also to discuss the question of boundary conditions in greater detail.

Consider the perturbations in the dark matter component in the wavelength range $\lambda_J \ll \lambda \ll d_H$ when the universe is radiation dominated[6]. Since $\lambda_J \ll \lambda$, we can ignore the pressure support. Further, in the right hand side of (4.159), we ignore δ_R because $< \delta_R > \approx 0$ at scales well inside the Hubble radius scales ($\lambda \ll ct$) due to rapid oscillations in (4.128). Therefore, the perturbation equation becomes

$$\ddot{\delta}_{\rm DM} + \frac{2\dot{a}}{a}\dot{\delta}_{\rm DM} \cong 4\pi G \rho_{\rm DM}\delta_{\rm DM}, \tag{4.172}$$

where the background universe is governed by the equation

$$\frac{\dot{a}^2}{a^2} = \frac{8\pi G}{3}\left(\rho_R + \rho_{\rm DM}\right). \tag{4.173}$$

Introducing the variable $x \equiv (a/a_{\rm eq})$ and using (4.173) in (4.172), we can recast the equation in the form

$$2x(1+x)\frac{d^2\delta_{\rm DM}}{dx^2} + (2+3x)\frac{d\delta_{\rm DM}}{dx} = 3\delta_{\rm DM}; \qquad x = \frac{a}{a_{\rm eq}}. \tag{4.174}$$

One solution to this equation can be written down by inspection:

$$\delta_{\rm DM} = 1 + \frac{3}{2}x. \tag{4.175}$$

In other words $\delta_{\rm DM} \approx$ constant for $a \ll a_{\rm eq}$ (no growth in the radiation dominated phase) and $\delta_{\rm DM} \propto a$ for $a \gg a_{\rm eq}$ (growth proportional to a in the matter dominated phase). Thus $\delta_{\rm DM}$ does not grow significantly in the radiation dominated phase even though $\lambda > \lambda_J$. A mode which enters the Hubble radius at $a = a_{\rm enter}$ can grow to a value $\delta_{\rm DM} = 1+(3/2)(a/a_{\rm eq})$ at some time $a > a_{\rm enter}$. This growth is by a factor

$$F = \frac{1+(3/2)(a/a_{\rm eq})}{1+(3/2)(a_{\rm enter}/a_{\rm eq})}. \qquad (4.176)$$

Assuming $a_{\rm enter} \ll a_{\rm eq}$ and $a \approx a_{\rm eq}$, the maximum growth factor is seen to be $F_{\rm max} = (5/2)$.

This statement, however, needs to be qualified. The above equations – and in fact all perturbation equations considered so far – are second order differential equations having two linearly independent solutions. A general solution can be found only when two initial conditions are given. We have been avoiding this problem so far by just choosing the 'growing' solution as *the* solution to the equation. (This procedure is justified as long as we are not interested in any transient phenomena). In this particular case a proper analysis will require matching both δ and $\dot{\delta}$ at the instant $t = t_{\rm enter}$. This would force us to choose a linear combination of solutions during the epoch $t_{\rm enter} < t < t_{\rm eq}$ rather than the purely 'growing' mode given above. It turns out that such an analysis changes the result slightly.

To analyze this situation correctly, we have to find the second solution. Given the growing solution, the decaying solution Δ can be easily found by the Wronskian condition $(Q'/Q) = -[(2+3x)/2x(1+x)]$ where $Q = \delta_{\rm DM}\Delta' - \delta'_{\rm DM}\Delta$. Writing the decaying solution as $\Delta = f(x)\delta_{\rm DM}(x)$ and substituting in this equation, we find

$$\frac{f''}{f'} = -\frac{2\delta'_{\rm DM}}{\delta_{\rm DM}} - \frac{2+3x}{2x(1+x)}. \qquad (4.177)$$

This can be solved to give

$$f = -\int \frac{dx}{x(1+3x/2)^2(1+x)^{1/2}}. \qquad (4.178)$$

Integrating, we find the decaying solution to be

$$\Delta = f\delta_{\rm DM} = \left(1+\frac{3x}{2}\right)\ln\left[\frac{(1+x)^{1/2}+1}{(1+x)^{1/2}-1}\right] - 3(1+x)^{1/2}. \qquad (4.179)$$

Thus the general solution to the perturbation equation, for a mode which is inside the Hubble radius, is the linear superposition

$$\delta_{\rm gen}(x) = A\delta_{\rm DM}(x) + B\Delta(x) = \begin{cases} A + B\ln(4/x) & (x \ll 1) \\ (3/2)Ax + (4/5)Bx^{(-3/2)} & (x \gg 1) \end{cases}. \tag{4.180}$$

Notice that, for $x \ll 1$, $\delta_{\rm gen}(x)$ has the same form as found earlier by solving the exact equations for the two component system.

The constants A and B have to be fixed by matching this solution to the growing solution, which was valid when the mode was bigger than the Hubble radius. Since the latter solution is given by $\delta(x) = x^2$ in the radiation dominated phase, we get the matching conditions:

$$\begin{aligned} x_{\rm enter}^2 &= \left[A\delta_{\rm DM}(x) + B\Delta(x)\right]_{x = x_{\rm enter}}, \\ 2x_{\rm enter} &= \left[A\delta'_{\rm DM}(x) + B\Delta'(x)\right]_{x = x_{\rm enter}}. \end{aligned} \tag{4.181}$$

This determines the constants A and B in terms of $x_{\rm enter} = (a_{\rm enter}/a_{\rm eq})$ which, in turn, depends on the wavelength of the mode through $a_{\rm enter}$.

As an example, consider a mode for which $x_{\rm enter} \ll 1$. The 'decaying' solution has the asymptotic form $\Delta(x) \simeq \ln(4/x)$ for $x \ll 1$. Using this, the matching conditions become

$$\begin{aligned} x_{\rm enter}^2 &= A\left(1 + \frac{3}{2}x_{\rm enter}\right) + B\ln\frac{4}{x_{\rm enter}}, \\ 2x_{\rm enter} &= \frac{3}{2}A - Bx_{\rm enter}^{-1}. \end{aligned} \tag{4.182}$$

Solving these to determine A and B (to the lowest order in $x_{\rm enter}$), we find

$$A = x_{\rm enter}^2\left[1 + 2\ln\left(\frac{4}{x_{\rm enter}}\right)\right]; \quad B = -2x_{\rm enter}^2, \tag{4.183}$$

so that the properly matched mode, inside the Hubble radius, is

$$\delta(x) = x_{\rm enter}^2\left[1 + 2\ln\left(\frac{4}{x_{\rm enter}}\right)\right]\left(1 + \frac{3x}{2}\right) - 2x_{\rm enter}^2\ln\frac{4}{x}. \tag{4.184}$$

During the radiation dominated phase, that is till $a \lesssim a_{\rm eq}$, ($x \lesssim 1$) this mode can grow by a factor

$$\frac{\delta(x \approx 1)}{\delta(x_{\rm enter})} = \frac{1}{x_{\rm enter}^2}\delta(x \simeq 1) \cong 5\ln\left(\frac{1}{x_{\rm enter}}\right) = 5\ln\left(\frac{a_{\rm eq}}{a_{\rm enter}}\right) = \frac{5}{2}\ln\left(\frac{t_{\rm eq}}{t_{\rm enter}}\right). \tag{4.185}$$

Since the time $t_{\rm enter}$ for a mode with wavelength λ is fixed by the condition $\lambda a_{\rm enter} \sim \lambda t_{\rm enter}^{1/2} \simeq d_H(t_{\rm enter}) \sim t_{\rm enter}$, it follows that $\lambda \propto t_{\rm enter}^{1/2}$. Hence,

$$\frac{\delta_{\rm final}}{\delta_{\rm enter}} = 5\ln\left(\frac{\lambda_{\rm eq}}{\lambda}\right) = \frac{5}{3}\ln\left(\frac{M_{\rm eq}}{M}\right) \tag{4.186}$$

for a mode with wavelength $\lambda \ll \lambda_{\rm eq}$. (Here, M is the mass contained in a sphere of radius λ.) The growth in the radiation dominated phase, therefore, is logarithmic. Notice that the matching procedure has brought in an amplification factor *which depends on the wavelength.*

As a second example of coupled density fluctuations, consider the perturbations in the baryon–dark matter system just after decoupling. This evolution is now governed by the equations

$$\ddot{\delta}_{\rm DM} + \frac{2\dot{a}}{a}\dot{\delta}_{\rm DM} = 4\pi G(\rho_B\delta_B + \rho_{\rm DM}\delta_{\rm DM}) \approx 4\pi G\rho_{\rm DM}\delta_{\rm DM}, \tag{4.187}$$

$$\ddot{\delta}_B + \frac{2\dot{a}}{a}\dot{\delta}_B = 4\pi G(\rho_B\delta_B + \rho_{\rm DM}\delta_{\rm DM}) \approx 4\pi G\rho_{\rm DM}\delta_{\rm DM}, \tag{4.188}$$

where we have used the fact that, just after $a_{\rm dec}$, $\rho_{\rm DM}\ \delta_{\rm DM} \gg \rho_B\delta_B$. Equation (4.187) represents the growth of perturbations in the dark matter component. From the previous analysis we know that it has the solution

$$\delta_{\rm DM} = ({\rm constant})a \equiv \alpha a. \tag{4.189}$$

Substituting this result in (4.188), we can rewrite it as

$$a^{3/2}\frac{d}{da}\left(\frac{1}{a^{1/2}}\frac{d\delta_B}{da}\right) + 2\frac{d\delta_B}{da} = \frac{3}{2}\alpha, \tag{4.190}$$

where we have also used the relation $a \propto t^{2/3}$. This equation has the growing solution

$$\delta_B = \alpha(a - a_0) = \delta_{\rm DM}(a)\left(1 - \frac{b}{a}\right), \tag{4.191}$$

where b is some constant. This solution shows that $\delta_B \to \delta_{\rm DM}$ for $a \gg b$, even if $\delta_B \approx 0$ at some $a = b = a_{\rm dec}$ (say). In other words, baryonic perturbations 'catch up' with dark matter perturbations after the decoupling. This result was also discussed earlier in section 2.

The two separate cases considered in this section and the four cases analyzed in the last section cover all the possibilities described earlier in section 4.2. Table 4.1 summarises the results of sections 4.4 and 4.5. The numbers in square brackets refer to the section in which the particular result is discussed.

4.6 Dissipation in dark matter and baryons

In the discussion of perturbations so far, we have taken the matter content of the universe to be an ideal fluid. This approximation breaks down

Table 4.1. Growth of perturbations

Epoch	δ_R	$\delta_{\rm DM}$	δ_B
$t < t_{\rm enter} < t_{\rm eq}$	grows as a^2	grows as a^2	grows as a^2
$\lambda > d_H$	[4(1), 4(3)]	[4(3)]	[4(3)]
$t_{\rm enter} < t < t_{\rm eq}$	oscillates	grows as ln a	oscillates
$\lambda < d_H$	[4(2), 4(3)]	[4(3), 5(2)]	[4(3)]
$t_{\rm eq} < t < t_{\rm dec}$	oscillates	grows as a	oscillates
$\lambda < d_H$	[4(3)]	[4(3), 5(1)]	[4(3)]
$t_{\rm dec} < t$	oscillates	grows as a	grows as a
$\lambda < d_H$	[4(3)]	[5(1)]	[5(1), 5(2)]

for wavelengths smaller than a particular critical value. At smaller wavelengths, energy is drained away by certain dissipative processes which we will now discuss.

The physical origin of dissipation is different in baryons and dark matter. In collisionless dark matter, dissipation occurs through a process called 'free streaming', while in baryons it arises due to coupling between radiation and matter.

Let us consider the dark matter component first[7]. If the dark matter is made up of weakly interacting particles, then these particles do not feel each other's presence via collisions (unlike an ordinary gas where collisions are significant). Each dark matter particle, therefore, moves along a geodesic in the spacetime. Perturbations modify the spacetime metric and consequently the geodesic orbits. We have been studying the response of dark matter particles to such perturbations by invoking an 'effective pressure' and treating dark matter as an ideal fluid. Such an approximation is valid only for sufficiently large wavelengths. At small scales, the 'free' geodesic motion of the particles will wipe out any structure, because the particles can freely propagate from an overdense region to an underdense region equalizing the densities. This process is called 'free streaming.'

Let $l_{\rm FS}(t)$ be the proper distance which a dark matter particle can travel in time t in the background spacetime; and let $\lambda(t)$ be the proper wavelength of a perturbation at time t. Then all modes, for which $l_{\rm FS}(t) > \lambda(t)$, will suffer dissipation due to free streaming. Since we know that $\lambda(t) \propto a(t)$; we only need to compute $l_{\rm FS}(t)$ to compare the two. This can be done as follows: The proper distance travelled by a particle in time t can be written as

$$l_{\rm FS}(t) = a(t)\int_0^t \frac{v(t')}{a(t')}dt'. \tag{4.192}$$

(This arises from the fact that $a\,dL = v\,dt$ defines the proper velocity $v(t)$.) During $0 < t < t_{\rm nr}$, the dark matter particles are relativistic and $v \simeq 1$; since $a(t) \propto t^{1/2}$, this gives

$$l_{\rm FS}(t) = a\int_0^t \frac{dt'}{a_{\rm nr}}\left(\frac{t_{\rm nr}}{t'}\right)^{1/2} = a(t)\left[\frac{2t_{\rm nr}^{1/2}t^{1/2}}{a_{\rm nr}}\right] = 2t \propto a^2 \quad (\text{for } t < t_{\rm nr}). \tag{4.193}$$

For $t_{\rm nr} < t < t_{\rm eq}$, $v \propto a^{-1}$ and we get

$$\begin{aligned} l_{\rm FS}(t) &= \left[\frac{l_{\rm FS}(t_{\rm nr})}{a_{\rm nr}} + \int_{t_{\rm nr}}^t \frac{dt'}{a(t')}\frac{a_{\rm nr}}{a(t')}\right]a(t) \\ &= \left[\frac{2t_{\rm nr}}{a_{\rm nr}} + \frac{2t_{\rm nr}}{a_{\rm nr}}\ln\frac{a}{a_{\rm nr}}\right]a = \frac{2t_{\rm nr}a}{a_{\rm nr}}\left[1 + \ln\frac{a}{a_{\rm nr}}\right] \qquad (t_{\rm nr} < t < t_{\rm eq}). \end{aligned} \tag{4.194}$$

Finally, for $t > t_{\rm eq}$, $a(t) \propto t^{2/3}$. So

$$\begin{aligned} l_{\rm FS}(t) &= \left[\frac{l_{\rm FS}(t_{\rm eq})}{a_{\rm eq}} + \int_{t_{\rm eq}}^t \frac{a_{\rm nr}}{a_{\rm eq}^2}\left(\frac{t_{\rm eq}}{t'}\right)^{4/3}dt'\right]a(t) \\ &= \left[\frac{2t_{\rm nr}}{a_{\rm nr}}\left(1 + \ln\frac{a_{\rm eq}}{a_{\rm nr}}\right) + \frac{3t_{\rm nr}}{a_{\rm nr}}\left(1 - \frac{a_{\rm eq}^{1/2}}{a^{1/2}}\right)\right]a(t). \end{aligned} \tag{4.195}$$

Thus we find that

$$\frac{l_{\rm FS}(t)}{a(t)} = \begin{cases} (2t_{\rm nr}/a_{\rm nr}^2)a = (2t/a) & t < t_{\rm nr} \\ (2t_{\rm nr}/a_{\rm nr})\left[1 + \ln(a/a_{\rm nr})\right] & t_{\rm nr} < t < t_{\rm eq} \\ (2t_{\rm nr}/a_{\rm nr})\left[(5/2) + \ln(a_{\rm eq}/a_{\rm nr})\right] & t_{\rm eq} \ll t \end{cases}. \tag{4.196}$$

(In arriving at the last equation, we have used the fact that, for $t \gg t_{\rm eq}$, the second term inside the square bracket in (4.195) becomes a constant.) We now have to determine the range of wavelengths for which the condition $\lambda(t) \leq l_{\rm FS}(t)$ is satisfied; or equivalently, the range for which $(\lambda/a) \leq (l_{\rm FS}(t)/a)$. Since (λ/a) is a constant independent of time we only have to consider the evolution of $(l_{\rm FS}/a)$. For $t < t_{\rm nr}$, $(l_{\rm FS}/a) \propto (t/a) \propto a$; during $(t_{\rm nr} < t < t_{\rm eq})$, $(l_{\rm FS}/a)$ grows only logarithmically; for $t > t_{\rm eq}$, $(l_{\rm FS}/a)$ grows still more slowly and $l_{\rm FS}$ saturates at the value

$$\lambda_{\rm FS} \equiv l_{\rm FS}(t_0) = \left(\frac{a_0}{a_{\rm nr}}\right)(2t_{\rm nr})\left(\frac{5}{2} + \ln\frac{a_{\rm eq}}{a_{\rm nr}}\right). \tag{4.197}$$

Since this is the largest value of $l_{\rm FS}$, all proper wavelengths $\lambda > \lambda_{\rm FS}$ will survive the process of free streaming. To be rigorous, we should evaluate

$\lambda_{\rm FS}$ at some $t = t_{\rm nl}$ at which nonlinear effects become important. In practice, this makes very little difference.

The above result for $\lambda_{\rm FS}$ can be understood as follows: When the dark matter is relativistic, it travels with the speed of light and covers a proper distance of $(2t_{\rm nr})$ by $t = t_{\rm nr}$. This distance corresponds today to the length $(2t_{\rm nr})\,(a_0/a_{\rm nr})$, which is to be identified as the free streaming scale. Notice that this argument gives the correct result up to a numerical factor.

To obtain numerical estimates, we need to identify the epoch $t_{\rm nr}$. We may take this to be the time at which $T_{\rm DM} \approx (m/3)$ where $T_{\rm DM}$ is the temperature of the dark matter and m is the mass of the dark matter particle. The temperature $T_{\rm DM}$, in general, will not be the same as the radiation temperature $T_R \equiv T$ because the dark matter could have decoupled earlier. We can however relate the ratio $(T/T_{\rm DM})$ to $\Omega_{\rm DM}$ and the mass of the wimp in the following way: If $n_{\rm DM}$ is the number density of dark matter particles, then the quantity

$$\left(\frac{T_{\rm DM}}{T}\right)^3 = \frac{n_{\rm DM}}{n_\gamma}, \tag{4.198}$$

is conserved during the expansion. But, since

$$\Omega_{\rm DM} = \left(\frac{m n_{\rm DM}}{\rho_c}\right)_{\rm now} = \frac{m n_\gamma}{\rho_c}\left(\frac{n_{\rm DM}}{n_\gamma}\right) \simeq 30\left(\frac{m}{1\,{\rm keV}}\right)\left(\frac{n_{\rm DM}}{n_\gamma}\right)h^{-2}, \tag{4.199}$$

we can write

$$R^3 \equiv \left(\frac{T_{\rm DM}}{T}\right)^3 = \left(\frac{n_{\rm DM}}{n_\gamma}\right) \cong \left(\frac{\Omega_{\rm DM}h^2}{30}\right)\left(\frac{m}{1\,{\rm keV}}\right)^{-1}. \tag{4.200}$$

Knowing this ratio, all other quantities can be determined. For example, $(a_{\rm nr}/a_0) = (T_0/T)_{\rm nr} = (T_0/T_{\rm DM})_{\rm nr}\,(T_{\rm DM}/T) = (3T_0/m)R$ etc. Substituting the numerical values, we get

$$\begin{aligned} \frac{a_{\rm nr}}{a_0} &= 7\times10^{-7}\left(\frac{m}{1\,{\rm keV}}\right)^{-1}\left(\frac{T_{\rm DM}}{T}\right), \\ t_{\rm nr} &= 1.2\times10^{7}\left(\frac{m}{1\,{\rm keV}}\right)^{-2}\left(\frac{T_{\rm DM}}{T}\right)^2, \\ \left(\frac{a_{\rm eq}}{a_{\rm nr}}\right) &= \left(\frac{m}{17\,{\rm keV}}\right)\left(\Omega h^2\right)^{-1}\left(\frac{T}{T_{\rm DM}}\right). \end{aligned} \tag{4.201}$$

Combining these with the value of R, it is easy to obtain the free streaming scale

$$\lambda_{\rm FS} \simeq 40\,{\rm Mpc}\left(\Omega_{\rm DM}h^2\right)^{-1}\left(\frac{T_{\rm DM}}{T_R}\right)^4 = 0.5\,{\rm Mpc}\left(\Omega_{\rm DM}h^2\right)^{1/3}\left(\frac{m}{1\,{\rm keV}}\right)^{-4/3}. \tag{4.202}$$

This result is of crucial importance. It shows that the length scale below which perturbations will be wiped out, $\lambda_{\rm FS}$, is essentially decided by the mass m (or, equivalently, $\Omega_{\rm DM}$) and the temperature $T_{\rm DM}$ of the dark matter. If, for example, the dark matter is made of neutrinos with $m \approx 30\,{\rm eV}$, $(T_{\rm DM}/T_R) \simeq 0.71$ and $\Omega_\nu h^2 \approx (m_\nu/91\,{\rm eV})$, then

$$\lambda_{\rm FS} \simeq 28\,{\rm Mpc}\left(\frac{m_\nu}{30\,{\rm eV}}\right)^{-1}. \tag{4.203}$$

This length scale contains a mass of

$$M_{\rm FS} \simeq 4\times 10^{15}\left(\frac{m_\nu}{30\,{\rm eV}}\right)^{-2} {\rm M}_\odot. \tag{4.204}$$

Thus, in a universe with neutrinos as dark matter, perturbations at all lower mass scales ($M < M_{\rm FS}$) will be wiped out; there will be very little small scale power.

If, on the other hand, the dark matter particle is much heavier with (say) $m \approx 1\,{\rm keV}$ and $\Omega_{\rm DM}h^2 \approx 1$, then

$$\lambda_{\rm FS} \approx 0.5\,{\rm Mpc}\left(\frac{m}{1\,{\rm keV}}\right)^{-4/3}, \tag{4.205}$$

containing only a mass of

$$M_{\rm FS} \approx 6\times 10^9\,{\rm M}_\odot. \tag{4.206}$$

In this case, power at scales above $10^9\,{\rm M}_\odot$ or so will survive dissipation due to free streaming. In general, a heavier dark matter candidate will let the power to survive at smaller scales. The free-streaming scale is shown in figure 4.1.

Let us next consider the damping of perturbations in the photon–baryon plasma which occurs for a different – and somewhat simpler – reason[8]. At $t \ll t_{\rm dec}$, photons and baryons are very tightly coupled due to Thomson scattering. The proper length corresponding to the photon mean free path at some time t is

$$l(t) = \frac{1}{X_e n_e \sigma} \simeq 1.3\times 10^{29}\,{\rm cm}\, X_e^{-1}(1+z)^{-3}(\Omega_B h^2)^{-1}, \tag{4.207}$$

where X_e is the electron ionization fraction. For wavelengths $\lambda \lesssim l$, the photon streaming will clearly damp any perturbation almost instantaneously.

The damping effect, however, is felt at even larger scales. For $\lambda > l$, dissipation occurs due to photon diffusion. Photons can slowly diffuse out of overdense to underdense regions, dragging the tightly coupled charged particles. Though not much decay occurs during one oscillation period of the perturbation, significant loss can take place within an expansion timescale. The characteristic distance an average photon can diffuse before $t = t_{\rm dec}$ can be computed as follows.

Consider a time interval Δt in which a photon suffers $N = (\Delta t / l(t))$ collisions. Between successive collisions it travels a proper distance $l(t)$, or equivalently, a coordinate distance $[l(t)/a(t)]$. Because of this random walk, it acquires a mean-square coordinate displacement:

$$(\Delta x)^2 = N \left(\frac{l}{a} \right)^2 = \frac{\Delta t}{l(t)} \frac{l^2}{a^2} = \frac{\Delta t}{a^2} l(t). \tag{4.208}$$

The total mean-square coordinate distance travelled by a typical photon until the time of decoupling is

$$x^2 \equiv \int_0^{t_{\rm dec}} \frac{dt}{a^2(t)} l(t) = \frac{3}{5} \frac{t_{\rm dec} l(t_{\rm dec})}{a^2(t_{\rm dec})} \tag{4.209}$$

where we have used the earlier expression for $l(t)$ and the fact that $a \propto t^{2/3}$. This corresponds to the proper distance

$$l_S = a(t_{\rm dec}) x = \left[\frac{3}{5} t_{\rm dec} l(t_{\rm dec}) \right]^{1/2} \simeq 3.5 \, {\rm Mpc} \left(\frac{\Omega}{\Omega_B} \right)^{1/2} (\Omega h^2)^{-3/4}. \tag{4.210}$$

If we assume that baryons are tightly coupled to photons before $t_{\rm dec}$, it follows that baryons will be dragged along with photons. Then all perturbations at wavelengths $\lambda < l_S$ will be wiped out. This length l_S corresponds to the mass

$$M_S \cong 6.2 \times 10^{12} \, {\rm M}_\odot \left(\frac{\Omega}{\Omega_B} \right)^{3/2} (\Omega h^2)^{-5/4}. \tag{4.211}$$

No baryonic perturbation carrying mass below M_S survives this damping process. The Silk mass is marked in figure 4.2.

It is easy to determine the scale dependence of l_S and M_S. For $t < t_{\rm eq}$, $l \sim a^3$, $t \sim a^2$ giving $l_S \sim a^{5/2}$ and $M_S \sim a^{9/2}$; for $t > t_{\rm eq}$, $l \sim a^3$, $t \sim a^{3/2}$ giving $l_S \sim a^{9/4}$ and $M_S \sim a^{15/4}$. Thus, M_S rises steeply with a. Notice that the dominant contribution to this damping arises around the time $t \simeq t_{\rm dec}$. At $t \ll t_{\rm dec}$, l_S is very small while for $t > t_{\rm dec}$ baryons are not coupled to the photons. It should also be clear that $l_S \gg l(t_{\rm dec})$ because $t_{\rm dec} \gg l(t_{\rm dec})$; thus the effect is felt at scales far larger than the mean free path.

The mean free path of the *electrons* is $l_{\rm elec} \simeq (n_\gamma \sigma)^{-1}$; this is much smaller than $l_{\rm photons} \simeq (n_e \sigma)^{-1}$ because $n_\gamma \gg n_e$. Hence any damping due to the random walk of electrons will be subdominant to the effects considered above.

4.7 The processed final spectrum

We have now assembled all the ingredients to evolve an initial perturbation at some $t = t_i \ll t_{\rm eq}$ to a final value at $t > t_{\rm dec}$. It is, as usual, best to consider the evolution of perturbations in dark matter and baryons separately.

Let $\delta_\lambda(t_i)$ denote the amplitude of the dark matter perturbation corresponding to some wavelength λ at the initial instant t_i. To each λ, we can associate a wavenumber $k \propto \lambda^{-1}$ and a mass $M \propto \lambda^3$; accordingly, we may label the perturbation as $\delta_M(t)$ or $\delta_k(t)$, as well, with the scalings $M \sim \lambda^3$, $k \sim \lambda^{-1}$. We are interested in the value of $\delta_\lambda(t)$ at some $t > t_{\rm dec}$.

To begin with, note that the process of free streaming will wipe out the perturbations at all scales smaller that $\lambda_{\rm FS}$ corresponding to a mass $M_{\rm FS}$. So we have the first result:

$$\delta_M(t) \approx 0 \qquad (\text{for } M < M_{\rm FS}; \lambda < \lambda_{\rm FS}). \tag{4.212}$$

Consider next the range of wavelengths $\lambda_{\rm FS} < \lambda < \lambda_{\rm eq}$. These modes enter the Hubble radius in the radiation dominated phase; however, their growth is suppressed in the *radiation dominated* phase by the rapid expansion of the universe; therefore, they do not grow significantly until $t = t_{\rm eq}$, giving $\delta_\lambda(t_{\rm eq}) \cong \delta_\lambda(t_{\rm enter})$. After matter begins to dominate, the amplitude of these modes grows in proportion to the scale factor a. Thus,

$$\delta_M(t) = \delta_M(t_{\rm enter}) \left(\frac{a}{a_{\rm eq}}\right) \qquad (\text{for } M_{\rm FS} < M < M_{\rm eq}; a > a_{\rm eq}). \tag{4.213}$$

Consider next the modes with $\lambda_{\rm eq} < \lambda < \lambda_H$ where $\lambda_H \equiv H^{-1}(t)$ is the Hubble radius at the time t when we are studying the spectrum. These modes enter the Hubble radius in the matter dominated phase and grow proportional to a afterwards. So,

$$\delta_M(t) = \delta_M(t_{\rm enter}) \left(\frac{a}{a_{\rm enter}}\right) \qquad (\text{for } M_{\rm eq} < M < M_H) \tag{4.214}$$

which may be rewritten as

$$\delta_M(t) = \delta_M(t_{\rm enter}) \left(\frac{a_{\rm eq}}{a_{\rm enter}}\right) \left(\frac{a}{a_{\rm eq}}\right). \tag{4.215}$$

But notice that, since $t_{\rm enter}$ is fixed by the condition $\lambda a_{\rm enter} \propto t_{\rm enter}$ $\propto \lambda t_{\rm enter}^{2/3}$, we have $t_{\rm enter} \propto \lambda^3$. Further $(a_{\rm eq}/a_{\rm enter}) = (t_{\rm eq}/t_{\rm enter})^{2/3}$, giving

$$\left(\frac{a_{\rm eq}}{a_{\rm enter}}\right) = \left(\frac{\lambda_{\rm eq}}{\lambda}\right)^2 = \left(\frac{M_{\rm eq}}{M}\right)^{2/3}. \tag{4.216}$$

Substituting (4.216) in (4.215), we get

$$\delta_M(t) = \delta_M(t_{\rm enter}) \left(\frac{\lambda_{\rm eq}}{\lambda}\right)^2 \left(\frac{a}{a_{\rm eq}}\right) = \delta_M(t_{\rm enter}) \left(\frac{M_{\rm eq}}{M}\right)^{2/3} \left(\frac{a}{a_{\rm eq}}\right). \tag{4.217}$$

Comparing (4.213) and (4.217) we see that the mode which enters the Hubble radius after $t_{\rm eq}$ has its amplitude decreased by a factor $M^{-2/3}$, relative to its original value.

Finally, consider the modes with $\lambda > \lambda_H$ which are still outside the Hubble radius at t and will enter the Hubble radius at some *future* time $t_{\rm enter} > t$. During the time interval $(t, t_{\rm enter})$, they will grow by a factor $(a_{\rm enter}/a)$. Thus

$$\delta_\lambda(t_{\rm enter}) = \delta_\lambda(t) \left(\frac{a_{\rm enter}}{a}\right) \tag{4.218}$$

or

$$\delta_\lambda(t) = \delta_\lambda(t_{\rm enter}) \left(\frac{a}{a_{\rm enter}}\right) = \delta_M(t_{\rm enter}) \left(\frac{M_{\rm eq}}{M}\right)^{2/3} \left(\frac{a}{a_{\rm eq}}\right) \quad (\lambda > \lambda_H). \tag{4.219}$$

(The last equality follows from the previous analysis.) Thus the behaviour of the modes is the same for the cases $\lambda_{\rm eq} < \lambda < \lambda_H$ and $\lambda_H < \lambda$; i.e. for all $\lambda > \lambda_{\rm eq}$. Combining all these pieces of information, we can state the final result as follows:

$$\delta_\lambda(t) = \begin{cases} 0 & (\lambda < \lambda_{\rm FS}) \\ \delta_\lambda(t_{\rm enter})(a/a_{\rm eq}) & (\lambda_{\rm FS} < \lambda < \lambda_{\rm eq}) \\ \delta_\lambda(t_{\rm enter})(a/a_{\rm eq})(\lambda_{\rm eq}/\lambda)^2 & (\lambda_{\rm eq} < \lambda) \end{cases}$$

or, equivalently

$$\delta_M(t) = \begin{cases} 0 & (M < M_{\rm FS}) \\ \delta_M(t_{\rm enter})(a/a_{\rm eq}) & (M_{\rm FS} < M < M_{\rm eq}) \\ \delta_M(t_{\rm enter})(a/a_{\rm eq})(M_{\rm eq}/M)^{2/3} & (M_{\rm eq} < M) \end{cases}. \tag{4.220}$$

Thus the amplitude at late times is completely fixed by the amplitude of the modes when they enter the Hubble radius. Of course, $\delta(t_{\rm enter})$ can be related to $\delta(t_i)$ for some $t_i < t_{\rm enter}$; but it is much more convenient to use $\delta_\lambda(t_{\rm enter})$ to characterize the fluctuations.

The specification of $\delta_\lambda(t_{\rm enter})$, or equivalently, the specification of $\delta_\lambda(t_i)$ at some t_i, is a fundamental, unsolved problem in cosmology. Any complete theory for structure formation must specify this function based on some physical considerations. In the absence of such a theory, we will have to make some reasonable assumption for this quantity and compare the results with observations.

Notice that the symbol $\delta_\lambda(t_{\rm enter})$ actually stands for the function $\delta(\lambda, t)$ evaluated at $t = t_{\rm enter}\ (\lambda)$. Thus $\delta_\lambda(t_{\rm enter}) \equiv \delta(\lambda, t_{\rm enter}(\lambda)) = F(\lambda)$, some function of λ. Since the range of wavelengths which are of astrophysical interest is not too large, we may attempt to approximate this function by a power law. Hence, we take

$$\delta_\lambda(t_{\rm enter}) = A\lambda^\alpha \propto k^{-\alpha} \propto M^{\alpha/3}. \tag{4.221}$$

It follows that

$$\delta_\lambda(t) = \delta_M(t) \propto \begin{cases} 0 & (\lambda < \lambda_{\rm FS};\, M < M_{\rm FS}) \\ \lambda^\alpha (a/a_{\rm eq}) \propto M^{\alpha/3}(a/a_{\rm eq}) & \\ & (\lambda_{\rm FS} < \lambda < \lambda_{\rm eq};\, M_{\rm FS} < M < M_{\rm eq}) \\ \lambda^{\alpha-2}(a/a_{\rm eq}) \propto M^{\alpha/3-2/3}(a/a_{\rm eq}) & (\lambda_{\rm eq} < \lambda;\, M_{\rm eq} < M) \end{cases}. \tag{4.222}$$

The total density contrast at any spatial location will be obtained by a superposition of modes with different wavelengths:

$$\delta(\mathbf{x}, t) = \int \frac{d^3\mathbf{k}}{(2\pi)^3} \delta_{\mathbf{k}}(t) \exp(i\mathbf{k}\cdot\mathbf{x}). \tag{4.223}$$

The strength of such a perturbation can be measured by the value of $|\delta(\mathbf{x}, t)|^2$. From (4.223), it follows that modes in the range $(\mathbf{k}, \mathbf{k} + d^3\mathbf{k})$ contribute an amount proportional to $d^3\mathbf{k}|\delta_k|^2$ to $|\delta(\mathbf{x}, t)|^2$. Writing the contribution as $d^3\mathbf{k}|\delta_k|^2 \simeq k^2\, dk|\delta_k|^2 \simeq d(\ln k)\ (k^3|\delta_k|^2)$ we see that each logarithmic interval contributes an amount $(k^3|\delta_k|^2)$ to $|\delta(\mathbf{x}, t)|^2$. This quantity is given by:

$$k^3|\delta_k(t)|^2 \propto M^{-1}|\delta_M(t)|^2 \propto \begin{cases} 0 & (M < M_{\rm FS}) \\ M^{2\alpha/3-1}(a/a_{\rm eq})^2 & (M_{\rm FS} < M < M_{\rm eq}) \\ M^{2\alpha/3-7/3}(a/a_{\rm eq})^2 & (M_{\rm eq} < M) \end{cases}. \tag{4.224}$$

(We shall see in the next chapter that this combination is physically very important.) At the time of entering the Hubble radius, this quantity has the dependence

$$(k^3|\delta_k|^2)_{t=t_{\rm enter}} \propto k^{3-2\alpha} \propto M^{2\alpha/3-1}. \tag{4.225}$$

This expression shows that the value $\alpha = (3/2)$ has a special significance. If $\alpha > (3/2)$, then $k^3|\delta_k|^2$ will have more amplitude at large M (at the time of entering the Hubble radius); if $\alpha < (3/2)$, then most of the power will be concentrated on small scales. For the special value of $\alpha = (3/2)$, neither small nor large scales dominate[9]. Such a spectrum (called a scale-invariant spectrum) is predicted by some models of the early universe. If

$\alpha = (3/2)$, then at any fixed a,

$$k^3|\delta_k(t)|^2 = \begin{cases} 0 & (M < M_{\rm FS}) \\ (\text{constant}) & (M_{\rm FS} < M < M_{\rm eq}) \\ M^{-(4/3)} & (M_{\rm eq} < M) \end{cases} . \quad (4.226)$$

Most of the power is in the region $M_{\rm FS} < M < M_{\rm eq}$.

It follows that the actual shape of the spectrum depends crucially on the ratio

$$\frac{M_{\rm FS}}{M_{\rm eq}} = 0.05(\Omega h^2)^4 \left(\frac{m}{100\,{\rm eV}}\right)^{-4} . \quad (4.227)$$

If hot relics, like neutrinos with $m \approx 30\,{\rm eV}$, constitute the dark matter then $(M_{\rm FS}/M_{\rm eq}) \approx 4(\Omega h^2)^4$ which is around the range of unity. Thus the spectrum will have a relatively sharp peak around $M_{\rm FS}$. If, on the other hand, the dark matter particle is a heavier (say 1 MeV or so) cold relic, then $M_{\rm FS} \ll M_{\rm eq}$, and the spectrum will be relatively flat between $M_{\rm FS}$ and $M_{\rm eq}$ (see figure 4.4).

In this context, it is important to point out an extra complication: The flatness of the spectrum between $M_{\rm FS}$ and $M_{\rm eq}$ is a direct consequence of our assumption that the modes which enter the Hubble radius before $t_{\rm eq}$ start growing only after $t_{\rm eq}$. As we saw in section 4.5(2), this result is not strictly true; there is a small, logarithmic growth in the interval $t_{\rm enter} < t < t_{\rm eq}$. Owing to this reason the spectrum will not be completely flat for $M_{\rm FS} < M < M_{\rm eq}$ but will be sloping gently downwards. We saw in section 4.5 that modes with $\lambda \ll \lambda_{\rm eq}$ can grow by a factor $G(\lambda) \propto 5\ln(\lambda_{\rm eq}/\lambda) \propto (5/3)\ln(M_{\rm eq}/M)$ between entering the Hubble radius and the onset of matter domination. Therefore, a more precise statement will be

$$k^3|\delta_k|^2 = \begin{cases} 0 & (M < M_{\rm FS}) \\ G^2(M) & (M_{\rm FS} < M < M_{\rm eq}) \\ M^{-4/3} & (M_{\rm eq} < M) \end{cases} \quad (4.228)$$

where $G(M) \propto \ln(M_{\rm eq}/M)$ for $M \ll M_{\rm eq}$ and $G(M) \simeq 1$ for $M \lesssim M_{\rm eq}$. The exact form of this function can be determined from the matching conditions discussed in section 4.6; $G(M)$ will be a gently decreasing function of M. Thus the cold dark matter spectrum will have: (i) maximum power around $M_{\rm FS}$ (ii) a slow, gently sloping spectrum from $M_{\rm FS}$ to $M_{\rm eq}$ and (iii) a steep $M^{-2/3}$ fall after $M_{\rm eq}$.

The exact form of the spectrum, for both hot and cold dark matter scenarios, can be found by numerical integration of the relevant equations.

Such calculations show that the final spectrum can be well approximated by the following functions[10]: For hot dark matter,

$$|\delta_k|^2 = Ak^{(4-2\alpha)} \exp\left[-4.61(k/k_{\rm FS})^{3/2}\right], \tag{4.229}$$

where $k_{\rm FS} = 0.16\,{\rm Mpc}^{-1}\,(m_\nu/30\,{\rm eV})$ is the free streaming scale and A is an unknown amplitude. For cold dark matter

$$|\delta_k|^2 = \frac{Ak^{(4-2\alpha)}}{(1+Bk+Ck^{3/2}+Dk^2)^2}, \tag{4.230}$$

with $B = 1.7(\Omega h^2)^{-1}\,{\rm Mpc}$, $C = 9\,{\rm Mpc}^{3/2}\,(\Omega h^2)^{-3/2}$ and $D = 1\,{\rm Mpc}^2\,(\Omega h^2)^{-2}$; A is, again, the unknown amplitude. It is easy to verify that these functions have the correct form for small and large k.

We have parameterized the spectrum by the form of $\delta(k,t)$ at $t = t_{\rm enter}(k)$. One can also specify the same function by the power law $|\delta(k,t)|^2 \propto k^n$ *at a given time* t. It is easy to verify that the index n is related to α by $n = (4-2\alpha)$. The scale-invariant spectrum $\alpha = (3/2)$

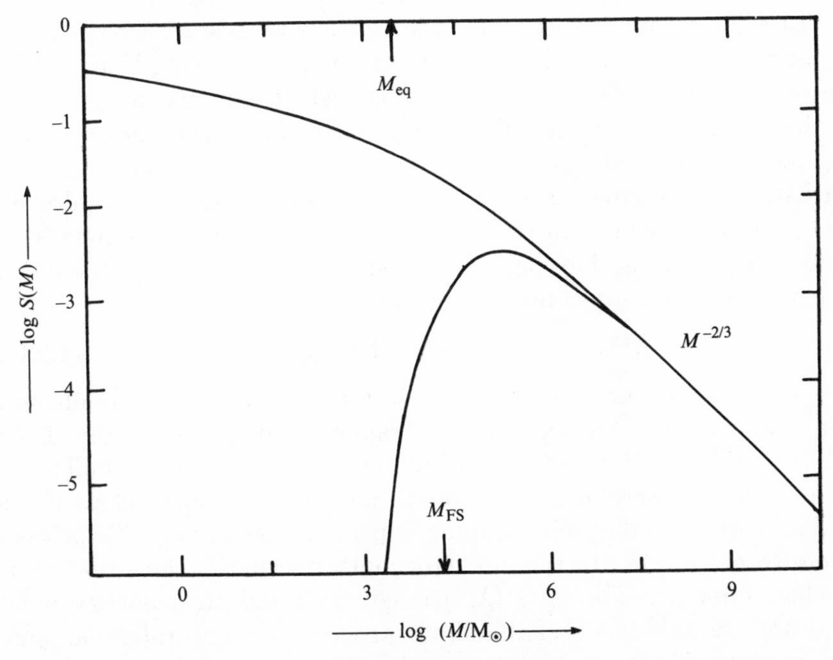

Fig. 4.4. The power in the perturbation spectrum $(k^{3/2}\delta_k)$ is plotted against the mass contained in the perturbation. The initial spectrum is assumed to be scale invariant. In scenarios with hot dark matter, the spectrum is peaked at $M_{\rm FS}$, while in scenarios with cold dark matter, there is a relatively flat, gently sloping-down portion, between $M_{\rm FS}$ and $M_{\rm eq}$. The power falls as $M^{-2/3}$ at large scales.

corresponds to the value $n = 1$. Note that the final processed spectrum is not a pure power law even if we start with a pure power law at some (very early) epoch.

As time goes on, the amplitudes grow in proportion to $a(t)$. Once δ becomes of order unity at some wavelength, the linear perturbation theory fails around that wavelength. For hot dark matter, the scale with mass $M \approx M_{\rm FS}$ reaches nonlinearity first. Thus the first structures to form will have masses around $10^{14}\,{\rm M}_\odot$ or so. Further evolution, involving fragmentation of these structures will generate power at smaller scales. For cold dark matter, the smallest scales near $M \approx M_{\rm FS}$ reach nonlinearity first; however, since the spectrum is relatively flat for $M_{\rm FS} < M < M_{\rm eq}$, there is considerable interference between the various scales. Larger and larger scales go nonlinear fairly quickly and disturb smaller scales. These nonlinear processes will be discussed in chapters 5 and 8.

The situation for baryonic perturbations is somewhat analogous. To begin with, the collisional damping wipes out all power in the scales smaller than the Silk mass ($M < M_S$), so that we only need to consider $M > M_S$. Since the baryons and photons are tightly coupled, even these perturbations do not grow until $t_{\rm dec}$; i.e., $\delta_M(t) = \delta_M(t_{\rm enter})$ for $t < t_{\rm dec}$. Just after decoupling, the baryonic perturbations start growing. Notice that the perturbations in the dark matter were growing from $t = t_{\rm eq}$ onwards. So, at $t = t_{\rm dec}$, the density contrast of dark matter is higher than that of baryons by the factor $(a_{\rm dec}/a_{\rm eq})$. From the known values of $a_{\rm dec}$ and $a_{\rm eq}$ we find that $(a_{\rm dec}/a_{\rm eq}) \cong 21(\Omega h^2)^{-1}$. So when baryonic perturbations start growing at $t \gtrsim t_{\rm dec}$, they are driven by the (already existing) inhomogeneities in the dark matter distribution. We have seen in section 4.5 that the baryonic density contrast equalizes with that of dark matter just after decoupling. Therefore

$$\delta_B(\lambda, t) \approx \delta_{\rm DM}(\lambda, t) \quad (\text{for } t \gtrsim t_{\rm dec}). \tag{4.231}$$

This analysis, of course, works for all λ at which $\delta_{\rm DM}(\lambda, t)$ is significant. In the cold dark matter scenario this range is fairly wide – say, from $10^5\,{\rm M}_\odot$ to $10^{14}\,{\rm M}_\odot$. Baryonic perturbations are now *generated* at all these scales even if they were originally absent; i.e., even at scales below $M_S \approx 10^{12}\,{\rm M}_\odot$ in which collisional damping wipes out the power, this process regenerates the power. In the hot dark matter scenario, the situation is somewhat different: The $\delta_{\rm DM}$ (λ, t) is concentrated on a narrow band around $M_{\rm FS} \approx 10^{14}\,{\rm M}_\odot$ and the above process only generates baryonic power in that region. At $M < M_S \approx 10^{12}\,{\rm M}_\odot$, all power is lost due to collisional damping. In the window ($10^{12}\,{\rm M}_\odot$ to $10^{13}\,{\rm M}_\odot$), there can be some small amount of baryonic power surviving from the early epochs; but it is not enhanced by the dark matter. Thus, in the hot dark matter scenario, most of the baryonic *and* dark matter power is concentrated around $M \approx M_{\rm FS} \approx 10^{14}\,{\rm M}_\odot$ for $t \gg t_{\rm dec}$. Further evolution of baryons is complicated by the fact that they can radiate and cool. This will be discussed in detail in chapter 8.

We emphasize that our analysis has only determined the *shape* of the power spectrum, given the value of α. The absolute value of $k^3|\delta_k|^2$ is not determined. Any theory which provides a physical mechanism for the generation of perturbations, will have to supply the values of α *as well as* the amplitude A. Alternatively, one may consider α and the amplitude to be two unknown parameters which have to be determined by comparison with observations.

Exercises

4.1 Some aspects of Newtonian approximation can be obtained by doing ordinary Newtonian gravity in a set of 'expanding' coordinates. Let us suppose that the separation between any two particles in the universe varies with time as $\mathbf{r} = a(t)\mathbf{x}$. The Lagrangian for the particle is taken to be $L = (1/2)m\dot{\mathbf{r}}^2 - m\Phi$ where $\nabla_r^2\Phi = 4\pi G\rho(\mathbf{x}, t)$.

(a) Show that the equations of motion imply the cosmological expansion law

$$\frac{d^2a}{dt^2} = -\frac{4\pi G}{3}\rho_b(t)a$$

if $\rho(\mathbf{x}, t) = \rho_b(t)$ is homogeneous.

(b) Show that, by a suitable canonical transformation, the Lagrangian can be written as

$$L = \frac{1}{2}ma^2\dot{x}^2 - m\phi$$

where

$$\nabla_x^2\phi = 4\pi Ga^2[\rho(\mathbf{x}, t) - \rho_b(t)].$$

(c) Show that the 'peculiar' velocity $\mathbf{v} = a\dot{\mathbf{x}}$ obeys the equation

$$\frac{d\mathbf{v}}{dt} + \mathbf{v}\frac{\dot{a}}{a} = -\frac{\nabla\phi}{a}.$$

(d) Show that the total Lagrangian for all the particles in the universe is

$$L = \sum_i \frac{1}{2}m_i a^2\dot{x}_i^2 - U,$$

$$U = -\frac{Ga^2}{2}\int d^3\mathbf{x}_1\, d^3\mathbf{x}_2 \frac{(\rho_1 - \rho_b)(\rho_2 - \rho_b)}{|\mathbf{x}_1 - \mathbf{x}_2|}.$$

What is the acceleration of the jth particle due to all other particles? Is it always finite?

4.2 Define the total mass M, gravitational energy U (for unit mass) and the kinetic energy K (for unit mass) by the relations:

$$M = \sum_j m_j, \quad MK = \sum_j \frac{1}{2} \frac{p_j^2}{m_j a^2},$$

$$MU = -\frac{1}{2} G a^2 \int d^3\mathbf{x}_1\, d^3\mathbf{x}_2 \frac{[\rho(\mathbf{x}_1 t) - \rho_b][\rho(\mathbf{x}_2, t) - \rho_b]}{|\mathbf{x}_1 - \mathbf{x}_2|}.$$

(a) Prove the 'cosmic virial theorem':

$$\frac{dK}{dt} + \frac{dU}{dt} + \frac{\dot{a}}{a}(2K + U) = 0.$$

This shows that the energy $(K+U)$ is conserved in an expanding universe for systems with $2K + U = 0$.
(b) Show that, in an expanding universe,

$$\frac{d}{dt}[a(K + U)] < 0.$$

So that, if at some t_0, $K \ll |U|$ and $U < 0$, then at all times $K < (-U)$.
(c) Show that $(d/dt)\ [a^2(K + U/2)] = -(a^2/2)\ (dU/dt)$. Argue why it is reasonable to assume $(dU/dt) < 0$. What bound does this imply on K?

4.3 The autocorrelation function for density contrast is defined to be

$$\xi(\mathbf{y}) = \int d^3\mathbf{x}\, \delta(\mathbf{x})\delta(\mathbf{x} + \mathbf{y})$$

where the integral is over a large volume in space. (We have suppressed the explicit time dependence.)
(a) Show that the quantity U, defined in the last problem, can be written as

$$U = -\frac{1}{2} G a^2 \rho_b \int d^3x \frac{\xi(x)}{x}.$$

(b) Prove that the peculiar velocity field, derived in the linear perturbation theory, has the mean-square value

$$< v^2 >= (Haf)^2 \int_0^\infty x\, dx\, \xi(x)$$

where $f = (a\, d\delta/\delta\, da)$. Thus conclude that, in the linear theory, $K = (2/3)(f^2/\Omega)|U|$.

4.4 Let the spatial Fourier transform of the velocity field $\mathbf{v}(\mathbf{x}, t)$ be $\mathbf{v}(\mathbf{k}, t)$. We can write $\mathbf{v}(\mathbf{k}, t) = \mathbf{v_1} + \mathbf{v_2}$ where $\mathbf{v_1} \cdot \mathbf{k} = 0$ and $\mathbf{v_2} \times \mathbf{k} = \mathbf{0}$. Show that $|\mathbf{v_1}| \propto a^{-1}$ in the linear theory. Because of this decay, we can ignore $\mathbf{v_1}$ and express $\mathbf{v}$ as a gradient.

4.5 We saw in chapter 2 that the evolution of the Friedmann model can be studied in terms of concentric shells of matter. Let the shell with radius $R(t)$ have mass M. Perturb the radius of the shell to $R(t) + L(t)$. For small L, show that

$$\frac{d^2L}{dt^2} = -\frac{2GM}{R^3}L = -\frac{8\pi G}{3}\rho(t)L; \quad \rho = \frac{3}{4\pi}\frac{M}{R^3},$$
$$\delta = \frac{\delta\rho}{\rho} = -3\frac{L}{R}.$$

Hence derive the perturbation equation for $\delta\rho$:

$$\frac{d^2}{dt^2}(\delta\rho)^{-1/3} = \frac{8\pi G}{3}(\delta\rho)^{2/3}.$$

Finally, show that this equation is identical to the equation for δ derived in the text, when the pressure term is neglected.

4.6 The solution for $\delta(z)$ obtained in the text for the matter dominated case is fairly messy. Plot this function for $\Omega = 0.01$, 0.1, and 0.9. Also plot the simpler function

$$F(z) = \frac{1 + (3/2)\Omega}{1 + (3/2)\Omega + (5/2)\Omega z}$$

for the same values. Convince yourself that this function is a good approximation.

4.7 For a collisional system (like an ordinary gas), the density perturbation $\delta\rho$, pressure perturbation δp and the velocity dispersion v^2 are related by $\delta p = v^2\delta\rho$. The Jeans length is defined in terms of this v^2. The purpose of this exercise is to show that similar gravitational instability occurs in the collisionless systems as well.

(a) For simplicity, consider the system in the Newtonian context. The perturbation in the collisionless dark matter, for example, is described by the equations

$$\frac{\partial f}{\partial t} + \mathbf{v}\cdot\frac{\partial f}{\partial \mathbf{x}} - \nabla\phi\cdot\frac{\partial f}{\partial \mathbf{v}} = 0; \quad \nabla^2\phi = 4\pi G\delta\rho,$$

$$f(\mathbf{x}, \mathbf{v}, t) = f_0(|\mathbf{v}|) + \delta f(\mathbf{x}, \mathbf{v}, t); \quad \delta\rho = m\int \delta f(\mathbf{x}, \mathbf{v}, t)\, d^3\mathbf{v}$$

where we have assumed the unperturbed background to be homogeneous with zero gravitational potential. Fourier transform $\delta\rho$, δf and ϕ in x and t using the modes $\exp(ikx + wt)$ for a wave propagating along the x-axis. Thus real w will indicate an instability. Show that

$$\delta f_{\mathbf{k}w}(v) = -\frac{4\pi G\delta\rho_{kw}}{k^2}\frac{2ikv_x}{w + ikv_x}\left(\frac{df_0}{dv^2}\right).$$

(b) Integrate $\delta f_{kw}(v)$ over v to obtain the dispersion relation between w and k:

$$-\frac{8\pi Gm}{k^2}\int d^3\mathbf{v}\left(\frac{k^2v_x^2 + ikv_xw}{k^2v_x^2 + w^2}\right)\left(\frac{df_0}{dv^2}\right) = 1.$$

Show that, in the long wavelength limit ($k \to 0$), w is real and is given by $w^2 = 4\pi G\rho$. Interpret this instability.

(c) Show that the critical wavelength at which this instability is triggered (corresponding to $w = 0$) is given by

$$k_J^2 = 4\pi G\rho < v^{-2} >$$

where

$$< \frac{1}{v^2} >= \frac{\int f_0 v^{-2}\, d^3\mathbf{v}}{\int f_0\, d^3\mathbf{v}} = \frac{\int_0^\infty f_0(v)\, dv}{\int_0^\infty f_0(v) v^2\, dv}.$$

Interpret this result.

4.8 The argument based on two evolving universes (given in the text) to determine the growth law for δ can be made more precise in the following way: Consider the radiation dominated phase, in which a spherical shell of radius R expands as

$$\frac{1}{2}\left(\frac{dR}{dt}\right)^2 = \frac{A}{R^2}; \quad A = \text{constant}.$$

(a) Argue that the evolution of a perturbed sphere can be described by an equation

$$\frac{1}{2}\left(\frac{dR'}{dt}\right)^2 = \frac{A}{(R')^2} + \epsilon; \qquad \epsilon \ll 1.$$

(b) Integrate this equation and show that, to leading order in ϵ, we have

$$\begin{aligned} R' &= (8A)^{1/4} t^{1/2} \left[1 + \frac{\epsilon}{16\sqrt{2}A^{3/2} t}(R')^4\right] \\ &\cong (8A)^{1/4} t^{1/2} \left[1 + \frac{\epsilon t}{(8A)^{1/2}}\right]. \end{aligned}$$

(c) Prove that this leads to the density contrast

$$\delta = \frac{\rho' - \rho}{\rho} = -\frac{4\epsilon}{(8A)^{1/2}} t \propto a^2.$$

4.9 The damping of the perturbations in the photon–baryon system (discussed in the text) can be derived more rigorously in the following manner: Consider the distribution function $f_t(x^\alpha, p^\beta)$ for the photons. Using t, x^j, $|\mathbf{p}| = \epsilon$ and the direction cosines γ^a of the vector $\mathbf{p}$ as the independent variables, we can write the evolution equation for f_t as

$$\frac{\partial f_t}{\partial t} + \frac{\partial f_t}{\partial x^a}\dot{x}^a + \frac{\partial f_t}{\partial \epsilon}\dot{\epsilon} + \frac{\partial f_t}{\partial \gamma^a}\dot{\gamma}^a = c_1 - c_2$$

where c_1 is the amount of photons scattered into the beam of radiation travelling in a particular direction and c_2 is the amount of photons scattered out of this beam.

(a) Separate f_t as $(f_0 + f)$ where f_0 is the homogeneous, isotropic background and f is a perturbation. Show that in the *rest frame* of the matter, f changes between two infinitesimally separated points, by the amount

$$df' = \delta t' \sigma n_e (g'(\epsilon') - f')$$

where the prime denotes quantities measured in the rest frame of matter, ϵ' is the energy of the photon and

$$g'(\epsilon') = \int \frac{d\Omega}{4\pi} f'(\epsilon', \gamma^{a\prime}).$$

(b) Prove that the quantities in the primed and unprimed coordinates are related by $\delta t' = (\epsilon'/\epsilon)\delta t$ and $\epsilon' \cong \epsilon(1 - \gamma_a v^a)$ to first order in v^a.
(c) Argue that, to the linear order in perturbation, $(\epsilon'/\epsilon)\ (g' - f') \approx (g' - f')$ and $(\partial f/\partial \gamma^a)\ (\dot{\gamma}^a) \approx 0$.
(d) Using the above facts, derive the equation

$$\frac{\partial I}{\partial t} + \frac{\gamma^a}{a}\frac{\partial I}{\partial x^\alpha} \cong \sigma n_e (\delta_R + 4\gamma_a v^a - I)$$

where the 'brightness function' I is defined through the relation:

$$\int dp\, p^3 f_t = \frac{\rho_b}{4\pi}\left[1 + I(\gamma^a, \mathbf{x}, t)\right]$$

and we have ignored the effects due to the perturbed gravitational field. Interpret the origin of each term in this equation.
(e) Fourier transform the variables, taking $\mathbf{k}$ along the z-axis, to obtain

$$\dot{I} + \frac{ik\mu}{a} I = \sigma_T n_e \left[\delta_R + 4\mu v - I\right]; \quad \mu = \hat{k}^\alpha \gamma_\alpha .$$

Also show that, to the same order of approximation, the equation of motion for matter is

$$\rho_m \left(\dot{v} + \frac{\dot{a}}{a} v\right) = \sigma_T n_e \rho_\gamma \left(P - \frac{4}{3} v\right); \quad P \equiv \frac{1}{2}\int_{-1}^{+1} d\mu\, I$$

and

$$\dot{\delta}_m = -\frac{ik}{a} v.$$

(f) Since $t_c \equiv (\sigma_T n_e)^{-1}$ is much smaller than the expansion timescale, these equations can be solved by a systematic expansion in t_c. Writing

$$I = \delta_R + 4\mu v + t_c \left(\dot{I} + \frac{ik\mu}{a} I\right)$$

and iterating systematically, show that to first order in t_c, we have the equations:

$$\dot{\delta}_R \cong \frac{-ik}{a}\left[\frac{4}{3} v - t_c \left(\frac{4}{3}\dot{v} + \frac{ik}{3a}\delta_R\right)\right],$$

$$\rho_m \dot{v} \cong \rho_\gamma \left[-\frac{4}{3}\dot{v} - \frac{ik}{3a}\delta_R + t_c \left(\frac{4}{3}\ddot{v} + \frac{2}{3}\frac{ik}{a}\dot{\delta}_R - \frac{4}{5}\frac{k^2}{a^2} v\right)\right].$$

These equations admit solutions with the time dependence $\exp(-\Gamma t)$ where

$$\Gamma = \frac{k^2 t_c}{6a^2}\left(1 - \frac{6}{5b} + \frac{1}{b^2}\right) \pm \frac{ik}{a\sqrt{3b}}; \quad b = 1 + \frac{3}{4}\frac{\rho_m}{\rho_\gamma}.$$

Demonstrate that the oscillations correspond to the acoustic vibrations with the correct adiabatic sound speed of the photon–baryon gas while the damping defines a critical wavelength λ_D where

$$\lambda_D^2 = \frac{4\pi^2}{6} t_c t.$$

Compare this result with the one derived in the text.

(g) If the viscous drag is completely ignored, we get the *zeroth* order approximation. Show that, in this limit

$$\delta_R = \frac{4}{3}\delta_m.$$

Thus the coupling between radiation and matter will preserve the adiabatic nature of baryonic perturbations.

4.10 The damping of perturbations due to 'free streaming' of collisionless particles can be viewed in the following manner:

(a) Let $f(\mathbf{x}, \mathbf{p}, t)$ be a perturbation in the wimp distribution function, with $f_k(\mathbf{p}, t)$ being the spatial Fourier transform. Show that a wimp at $(\mathbf{x}, \mathbf{p})$ at time t must have come from $(\mathbf{x}_0, \mathbf{p}_0)$ at time t_0 where

$$\mathbf{p}_0 a(t_0) = \mathbf{p}a(t); \quad \mathbf{x}_0 = \mathbf{x} - \int_{t_0}^{t} \frac{dt'}{a(t')} \frac{\mathbf{p}a(t)/a(t')}{\sqrt{p^2 a^2(t)/a^2(t') + m^2}}.$$

Hence show that

$$f_{\mathbf{k}}(\mathbf{p}, t) = f_{\mathbf{k}}(\mathbf{p}a(t)/a(t_0)) \times \exp\left[-i\mathbf{k}\cdot\hat{\mathbf{p}} \int_{a_0}^{a} \frac{dz}{z^2 H(z)[1 + m^2 z^2/p^2 a^2(t))]^{1/2}}\right].$$

(b) Assume that the perturbation in the distribution function at t_0 is independent of $\mathbf{p}$. Prove that the density contrast is

$$\delta(\mathbf{k}, t) = \frac{f_k(t_0)}{3\zeta(3)} \int_0^\infty \frac{dy\, y^2}{e^y + 1} \frac{\sin 2\phi(y, t)}{\phi(y, t)}$$

where,

$$\phi(y,t) \cong \begin{cases} (ka_d T_d t_0/ma_0^2) y \left\{sh^{-1}\left[ma/a_D T_D y\right] - sh^{-1}\left[ma/a_D T_D y\right]\right\} & (t < t_{nr} < t_{eq}) \\ (ka_D T_D t_0/ma_0^2) y \ln\left[2ma/a_D T_D y\right] & (t_{nr} < t < t_{eq}). \\ (3ka_D T_D t_0/2ma_0^2) y \left[1 - (1+\Omega z)^{1/2}(1+\Omega z_0)^{-1/2}\right] & (t_{nr} < t_{eq} < t) \end{cases}$$

(c) Evaluate the region over which ϕ makes a significant contribution to $\delta(\mathbf{k}, t)$ in each case and identify the free-streaming scales at each epoch. Compare the result with that given in the text.

4.11 As a simple example of the gauge ambiguities which arise due to the choice of coordinates, consider the following situation: Suppose we use the 'Newtonian' coordinate system, in which the metric is

$$ds^2 = (1 - 2\phi)\, dT^2 - (1 + 2\phi)(dX^2 + dY^2 + dZ^2)$$

with $\phi \ll 1$. Transform this metric to the comoving coordinate system:

$$ds^2 = dt^2 - [\delta_{ik} + h_{ik}(\mathbf{x}, t)]a^2(t)(dx^2 + dy^2 + dz^2)$$

where $\mathbf{X} = \mathbf{x}a(t)$. (How are t and T related?) Show that h_{ik} will have a term of the form $(\phi_{,ik}t^2)$ which arises from the coordinate transformation. Does this represent growth of fluctuations?

4.12 Consider the perturbed Friedmann metric, written in the form:

$$ds^2 = a^2(\eta)\left\{d\eta^2 - \left[\delta_{ik} + a^{-2}h_{ik}\right] dx^i\, dx^k\right\}.$$

Raising and lowering of indices will be done with the background metric so that $h^i_k = a^{-2}h_{ik}$ etc. Show that linearized Einstein equations for $\{0j, 00, jk\}$ components reduce to

$$u^i = \frac{1}{16\pi G}(\dot{h}^{,i} - \dot{h}^{i,j}_j)a^{-2}(\rho + p)^{-1},$$

$$\frac{\delta\rho}{\rho} = \frac{1}{16\pi G\rho a^2}\left(h^{i,j}_{j,i} - h^{,i}_{,i} + 2\frac{\dot{a}}{a}\dot{h}\right)$$

and

$$\left(h^{i,k}_{j,i} + h^{k,i}_{i,j} - h^{,k}_{,j} - h^{k,i}_{j,i}\right) + \ddot{h}^k_j + 2\frac{\dot{a}}{a}\dot{h}^k_j = 0$$

where $\dot{a} = (\partial a/\partial\eta)$ etc., $h = h^i_i$ and u^α is the four-velocity of matter, with $u^0 = a^{-1}$. Also show that the trace of the Einstein equations gives

$$\frac{1}{2}\left(h^{i,j}_{j,i} - h^{,i}_{,i}\right)[1 + 3(dp/d\rho)] + \ddot{h} + \dot{h}\frac{\dot{a}}{a}\left(2 + 3\frac{dp}{d\rho}\right) = 0.$$

When do these equations reduce to the Newtonian equations?

5
Statistical properties

5.1 Introduction

The linear perturbation theory developed in the last chapter is a powerful tool and can be used to make definite predictions about the large scale structures. However, in order to compare these predictions with actual observations, it is necessary to understand certain statistical features of the perturbations. These statistical aspects of the problem of structure formation are discussed in this chapter[1]. Sections 5.2 and 5.3 introduce the basic notion of the probability functional which is needed for characterizing the density contrast. The concept of filter functions is discussed in section 5.4 and several important properties of the two-point correlation function are studied in sections 5.5 and 5.6. The remaining sections discuss the applications of these concepts in some specific scenarios.

5.2 Probability functionals

The analysis in the previous chapter allows us to calculate the value of the density contrast $\delta_k(t)$ at some time t, given its value $\delta_k(t_0)$ at some initial time t_0, for each mode labelled by k. Since this analysis was based on linear theory, it is valid only if $\delta_k(t) \ll 1$. At a sufficiently early epoch in the past, this condition will be satisfied virtually for all values of k; however, as time goes on, the amplitudes increase (in proportion with $a(t)$ in the matter dominated phase) and soon δ_k will become comparable to unity for some range of values of k. It is usual to say that a mode labelled by k has become 'nonlinear' at some time t_k if $\delta_k(t_k)$ becomes unity at $t = t_k$. The detailed form of $t(k)$ – which contains the information as to which mode is becoming nonlinear at what time – depends on the spectrum $\delta_k(t)$.

We saw in the last chapter that if the universe is dominated by hot dark matter, then $\delta_k(t)$ has a fairly narrow peak at, say, $k \cong k_{\rm FS}$ with some width σ. This scale will go nonlinear at some time $t = t_{\rm nl}$. Since the peak is reasonably sharp, modes which are far from $k_{\rm FS}$ (more precisely, modes for which $| k - k_{\rm FS} | > \sigma$) will become nonlinear only much later. During this time, astrophysical processes due to the modes which have already become nonlinear will turn out to be significant.

If, on the other hand, the universe is dominated by cold dark matter, then $\delta_k(t)$ is almost flat and gently sloping downwards in the range $k_{\rm FS} < k < k_{\rm eq}$, where $k_{\rm FS} \ll k_{\rm eq}$. Since δ_k is maximum around $k \simeq k_{\rm FS}$, it is this scale which will go nonlinear first. However, since the spectrum is almost flat, the whole range of modes (from $k_{\rm FS}$ to $k_{\rm eq}$) will go nonlinear one after another in a rather continuous manner.

In both the scenarios, the amplitude δ_λ *decreases* with increasing λ for $\lambda > \lambda_{\rm eq}$. Therefore, *larger* wavelengths in the range $\lambda > \lambda_{\rm eq}$ will become nonlinear at *later* times. It follows that modes with sufficiently large λ are yet to become nonlinear in the present day universe. Thus, we may imagine the mass distribution in the present day universe to be made of two components: (i) Small scale, highly nonlinear structures which have arisen out of modes which had become nonlinear earlier and (ii) Large scale structures with $\delta < 1$ which are yet to become nonlinear.

The evolution of the modes in the linear regime is known from the discussion in the last chapter. The nonlinear evolution, leading to the formation of small scale structures is not so well understood; but – in principle – this can always be done, say, by evolving the system numerically in a computer. Thus, given an initial density perturbation $\delta(\mathbf{x}, t_0)$ at some t_0, we can compute – in principle – the final mass distribution, including the distribution of condensed objects like galaxies etc.

In comparing the result of such a theoretical calculation with the observed universe, there arises an interesting subtlety. It would be quite meaningless to hope that the theory will be able to produce the properties of *a particular* galaxy located at a particular coordinate $\mathbf{x}$; or, more generally, we do not expect the theory to predict the density contrast of matter $\delta(\mathbf{x})$ at any *specified* location $\mathbf{x}$. Such a prediction – even if the theoretical, nonlinear calculations could be carried out – will require an extremely fine-tuned initial density contrast $\delta(\mathbf{x}, t_i)$, which is not known. The question, therefore, arises as to how one can compare the theory with observations.

Such a comparison can be attempted in the following manner: Instead of predicting the exact mass distribution of our universe, we shall predict the average, statistical properties of the mass distribution. (This is the procedure adopted in statistical mechanics as well, again because of the non-availability of information about the initial conditions). In doing so, we can make use of the following fact: Even though our present universe is not strictly homogeneous (in the sense that $\nabla\rho \neq 0$), it does appear to be 'statistically homogeneous'. That is, it is possible to divide the universe into sufficiently large regions $\mathcal{R}_1, \mathcal{R}_2 ... \mathcal{R}_N$ (each of volume V, say) such that the mass distribution in each of the volumes is the same on the average and has identical statistical properties.

Let $\rho_1(\mathbf{x}), \rho_2(\mathbf{x}), ... \rho_N(\mathbf{x})$ be the actual density functions in each of these regions. We shall assume that each region is identical; say, a cube of size L. We can then set up identical coordinate systems in each region. The coordinate $\mathbf{x}$ will correspond to points with the same coordinates in each region. These functions, of course, will be different. However, we would like to interpret them as realizations of a single random variable $\rho(\mathbf{x})$, whose statistical properties can be determined.

These statistical properties can be specified, if the probability for any particular realization of $\rho(\mathbf{x})$ is given. This probability, denoted by $P[\rho(\mathbf{x})]$, is a 'functional' of $\rho(\mathbf{x})$. The concept of a 'functional' involves an elementary generalization of the notion of a function. A function $f(z)$ is a rule by which a unique number $f(z)$ is assigned to each value of z; similarly, the functional $P[\rho(\mathbf{x})]$ assigns a unique number to each function $\rho(\mathbf{x})$. As an example, consider the functional:

$$G[\rho(\mathbf{x})] = \exp\left[-\int d^3\mathbf{x}\, d^3\mathbf{y} \frac{\rho(\mathbf{x})\,\rho(\mathbf{y})}{|\mathbf{x}-\mathbf{y}|}\right]. \tag{5.1}$$

Given *any* function $\rho(\mathbf{x})$, one can – in principle – evaluate the value of G, which will be a pure number. This functional generalizes the notion of an ordinary Gaussian function $f(z) = \exp(-\sigma z^2)$.

Any well-behaved function $\rho(\mathbf{x})$ can be completely characterized by its Fourier transform $\rho(\mathbf{k})$. Therefore, any functional $G[\rho(\mathbf{x})]$ can be re-expressed in terms of $\rho(\mathbf{k})$. For example, the functional in (5.1) may be written as

$$\begin{aligned}\mathcal{G}[\rho(\mathbf{k})] &= \exp\left[-\int d^3\mathbf{x}\, d^3\mathbf{y} \int \frac{d^3\mathbf{k}}{(2\pi)^3}\frac{d^3\mathbf{q}}{(2\pi)^3}\frac{\rho(\mathbf{k})\rho^\star(\mathbf{q})}{|\mathbf{x}-\mathbf{y}|}\mathrm{e}^{i(\mathbf{k}\cdot\mathbf{x}-\mathbf{q}\cdot\mathbf{y})}\right] \\ &= \exp\left[-\int \frac{d^3\mathbf{k}}{2\pi^2}\frac{|\rho_k|^2}{k^2}\right].\end{aligned} \tag{5.2}$$

The notation is changed from G to $\mathcal{G}$ to emphasize the fact that G and $\mathcal{G}$ are different functionals of their respective arguments. We will not bother to do this when no confusion is likely to arise. Of course, the numerical value of $\mathcal{G}[\rho(\mathbf{k})]$ and $G[\rho(\mathbf{x})]$ will be the same if $\rho(\mathbf{k})$ and $\rho(\mathbf{x})$ are related by the Fourier transform.

The probability functional $P[\rho(\mathbf{x})]$ may also be thought of in a different manner which is physically more transparent. Let us suppose that the region of interest is divided into a large number (N) of small cells. Then the function $\rho(\mathbf{x})$ can be approximated by its value at the centre of each cell: $\mathbf{x}_1, \mathbf{x}_2, \ldots \mathbf{x}_N$. This description will be exact in the limit of $N \to \infty$, with the cell-size shrinking to zero. Thus we have approximated a random *function* $\rho(\mathbf{x})$ by a finite (though large) number of random variables $\{\rho_1, \rho_2, \cdots \rho_N\}$ where ρ_i corresponds to the density in the ith cell. The probability functional $P[\rho(\mathbf{x})]$ now becomes the joint probability distribution $P[\rho_1, \rho_2, \cdots \rho_N]$ for the simultaneous occurrence of the values $\{\rho_1, \rho_2, \cdots \rho_N\}$. The same approach can be followed in the Fourier space as well. Instead of considering the *functional* $P[\rho(\mathbf{k})]$, we may consider a *function* of N variables $P[\rho_1, \rho_2, \cdots \rho_N]$ where $\rho_a = \rho(k_a)$. (That is, we have introduced small cells in the $\mathbf{k}$-space centred at $\mathbf{k}_1, \mathbf{k}_2, \cdots \mathbf{k}_N$). Such a partial discretisation of $\mathbf{k}$-space does arise naturally if our region is bounded by (say) a large cubical box of size L, with $\rho(\mathbf{x})$ vanishing on the boundary of the box. Then the Fourier transform of $\rho(\mathbf{x})$ becomes a Fourier *series* of the form

$$\rho(\mathbf{x}) = \frac{1}{V} \sum_{\mathbf{n}} \rho_{\mathbf{n}} \exp(i\mathbf{k}_n \cdot \mathbf{x}), \tag{5.3}$$

where the allowed values of $\mathbf{k}_n$ are $[(2\pi n_x/L), (2\pi n_y/L), (2\pi n_z/L)]$ with (n_x, n_y, n_z) being integers. We now have an infinite number of discrete variables rather than one continuous function. Again, the probability functional can be expressed as a *function* of these variables: $P[\rho_1, \rho_2, \cdots]$.

One basic distinction between the representation in $\mathbf{x}$-space and $\mathbf{k}$-space should be noted. Since $\rho(\mathbf{x})$ is real, its Fourier transform $\rho(\mathbf{k})$ satisfies the constraint $\rho^\star(\mathbf{k}) = \rho(-\mathbf{k})$. Thus, $\rho(\mathbf{k})$ with $\mathbf{k}$ confined to the positive quadrant ($k_x \geq 0, k_y \geq 0, k_z \geq 0$) contains the full information about $\rho(\mathbf{x})$. Therefore, only the modes with positive $\mathbf{k}$-values are the independent degrees of freedom. In converting the sums over $\mathbf{k}_n$ to integrals over $\mathbf{k}$, this point should be borne in mind.

The discussion so far has been confined to a given instant in time. As time goes on, the probability functionals will change; this is to be expected because the statistical properties – like the mean density $< \rho(\mathbf{x}) >$ – do change with time. Thus, to be precise, we should write the probabilities as $P[\rho(\mathbf{x}); t]$ where t is the time at which we require the probability.

5.3 Gaussian probability functional

The quantity $P[\rho(\mathbf{x}); t]$ contains complete information about the statistical properties of the mass distribution. Its exact form can be determined only if the dynamical equations are solved; in the absence of such a solution, we can only hope to obtain some of its properties by general considerations.

To determine these properties it is more convenient to work with the dimensionless density contrast $\delta(\mathbf{x}) = [(\rho(\mathbf{x}) - \overline{\rho})/\overline{\rho}]$ where $\overline{\rho} =< \rho(\mathbf{x}) >$. This random variable, of course, has zero mean. Let us suppose that the space is discretized into cells and the probability is given as a function of a set of variables $\{\delta_1, \delta_2, \cdots \delta_N\}$ by $P = P[\delta_1, \delta_2, \cdots \delta_N; t]$. It would be very convenient if this probability can be expressed as the product of N individual distributions:

$$P\left[\delta_1, \delta_2, \cdots \delta_N; t\right] = \prod_{a=1}^{N} f_a(\delta_a; t). \tag{5.4}$$

The validity of such a decomposition would imply that each of the variables δ_a is evolving independently of the other variables. It is, therefore, clear that such a decomposition is impossible. Even if this form is postulated at some time $t = t_i$, the gravitational influence of particles on each other will destroy the mutual independence very quickly.

However, such a decomposition can be *approximately* preserved if we work in the **k**-space rather than in the **x**-space. Suppose we represent the probability as the product

$$P\left[\delta(\mathbf{k}); t\right] = P\left[\delta_1, \delta_2, \cdots; t\right] = \prod_{a} g_a(\delta_a; t), \tag{5.5}$$

where δ_a stands for the Fourier mode corresponding to the wavevector $\mathbf{k}_a$. For simplicity of notation, we will omit the subscript **k** and write

$$P\left[\delta(\mathbf{k}); t\right] = \prod_{\mathbf{k}} g_{\mathbf{k}}(\delta_{\mathbf{k}}; t), \tag{5.6}$$

with

$$\delta_{\mathbf{k}} = \int_V \delta(\mathbf{x}) \mathrm{e}^{-i\mathbf{k}\cdot\mathbf{x}}\, d^3\mathbf{x}. \tag{5.7}$$

It is understood that $\mathbf{k} = (2\pi/L)\mathbf{n}$ where $\mathbf{n}$ is a vector with integer components and $V = L^3$ is the normalization volume. This decomposition is preserved, at least during the linear regime, when each mode **k** evolves independently. Thus, we expect this decomposition to become invalid only in the nonlinear regime and that too (probably) *only* for those modes which have already become nonlinear.

Thus the probability functionals are much easier to handle in the **k**-space than in the **x**-space, because in the **k**-space each mode evolves independently at least during most of the evolution. The problem of finding the *functional* $P\left[\delta(\mathbf{x});\ t\right]$ reduces to the problem of determining the form of the *function* $g_{\mathbf{k}}(\delta_{\mathbf{k}}; t_i)$, at a sufficiently early time $t = t_i$. Once this form is known, $g_{\mathbf{k}}(\delta_{\mathbf{k}}; t)$ at any later time t can be found by linear evolution provided the mode labelled by k has not become nonlinear at this time t.

One of the simplest choices for $g_{\mathbf{k}}(\delta_{\mathbf{k}}; t_i)$ is a Gaussian. We shall see in chapter 10 that some models for the early universe predict such a form. This choice can also be motivated by the following suggestive argument, though it should not be considered as a valid proof: From (5.7), it is clear that $\delta_{\mathbf{k}}$ is obtained by adding up $\delta(\mathbf{x})$ at various spatial locations.

Further, the statistical description was invoked because we do not possess complete knowledge about $\delta(\mathbf{x})$. It is therefore (probably) reasonable to assume that $\delta(\mathbf{x})$ at different locations in space is uncorrelated (or only very weakly correlated) in the *very* early universe. Since $\delta_{\mathbf{k}}$ is a sum of large number of uncorrelated random variables, central limit theorem would imply that the probability distribution for $\delta_{\mathbf{k}}$ is a Gaussian. In what follows, we shall assume that $g_{\mathbf{k}}$ is a Gaussian.

It should be noted that, since $\delta(\mathbf{x})$ is real, $\delta_{\mathbf{k}}$ is complex but satisfies the constraint $\delta_{\mathbf{k}}^{\star} = \delta_{-\mathbf{k}}$. If we separate out $\delta_{\mathbf{k}}$ into real and imaginary parts as $\delta_{\mathbf{k}} = a_{\mathbf{k}} + ib_{\mathbf{k}}$, (with the constraints $a_{\mathbf{k}} = a_{-\mathbf{k}}$ and $b_{\mathbf{k}} = -b_{-\mathbf{k}}$), then the Gaussian probability distribution for $\delta_{\mathbf{k}}$ can be stated as

$$g_{\mathbf{k}}\left[a_{\mathbf{k}}, b_{\mathbf{k}}; t\right] da_{\mathbf{k}}\, db_{\mathbf{k}} = \frac{1}{(2\pi\mu_{\mathbf{k}}^2(t))} \exp\left(-\frac{a_{\mathbf{k}}^2 + b_{\mathbf{k}}^2}{2\mu_{\mathbf{k}}^2(t)}\right) da_{\mathbf{k}}\, db_{\mathbf{k}}, \tag{5.8}$$

where $\mu_{\mathbf{k}}^2(t)$ is the variance in $a_{\mathbf{k}}$ and $b_{\mathbf{k}}$. This single function $\mu_{\mathbf{k}}(t)$ characterizes the distribution. If the background space is homogeneous and isotropic, then $\mu_{\mathbf{k}}(t) = \mu(|\mathbf{k}|, t) = \mu_k(t)$ can only depend on the magnitude of $\mathbf{k}$.

Writing $\delta_{\mathbf{k}} = r_{\mathbf{k}} \exp(i\phi_{\mathbf{k}})$, one can express the same probability distribution as a probability distribution for $r_{\mathbf{k}}$ and $\phi_{\mathbf{k}}$:

$$\begin{aligned} g_{\mathbf{k}}(r_{\mathbf{k}},\ \phi_{\mathbf{k}};\ t)\, dr_{\mathbf{k}}\, d\phi_{\mathbf{k}} &= \frac{1}{2\pi\mu_k^2} \exp\left(-\frac{1}{2}\frac{r_{\mathbf{k}}^2}{\mu_k^2}\right) r_{\mathbf{k}}\, dr_{\mathbf{k}}\, d\phi_{\mathbf{k}} \\ &= \frac{2(r_{\mathbf{k}}\, dr_{\mathbf{k}})}{\sigma_k^2}\left(\frac{d\phi_{\mathbf{k}}}{2\pi}\right) \exp\left(-\frac{r_{\mathbf{k}}^2}{\sigma_k^2}\right); \qquad \sigma_k^2 = 2\mu_k^2. \end{aligned} \tag{5.9}$$

This shows that the $\phi_{\mathbf{k}}$ are distributed uniformly in the allowed range of $(0, 2\pi)$ while $r_{\mathbf{k}}^2$ obeys an exponential distribution with $< r_{\mathbf{k}}^2 >^{1/2} = \sigma_k = \sqrt{2}\mu_k$. In other words, the phases of the $\delta_{\mathbf{k}}$ are completely random in the interval $(0, 2\pi)$. Using this fact, we can characterize the Gaussian distribution in a more convenient way. Consider the average,

$$\begin{aligned} < \delta_{\mathbf{k}}\delta_{\mathbf{p}}^{\star} > &= \int \frac{d\phi_{\mathbf{k}}\, d\phi_{\mathbf{p}}}{4\pi^2} e^{i(\phi_{\mathbf{k}} - \phi_{\mathbf{p}})} \\ &\times \int \frac{(2r_{\mathbf{k}}\, dr_{\mathbf{k}})(2r_{\mathbf{p}}\, dr_{\mathbf{p}})}{\sigma_k^2\, \sigma_p^2} r_{\mathbf{k}}\, r_{\mathbf{p}} \exp\left[-\left(\frac{r_{\mathbf{k}}^2}{\sigma_k^2} + \frac{r_{\mathbf{p}}^2}{\sigma_p^2}\right)\right] \\ &= \eta_{\mathbf{kp}} < r_{\mathbf{k}}^2 > = \eta_{\mathbf{kp}}\sigma_{\mathbf{k}}^2, \end{aligned} \tag{5.10}$$

where $\eta_{\mathbf{kp}}$ is the Kronecker delta function. The Gaussian probability distribution we are considering can now be completely specified by its first two moments:

$$< \delta_{\mathbf{k}} > = 0; \quad < |\delta_{\mathbf{k}}|^2 > = \sigma_k^2; \quad < \delta_{\mathbf{k}}\delta_{\mathbf{p}} > = 0 \quad (\text{for } k \neq p). \tag{5.11}$$

The quantity σ_k^2 is usually called the power spectrum of the fluctuations.

The probability for a given configuration $\{\delta_{\mathbf{k}}\}$ is obtained by multiplying the probabilities for each independent mode; i.e., for modes with positive $\mathbf{k}$:

$$P\left[\{\delta_{\mathbf{k}}\},t\right] = \prod_{\mathbf{k}>0} g_{\mathbf{k}}(\delta_{\mathbf{k}},t) = N\exp - \sum_{\mathbf{k}>0} \frac{1}{2\mu_k^2(t)}(a_{\mathbf{k}}^2 + b_{\mathbf{k}}^2). \tag{5.12}$$

This can be rewritten with the sum ranging over *all* $\mathbf{k}$ by using the fact that $a_{-\mathbf{k}}^2 = a_{\mathbf{k}}^2$, $b_{-\mathbf{k}}^2 = (-b_{\mathbf{k}})^2 = b_{\mathbf{k}}^2$ and $\mu_{-\mathbf{k}}^2 = \mu_{\mathbf{k}}^2$. Then

$$\begin{aligned} P\left[\{\delta_{\mathbf{k}}\},t\right] &= N\exp -\frac{1}{2}\sum_{\text{all }\mathbf{k}} \frac{1}{(2\mu_k^2(t))}(a_{\mathbf{k}}^2 + b_{\mathbf{k}}^2) \\ &= N\exp -\frac{1}{2}\sum_{\mathbf{k}} \frac{|\delta_{\mathbf{k}}|^2}{\sigma_k^2(t)}. \end{aligned} \tag{5.13}$$

Having determined the form of the probability distribution in $\mathbf{k}$-space we can switch back to the $\mathbf{x}$-space by using the relation:

$$a_{\mathbf{k}}^2 + b_{\mathbf{k}}^2 = |\delta_{\mathbf{k}}|^2 = \int \delta(\mathbf{x})\delta(\mathbf{y})\mathrm{e}^{-i\mathbf{k}\cdot(\mathbf{x}-\mathbf{y})}\, d\mathbf{x}\, d\mathbf{y}. \tag{5.14}$$

Substituting this in (5.13), we get

$$\begin{aligned} P\left[\delta(\mathbf{x}),t\right] &= N\exp -\frac{1}{2}\sum_{\mathbf{k}}\int d\mathbf{x}\, d\mathbf{y}\; \delta(\mathbf{x})\delta(\mathbf{y})\,\frac{\mathrm{e}^{-i\mathbf{k}\cdot(\mathbf{x}-\mathbf{y})}}{\sigma_k^2} \\ &= N\exp\left[-\frac{1}{2}\int d\mathbf{x}\, d\mathbf{y}\; \delta(\mathbf{x})F(\mathbf{x}-\mathbf{y})\delta(\mathbf{y})\right], \end{aligned} \tag{5.15}$$

where

$$F(\mathbf{x}-\mathbf{y}) = \sum_{\mathbf{k}} \frac{\mathrm{e}^{-i\mathbf{k}\cdot(\mathbf{x}-\mathbf{y})}}{\sigma_k^2} = \frac{V}{(2\pi)^3}\int \frac{d^3\mathbf{k}}{\sigma_{\mathbf{k}}^2(t)}\exp[-i\mathbf{k}\cdot(\mathbf{x}-\mathbf{y})]. \tag{5.16}$$

In arriving at the last equality, we have taken the continuum limit of the Fourier series by replacing $\sum_{\mathbf{k}}$ by $\left[V/(2\pi)^3\right]$. The form of $P\left[\delta(\mathbf{x}),t\right]$ shows that $\delta(\mathbf{x})$ at different points in space are *not* independent; i.e., $P\left[\delta(\mathbf{x}),t\right]$ cannot be expressed as a product over probability functions $f_{\mathbf{x}}(\delta_{\mathbf{x}})$ at each point in space. In this respect, the $\mathbf{x}$-space and $\mathbf{k}$-space differ significantly.

The value of the density contrast $\delta(\mathbf{x})$ at any specified point $\mathbf{x}$ is also a random variable. We can, therefore, ask for the probability that this random variable has some specified value q. Since

$$\delta(\mathbf{x}) = \frac{1}{V}\sum \delta_{\mathbf{k}}\exp(i\mathbf{k}\cdot\mathbf{x}) \tag{5.17}$$

is the sum of several random variables, each of which is a Gaussian with zero mean, this sum will also be a Gaussian variable with zero mean.

Hence the probability distribution for q will be of the form

$$P[q] = \frac{1}{(2\pi\Delta^2)^{1/2}} \exp\left(-\frac{q^2}{2\Delta^2}\right), \tag{5.18}$$

where the dispersion $\Delta^2 =< q^2 >=< \delta^2(\mathbf{x}) >$ is yet to be determined. Though this dispersion can be computed from (5.15), it is much easier to use (5.11):

$$\begin{aligned} < \delta^2(\mathbf{x}) > &= \frac{1}{V^2} \sum_{\mathbf{k},\mathbf{p}} <\delta_\mathbf{k}\delta^\star_\mathbf{p}> \exp[i(\mathbf{k}-\mathbf{p})\cdot\mathbf{x}] = \frac{1}{V^2}\sum_\mathbf{k} \sigma_k^2 \\ &= \frac{V^{-1}}{(2\pi)^3}\int d^3\mathbf{k}\,\sigma_k^2 = V^{-1}\int_0^\infty \sigma_k^2 \frac{k^2\,dk}{2\pi^2} \equiv \int_0^\infty \Delta_k^2 \frac{dk}{k}, \end{aligned} \tag{5.19}$$

where

$$\Delta_k^2 \equiv \frac{V^{-1}}{2\pi^2} k^3 \sigma_k^2 \tag{5.20}$$

is a measure of the dispersion. In arriving at (5.19), we have also used the fact that $\sigma_k^2 = \sigma^2(|\mathbf{k}|)$ when the background is homogeneous and isotropic. The distribution of $q = \delta(\mathbf{x})$ shows that the density contrast at any location may fluctuate by the amount $(\pm\Delta)$, while larger fluctuations are relatively unlikely.

Because of the homogeneity, the distribution for the density contrast $\delta(\mathbf{x})$ in (5.18) is independent of the location $\mathbf{x}$. To obtain some information about the spatial distribution of $\delta(\mathbf{x})$, we can construct the two-point correlation function:

$$\begin{aligned} \xi(\mathbf{x}) \equiv < \delta(\mathbf{y}+\mathbf{x})\delta(\mathbf{y}) > &= \frac{1}{V^2}\sum_{\mathbf{k},\mathbf{p}} <\delta_\mathbf{k}\delta^\star_\mathbf{p}> \exp\left[i\mathbf{k}\cdot(\mathbf{x}+\mathbf{y}) - i\mathbf{p}\cdot\mathbf{y}\right] \\ &= \frac{1}{V^2}\sum_\mathbf{k} \sigma_k^2 \exp(i\mathbf{k}\cdot\mathbf{x}) = \int \frac{V^{-1}\,d^3\mathbf{k}}{(2\pi)^3}\sigma_k^2 e^{i\mathbf{k}\cdot\mathbf{x}}. \end{aligned} \tag{5.21}$$

The last equation shows that the correlation function $\xi(\mathbf{x})$ is the Fourier transform of the power spectrum σ_k^2. Inverting this relation, the power spectrum can be expressed as the Fourier transform of the correlation function:

$$V^{-1}\sigma_k^2 = \int \xi(\mathbf{x}) e^{-i\mathbf{k}\cdot\mathbf{x}}\, d^3\mathbf{x}. \tag{5.22}$$

Since σ_k^2 completely specifies the statistical properties, it follows that $\xi(\mathbf{x})$ also contains the same amount of information. In particular, from (5.19) and (5.21) it follows that $\Delta^2 = \xi(0)$. Thus $\xi(0)$ determines the distribution of $\delta(\mathbf{x})$; also note that $\xi(0) = \Delta^2$ must be positive.

Another simple property of $\xi(\mathbf{x})$ can be derived directly from its definition. Consider the integral

$$I = \int d^3\mathbf{x}\, \xi(\mathbf{x}) = \int d^3\mathbf{x} < \delta(\mathbf{y}+\mathbf{x})\delta(\mathbf{y}) > . \tag{5.23}$$

Since the processes of statistical averaging and spatial integration can be interchanged, this integral can be written as

$$I =< \delta(\mathbf{y}) \int d^3\mathbf{x}\, \delta(\mathbf{y}+\mathbf{x}) >= 0, \tag{5.24}$$

where we have used the fact that the integral of $\delta(\mathbf{x})$ over all space vanishes. Since $\xi(\mathbf{x}) = \xi(|\mathbf{x}|) \equiv \xi(x)$, in an isotropic universe,

$$I = \int_0^\infty 4\pi x^2 \xi(x)\, dx = 0 \tag{5.25}$$

which implies that $\xi(x)$ should change sign at some $x = x_0$.

A physical interpretation for the correlation function can be given along the following lines: Suppose $\delta(\mathbf{x})$ is such that there are regions with typical size L on which $\delta > 0$ and similar regions with $\delta < 0$. If $|\mathbf{x}| \ll L$ then $(\mathbf{x}+\mathbf{y})$ and $(\mathbf{y})$ in (5.21) will both lie in an overdense or underdense region most of the time; in either case, the product $\delta(\mathbf{x}+\mathbf{y})\delta(\mathbf{y})$ will be positive, making $\xi(\mathbf{x})$ positive. On the other hand, if $|\mathbf{x}| \gtrsim L$, then $(\mathbf{x}+\mathbf{y})$ and $(\mathbf{y})$ are likely to be in regions with $\delta > 0$ and $\delta < 0$, making $\xi(\mathbf{x})$ negative. So $\xi(\mathbf{x})$ will be positive for small $|\mathbf{x}|$ and will become negative for $x \gtrsim L$. Thus the first zero of $\xi(x)$ will give the typical size of overdense and underdense regions.

5.4 Spatial averages and filter functions

The probability distributions discussed in the last section can be related to the actual density distribution in the universe by invoking the assumption of large scale statistical homogeneity. Consider a large volume of the universe, in which the density contrast is $\delta(\mathbf{x})$. The spatial average of $\delta^2(\mathbf{x})$ will be

$$\begin{aligned} < \delta^2(\mathbf{x}) >_V &= \int \frac{d^3\mathbf{x}}{V} \delta^2(\mathbf{x}) = \int \frac{d^3\mathbf{x}}{V} \int \frac{d^3\mathbf{k}\, d^3\mathbf{q}}{(2\pi)^3(2\pi)^3} \delta_{\mathbf{k}} \delta^\star_{\mathbf{q}}\, e^{i(\mathbf{k}-\mathbf{q})\cdot\mathbf{x}} \\ &\cong V^{-1} \int \frac{d^3\mathbf{k}}{(2\pi)^3} |\delta_k|^2 = \int_0^\infty \frac{V^{-1}}{2\pi^2} k^3 |\delta_k|^2 \frac{dk}{k}, \end{aligned} \tag{5.26}$$

where the third equality is valid for sufficiently large V. If the volume V is large enough to contain several statistically homogeneous regions, in each of which we have a realization of the random variable $\delta(\mathbf{x})$, then we may assume that the spatial averaging is the same as the statistical averaging used in (5.19). A comparison of (5.19) and (5.26) then allows us to identify the parameter σ_k^2 of the probability distribution with $|\delta_k|^2$ of the density contrast. Once this identification is made, all the statistical properties can be determined in terms of $|\delta_k|^2$. The statistical averaging can be replaced by the more convenient spatial integration.

In particular, the two-point correlation function can now be directly related to $|\delta_k|^2$ as

$$\xi(\mathbf{x}) = \int \frac{V^{-1}\, d^3\mathbf{k}}{(2\pi)^3} |\delta_k|^2 e^{i\mathbf{k}\cdot\mathbf{x}} = \int \frac{d^3\mathbf{y}}{V} \delta(\mathbf{y}+\mathbf{x})\delta(\mathbf{y}), \tag{5.27}$$

where the last equality follows from replacing statistical averaging by spatial averaging in equation (5.21).

Using this relation, one can provide direct probabilistic interpretation of $\xi(\mathbf{x})$. Writing $\delta = (\rho - \rho_b)/\rho_b$, this equation becomes

$$\xi(\mathbf{x}) = \frac{1}{\rho_b^2} \int \frac{d^3\mathbf{y}}{V} \left[\rho(\mathbf{y}+\mathbf{x}) - \rho_b\right] \left[\rho(\mathbf{y}) - \rho_b\right] = \frac{1}{\rho_b^2} \int \frac{d^3\mathbf{y}}{V} \rho(\mathbf{y}+\mathbf{x})\rho(\mathbf{y}) - 1. \tag{5.28}$$

Shifting the integration variable to $(\mathbf{y}+\mathbf{x}_1)$, writing $(\mathbf{x}_1+\mathbf{x})$ as $\mathbf{x}_2$ and rearranging the terms, we get:

$$\int \frac{d^3\mathbf{y}}{V} \rho(\mathbf{y}+\mathbf{x}_1)\rho(\mathbf{y}+\mathbf{x}_2) = \rho_b^2 \left[1 + \xi(\mathbf{x}_1 - \mathbf{x}_2)\right]. \tag{5.29}$$

If $\rho(\mathbf{x})$ denotes the smoothed out density of some class of objects – like galaxies, say – then the probability P_{12} of finding two galaxies simultaneously with the separation $(\mathbf{x}_1 - \mathbf{x}_2)$ will be proportional to

$$P_{12} \propto \int d^3\mathbf{y}\, \rho(\mathbf{y}+\mathbf{x}_1)\rho(\mathbf{y}+\mathbf{x}_2) \propto \rho_b^2 \left[1 + \xi(\mathbf{x}_1 - \mathbf{x}_2)\right]. \tag{5.30}$$

If these two events are independent, i.e. if the existence of a galaxy at $\mathbf{x}_1$ has no effect on the probability to find a galaxy at $\mathbf{x}_2$, then the same probability would have been proportional to $P_{\rm ind} \propto \rho_b^2$. Hence the 'excess' probability over random occurrence is given by

$$(P_{12}/P_{\rm ind}) = 1 + \xi(\mathbf{x}_1 - \mathbf{x}_2). \tag{5.31}$$

The correlation function, therefore, is a direct measure of the excess probability (over the random) of finding two objects at $\mathbf{x}_1$ and $\mathbf{x}_2$.

If $\delta(\mathbf{x})$ is a Gaussian variable, any other linear function of δ will also be a Gaussian random variable. One such function is the excess mass contained within a sphere of radius R, centred at some point $\mathbf{x}$:

$$\delta M_R(\mathbf{x}) \equiv \rho_b \int_{|\mathbf{y}|<R} \delta(\mathbf{x}+\mathbf{y})\, d^3\mathbf{y} = \rho_b \int_{\text{all}\,\mathbf{y}} \delta(\mathbf{x}+\mathbf{y})\theta(R - |\mathbf{y}|)\, d^3\mathbf{y}, \tag{5.32}$$

where $\theta(z) = 1$ for $z > 0$ and zero otherwise. For future applications, it is convenient to generalize this definition somewhat and choose:

$$\delta M_R(\mathbf{x}) = \rho_b \int d^3\mathbf{y}\, \delta(\mathbf{x}+\mathbf{y})W(\mathbf{y}), \tag{5.33}$$

where $W(\mathbf{y}) \cong 1$ for $|\mathbf{y}| < R$ and $W(\mathbf{y}) \cong 0$ for $|\mathbf{y}| > R$. This function $W(\mathbf{y})$ is a smoothed out version of the step function θ; because of its smoother behaviour at $|\mathbf{y}| = R$, this function has better mathematical properties. One example of the 'window function' $W(\mathbf{y})$ is the Gaussian:

$$W(\mathbf{y}) = \exp\left(-\frac{y^2}{2R^2}\right), \tag{5.34}$$

which will be used whenever we need a specific form for W.

Notice that the effect of the window function is to average out the fluctuations in $\delta(\mathbf{x})$ at scales smaller than R. This is clearly seen in the Fourier transform of (5.33):

$$\delta M_R(\mathbf{k}) = \rho_b\, W(\mathbf{k})\delta_{\mathbf{k}} = (2\pi)^{3/2}R^3\delta_{\mathbf{k}}\exp\left(-\frac{k^2R^2}{2}\right), \tag{5.35}$$

where we have taken W to be a Gaussian. The amplitude δ_k at $k \gg R^{-1}$ (i.e. for wavelengths $\lambda \ll R$) is exponentially damped by the smoothening process.

Since $\delta(\mathbf{x})$ is a Gaussian variable, so is δM_R. The dispersion in the relative mass excess $(\delta M_R/M_R)$ can be computed exactly as in the case of (5.19) once we define the mean mass M_R inside the sphere of radius R. When $W_R(x) = \theta(R-x)$, the mass contained inside the radius R is precisely $(4\pi/3)R^3\rho_b$. For a Gaussian window function, this notion is not so clearly defined. One reasonable choice for the average mass M sampled by the window function is:

$$M_R \equiv \int d^3y\, \rho_b W(\mathbf{y}) = (2\pi)^{3/2}R^3\rho_b \equiv V_w\rho_b, \tag{5.36}$$

where V_w is the 'effective volume' of the Gaussian. With this definition, the Fourier transform of $(\delta M/M)_R$ becomes

$$\left(\frac{\delta M}{M}\right)_R = \int \frac{d^3\mathbf{k}}{(2\pi)^3}\left(\frac{\delta_k W_k}{V_w}\right)\exp(i\mathbf{k}\cdot\mathbf{x}). \tag{5.37}$$

Therefore, we find, in analogy with (5.19),

$$< (\delta M/M)_R^2 >= \frac{1}{V_w^2}\int_0^\infty \Delta_k^2 W_k^2 \frac{dk}{k}, \tag{5.38}$$

where Δ_k^2 is defined in (5.20). With the identification $\sigma_k^2 = |\delta_k|^2$, it follows that the fluctuation in the mass excess is governed by the combination

$k^3|\delta_k|^2$. More explicitly,

$$< (\delta M/M)_R^2 >= \int_0^\infty \frac{dk}{k} \left[\frac{W_k^2}{V_w^2}\right] \left[\frac{V^{-1}}{2\pi^2} k^3 |\delta_k|^2\right]. \tag{5.39}$$

This integral can be easily evaluated when W is a Gaussian and $|\delta_k|^2 = VAk^n$ is a power law; we get

$$< (\delta M/M)_R^2 >\equiv \sigma_M^2(R) = \frac{A}{2}\Delta^2 \left(k = R^{-1}\right) \Gamma\left(\frac{n+3}{2}\right). \tag{5.40}$$

The fluctuation in mass, in a region of size R, directly probes the quantity $k^3|\delta_k|^2$ at wavelength of the order of R. Further, for a power law, $\sigma_M^2(R) \propto R^{-(n+3)}$.

If we had used the step function itself as the window function, the corresponding result would have been

$$< (\delta M/M)_R^2 >= \int_0^\infty \Delta_k^2 . 9 \left[\frac{\sin kR}{(kR)^3} - \frac{\cos kR}{(kR)^2}\right]^2 \frac{dk}{k}, \tag{5.41}$$

where we have used the following properties of the step function:

$$W_k = 4\pi R^3 \left[\frac{\sin kR}{(kR)^3} - \frac{\cos kR}{(kR)^2}\right]; \quad V_w = \frac{4\pi}{3} R^3. \tag{5.42}$$

If $|\delta_k|^2 = VAk^n$ with $n < 1$, numerical integration of (5.41) gives $(\delta M/M) \simeq \mathcal{O}(1)\Delta(k = R^{-1})$. However, notice that $|W_k|^2 \propto k^{-4}$ for $k \gtrsim R^{-1}$; hence the integral converges only for $n < 1$. This difficulty does not arise in the case of the Gaussian window function.

We shall now discuss several useful scaling relations for a power-law spectrum with $|\delta_k|^2 \propto k^n, \sigma_M^2(R) \propto R^{-(n+3)} \propto M^{-(n+3)/3}$. Let us begin by asking what restrictions can be put on the index n.

The integrand defining σ^2 behaves as $k^2|\delta_k|^2 dk$ near $k = 0$. (Note that $W_k \simeq 1$ for small k in any window function.) Hence the finiteness of σ^2 will require the condition $n > -3$. The behaviour of the integrand for large values of k depends on the window function W_k. If we take the window function to be a Gaussian, then the convergence is ensured for all n. This might suggest that n can be made as large as one wants; that is, we can keep the power at small k (i.e., large wavelengths) to be as small as we desire. This result, however, is not quite true for the following reason: As the system evolves, small scale nonlinearities will develop in the system which can affect the large scales. If the large scales have too little power intrinsically (i.e. if n is large), then the long wavelength power will soon be dominated by the 'tail' of the short wavelength power arising from the nonlinear clustering. Thus there will be an *effective* upper bound on n.

The actual value of this upper-bound depends, to some extent, on the details of the small scale physics. It is, however, possible to argue that the *natural* value for this bound is $n = 4$. The argument runs[1] as follows: Let us suppose that a large number of particles, each of mass m, are distributed in space in such a way that there is very little power at large wavelengths. (That is, $|\delta_k|^2 \propto k^n$ with $n \gg 4$ for small k.) As time goes on, the particles influence each other gravitationally and will start clustering. The density $\rho(\mathbf{x}, t)$ due to the particles in some region will be

$$\rho(\mathbf{x}, t) = \sum_i m\delta[\mathbf{x} - \mathbf{x}_i(t)], \tag{5.43}$$

where $\mathbf{x}_i(t)$ is the position of the ith particle at time t and the summation is over all the particles in some specified region. The density contrast in the Fourier space will be

$$\delta_{\mathbf{k}}(t) = \frac{1}{N}\sum_i \exp[i\mathbf{k}\cdot\mathbf{x}_i(t)] - 1, \tag{5.44}$$

where N is the total number of particles in the region. For small enough $|\mathbf{k}|$, we can expand the right hand side in a Taylor series obtaining

$$\delta_{\mathbf{k}}(t) = i\mathbf{k}\cdot\left\{\frac{1}{N}\sum_i \mathbf{x}_i(t)\right\} - \frac{k^2}{2}\left\{\frac{1}{N}\sum_i x_i^2(t)\right\} + \cdots. \tag{5.45}$$

If the motion of the particles is such that the centre-of-mass of the region under consideration does not change, then $\sum \mathbf{x}_i$ will vanish; under this (reasonable) condition, $\delta_{\mathbf{k}}(t) \propto k^2$ for small k. Note that this result follows, essentially, from the three assumptions: small-scale graininess of the system, conservation of mass and conservation of momentum. This will lead to a long wavelength tail with $|\delta_k|^2 \propto k^4$ which corresponds to $n = 4$. Thus, for all practical purposes, $-3 < n < 4$.

The value $n = 4$ corresponds to $\sigma_M^2(R) \propto R^{-7} \propto M^{-7/3}$. For comparison, note that purely Poisson fluctuations will correspond to $(\delta M/M)^2 \propto (1/M)$; i.e. $\sigma_M^2(R) \propto M^{-1} \propto R^{-3}$ with an index of $n = 0$.

Some more properties of the power spectra with different values of n can be obtained if the nonlinear effects are taken into account. We will develop this model in detail in chapter 8; but some of the results can be anticipated by simple scaling arguments of the following kind.

We know that, in the matter-dominated phase, linear perturbations grow as $\delta_k(t) \propto a(t) \propto t^{2/3}$. Hence $\sigma_M^2(R) \propto t^{4/3}R^{-(3+n)}$. We may assume that the perturbations at some scale R become nonlinear when $\sigma_M(R) \simeq 1$. It follows that the time t_R, at which a scale R becomes nonlinear, satisfies the relation

$$t_R \propto R^{3(n+3)/4} \propto M^{(n+3)/4}. \tag{5.46}$$

For $n > -3, t_R$ is an increasing function of M; small scales become nonlinear at earlier times. The proper size L of the region which becomes nonlinear is

$$L \propto Ra(t_R) \propto Rt_R^{2/3} \propto R^{(5+n)/2} \propto M^{(5+n)/6}. \tag{5.47}$$

Further, the objects which are formed at $t = t_R$ will have density ρ of the order of the background density $\overline{\rho}$ of the universe at t_R. Since $\overline{\rho} \propto t^{-2}$, we get

$$\rho \propto t_R^{-2} \propto R^{-3(3+n)/2} \propto M^{-(3+n)/2}. \tag{5.48}$$

Combining (5.47) and (5.48) we get $\rho \propto L^{-\beta}$ with

$$\beta = \frac{3(3+n)}{(5+n)}. \tag{5.49}$$

In the nonlinear case, one may interpret the correlation function ξ as $\xi(L) \propto \rho(L)$; this would imply $\xi(x) \propto x^{-\beta}$. We shall see later that such a behaviour is to be expected. The gravitational potential due to these bodies is

$$\phi \simeq G\rho(L)L^2 \propto L^{(1-n)/(5+n)} \propto M^{(1-n)/6}. \tag{5.50}$$

The same scaling, of course, can be obtained from $\phi \propto (M/L)$. This result shows that the binding energy of the structures increases with M for $n < 1$. In that case, the substructures will be rapidly erased as larger and larger structures become nonlinear.

For $n = 1$, the gravitational potential is independent of the scale, and $\rho \propto L^{-2}$. This value of n corresponds to the Harrison–Zeldovich spectrum discussed in the last chapter. This can be seen directly as follows: In the radiation dominated phase, $\delta \propto a^2 \propto t$. So the power $k^3|\delta_k|^2$ in the perturbation at the time of entering the Hubble radius will be proportional to $k^{(n+3)}t^2_{\text{enter}}(k)$. But since $k^{-1}a(t_{\text{enter}}) \propto d_H(t_{\text{enter}}) \propto t_{\text{enter}}$, we have $t_{\text{enter}}(k) \propto k^{-2}$. Hence the power at the time of entering the Hubble radius is $k^{(n+3)}k^{-4} \propto k^{(n-1)}$. This is independent of the scale for $n = 1$.

Finally, we derive an (approximate) relation between the correlation function $\xi(\mathbf{x})$ and the relative mass excess $(\delta M/M)_R$ which is of considerable practical importance. Consider the integral

$$\begin{aligned} J_3(R) &\equiv \frac{1}{4\pi}\int_{R^3} d^3\mathbf{x}\,\xi(\mathbf{x}) = \int_0^R r^2\xi(r)\,dr \\ &= \frac{1}{4\pi}\int d^3\mathbf{x}\,\xi(\mathbf{x})\theta(R - |\mathbf{x}|) = \frac{1}{4\pi}\int \frac{d^3\mathbf{k}}{(2\pi)^3}\xi(\mathbf{k})W_k. \end{aligned} \tag{5.51}$$

Using now the relation that $\xi(\mathbf{k}) = |\delta_k|^2 V^{-1}$ we get

$$\begin{aligned} J_3(R) &= R^3 \int \frac{d^3\mathbf{k}}{(2\pi)^3} \frac{|\delta_k|^2}{V} \left[\frac{\sin kR}{(kR)^3} - \frac{\cos kR}{(kR)^2} \right] \\ &\cong R^3 \int_0^\infty \frac{dk}{k} \Delta_k^2 \left[\frac{\sin kR}{(kR)^3} - \frac{\cos kR}{(kR)^2} \right] \cong \frac{R^3}{3} \int_0^{R^{-1}} \frac{dk}{k} \Delta_k^2. \end{aligned} \tag{5.52}$$

In arriving at the last equality, we have ignored the contribution from wavelengths smaller than R. For the $kR \ll 1$ range, the expression in square brackets has the value (1/3). In the same approximation,

$$\left\langle \left(\frac{\delta M}{M} \right)_R^2 \right\rangle = \int_0^\infty 9\Delta^2(k) \left[\frac{\sin kR}{(kR)^3} - \frac{\cos kR}{(kR)^2} \right]^2 \frac{dk}{k} \cong \int_0^{R^{-1}} \Delta^2(k) \frac{dk}{k}. \tag{5.53}$$

Combining these results we find:

$$\sigma_M^2(R) = \left\langle \left(\frac{\delta M}{M} \right)_R^2 \right\rangle = \frac{3J_3(R)}{R^3}. \tag{5.54}$$

Thus the integral $J_3(R)$ is a direct measure of mass fluctuations in the scale R.

5.5 Normalization of the fluctuation spectrum

It was mentioned in chapter 4 that the spectrum of fluctuations can be characterized by a power law $|\delta_k|^2 = VAk^n$ where n is the spectral index and A is the amplitude at very early epochs. (The normalization volume V is added purely for convenience.) The values of these two parameters should emerge from the physical model which describes the production of the initial spectrum. In the absence of any reliable theoretical predictions for A and n, it is best to treat them as free parameters which have to be determined by comparison with observations.

Since the linear theory is conceptually and mathematically much better understood than the nonlinear evolution which follows it, the parameters A and n can be determined most reliably if some predictions from linear theory can be tested directly. It turns out that the density perturbations at the time of decoupling, when the linear theory is valid, lead to an anisotropy in the MBR temperature. If this anisotropy in MBR is detected, it would provide the most reliable way of determining A, and possibly n. However, as we shall see in the next chapter, no such anisotropies have been detected.* The linear theory also predicts a pattern of large scale velocity fields in the universe. In principle, this feature

---

* COBE has now detected this anisotropy. See Appendix C regarding the implications for normalization.

also allows a comparison between theory and observation; in practice, as we shall see in chapter 7, the method also suffers from several sources of error.

Because of these difficulties, it is usual to try to determine the value of A (for a given n) by comparison with the observed galaxy–galaxy correlation function $\xi_{\rm gg}(r)$. We have seen that the mean square fluctuation $\sigma_M^2(R)$ in the mass contained within a radius R, and the integral over $\xi(r)$ are related by

$$\begin{aligned} J_3(R) \equiv \int_0^R r^2 \xi(r)\, dr &= \frac{R^3}{3} \left\langle \left(\frac{\delta M}{M} \right)_R^2 \right\rangle \\ &= \frac{V^{-1}}{2\pi^2} \int_0^\infty \frac{|\delta_k(t_0)|^2}{k} [\sin kR - kR \cos kR]\, dk. \end{aligned} \tag{5.55}$$

That is

$$\int_0^R r^2 \xi(r)\, dr = \frac{V^{-1}}{2\pi^2} \int_0^\infty \frac{dk}{k} |\delta_k(t_0)|^2 [\sin kR - kR \cos kR]. \tag{5.56}$$

The left hand side can be determined from the observed form of $\xi(r)$; for a given spectral index n, the right hand side is proportional to A. Comparing the two expressions, we can determine A.

Notice that the right hand side is evaluated by using expressions from the linear theory. The integral on the left hand side, however, receives contributions from $\xi(r)$ at all values of r in the range $(0, R)$. For small r, the galaxy distribution will be strongly influenced by the nonlinear evolution which we still have not worked out. One may, therefore, feel that the comparison between the two sides of this equation is meaningless. Fortunately, this is not the case. A comparison does turn out to be meaningful provided R is chosen large enough such that ξ *at* R obeys the linear theory. To demonstrate this fact, we shall prove[2] that the time evolution of the quantity

$$\sigma = a^3 \int_{R^3} d^3\mathbf{x}\, (\rho - \rho_b) \propto \left(\frac{\Delta M}{M} \right)_R \tag{5.57}$$

is determined by the linear perturbation equation

$$\ddot{\sigma} + 2\frac{\dot{a}}{a}\dot{\sigma} = 4\pi G \rho_b \sigma, \tag{5.58}$$

provided R is sufficiently large. Since this is precisely the equation that arises for σ in the linear theory, we can compare the value of $J_3(R)$ with the prediction of the linear theory provided R is sufficiently large. To prove this claim we start with the fundamental equations:

$$\dot{\mathbf{v}} + a^{-1}(\mathbf{v} \cdot \nabla)\mathbf{v} + \left(\frac{\dot{a}}{a} \right) \mathbf{v} = -\frac{1}{a} \nabla \phi, \tag{5.59}$$

$$\dot{\rho} + 3\frac{\dot{a}}{a}\rho + \frac{1}{a}\nabla \cdot (\rho \mathbf{v}) = 0, \tag{5.60}$$

and

$$\nabla^2 \phi = 4\pi G a^2 \left[\rho - \rho_b\right]. \tag{5.61}$$

The mass excess σ in a volume V changes with time as

$$\dot{\sigma} = -a^2 \int_\Sigma \rho \mathbf{v} \cdot d\mathbf{S}, \tag{5.62}$$

where $d\mathbf{S}$ is the element of a surface Σ enclosing the volume V. Differentiating this expression further and using (5.59) and (5.60) to eliminate $\dot{\rho}$ and $\dot{\mathbf{v}}$, we get

$$\begin{aligned}
\ddot{\sigma} &= \frac{2\dot{a}}{a}\dot{\sigma} - a^2 \int d\mathbf{S} \cdot \left[\rho \frac{\partial \mathbf{v}}{\partial t} + \mathbf{v}\frac{\partial \rho}{\partial t}\right] \\
&= -\frac{2\dot{a}}{a}\dot{\sigma} - a^2 \int d\mathbf{S} \cdot \left[-\frac{\rho}{a}\nabla\phi - \frac{\rho}{a}(v^i \partial_i)\mathbf{v} - \frac{\mathbf{v}}{a}\partial_i(\rho v^i)\right] \\
&= -\frac{2\dot{a}}{a}\dot{\sigma} + a \int d\mathbf{S} \cdot \rho \nabla \phi + a \int d\mathbf{S} \cdot \frac{\partial}{\partial x^i}(\rho v^i \mathbf{v}).
\end{aligned} \tag{5.63}$$

This expression can be rewritten as

$$\ddot{\sigma} + \frac{2\dot{a}}{a}\dot{\sigma} = a\rho_b \int \nabla\phi \cdot d\mathbf{S} + a\int (\rho - \rho_b)\nabla\phi \cdot d\mathbf{S} + a\int \frac{\partial^2}{\partial x^i \partial x^j}(\rho v^i v^j)\, d^3x. \tag{5.64}$$

If R is much larger than the coherence length of the mass distribution, then $(\rho - \rho_b)$ will fluctuate in sign many times within this volume; therefore, the second term on the right hand side can be ignored, compared to the first. Using (5.62) one can estimate the magnitude of the third term to be $R^3(\rho v^2/R^2)$ while $\dot{\sigma} \sim \rho v R^2$. So

$$a\int d^3x \frac{\partial^2}{\partial x^i \partial x^j}(\rho v^i v^j) \cong R\rho v^2 \cong \frac{1}{\rho R^3}\dot{\sigma}^2 \simeq \frac{G\rho\sigma^2}{\rho V} \simeq (G\rho\sigma)\left(\frac{\sigma}{\rho V}\right). \tag{5.65}$$

Therefore, if $|\sigma| \lesssim \rho V$, then this term is small compared to the first term $(G\rho\sigma)$. Therefore, for sufficiently large R (i.e. if linear theory is valid at R), $\sigma_R(t)$ satisfies the linear equation. Hence, we are justified in using J_3 for determining the normalization constant A provided R is large enough.

The value of J_3 can be determined[3] from the galaxy surveys. The most reliable value is from the Center for Astrophysics (CfA) redshift survey:

$$J_3(R) = \begin{cases} 270h^{-3}\ \text{Mpc}^3 & (\text{at } 10h^{-1}\ \text{Mpc}) \\ 600h^{-3}\ \text{Mpc}^3 & (\text{at } 30h^{-1}\ \text{Mpc}) \end{cases}. \tag{5.66}$$

Using this, we get

$$\left(\frac{\Delta M}{M}\right)_R \simeq \begin{cases} 0.9 & (\text{at } 10h^{-1}\,\text{Mpc}) \\ 0.25 & (\text{at } 30h^{-1}\,\text{Mpc}) \end{cases} . \tag{5.67}$$

This value, in turn, fixes A once the spectral index n is given.

While this is the best method currently available, it suffers from some serious shortcomings. Since $(\Delta M/M) \simeq 1$ near $8h^{-1}$ Mpc or so, one would have preferred a value of R significantly larger than $10h^{-1}$ Mpc for a reliable estimate based on linear theory. But at depths larger than $10h^{-1}$ Mpc, the uncertainties in ξ increase substantially. It is possible to tackle this problem of normalization using some results from the study of the nonlinear evolution. We will discuss this aspect in chapter 9.

The second problem with the above approach is that it attempts to estimate the distribution of *matter* in the universe using the distribution of *galaxies*. The correlation function for the mass distribution *need* not, however, be the same as the correlation function for galaxies. For example, suppose that galaxies formed preferentially in the high density regions of the mass distribution. We will see later in section 5.7 that regions of high density in a Gaussian distribution will be more strongly correlated than the underlying distribution. Then the galaxy–galaxy correlation function will be a biased indicator of the mass correlation function.

If the biasing occurs due to the fact that galaxies have formed in a high-density environment, then we may use an approximately linear relation,

$$\xi_{\text{gg}} = b^2 \xi_{\text{mass}}, \tag{5.68}$$

where b is called the biasing parameter[4]. In that case, we have

$$(\delta\rho/\rho)_{\text{galaxy}} = b(\delta\rho/\rho)_{\text{mass}}. \tag{5.69}$$

This procedure introduces an additional free parameter b in the theory, which also needs to be determined by comparison with the observations.

The method based on galaxy–galaxy correlation faces yet another difficulty in the case of hot dark matter scenarios. In these scenarios, the first structures which form have masses of about $10^{14}\,\text{M}_\odot$, which corresponds to clusters. Galaxies are produced by secondary processes involving fragmentation of the cluster-sized objects. Hence, in these models, galaxies will be highly distorted indicators of the original mass distribution. It is, therefore, necessary to normalize the amplitude of the perturbations at an earlier epoch, at redshift $z = z_{\text{nl}}$, when the linear theory is definitely valid. Since z_{nl} is quite uncertain, the normalization is difficult to adopt in practice.

5.6 The time evolution of the correlation function

The discussion in chapter 1 showed that the two-point correlation function for galaxies is observationally well-determined. It is, therefore, of interest to see what theoretical predictions[5] can be made about the quantity $\xi(\mathbf{x}, t)$.

The behaviour of $\xi(\mathbf{x}, t)$ in the linear regime is easy to determine. From (5.21), it follows that

$$\xi(\mathbf{x}, t) \propto a^2(t) \int \frac{d^3\mathbf{k}}{(2\pi)^3} |\delta_k|^2 \mathrm{e}^{i\mathbf{k}\cdot\mathbf{x}} \tag{5.70}$$

where we have used the fact that $\delta_k \propto a(t)$ in the matter dominated universe. If $|\delta_k|^2 \propto k^n$, then the integral is proportional to $x^{-(n+3)}$. Therefore, in the linear regime

$$\xi(\mathbf{x}, t) \propto a^2(t) x^{-(n+3)} \propto t^{4/3} x^{-(n+3)} \qquad (x \gg 1). \tag{5.71}$$

It is possible to extrapolate this result into the nonlinear regime in the following manner: We shall first derive an exact equation satisfied by ξ in the nonlinear regime and show that this equation admits power law solutions similar to (5.71). These two solutions can then be connected up using the scale invariance of a particular solution of the collisionless Boltzmann equation.

The exact equation satisfied by $\xi(\mathbf{x}, t)$ arises from the conservation of particle pairs. The mean number of neighbours which a particle located at some point $\mathbf{x}$ will have is given by

$$N(\mathbf{x}, t) = na^3 \int_0^x 4\pi r^2 \, dr \, [1 + \xi(r, t)] \, . \tag{5.72}$$

(This follows directly from the probabilistic interpretation for $\xi(r, t)$.) The mean flux of neighbours out of the surface $x =$ constant, is $4\pi^2 x^2 n(1 + \xi)v$. Hence the conservation of neighbours implies the relation

$$\frac{\partial N}{\partial t} + 4\pi a^2 x^2 n(1 + \xi)v = 0. \tag{5.73}$$

Since na^3 is constant in time, this is equivalent to:

$$\int_0^x 4\pi r^2 \, dr \left(\frac{\partial \xi}{\partial t}\right) = -\frac{4\pi x^2}{a}(1 + \xi)v. \tag{5.74}$$

Differentiating this relation with respect to x, we get

$$\frac{\partial \xi}{\partial t} + \frac{1}{ax^2} \frac{\partial}{\partial x} \left[x^2(1 + \xi)v\right] = 0.$$

At small scales, where nonlinear effects dominate, we expect $v \cong -\dot{a}x$ so as to cancel the Hubble expansion and keep the structures as bound. In this limit, this equation becomes

$$\frac{\partial}{\partial t}(1+\xi) = \frac{\dot{a}}{x^2 a}\frac{\partial}{\partial x}\left[x^3(1+\xi)\right], \tag{5.75}$$

or, if $\xi \gg 1$,

$$\left[a\frac{\partial}{\partial a} - x\frac{\partial}{\partial x} - 3\right]\xi = 0. \tag{5.76}$$

This equation has the power law solution with some index m:

$$\xi = a^{3-m}x^{-m} \propto t^{(2/3)(3-m)}x^{-m}; \quad (x \ll 1). \tag{5.77}$$

We have thus obtained power law solutions for $\xi(x,t)$ for both large x and small x, with different indices m and n. To connect up the indices m and n, we use the fact that the distribution function $f(\mathbf{x},\mathbf{p},t)$ of galaxies satisfies the Boltzmann equation

$$\frac{\partial f}{\partial t} + \dot{x}^i\frac{\partial f}{\partial x^i} + \dot{p}^i\frac{\partial f}{\partial p^i} = \frac{\partial f}{\partial t} + \frac{p^i}{ma^2}\frac{\partial f}{\partial x^i} - m\frac{\partial \phi}{\partial x^i}\frac{\partial f}{\partial p^i} = 0, \tag{5.78}$$

with

$$\nabla^2\phi = 4\pi G m\, a^{-1}\int f\, d^3p; \quad a \propto t^{2/3}. \tag{5.79}$$

It can be seen, by direct substitution, that this equation admits solutions of the form

$$f(\mathbf{x},\mathbf{p},t) = t^{-3\beta}F(\mathbf{x}/t^{\alpha}, \mathbf{p}/t^{\beta}), \tag{5.80}$$

with $\beta = \alpha + (1/3)$. Hence the density must have the form $\rho(\mathbf{x},t) = A(\mathbf{x}/t^{\alpha})$ and the correlation function should have the form determined by the relation $[1+\xi(x,t)]\,d^3\mathbf{x} = t^{3\alpha}B(\mathbf{x}/t^{\alpha})$. From this it follows that the correlation function must have the form

$$\xi(x,t) = G(x/t^{\alpha}). \tag{5.81}$$

This form is consistent with the result (5.71) for large scales (linear limit) only if

$$\alpha = \frac{4}{3(3+n)} = \frac{4}{9+3n}. \tag{5.82}$$

On the other hand, the result (5.77) is consistent with (5.81) only if

$$\alpha = \frac{2}{m} - \frac{2}{3}. \tag{5.83}$$

Combining (5.82) and (5.83) we get a relation between the power law index at small and large scales:

$$m = \left(\frac{9+3n}{5+n}\right). \tag{5.84}$$

Notice that this result connects up the index of the nonlinear, small scale regime (where $\xi \propto x^{-m}$) with the spectral index of the perturbations ($|\delta_k|^2 \propto k^n$). This is one of the few exact results available in the nonlinear limit. Observations suggest that $m \simeq 1.8 = (9/5)$ implying that $n \simeq 0$, near galactic scales.

5.7 Correlation of high density regions

The approach taken so far treats $\delta(\mathbf{x})$ and consequently the density $\rho(\mathbf{x})$ as a Gaussian random field. Any sufficiently large volume of the universe will provide a realization of this random field. Let us consider one such realization. The density $\rho(\mathbf{x})$ in this realization will vary from place to place, with some regions having densities far higher than the average. Since such high density regions are the more likely sites for gravitational condensation to occur, it is important to know how they are distributed in space. In particular, one would like to know the correlation function for these high density regions.

To study the properties of the high density regions, we will need a preliminary result, which we will derive first. Let $(q_1, q_2, \cdots, q_N)$ be a set of N Gaussian random fields with zero mean and a specified correlation function $< q_i q_j >= F_{ij}$. (This, of course, implies that the q_i are *not* independent random variables.) Let $P[q_1, q_2, \cdots, q_N]$ be the probability that these random variables simultaneously take the values $\{q_1, \cdots, q_N\}$. To determine P we will proceed in the following manner: consider the integral

$$\begin{aligned} W[J_1, \cdots, J_N] &\equiv \int_{-\infty}^{+\infty} dq_1 \cdots dq_N \, P(q_1, \cdots, q_N) \exp\left[\sum_{i=1}^{N} q_i J_i\right] \\ &\equiv \left\langle \exp \sum_{i=1}^{N} q_i J_i \right\rangle . \end{aligned} \tag{5.85}$$

This integral can be expressed in terms of F_{ij} as

$$W[J_1, \cdots, J_N] = \left\langle \exp \sum_{i=1}^{N} q_i J_i \right\rangle = \exp\left[\frac{1}{2} \sum_{i,k=1}^{N} F_{ik} J_i J_k\right] . \tag{5.86}$$

To prove this result, note that differentiating W with respect to J_i has the effect of bringing q_i into the integrand; therefore

$$\langle q_1 q_2 \cdots q_k \rangle = \left(\frac{\partial^k W}{\partial J_1 \partial J_2 \cdots \partial J_k} \right)_{J=0} . \tag{5.87}$$

Clearly, $< q_i >= 0$, $< q_i q_j >= F_{ij}$ and all other higher order correlation functions are given in terms of F_{ij} by suitable products. In other words, *all* derivatives with respect to J_i, evaluated at $J_i = 0$ match on both sides of equation (5.86). Hence these two functions are the same.

Since F_{ij} is a symmetric matrix, it can be diagonalized by an orthogonal matrix M; i.e., we can express F on the form $F_{ij} = (M^T \Lambda M)_{ij}$ where M is made of the eigenvectors of F and $\Lambda = \text{dia}(\lambda_1, \cdots, \lambda_N)$ where λ_i are the eigenvalues. Once F is expressed in the diagonal form, (5.85) can be inverted easily to obtain

$$\begin{aligned} P[q_1, \cdots, q_N] &= \frac{1}{(2\pi)^{N/2}(\lambda_1 \cdots \lambda_N)^{1/2}} \exp\left[-\frac{1}{2}\sum_{i,j=1}^{N} \frac{(M_{ij}q_j)^2}{\lambda_i}\right] \\ &= \frac{1}{(2\pi)^{N/2}(\det F)^{1/2}} \exp\left(-\frac{1}{2} q^T F^{-1} q\right) \end{aligned} \tag{5.88}$$

where F^{-1} is the inverse of the matrix F. Thus the probability $P[q_1, \cdots, q_N]$ can be completely expressed in terms of the correlation matrix F_{ij}.

Consider now a Gaussian random field $m(\mathbf{x})$ obtained by smoothening $\delta(\mathbf{x})$ by a window function W:

$$m(\mathbf{x}) = \int \delta(\mathbf{x} + \mathbf{y}) W(\mathbf{y})\, d\mathbf{y}; \quad m_{\mathbf{k}} = \delta_{\mathbf{k}} W_{\mathbf{k}}. \tag{5.89}$$

The correlation function for the variable $m(\mathbf{x})$ can be determined exactly as in (5.21):

$$\xi_R(\mathbf{x}) \equiv < m(\mathbf{x} + \mathbf{y}) m(\mathbf{y}) > = \int \frac{d^3k}{(2\pi)^3} |\sigma_k W_k|^2 e^{i\mathbf{k}\cdot\mathbf{x}}. \tag{5.90}$$

(We have now set $V = 1$ for simplicity.) In particular, the dispersion in the field is

$$< m(\mathbf{x})^2 > = \xi_R(0) = \int \frac{d^3\mathbf{k}}{(2\pi)^3} |\sigma_k W_k|^2 \equiv \mu^2. \tag{5.91}$$

Since the probability that the field has a value m at any location is given by the Gaussian distribution

$$P(m) = \frac{1}{(2\pi\mu^2)^{1/2}} \exp\left(-\frac{1}{2}\frac{m^2}{\mu^2}\right) \tag{5.92}$$

the probability that the field at any given location – say, $\mathbf{x}_1$ – is *larger* than ν times the dispersion μ is given by the integral

$$\begin{aligned} P_1(m > \nu\mu) &= \int_{\nu\mu}^{\infty} dm\, P(m) = \int_{\nu}^{\infty} \frac{dt}{\sqrt{2\pi}} \exp\left(-\frac{1}{2}t^2\right) \\ &\simeq \frac{1}{\sqrt{2\pi}} \exp\left(-\frac{1}{2}\nu^2\right), \end{aligned} \tag{5.93}$$

where the last approximation is valid if $\nu \gg 1$.

Consider now two fixed locations in space, $\mathbf{x}_1$ and $\mathbf{x}_2$. Let the joint probability that the fields at these points are m_1 and m_2 be $P(m_1, m_2)$. Since $m(\mathbf{x}_1)$ and $m(\mathbf{x}_2)$ are two Gaussian variables with dispersion μ^2 and correlation $\xi_R(\mathbf{x}_1 - \mathbf{x}_2) \equiv \xi_R(\mathbf{r})$, we can use the general result obtained in (5.88) and determine $P(m_1, m_2)$:

$$P(m_1, m_2) = \frac{1}{2\pi}\left[\xi_R^2(0) - \xi_R^2(r)\right]^{-1/2} \exp\left\{-\frac{\left[\xi_R(0)(m_1^2 + m_2^2) - 2\xi_R(r)m_1 m_2\right]}{2\left[\xi_R^2(0) - \xi_R^2(r)\right]}\right\}. \tag{5.94}$$

We can now determine the probability P_2 that the values m_1 and m_2 are both (simultaneously) larger than $(\nu\mu)$. This is given by

$$P_2 = \int_{\nu\mu}^{\infty} dm_1 \int_{\nu\mu}^{\infty} dm_2 \, P(m_1, m_2). \tag{5.95}$$

If the two variables were independent, then this probability would have been just P_1^2. We define the correlation function $\xi_\nu(r)$ for regions with density higher than $\nu\mu$ by the relation

$$P_2 = P_1^2\left[1 + \xi_\nu(r)\right]. \tag{5.96}$$

(This is the usual probabilistic, interpretation of the correlation function; see equation (5.31).) From the expression for $P(m_1, m_2)$, we find

$$P_2 = \int_\nu^\infty dt_1 \int_\nu^\infty dt_2 \, N \exp\left\{-\frac{1}{2}\frac{1}{(1 - A^2)}\left[t_1^2 + t_2^2 - 2At_1\, t_2\right]\right\}, \tag{5.97}$$

where $t_i = (m_i/\mu)$, $A = (\xi_R(r)/\xi_R(0))$, $N = (2\pi)^{-1}(1 - A^2)^{-1/2}$ and we have used the fact $\mu^2 = \xi_R(0)$. This integral can be expressed in terms of the error functions leading to an exact form for $\xi_\nu(r)$. The key result can, however, be understood by considering the limit $A \ll 1, \nu \gg 1$ (with $A\nu^2$ arbitrary). In that case, the integral over t_1 becomes, on completing the square:

$$\begin{aligned}\int_\nu^\infty dt_1 \exp\left[-\frac{1}{2}\frac{1}{(1 - A^2)}(t_1 - At_2)^2\right] &\simeq \int_{(\nu - At_2)}^\infty dz \exp\left(-\frac{1}{2}z^2\right) \\ &\simeq \exp\left[-\frac{1}{2}(\nu - At_2)^2\right] \\ &\simeq \exp\left(-\frac{1}{2}\nu^2\right)\exp(A\nu t_2),\end{aligned} \tag{5.98}$$

where we ignored A^2 and higher terms and used the result

$$\int_\nu^\infty dz \exp\left(-\frac{1}{2}z^2\right) \simeq \exp\left(-\frac{1}{2}\nu^2\right) \quad (\text{for } \nu \gg 1). \tag{5.99}$$

The remaining integral is, in the same approximation:

$$\begin{aligned} P_2 &\cong \frac{1}{2\pi}\int_\nu^\infty dt_2 \, \exp\left(-\frac{1}{2}\nu^2\right)\exp(A\nu t_2) \\ &\cong \frac{1}{(2\pi)}\exp\left(-\frac{1}{2}\nu^2\right)\exp\left[-\frac{1}{2}\nu^2(1-2A)\right] \\ &= (2\pi)^{-1}\exp(-\nu^2)\exp(A\nu^2). \end{aligned} \tag{5.100}$$

Combining this relation with (5.93) and using (5.96), we get

$$\xi_\nu(r) \cong \exp(\nu^2 A) - 1 = \exp\left[\nu^2\frac{\xi_R(r)}{\xi_R(0)}\right] - 1 = \exp\left[\frac{\nu^2}{\mu^2}\xi_R(r)\right] - 1. \tag{5.101}$$

This result is valid for arbitrary $A\nu^2$ (provided $A \ll 1$ and $\nu \gg 1$) and shows that $\xi_\nu(r)$ can be much higher than $\xi_R(r)$. In other words, high density regions are more strongly correlated than the underlying distribution. Even for small values of $A\nu^2$, $\xi_\nu(r)$ is enhanced with respect to $\xi_R(r)$. When $\nu \gg 1$, $A \ll 1$ such that $A\nu^2 \ll 1$, we can write

$$\xi_\nu(r) \cong \nu^2\frac{\xi_R(r)}{\xi_R(0)} = \frac{\nu^2}{\Delta_R^2}\xi_R(r). \tag{5.102}$$

For $\nu > \Delta_R, \xi_\nu > \xi_R$; if $A\nu^2$ is not small, the enhancement can be still higher.

The result shows that high density regions of a Gaussian field are more strongly correlated than the background[6]. Higher and higher thresholds (indicated by larger values of ν) will be more and more strongly clustered. This result may be understood based on the following physical picture: Consider the superposition of a 'noise' $f_N(\mathbf{x})$ and a 'signal' $f_S(\mathbf{x})$ which have the following properties: The noise is assumed to have a small coherence length L_N with no correlation on higher scales; the signal, on the other hand, has a relatively smaller amplitude but much larger coherence length L_S. The *distribution* function for the combined field, $f(\mathbf{x}) = f_N(\mathbf{x}) + f_S(\mathbf{x})$, is essentially that of the noise $P_N[f_N]$ because f_S is relatively steady. But the *correlations* are that of f_S for $r > L_N$. The probability to find a region with the value of f higher than $\nu\sigma$ is the same as the probability that f_N exceeds $[\nu\sigma - f_S(\mathbf{x})]$. This probability will be

$$\begin{aligned} P[f > \nu\sigma] &\cong P_N[f_N > (\nu\sigma - f_S)] \\ &\cong \mathcal{P}_N(\nu)\left[1 - \frac{1}{\sigma}\frac{d\ln\mathcal{P}_N(\nu)}{d\nu}f_S(\mathbf{x})\right], \end{aligned} \tag{5.103}$$

where $\mathcal{P}_N(\nu)$ is the probability of exceeding a threshold ν in the absence of the signal. Since $\delta(\mathbf{x}) = (f - \overline{f})/\overline{f}$ is proportional to the relative probabilities, this equation suggests that the system behaves as though the effective density contrast is

$$\delta_{\rm eff}(\mathbf{x}) = -\frac{1}{\sigma}\frac{d\ln \mathcal{P}_N(\nu)}{d\nu} f_S(\mathbf{x}). \tag{5.104}$$

Since $\xi \propto \delta^2$ this leads to amplification by the factor

$$A = \left[\frac{1}{\sigma}\frac{d\ln \mathcal{P}_N(\nu)}{d\nu}\right]^2. \tag{5.105}$$

This result is quite general and is valid whenever the Taylor expansion in (5.103) is valid. When the noise is Gaussian the amplification A reduces to the expression found earlier.

5.8 Mass functions

Gravitationally bound objects in the universe, like galaxies, span a large dynamic range in mass. Let $f(M)\,dM$ be the number density of bound objects in the mass range $(M, M+dM)$ (usually called the 'mass function') and let $F(M)$ be the number density of objects with masses *greater* than M.

Since the formation of gravitationally bound objects is an inherently nonlinear process, it might seem that the linear theory developed so far cannot be used to determine $F(M)$. This, however, is not entirely true. In any one realization of the linear density field $\delta_R(\mathbf{x})$, (filtered using a window function of scale R), there will be regions with high density (i.e. regions with $\delta_R > \delta_c$ where δ_c is some critical value slightly greater than unity, say). It seems reasonable to assume that such regions will eventually condense out as bound objects. Though the dynamics of that region will be nonlinear, the process of condensation is unlikely to change the mass contained in that region significantly. Therefore, if we can estimate the mean number of regions with $\delta_R > \delta_c$ in a Gaussian random field, we will be able to determine $F(M)$.

An approximate way of achieving this is as follows[7]: Let us consider a density field $\delta_R(\mathbf{x})$ smoothed by a window function W_R of scale radius R. We have seen earlier that the probability that this field will have a value δ at any chosen point is

$$P(\delta, t) = \left[\frac{1}{2\pi\sigma^2(R,t)}\right]^{1/2} \exp\left(-\frac{\delta^2}{2\sigma^2(R,t)}\right), \tag{5.106}$$

where

$$\sigma^2(R,t) = \int \frac{d^3k}{(2\pi)^3} |\delta_k(t)|^2 W_k^2(R). \tag{5.107}$$

(To be precise we should always write $\sigma^2(R,t)$ since $\sigma^2 \propto t^{4/3}$ is a function of time; we will, however, suppress this time-dependence when it is not likely to cause any confusion. We have also set $V = 1$.) As a first approximation, we may assume that the region with $\delta > \delta_c\ (t, t_i)$ (when smoothed on the scale R at time t_i) will form a gravitationally bound object with mass $M \propto \overline{\rho} R^3$ by the time t. (The precise form of the M–R relation depends on the window function used; for a step function $M = (4\pi/3)\ \overline{\rho} R^3$, while for a Gaussian $M = (2\pi)^{3/2} \overline{\rho} R^3$; see (5.36).) Here $\delta_c(t, t_i)$ is a critical value density contrast needed at time t_i so that $\delta_c \simeq 1$ by the time t. We will see in chapter 8 that $\delta_c(t, t_i) \cong (t_i/t)^{2/3}$. Therefore, the fraction of bound objects with mass greater than M will be

$$
\begin{aligned}
F(M) &= \int_{\delta_c(t,t_i)}^{\infty} P(\delta, R, t_i)\, d\delta = \frac{1}{\sqrt{2\pi}} \frac{1}{\sigma(R, t_i)} \int_{\delta_c}^{\infty} \exp\left(-\frac{\delta^2}{2\sigma^2(R, t_i)}\right) d\delta \\
&= \frac{1}{2} \operatorname{erfc}\left(\frac{\delta_c(t, t_i)}{\sqrt{2}\sigma(R, t_i)}\right),
\end{aligned}
\tag{5.108}
$$

where $\operatorname{erfc}(x)$ is the complementary error function. The mass function $f(M)$ is just $(\partial F/\partial M)$; the (comoving) number density $N(M, t)$ can be found by dividing this expression by $(M/\overline{\rho})$. Carrying out these operations we get

$$
N(M, t)\, dM = -\left(\frac{\overline{\rho}}{M}\right)\left(\frac{1}{2\pi}\right)^{1/2}\left(\frac{\delta_c}{\sigma}\right)\left(\frac{1}{\sigma}\frac{d\sigma}{dM}\right)\exp\left(-\frac{\delta_c^2}{2\sigma^2}\right) dM.
\tag{5.109}
$$

Given the power spectrum $|\delta_k|^2$ and a window function W_R one can explicitly compute the right hand side of this expression.

There is, however, one fundamental difficulty with the equation (5.108). The integral of $f(M)$ over all M should give unity; but it is easy to see that, for the expression in (5.108),

$$
\int_0^{\infty} f(M)\, dM = \int_0^{\infty} dF = \frac{1}{2}.
\tag{5.110}
$$

This arises because we have not taken into account the underdense regions correctly.

To see the origin of this difficulty more clearly, consider the interpretation of (5.108). If a point in space has $\delta > \delta_c$ when filtered at scale R, then that point should correspond to a system with mass greater than $M(R)$; this is taken care of correctly by equation (5.108). However, consider those points which have $\delta < \delta_c$ under this filtering. There is a *non-zero* probability that such a point will have $\delta > \delta_c$ when the density field is filtered with a radius $R_1 > R$. Therefore, to be consistent with the interpretation in (5.108), such points should *also* correspond to a region with mass greater than M. But (5.108) ignores these points completely and thus *underestimates* $F(M)$ (by a factor $(1/2)$).

To correct this mistake, we should replace (5.108) by the relation

$$F(M) = \int_{\delta_c}^{\infty} P(\delta, R)\, d\delta + \int_{-\infty}^{\delta_c} C(\delta_c, \delta)\, d\delta, \tag{5.111}$$

where the second term represents the probability P_u that a point which has $\delta < \delta_c$ at the filter scale R has the density $\delta > \delta_c$ at a larger filter scale $R_1 > R$. For a sequence of filter scales $R_1, R_2, \cdots R_n$, we obtain a sequence of Gaussian random fields parametrized by the dispersions $\Delta_1, \Delta_2, \cdots \Delta_n$. The probability that a point *remains underdense* (i.e. $\delta < \delta_c$) for all these filter scales is given by

$$P_{\rm survive} \equiv P_s = \int_{-\infty}^{\delta_c} d\delta_1 \int_{-\infty}^{\delta_c} d\delta_2 \cdots \int_{-\infty}^{\delta_c} d\delta_n\, P_J(\delta_1, \delta_2 \cdots \delta_n), \tag{5.112}$$

where $P_J[\delta_i]$ is the joint probability distribution that the Gaussian variables δ_i take the set of values simultaneously. Obviously, $(1 - P_s)$ gives the probability that a point becomes overdense somewhere along the sequence of filterings $(R_1, \cdots R_n)$.

The Gaussian variables obtained by different filtering scales, unfortunately, are not independent. We can see that

$$\begin{aligned} < \delta_a \delta_b > &= \int \frac{d^3\mathbf{k}}{(2\pi)^3} \frac{d^3\mathbf{p}}{(2\pi)^3} W_k(R_a) W_p^\star(R_b) < \delta_k \delta_p^\star > \mathrm{e}^{i(\mathbf{k}-\mathbf{p})\cdot\mathbf{x}} \\ &= \int \frac{d^3\mathbf{k}}{(2\pi)^3} W_k(R_a) W_k^\star(R_b) \sigma_k^2, \end{aligned} \tag{5.113}$$

is, in general, non-zero. Hence, calculating (5.112) is a non-trivial task.

We can look upon this process in a different, but equivalent, manner. Consider any one fixed location in space. When the filtering scale is some large value R_1 (with a dispersion Δ_1), let us assume that this point has a density contrast δ_1. When we reduce the scale to R_2, we will have a *new* probability distribution for δ; let the value of density contrast at our chosen point now be δ_2. As we go through a sequence of filtering scales, R_1, $R_2, \cdots R_n$ (in *decreasing* order), the density contrast performs a random walk through the points $(\delta_1, \delta_2 \cdots \delta_n)$. Suppose the *first* instance when δ crosses the value δ_c occurs at the kth step. Then we will attribute the chosen point to a mass $M_k \propto \overline{\rho} R_k^3$. Notice that, since $\delta < \delta_c$ for all the *higher* filtering scales – i.e. for all $(R_1, R_2 \cdots R_{k-1})$ – this point *does not* belong to any higher mass. (This takes care of the original difficulty in (5.108).) The random walk concept merely translates into a pictorial form the content of (5.112). This random walk problem is equally difficult to solve because the steps are not independent. In fact the answer will clearly depend on the correlation between the steps; and from (5.113) it follows that the answer will critically depend on the form of the window function.

Since no result which is independent of the form of the window function is possible, one might consider window functions for which the analysis is the simplest. This happens for window functions which are sharply truncated in **k**-space; that is for $W_k(R) = \theta(R^{-1} - k)$ which acts as a low-pass-filter in **k**-space. From (5.113) it follows that, for this window function,

$$< \delta_a \delta_{a+1} >= \sigma_a^2; \quad < \delta_a \delta_b >= \sigma_a^2 \quad (\text{for } a \leq b). \tag{5.114}$$

The step lengths of the random walks are $l_1 \equiv (\delta_2 - \delta_1)$, $l_2 = (\delta_3 - \delta_2)$, $\cdots l_a = (\delta_{a+1} - \delta_a)$ etc. Each of these is a Gaussian variable with the dispersion

$$< l_a^2 >=< (\delta_{a+1} - \delta_a)^2 >= \sigma_{a+1}^2 + \sigma_a^2 - 2 < \delta_{a+1}\delta_a >= \sigma_{a+1}^2 - \sigma_a^2, \tag{5.115}$$

and *zero* cross correlation:

$$\begin{aligned} < l_a l_b > &=< (\delta_{a+1} - \delta_a)(\delta_{b+1} - \delta_b) > \\ &=< \delta_{a+1}\delta_{b+1} > - < \delta_{a+1}\delta_b > - < \delta_a b_{b+1} > + < \delta_a \delta_b > \\ &= \sigma_{a+1}^2 - \sigma_{a+1}^2 - \sigma_a^2 + \sigma_a^2 = 0 \end{aligned} \tag{5.116}$$

for $(a+1) < b$; the other cases can be considered in a similar manner and can be shown to vanish. In other words, a sharp filter in the **k**-space produces a random walk in which each step l_a is independent and is drawn from a Gaussian variable with dispersion $(\sigma_{a+1}^2 - \sigma_a^2)$. In the continuum limit, this random walk is described by a diffusion equation. The probability $P(\delta, \sigma^2)$ that the particle is at $(\delta, \delta + d\delta)$ when the dispersion is σ^2 obeys the diffusion equation

$$\frac{\partial P}{\partial \sigma^2} = \frac{1}{2}\frac{\partial^2 P}{\partial \delta^2}. \tag{5.117}$$

We are interested in the probability that the trajectory reaches (δ, σ) without exceeding δ_c earlier, i.e. at smaller σ. This is equivalent to solving (5.117) with the boundary condition that there exists an absorbing barrier at $\delta = \delta_c$. This is straightforward and the answer is[8]

$$P(\delta, \sigma^2) = \frac{1}{\sigma\sqrt{2\pi}}\left[\exp\left(-\frac{\delta^2}{2\sigma^2}\right) - \exp\left(-\frac{(\delta - 2\delta_c)^2}{2\sigma^2}\right)\right]. \tag{5.118}$$

Integrating this expression from δ_c to ∞ and differentiating with respect to M, we get

$$dF(M) = \sqrt{\frac{2}{\pi}}\frac{\delta_c}{\sigma_2}\left(-\frac{\partial \sigma}{\partial M}\right)\exp\left(-\frac{\delta_c^2}{2\sigma^2}\right)dM, \tag{5.119}$$

or

$$N(M)\,dM = -\frac{\overline{\rho}}{M}\left(\frac{2}{\pi}\right)^{1/2}\frac{\delta_c}{\sigma^2}\left(\frac{\partial\sigma}{\partial M}\right)\exp\left(-\frac{\delta_c^2}{2\sigma^2}\right)dM, \tag{5.120}$$

which is precisely *twice* the value obtained by using (5.108). Of course, the normalization problem is solved automatically.

This result can be expressed in an explicit form for power law spectra with $|\delta_k|^2 \propto k^n$. In that case $\sigma^2 = c^2 M^{-(3+n)/3} t^{4/3}$ where c is some constant. Since the limit of linear theory occurs when $\sigma \simeq 1$, the characteristic mass scale $M_{\rm nl}$ which goes nonlinear at time t obeys the scaling relation $M_{\rm nl} \propto t^{4/(3+n)}$. Therefore, $c^2 = M_{\rm nl}^{(3+n)/3}(t_0)t_0^{-4/3}$ where t_0 is the epoch at which we need the $N(M)$; say, the present epoch. Since $\delta_c(t_0, t_i) \cong (t_i/t_0)^{2/3}$,

$$\frac{\delta_c(t_0,t_i)}{\sigma(M,t_i)} = \left(\frac{t_i}{t_0}\right)^{2/3}\frac{M^{(3+n)/6}}{t_i^{2/3}}\,\frac{t_0^{2/3}}{M_{\rm nl}^{(3+n)/6}(t_0)} = \left[\frac{M}{M_{\rm nl}(t_0)}\right]^{(3+n)/6} \tag{5.121}$$

where $M_{\rm nl}(t_0)$ is the scale which is becoming nonlinear today. The expression for $N(M, t_0)$ now becomes

$$\begin{aligned} N(M,t_0)\,dM = {} & \frac{\overline{\rho}}{\sqrt{2\pi}}\left(1+\frac{n}{3}\right)\left(\frac{M}{M_{\rm nl}}\right)^{(3+n)/6} \\ & \times \exp\left[-\frac{1}{2}\left(\frac{M}{M_{\rm nl}}\right)^{(3+n)/3}\right]\frac{dM}{M^2}. \end{aligned} \tag{5.122}$$

A different choice of the window function, in general, will give a different result especially in the low mass limit.

Exercises

5.1 Consider a distribution of points along a line according to the following prescription: (i) The points are in non-overlapping clumps of length a, with average density $\overline{n}$ and constant density n_c within the clumps. (ii) The clumps are distributed randomly on the line.
(a) Show that the quantity $< n(x+y)n(y) >$ is $\overline{n}^2$ for $x > a$; conclude that the correlation function $\xi(x) = 0$ for $x > a$.
(b) Show that, for $x < a$, the correlation function is $\zeta(x) = (n_c/\overline{n} - 1)(1 - x/a)$.
(c) How will the correlation function behave if the clumps themselves are grouped together in superclumps, with rules similar to (i) and (ii) above?

5.2 Assume that a spherical cluster of galaxies with n members has a radius $r_n = Bn^{\alpha}$. Let the number of such clusters with n members be $An^{-\beta}$. The mean density in any one cluster is $\overline{n} = (n/v_n)$, where $v_n = (4\pi/3)r_n^3$.

(a) Calçulate $< n(\mathbf{y})\, n(\mathbf{y}+\mathbf{x}) >$ retaining only the contribution from pairs lying in the same cluster and when one of the cluster members lies within the distance $(r_n - |\mathbf{x}|)$ of the cluster centre. Show that

$$< n(\mathbf{y}) n(\mathbf{y}+\mathbf{x}) > \simeq \sum_n \overline{n}^2 \frac{(r_n - |\mathbf{x}|)^3}{r_n^3} \frac{A n^{-\beta} v_n}{V}.$$

(b) Convert the sums to integrals and show that

$$n_0^2[1+\xi(x)] \simeq \frac{3}{4\pi} \frac{(\beta-2)}{\alpha} \frac{A}{B} n_0 x^{-\mu} I,$$

where $\mu = 3 + (\beta - 3)\alpha^{-1}$ and

$$I = \int_1^\infty t^{-(\mu+4)} (t-1)^3\, dt; \quad n_0 = (N/V).$$

Interpret this result.

(c) What is the range of α and β for which our approximations are valid?

5.3 Consider a set of particles inside a large volume V, with positions $\mathbf{x}_i(t)$, velocities $\mathbf{v}_i(t)$ etc. The density $\rho(\mathbf{x},t)$ due to these particles is given by

$$\rho(\mathbf{x},t) = \frac{1}{a^3(t)} \sum_i m_i \delta_D(\mathbf{x} - \mathbf{x}_i(t)),$$

where $\delta_D(z)$ is the Dirac delta function.

(a) Show that the density contrast, in Fourier space, is

$$\delta_k(t) = \frac{1}{M} \sum_i m_i \exp[i\mathbf{k}\cdot\mathbf{x}_i(t)]; \quad M = \sum_j m_j.$$

(b) Using the results

$$\ddot{\mathbf{x}} + 2\frac{\dot{a}}{a}\dot{\mathbf{x}} = -\frac{\nabla\phi}{a^2}; \quad \nabla^2\phi = 4\pi G a^2 \left[\rho(\mathbf{x}) - \rho_b\right]$$

where $\rho_b = [M/Va^3(t)]$, show that $\delta_{\mathbf{k}}(t)$ satisfies the equation

$$\ddot{\delta}_{\mathbf{k}} + 2\frac{\dot{a}}{a}\dot{\delta}_{\mathbf{k}} = 4\pi G \rho_b \delta_{\mathbf{k}} + A - B$$

with

$$A = 2\pi G \rho_b \sum_{\mathbf{k}'\neq 0,\mathbf{k}} \delta_{\mathbf{k}} \delta_{\mathbf{k}-\mathbf{k}'} \left\{ \frac{\mathbf{k}\cdot\mathbf{k}'}{k'^2} + \frac{\mathbf{k}\cdot(\mathbf{k}-\mathbf{k}')}{|\mathbf{k}-\mathbf{k}'|^2} \right\},$$

$$B = \frac{1}{M} \sum_j m_j (\mathbf{k}\cdot\dot{\mathbf{x}}_j)^2 \exp[i\mathbf{k}\cdot\mathbf{x}_j(t)].$$

(c) Derive the minimal k^4 fluctuations, discussed in the text, by estimating the effects of A and B.

(d) Consider a bunch of particles which form a gravitationally bound cluster. As the cluster becomes more and more tightly bound, the magnitude of B increases. It might seem, at first sight, that this effect can drive fluctuations at scales much bigger than the cluster. Show that no such unphysical result can occur.

5.4 Though the choice of window function is arbitrary, some window functions will have properties not shared by others. Consider, for example, the property that *both* $W(\mathbf{x})$ and its Fourier transform $\tilde{W}(\mathbf{k})$ should be positive definite. The Gaussian window function clearly satisfies this condition. Show that

$$W_1(r) = (1 + r^2)^{-n}$$

with $n > 2$ also has this property. On the other hand, prove that

$$W_2(r) = (1 + r^{2n})^{-1}$$

does not have, even though $W_1(r)$ and $W_2(r)$ have identical asymptotic behaviour. Give an intuitive explanation for this difference.

6
Microwave background radiation

6.1 Introduction

The inhomogeneities in the universe lead to two crucial effects: (i) they induce anisotropies in the distribution of relic background of photons present in the universe; (ii) they produce deviations from the pure Hubble velocity for distant galaxies. Both these effects can be studied using the formalism developed in chapters 4 and 5. This chapter and the next are devoted to the study of these effects.

The temperature anisotropies in the microwave background radiation (MBR, for short) are discussed in this chapter. Anisotropies due to peculiar velocities and fluctuations in the gravitational potential are derived in section 6.3 and discussed in detail in sections 6.4 and 6.5. Intrinsic anisotropies of the radiation field are taken up in section 6.6. Damping of anisotropies and the distortions which arise because of the astrophysical processes are studied in sections 6.7 and 6.8.

6.2 Processes leading to distortions in the MBR

We have seen in chapter 2 that the photons in the universe decoupled from matter at a redshift of about 10^3. These photons have been propagating freely in spacetime since then and can be detected today. In an *ideal* Friedmann universe, a *comoving* observer will see these photons as a blackbody spectrum at some temperature T_0. The deviations in the metric from that of the Friedmann universe, motion of the observer with respect to the comoving frame, the astrophysical processes which take place along the trajectory of the photon, can all lead to potentially observable effects in this radiation. These effects can be of two kinds: (i) The spectrum may not be strictly blackbody; that is, the radiation intensity $I(\omega)$ at different frequencies, coming to us from a particular direction in the sky, may not correspond to a Planck spectrum with a single temperature T_0. (ii) The spectrum may be Planckian in any given direction;

but the temperature of the radiation may be different in *different* directions; that is $T_0 = T_0(\theta, \phi)$ where (θ, ϕ) are the angular coordinates on the sky.

No deviations of either kind have been seen* in the MBR except for the so called 'dipole anisotropy' which is most likely to be due to our motion with respect to the comoving coordinates. There exist tight observational bounds on both the spectral deviations and the temperature anisotropies. This negative result is extremely significant. The processes leading to structure formation *must* have led to *some* anisotropies in the temperature distribution. Therefore, the observational upper bounds on the distortions of MBR serve as a powerful constraint on the models describing the structure formation. Between the two forms of distortions mentioned above, the bounds on $(\delta T/T)$ at different angular scales turn out to be more restrictive than the bounds on spectral deviations. Because of this reason, we shall concentrate on the angular anisotropies.

Angular anisotropies, leading to the directional dependence $T = T(\theta, \phi)$ of the temperature, arise primarily due to the following reasons: (1) If the observer is moving with respect to the comoving frame, then the photons reaching the observer from different directions with respect to the direction of motion will be redshifted (or blueshifted) by different amounts. This will lead to a dependence of T on $\cos\theta$, where θ is the angle between the direction of motion and direction of the photon. (2) If the matter which scattered the radiation in our direction had a peculiar velocity (with respect to the comoving frame) when the scattering occurred, then an effect similar to (1) can occur. Since the peculiar velocity of matter will be different at different locations in the last scattering surface, corresponding to different directions in the sky today, this will lead to an anisotropy. (3) If the local gravitational potential on the last scattering surface was different at different locations, then the photons will be climbing out of different gravitational potential wells and hence, will experience different amounts of redshift. This will lead to an angular dependence for the observed temperature. (4) The energy density of radiation in the last scattering surface can have an intrinsic inhomogeneity $\delta_R = (\delta\rho_R/\rho_R)$. This will appear as a temperature fluctuation $(\delta T/T)$. (5) Processes which take place along the path of the photon reaching us can also produce a $(\delta T/T)$. For example, consider a photon which travels through a large mass concentration which is collapsing. Since the depth of the gravitational potential well of a collapsing structure is increasing with time, the blueshift suffered by the photon as it travels towards the centre of the condensation will be lower than the redshift which occurs while it

* The analysis of COBE data now shows that MBR has an anisotropy of $(\Delta T/T) \simeq (1.1 \pm 0.2) \times 10^{-5}$. This result and its implications are discussed in Appendix C.

is emerging from the centre. Similarly, photons travelling through hot gas in a cluster will be affected by Compton scattering with the charged particles in the gas. These two processes can cause temperature distortions in specific directions. (6) Lastly, one must also consider effects which can *wipe out* any $(\delta T/T)$ produced by some of the processes mentioned above. Such an effect can arise primarily due to two reasons: Firstly, if the matter in the universe was re-ionized at any redshift $z_{\rm ion}$ (which is lower than $z_{\rm dec} \simeq 10^3$), then the photons would have interacted with those charged particles at $z \lesssim z_{\rm ion}$. Thus the 'last' scattering was not at $z \simeq z_{\rm dec}$ but at $z \simeq z_{\rm ion}$ and some of the anisotropies induced at $z \simeq z_{\rm dec}$ would be wiped out. Secondly, it should be noted that the decoupling was not an instantaneous event but took an interval of $\Delta z \simeq 80$ in the redshift space. This allows the photon to perform a random walk during the interval $(z_{\rm dec}, z_{\rm dec} + \Delta z)$ which can wipe out the original temperature fluctuations at small scales.

The physical processes described above operate at different characteristic length scales. For example, the effect in (3) will be governed by a (proper) length scale L over which the gravitational potential ϕ varies on the $z \simeq 10^3$ surface. Such a proper length L will subtend an angle $\theta = [L/d_A(z)]$ in the sky where $d_A(z)$ is the angular diameter distance discussed in chapter 2. For $z \gg 1$, $d_A(z) \simeq 2H_0^{-1}(\Omega z)^{-1}$ and we get

$$\theta(L) \cong \left(\frac{\Omega}{2}\right)\left(\frac{Lz}{H_0^{-1}}\right) = 34.4''(\Omega h)\left(\frac{\lambda_0}{1\,{\rm Mpc}}\right). \tag{6.1}$$

(As always, we quote the numerical values with the length scales by extrapolating them to the present epoch. Thus $\lambda_0 = L(1+z_{\rm dec}) \simeq L z_{\rm dec}$ is the proper length *today* which would have been L at the redshift of $z_{\rm dec}$. We will not bother to indicate this fact with a subscript '0' when no confusion is likely to arise.) In particular, consider the angle subtended by the region which has the size equal to that of the Hubble radius at $z_{\rm dec}$; that is, we take L to be $d_H(z_{\rm dec}) = H^{-1}[z = z_{\rm dec}] = H_0^{-1}(\Omega z_{\rm dec})^{-1/2} z_{\rm dec}^{-1}$ so that $\lambda = d_H(z_{\rm dec})\,[1 + z_{\rm dec}] \cong d_H(z_{\rm dec}) z_{\rm dec} \cong H_0^{-1}(\Omega z_{\rm dec})^{-1/2}$. Then:

$$\theta_H \equiv \theta(d_H) \cong 0.87^\circ \Omega^{1/2}\left(\frac{z_{\rm dec}}{1100}\right)^{-1/2} \simeq 1^\circ. \tag{6.2}$$

Therefore, angular separation of more than one degree in the sky would correspond to regions which were bigger than the Hubble radius at the time of decoupling. This fact makes the angle of about one degree a natural dividing line between what may be called 'small-angles' $(\theta < 1^\circ)$ and 'large-angles' $(\theta > 1^\circ)$. Since most of the non-gravitational processes operate at proper lengths smaller than Hubble radius, they predominantly produce anisotropies at small angles. On the other hand, fluctuations in gravitational field can cause anisotropies at large angles.

A linear scale of 100 Mpc today has evolved from a region of size $100(1+z_{\rm dec})^{-1}$ at recombination; this region subtends about (Ωh) degrees in the sky. Even a scale of 1000 Mpc corresponds to only about 10 degrees. Therefore, observations of $(\delta T/T)$ at these angular scales probe the density fluctuations at length scales much larger than those available for astronomical investigations today, say, in the study of galaxy–galaxy clustering. Further, we are now probing very large scales at which density perturbations are small and hence can be analyzed using the linear theory with complete confidence.

In the next few sections we shall discuss in detail each of the processes mentioned above. However, before doing that, it is worthwhile to review the observational results which are currently available[1].

These results are usually quoted in a different manner for large and small angles. The procedure adopted in most of the experiments measuring $(\delta T/T)$ at *large* angles can be summarized as follows: Consider an expansion of the relative anisotropy in the sky temperature $(\delta T(\theta,\phi)/T)$ in terms of the complete set of functions $Y_{lm}(\theta,\phi)$:

$$\left(\frac{\delta T}{T}\right) \equiv S(\theta,\phi) = \sum_l a_{lm} Y_{lm}(\theta,\phi). \tag{6.3}$$

Since S is generated by the inhomogeneities in the universe, we will see later that it can be expressed as a linear superposition of δ_k, each of which is a Gaussian random variable with zero mean. It follows that S and a_{lm} are also Gaussian random variables. Since the mean value of S is zero, full statistical information about S is contained in the two-point correlation function $C(\alpha) =< S(\mathbf{n})S(\mathbf{m}) >$ where $\mathbf{n}$ and $\mathbf{m}$ are unit vectors denoting two different directions (θ,ϕ) and (θ',ϕ') and $\cos\alpha = \mathbf{n}\cdot\mathbf{m}$. We can express this correlation function as:

$$\begin{aligned} C(\alpha) =< S(\mathbf{n})S(\mathbf{m}) > &= \sum_{lm}\sum_{l'm'} < a_{lm} a^\star_{l'm'} > Y_{lm}(\theta,\phi) Y^\star_{l'm'}(\theta',\phi') \\ &= \frac{1}{4\pi}\sum_l (2l+1) C_l\, P_l(\cos\alpha) \end{aligned} \tag{6.4}$$

where we have set $< a_{lm}a^\star_{l'm'} >= C_l \delta_{ll'}\delta_{mm'}$ and used the addition theorem for spherical harmonics. Observational limits on large angle anisotropies are expressed as bounds on the coefficients $C_l =< |a_{lm}|^2 >$ for various l or as bounds on $C(\alpha)$ over a specified range of the angle α. Given the observed values of $S(\theta,\phi)$ on a large region of the sky, one can determine the values of $< |a_{lm}|^2 >$ by angular averaging.

The results available now can be summarized[2] as follows: (1) Analysis of $S(\theta,\phi)$ shows a statistically significant detection of a dipole $(l=1)$ component in the anisotropy; $C_1^{1/2} \simeq 3\times 10^{-3}$. We shall see in a later section that such an anisotropy is expected because of the velocity $\mathbf{v}_{\text{obs}}$ of the observer. (It is, of course, *possible* that $S(\theta, \phi)$ has other sources of inherent dipole anisotropy; however, most physical mechanisms which produce a C_1 are also likely to produce C_2, C_3 etc. of comparable magnitude. Since this is not observed (see below), it seems reasonable to attribute C_1 to our peculiar velocity.) (2) The best *upper bounds* on the quadrupole anisotropy $C_2^{1/2}$ are

$$C_2^{1/2} \leq \begin{cases} 1.1 \times 10^{-4} & (\text{at } \lambda = 3\,\text{mm}) \\ 4.8 \times 10^{-5} & (\text{at } \lambda = 8\,\text{mm}) \\ 1.2 \times 10^{-4} & (\text{at } \lambda = 12\,\text{mm}) \end{cases} . \tag{6.5}$$

The bounds on $C_l^{1/2}$ with $l > 2$ are somewhat smaller but of the same order. (3) The bounds on $C(\alpha)$ are slightly better. In the entire angular range $10° < \alpha < 180°$, $C^{1/2}(\alpha)$ is less than 2×10^{-5}.

Similar results are available[3] from the COBE satellite experiment: The overall bound on the root-mean-square fluctuation of temperature – after the dipole term due to our motion has been subtracted out – is $(\Delta T/T) < 3 \times 10^{-5}$ at 90 per cent confidence level.* A similar bound, $C(\alpha) < 3 \times 10^{-5}$ applies to $C(\alpha)$ for $15° < \alpha < 165°$.

The observational situation is more complicated at small angles because we are now interested in $C(\alpha)$ at some specific value of α. Suppose two antennas, separated by a fixed angle α, measure the temperatures T_1 and T_2 simultaneously; the quantity $[(T_1 - T_2)/T_0]^2$ can then be measured and averaged over a large region of the sky by allowing the antennas to scan the sky. Simple algebra shows that

$$\left(\frac{\Delta T}{T}\right)^2 \equiv \left\langle \frac{(T_1 - T_2)^2}{T_0^2} \right\rangle = 2\left[C(0) - C(\alpha)\right]. \tag{6.6}$$

In most of the experiments, however, it is usual to use three antennas and measure the quantity $\langle [T_2 - (T_1 + T_3)/2]^2 \rangle / T_0$, where T_2 corresponds to temperature measured by the central beam. In this case,

$$\left(\frac{\Delta T}{T}\right)^2 = \frac{3}{2}C(0) + \frac{1}{2}C(2\alpha) - 2C(\alpha). \tag{6.7}$$

The results are quoted as bounds on the measured value of $(\Delta T/T)$.

---

* The analysis of COBE data now shows that MBR has an anisotropy of $(\Delta T/T) \simeq (1.1 \pm 0.2) \times 10^{-5}$. This result and its implications are discussed in Appendix C.

The best upper bounds on intermediate and small scales are[4] as follows:

$$\left(\frac{\Delta T}{T}\right) \leq \begin{cases} 3 \times 10^{-5} & (\alpha = 8^\circ,\ \lambda = 2.9\,\text{cm}) \\ 4.8 \times 10^{-5} & (\alpha = 6^\circ,\ \lambda = 500 - 3000\,\mu\text{m}) \\ 3 \times 10^{-5} & (\alpha = 1^\circ,\ \lambda = 3\,\text{mm}) \end{cases} \tag{6.8}$$

and

$$\left(\frac{\Delta T}{T}\right) \leq \begin{cases} 1.5 \times 10^{-5} & (\alpha = 7.2',\ \lambda = 1.5\,\text{cm}) \\ 3 \times 10^{-5} & (\alpha = 4.5',\ \lambda = 1.5\,\text{cm}) \\ 5 \times 10^{-5} & (\alpha = 50'',\ \lambda = 6\,\text{cm}) \\ 1.1 \times 10^{-4} & (\alpha = 30'',\ \lambda = 6\,\text{cm}) \end{cases}. \tag{6.9}$$

There have been several other measurements at very small angles ($10''$ to $40''$) using VLA; the bounds are usually in the range of a few times 10^{-4}. It should be clear that the typical values of anisotropy bounds are in the range 2×10^{-5} to 10^{-4}. Of these bounds, the limit of 4.8×10^{-5} at 6° and 1.5×10^{-5} at $7.2'$ turns out to be extremely restrictive.

6.3 Propagation of light in a perturbed universe

The effects due to the peculiar velocities and the gravitational potential (processes 1, 2 and 3 mentioned in the last section) can be derived[5] together from the redshift of a photon travelling in a perturbed Friedmann metric. Consider a spacetime with the line interval

$$ds^2 = dt^2 - a^2(t)[\delta_{ij} - h_{ij}(t, \mathbf{x})]\, dx^i\, dx^j\,, \tag{6.10}$$

where h_{ij} represent the perturbations around a $k = 0$ Friedmann universe. These perturbations are assumed to be small in the following sense: If we replace h_{ij} by ϵh_{ij} in any expression, we will ignore terms of order ϵ^2 and higher.

Consider a photon which was emitted at $t = t_e$ and reaches an observer at $t = t_0$ travelling along a null geodesic. Let the *unit* vector pointing along this null ray, *from* the observer *to* the source be n^i. Assuming the observer is at the origin, the location of the photon at some time t will be at

$$x^i = n^i \eta(t); \quad \eta(t) = \int_t^{t_0} \frac{dt'}{a(t')}. \tag{6.11}$$

(This is the path of propagation in the unperturbed metric; as we shall see, we do not need the perturbed trajectory.) We are interested in calculating the redshift suffered by this photon, to first order in h_{ij}.

This can be done by following exactly the same procedure as used in chapter 2. Consider two observers located at x^i and $x^i + \delta x^i$ along the path of the ray. The proper separation between these observers at any time t will be

$$\delta l = (-g_{ij}\delta x^i \delta x^j)^{1/2} \propto a\left(1 - \frac{1}{2}h_{ij}n^i n^j\right), \tag{6.12}$$

to first order in h_{ij}. Therefore, the rate of change of δl (in the lowest order in h_{ij}) will be

$$\begin{aligned} v = \frac{d\delta l}{dt} &= (\delta l)\frac{d}{dt}\{\ln[\delta l]\} \cong \left(\frac{\dot{a}}{a} - \frac{1}{2}\dot{h}_{ij}n^i n^j\right)(\delta l) \\ &= \left(\frac{\dot{a}}{a} - \frac{1}{2}\dot{h}_{ij}n^i n^j\right)(\delta t). \end{aligned} \tag{6.13}$$

In arriving at the last equality, we have used the fact that $\delta l = c\delta t = \delta t$ is the time taken by the photon to travel the proper distance δl. Since the proper velocity of one observer with respect to the second is v, the redshift of the photon will be

$$\frac{\delta\omega}{\omega} = -v = -\frac{\dot{a}}{a}\delta t + \frac{1}{2}\dot{h}_{ik}n^i n^k \delta t. \tag{6.14}$$

(This is precisely the argument used in chapter 2; it works because, in a region surrounding two infinitesimally separated observers, the laws of special relativity remain valid.) The first term represents the standard redshift $\omega \propto a^{-1}$ due to the Friedmann background; since we are interested in the deviations from this result arising due to the perturbations, it is best to rewrite equation (6.14) as follows:

$$\frac{\delta\omega}{\omega} + \frac{\dot{a}}{a}\delta t = \frac{\delta\omega}{\omega} + \frac{\delta a}{a} = \frac{\delta(\omega a)}{(\omega a)} = \frac{1}{2}\dot{h}_{ik}n^i n^k \delta t. \tag{6.15}$$

Further, since the deviations in temperature δT will be proportional to $\delta\omega$, we can write

$$\frac{\delta(Ta)}{(Ta)} = \frac{1}{2}\dot{h}_{ik}n^i n^k \delta t. \tag{6.16}$$

Integrating this expression *along the path of the ray* from t_e to t_0, we find

$$\ln\left(\frac{T_{\rm ob}a_{\rm ob}}{T_{\rm em}a_{\rm em}}\right) = \frac{1}{2}\int_{t_{\rm em}}^{t_{\rm ob}} \dot{h}_{ik}\left[t, x^i(t)\right] n^i n^k \, dt. \tag{6.17}$$

To avoid possible misunderstanding, we stress the following point: The quantity $\dot{h}_{ik}\ [t, x^i(t)]$ is obtained by taking the partial derivative of $h_{ik}(t, x^i)$ with respect to t (at constant x^i), *followed* by the replacement of x^i by the function $x^i(t) = n^i\eta(t)$. We cannot, therefore, write the result of the integration as $(h_{ik}^{(2)} - h_{ik}^{(1)})$.

Several further simplifications can be made. In the absence of perturbations, $T \propto a^{-1}$; blackbody photons emitted at an epoch when the temperature of the universe was T_{em} would have appeared as a blackbody radiation with temperature $(T_{\text{em}}a_{\text{em}}/a_{\text{ob}}) \equiv T_0$ today. The deviation ΔT between the observed temperature T_{ob} and the expected temperature T_0 will be a small quantity caused by h_{ik}. Therefore, we can replace the left hand side of (6.17) by

$$\ln\left(\frac{T_{\text{ob}}}{T_0}\right) = \ln\left(\frac{T_0 + \delta T}{T_0}\right) \simeq \left(\frac{\delta T}{T_0}\right). \tag{6.18}$$

Further notice that, on the right hand side, h_{ik} is already a first order quantity. So in evaluating the integral we only need the trajectory $x^i(t)$ of the photon to zeroth order accuracy; i.e. we only need the trajectory (6.11) in the unperturbed metric. Therefore, we can write the final answer as:

$$\left(\frac{\delta T}{T_0}\right) = \frac{1}{2}n^i n^k \int_{t_e}^{t_0} dt \left[\frac{\partial}{\partial t} h_{ik}(t, n^j \eta(t))\right]. \tag{6.19}$$

Once the form of the perturbations h_{ik} is known, we can compute $(\delta T/T_0)$.

The form of h_{ik} can be determined by linearizing Einstein's equations and solving them. Since we are working in the linear order, $h_{ik}(\mathbf{x}, t)$ can be expressed in terms of the independently evolving Fourier components $h_{ik}(\mathbf{p}, t)$. Each component contributes independently to $(\delta T/T)$ in (6.19). We will see later that the significant effects arise from the growing modes which are bigger than the Hubble radius at z_{dec}. For these modes we can ignore forces due to pressure gradients. In that case, the linearized equations governing the evolution of h_{ik} are the following:

$$\ddot{h} + 2\frac{\dot{a}}{a}\dot{h} = 8\pi G(\rho - \rho_b), \tag{6.20}$$

$$h_{ik,j}^{\ ,j} + h_{,ik} - h_{i,jk}^{j} - h_{k,ji}^{j} = 0, \tag{6.21}$$

and

$$\dot{h}_{,i} = \dot{h}_{i,k}^{k} \tag{6.22}$$

where $h = \text{Tr}\,(h_k^i)$. Further, the conservation of mass density $\partial(\sqrt{-g}\rho)/\partial t = 0$ implies the relation:

$$\dot{h} = 2\dot{\delta}; \quad \delta = (\rho - \rho_b)/\rho_b. \tag{6.23}$$

Combining (6.23) and (6.20) we see that δ satisfies the standard perturbation equation with $\delta \propto a$ in the matter dominated era. The solution to (6.22) can be expressed as

$$\frac{\partial h_{ik}}{\partial t} = -\frac{1}{4\pi}\frac{\partial^2}{\partial x^i \partial x^k}\int \frac{d^3x'}{|\mathbf{x} - \mathbf{x}'|}\left(\frac{\partial h}{\partial t}\right). \tag{6.24}$$

Taking the time derivative of (6.21) and using (6.24), we find that the equation (6.21) is identically satisfied. Thus (6.24), along with (6.23) completely determines h_{ik} in terms of δ.

We shall now rewrite these expressions in a more convenient form. Since the 'Newtonian' potential due to the density contrast δ is

$$\phi = -Ga^2\rho_b \int \frac{\delta(\mathbf{x}', t)\, d^3\mathbf{x}'}{|\mathbf{x} - \mathbf{x}'|}, \tag{6.25}$$

it follows that

$$\frac{\partial(\phi a)}{\partial t} = -G\rho_b a^3 \int \frac{\dot{\delta}(\mathbf{x}', t)\, d^3\mathbf{x}'}{|\mathbf{x} - \mathbf{x}'|}, \tag{6.26}$$

where we have used the fact that $\rho_b a^3$ is constant in time. From the linear perturbation theory in the matter dominated universe, we know that $\delta(t)$ grows as $a(t)$ and has the form $\delta(t) = a(t) f(\mathbf{x})$. Therefore $\phi(t, \mathbf{x})$ has the form

$$\phi(t, \mathbf{x}) = -Ga^2\rho_b \int \frac{a(t) f(\mathbf{x}')\, d^3x'}{|\mathbf{x} - \mathbf{x}'|} = -G\rho_b a^3 \int \frac{f(\mathbf{x}')}{|\mathbf{x} - \mathbf{x}'|} d^3x'. \tag{6.27}$$

Since $\rho_b a^3$ is constant in time, it follows that $\phi(t, \mathbf{x}) \equiv \phi(\mathbf{x})$ is also constant in time with *no* explicit time dependence. On the other hand,

$$\begin{aligned} \frac{\partial}{\partial t}(\phi a) = \dot{a}\phi &= \frac{\partial}{\partial t}\left[-G\rho_b a^3 \int \frac{\delta(\mathbf{x}', t)\, d^3\mathbf{x}'}{|\mathbf{x} - \mathbf{x}'|}\right] \\ &= -G\rho_b a^3 \int \frac{\dot{\delta}\, d^3\mathbf{x}'}{|\mathbf{x} - \mathbf{x}'|} = -\frac{1}{2} G\rho_b a^3 \int \frac{\dot{h}\, d^3x'}{|\mathbf{x} - \mathbf{x}'|}, \end{aligned} \tag{6.28}$$

where we have used the constancy of $\rho_b a^3$ and the relation (6.23). Therefore

$$\int \frac{\dot{h}\, d^3x'}{|\mathbf{x} - \mathbf{x}'|} = -\frac{2\dot{a}}{(G\rho_b a^3)}\phi. \tag{6.29}$$

Using this result in (6.24) we get

$$n^i n^k \frac{\partial h_{ik}}{\partial t} = \frac{\dot{a} n^i n^k}{(2\pi G\rho_b a^3)} \frac{\partial^2 \phi}{\partial x^i \partial x^k} \equiv \frac{\dot{a}}{Q}\phi_{ik} n^i n^k, \tag{6.30}$$

with the notation $\phi_{ik} = (\partial^2\phi/\partial x^i \partial x^k)$ and $Q \equiv 2\pi G\rho_b a^3$. The integrand in (6.19) involves this quantity evaluated along the photon path $x^i = n^i \eta(t)$. Consider any function $f(x^i)$ evaluated along a path $x^i = n^i\eta(t)$. Clearly, $(df/d\eta) = (\partial f/\partial x^i) n^i$. Therefore, along the path, we can write

$$\phi_{ik} n^i n^k = \left(\frac{d^2\phi}{d\eta^2}\right), \tag{6.31}$$

where $\phi = \phi(x^i = n^i\eta) = \phi(\eta)$ is now treated as a function of η. From (6.11), it follows that $dt = -a\,d\eta$ with the minus sign arising from the lower limit of the integral; converting the variable from t to η, (6.19) becomes

$$\frac{\delta T}{T_0} = \frac{1}{2Q}\int dt\,\dot{a}\left(\frac{d^2\phi}{d\eta^2}\right) = \frac{1}{2Q}\int d\eta\left(\frac{da}{d\eta}\right)\left(\frac{d^2\phi}{d\eta^2}\right). \tag{6.32}$$

Integrating by parts, this becomes

$$\frac{\delta T}{T_0} = \left[\frac{1}{2Q}\left(\frac{da}{d\eta}\right)\left(\frac{d\phi}{d\eta}\right)\right]_{\eta_1}^{\eta_1} - \frac{1}{2Q}\int d\eta\frac{d^2a}{d\eta^2}\left(\frac{d\phi}{d\eta}\right). \tag{6.33}$$

In the matter dominated case, $a \propto t^{2/3}$ so that $\eta \propto t^{1/3}$ and $a \propto \eta^2$. Therefore, $(d^2a/d\eta^2)$ in the second term is a constant. Further notice that, in the expression

$$\frac{1}{2Q}\left(\frac{d^2a}{d\eta^2}\right)\left(\frac{d\phi}{d\eta}\right)d\eta = \frac{1}{(4\pi G\rho_b a^3)}\left[a\frac{d}{dt}(a\dot{a})\right]\dot{\phi}\,dt, \tag{6.34}$$

a appears three times both in the numerator and denominator. Thus the proportionality constant in the $a \propto t^{2/3}$ relation is irrelevant. Taking just $a = t^{2/3}$, we find the right hand side of (6.34) to be

$$\frac{\dot{\phi}\,dt}{(4\pi G\rho_b t^2)}\left[t^{2/3}\frac{d}{dt}\left(t^{2/3}\frac{2}{3}t^{-1/3}\right)\right] = \frac{2}{9}\frac{\dot{\phi}\,dt}{(4\pi G\rho_b t^2)} = \frac{1}{3}\dot{\phi}\,dt. \tag{6.35}$$

In arriving at the last step, we have used the relation $6\pi G\rho_b t^2 = 1$ for the matter dominated phase. Consider now the first term in equation (6.33), which can be written as

$$-\frac{1}{(4\pi G\rho_b a^3)}(a\dot{a})\left(\frac{d\phi}{d\eta}\right) = -\frac{(\dot{a}/a)}{(4\pi G\rho_b)}\frac{1}{a}n^i\left(\frac{\partial\phi}{\partial x^i}\right). \tag{6.36}$$

Note that if the growing solution to the perturbation equation is δ then, since $\delta \propto a$, $(\dot{a}/a) = (\dot{\delta}/\delta)$. Using this fact, we can relate this term to the peculiar velocities generated by the density inhomogeneities. We have seen in chapter 3 that the peculiar velocity v^i induced by the potential ϕ is

$$v^i = -\frac{(\dot{\delta}/\delta)}{(4\pi G\rho_b)}\left(\frac{1}{a}\frac{\partial\phi}{\partial x^i}\right). \tag{6.37}$$

Therefore the first term can be expressed as

$$-\frac{(\dot{a}/a)}{(4\pi G\rho_b)}\left(\frac{1}{a}\right)n^i\left(\frac{\partial\phi}{\partial x^i}\right) = \mathbf{v}\cdot\mathbf{n}. \tag{6.38}$$

Using this result and (6.35) in (6.33) we find that the temperature fluctuation becomes

$$\frac{\delta T}{T_0} = \mathbf{n}\cdot(\mathbf{v}_{\rm ob}-\mathbf{v}_{\rm em}) - \frac{1}{3}\int \dot{\phi}\,dt = \mathbf{n}\cdot(\mathbf{v}_{\rm ob}-\mathbf{v}_{\rm em}) - \frac{1}{3}\left[\phi(0)-\phi(\mathbf{x}_{\rm em})\right]. \tag{6.39}$$

Each of these terms has a direct interpretation: $(\mathbf{n}\cdot\mathbf{v}_{\rm ob})$ is the Doppler shift due to the motion of the observer with respect to the comoving frame; $(-\mathbf{n}\cdot\mathbf{v}_{\rm em})$ is the corresponding Doppler effect in the emitting surface; the $(1/3)[\phi(0)-\phi(x_{\rm em})]$ is caused by the local variations in the gravitational potential at the source and the observer. The term $(1/3)$ $\phi(0)$ merely adds a constant to $(\delta T/T_0)$ and will not introduce any directional dependence in $(\delta T/T_0)$. The other three terms, of course, depend on $\mathbf{n}$ and hence vary from direction to direction.

A cautionary note regarding the interpretation in the last paragraph should be made. The above analysis was done in the gauge in which $g_{00}=1$ and $g_{0j}=0$, and the contribution in (6.19) was from $\dot{h}_{ik}$. It is possible to choose other gauges, for example, one with $g_{00}=(1+2\phi_N)$ and $\dot{h}_{ik}=0$. Detailed computation will show that the numerical value of $(\delta T/T)$ as measured by any specified observer will be the same (as to be expected); but the separation of the contribution, as due to Doppler shift and gravitational potential, depends on the gauge which is chosen. Thus the factor $(1/3)$ in the second term has no invariant significance.

An order of magnitude estimate of these terms can be easily made. Since the perturbation in the gravitational potential is

$$\begin{aligned}\delta\phi &\simeq (2G\delta M/L) \simeq (8\pi G/3)(\delta\rho L^2) \simeq (8\pi G\rho_b/3)(\delta\rho/\rho_b)L^2\\ &\simeq (\delta\rho/\rho_b)(HL)^2,\end{aligned} \tag{6.40}$$

and $v\simeq H^{-1}(\delta\phi/L)$, the two terms contribute the amounts

$$\left(\frac{\delta T}{T}\right)_{\rm Doppler} \simeq \left(\frac{\delta\rho}{\rho}\right)(LH) \simeq \left(\frac{\delta\rho}{\rho}\right)\left(\frac{L}{ct}\right) \tag{6.41}$$

and

$$\left(\frac{\delta T}{T}\right)_{\rm potential} \simeq \frac{1}{3}\left(\frac{\delta\rho}{\rho}\right)\left(\frac{L}{ct}\right)^2, \tag{6.42}$$

where the expressions in the right hand side are to be evaluated on the $z\approx z_{\rm dec}$ surface. Since the angle $\theta(L)$ subtended on the sky by a scale L is proportional to L, it is clear that the effect due to potential will be dominant at large angles $(\theta>\theta_H)$ while the Doppler term will dominate at small angles.

Most of the conclusions derived above will be valid even in a universe with $\Omega\neq 1$. The magnitude of the radius of curvature of the universe (today) is $l_{\rm curv}=H_0^{-1}|\Omega-1|^{-1/2}$. This length subtends an angle $\theta_{\rm curv}\simeq$

$(\Omega/2)|(\Omega-1)|^{-1/2}$ in the sky. For $\theta \ll \theta_{\rm curv}$, we can ignore the effects due to the curvature.

We shall now work out the detailed form of each of the terms contributing to $(\Delta T/T)$, starting with the $(1/3)\ \phi(\mathbf{x}_{\rm em})$ term.

6.4 Anisotropy due to variations in the potential

The contribution arising from the variations in the gravitational potential at the last scattering surface, called the Sachs–Wolfe effect, is given by the $(1/3)\phi(\mathbf{x}_{\rm em})$ term in (6.39). Strictly speaking, we should write $\phi(\mathbf{x}_{\rm em}, t_{\rm em})$; however since ϕ is independent of time in the matter dominated phase (in a universe with $\Omega = 1$), $\phi(\mathbf{x}_{\rm em}, t_{\rm em}) = \phi(\mathbf{x}_{\rm em}, t_0)$ and can be evaluated at the present epoch. Since the scale of a_0 is arbitrary, we shall choose for convenience $a_0 = a(t_0) = 1$. Then using (6.25), we get

$$\left(\frac{\delta T}{T}\right) = -\frac{1}{3} G\rho_b(t_0) \int d^3\mathbf{x}' \frac{\delta(\mathbf{x}', t_0)}{|\mathbf{x}_{\rm em} - \mathbf{x}'|} \tag{6.43}$$

where

$$\mathbf{x}_{\rm em} = \hat{\mathbf{n}} \int_{t_{\rm em}}^{t_0} \frac{dt}{a(t)} \cong \hat{\mathbf{n}} \int_0^{t_0} \frac{dt}{a(t)} = \hat{\mathbf{n}}\eta \tag{6.44}$$

since $t_{\rm em} \ll t_0$. Fourier transforming the expression and using $G\rho_b = (3H^2/8\pi)$ we get

$$\left(\frac{\delta T}{T}\right) = -\frac{1}{2} H_0^2 \int \frac{d^3\mathbf{k}}{(2\pi)^3} \frac{\delta_k}{k^2} \exp(-i\mathbf{k}\cdot\mathbf{x}_{\rm em}) \equiv \int \frac{d^3\mathbf{k}}{(2\pi)^3} \Phi_{\mathbf{k}} \exp(-i\mathbf{k}\cdot\mathbf{x}_{\rm em}), \tag{6.45}$$

where $\Phi_{\mathbf{k}} = -\left[(H)^2/2\right]\left[\delta_k/k^2\right]$ is proportional to the Fourier transform of ϕ. The vector $\mathbf{x}_{\rm em}$ has fixed magnitude; therefore $(\delta T/T)$ is essentially a function of the two angles (θ, ϕ) on the sky.

Though this expression gives the value of the temperature in any direction in the sky, it is not yet in a form which is suitable for comparison with observations. It is nearly impossible to measure – with any useful level of accuracy – the temperature along a *single* direction. What can be achieved is some kind of an average over all (θ, ϕ). It is, therefore, necessary to derive from $(\delta T(\theta, \phi)/T)$ some characteristic parameters which can be compared with observations.

We saw earlier that the observational results for large angles are quoted as bounds on the coefficients C_l which occur in the expansion of the angular correlation function $C(\alpha)$ in terms of the $P_l(\cos\alpha)$. By computing the angular correlation function from (6.45) we can determine the values of C_l in term of the power spectrum for the inhomogeneities[6]. This will allow us to compare the theory with observation. Expanding $(\delta T/T)$ in terms of $Y_{lm}(\theta, \phi)$ and using the orthonormality of $Y_{lm}(\theta, \phi)$ we can express the expansion coefficients a_{lm} in terms of $\Phi_{\mathbf{k}}$. Averaging this expression, it is straightforward to show that

$$< |a_{lm}|^2 >= C_l = \frac{V^{-1}H_0^4}{(2\pi)} \int_0^\infty dk \frac{|\delta_k|^2}{k^2} |j_l(k\eta)|^2 \tag{6.46}$$

where j_l is the spherical Bessel function. Using the expression on the right hand side of (6.46), one can theoretically predict the value of C_l provided the power spectrum $|\delta_k|^2$ is known. This expression (6.46) is in a form which allows direct comparison between theory and observation.

Let us compute the values for C_l predicted by the theory when the spectrum is a power law with $|\delta_k|^2 = VAk^n$ where A is the amplitude and n is the spectral index. In this case, the integral in (6.46) can be evaluated exactly provided $n < 3$; we get:

$$C_l = \left\{ \frac{AH_0^{n+3}}{16} \frac{\Gamma(3-n)}{\Gamma^2[(4-n)/2]} \right\} \frac{\Gamma[(2l+n-1)/2]}{\Gamma[(2l+5-n)/2]}. \tag{6.47}$$

This expresses the C_l for various l in terms of the (unknown) amplitude A of the power spectrum and the index n. In principle, a positive measurement of the C_l will allow us to determine n and A.

To make a concrete prediction, we need an independent estimate of the normalization constant A. For scenarios with cold dark matter, we can use the procedure described in the last chapter, and relate A to the mass fluctuations. In fact, we can directly connect up C_l and $(\Delta M/M)_R^2$ in the following way: We have seen in chapter 5 that the fluctuation $\sigma_M^2(R) = (\Delta M/M)_R^2$ in the mass, when smoothed by a window function of radius R, is given by the value of $[(k^3|\delta_k|^2)/2\pi^2]$ at $k \simeq R^{-1}$; i.e. $\sigma_M^2(R) \propto AR^{-(n+3)}$. Comparing with (6.47), we see that $C_l \propto \sigma_M^2(H_0^{-1})$; that is, C_l is proportional to the value of $(\Delta M/M)^2$ at $R = H_0^{-1}$, which probes the mass fluctuation at the scale of the Hubble radius today. As a specific example, consider the value of C_2 for the $n = 1$ (scale-invariant) spectrum. The mass fluctuation inside a sphere of radius R can be defined, using a spherical top-hat window function, to be

$$\begin{aligned} \sigma_M^2(R) &= \int_0^\infty \frac{(Ak)k^3}{2\pi^2}.9 \left[\frac{\sin kR}{k^3R^3} - \frac{\cos kR}{k^2R^2}\right]^2 \frac{dk}{k} \\ &= \frac{9A}{2\pi^2R^4} \int_0^\infty \frac{dx}{x^3} [\sin x - x\cos x]^2 \simeq \frac{A}{8\pi^2R^4}. \end{aligned} \tag{6.48}$$

In arriving at the last step we have approximated the integrand by its value near zero and truncated the integral at $x = 1$. Evaluating C_2 from (6.47) and using the above result, we can express C_2 in terms of σ_M:

$$C_2 = \frac{AH_0^4}{24\pi} = \frac{H_0^4}{24\pi}(8\pi^2R^4\sigma_M^2) = \frac{\pi}{3}(H_0R)^4 \left(\frac{\Delta M}{M}\right)_R^2, \tag{6.49}$$

Since $(\Delta M/M)^2 \propto R^{-(n+3)} \propto R^{-4}$ (for $n = 1$), the right hand side is independent of the scale R at which it is evaluated. Using the normalization that $(\Delta M/M) \approx 0.25$ at $R \approx 30h^{-1}$ Mpc, we get

$$C_2^{1/2} \simeq 2 \times 10^{-5}, \tag{6.50}$$

which is marginally less than the observational bounds quoted above. Similar calculations can be carried out[7] for other values of l and other forms of spectrum.

The above result assumes that no biasing has been invoked. If the galaxy–galaxy correlation function $\xi_{\rm gg}$ is taken to be $b^2\xi_{\rm true}$, then $C_2^{1/2}$ is reduced further by the factor b^{-1}. Since b can be as high as 2.5, the predicted value of $C_2^{1/2}$ is further lowered.

The situation is more complicated for the hot dark matter scenarios because the normalization of the spectrum is not straightforward. Since galaxies form through fragmentation of larger structures in these models, it may not be proper to normalize the spectrum using the value of ξ_{gg} observed today. It is conventional to set the correlation function to unity at some past epoch with redshift z_{nl}. In that case, the corresponding calculation will give

$$C_2^{1/2} \simeq 1.6 \times 10^{-6} h^{-2}(1 + z_{\rm nl}). \tag{6.51}$$

This is not a strong constraint either; if $h = 0.5$, the observational bounds are satisfied for $z_{\rm nl} \lesssim 7$.

Note that our result for C_l allows us to express all C_l with $l \neq 2$ in terms of C_2:

$$C_l = C_2 \frac{\Gamma[l + (1/2)(n-1)]}{\Gamma[l + (1/2)(5-n)]} \frac{\Gamma[(9-n)/2]}{\Gamma[(3+n)/2]}. \tag{6.52}$$

In particular $C_1 = (3/2)C_2$. Given the bounds on C_2, the observed value of C_1 cannot be generated by this power spectrum. This is one of the reasons for attributing C_1 entirely to Doppler effect which will be discussed in the next section.

The expression (6.45) also allows us to determine the characteristic angular scales at which each mode contributes. Since the contribution to $< (\delta T/T)^2 >$ from an interval $(k, k + d^3k)$ is proportional to $d^3k|\delta_k|^2 k^{-4}$ $\propto (dk/k)|\delta_k|^2 k^{-1}$, each logarithmic interval in k contributes to $(\delta T/T)$ the amount $[|\delta_k|^2 k^{-1}]^{1/2}$ at the time of decoupling. We have seen in chapter 4 that $k^3|\delta_k|^2$ varies as $\lambda^{2\alpha-3}$ for $\lambda < d_H(t_{\rm dec})$ and as $\lambda^{2\alpha-5}$ for $\lambda > d_H(t_{\rm dec})$. (The index α is related to n by $2\alpha = (4 - n)$.) Therefore,

$$\left(\frac{\delta T}{T}\right) \propto k^{-2}\left[k^3|\delta_k|^2\right]^{1/2} \simeq \begin{cases} \lambda^{\alpha+1/2} \propto \theta^{\alpha+1/2} & (\lambda < d_H) \\ \lambda^{\alpha-3/2} \propto \theta^{\alpha-3/2} & (\lambda > d_H) \end{cases}. \tag{6.53}$$

In arriving at the last relation we have used the fact that $\theta \propto \lambda$ where $\theta(\lambda)$ is the angle subtended in the sky by a proper wavelength λ. The length d_H corresponds to an angle of $\theta_H \simeq 1°$ (see (6.2)). In particular, for the scale invariant spectrum with $\alpha = (3/2)$, the angular dependence will be

$$\left(\frac{\delta T}{T}\right) \propto \begin{cases} \theta^2 & (\theta < 1°;\ \text{small angles}) \\ \text{constant} & (\theta > 1°;\ \text{large angles}) \end{cases} . \tag{6.54}$$

This effect is insignificant at small angles and dominates at $\theta \gtrsim 1°$.

In obtaining the above result, we have identified the $h_{\alpha\beta}$ which appears in (6.19) with the $h_{\alpha\beta}$ produced by the matter distribution. It is, however, possible to add to this $h_{\alpha\beta}$ any solution of the homogeneous part of the linearized Einstein equations. From (6.20), (6.21) and (6.22), it follows that we can choose these solutions to satisfy the conditions

$$h^{\alpha}_{\beta,\alpha} = 0; \quad h = 0; \quad h^{\alpha\beta,\mu}_{\ \ \ \ ,\mu} = 0. \tag{6.55}$$

These equations describe the transverse, traceless, gravitational wave modes propagating in the Friedmann universe. Given any specific origin for these waves, we can determine $h_{\alpha\beta}$ and – through (6.19) – the temperature anisotropies which are produced[8].

An order of magnitude estimate of this effect can be made as follows: The energy density of the gravitational waves is given by $\rho_g \simeq (32\pi G)^{-1} < [\dot{h}^2_{\alpha\beta}] > \sim (32\pi G)^{-1}(\omega^2 h^2)$ where ω is the frequency and h is the amplitude. Hence,

$$\left\langle \left(\frac{\delta T}{T}\right)^2 \right\rangle \simeq \frac{1}{4} < \omega h H_0^{-1} >^2 \simeq \frac{1}{4H_0^2}\omega^2 h^2 \simeq \frac{8\pi G\rho_g}{H_0^2} \simeq 3\Omega_g. \tag{6.56}$$

The bound $(\delta T/T) < 10^{-4}$ will therefore imply $\Omega_g < 3 \times 10^{-9}$. Thus the energy density of the stochastic background of the gravitational waves is severely constrained.

More specific bounds can be imposed if the spectral distribution of the gravitational waves is known. For example, a gravitational wave with $\lambda \sim H_0^{-1}$ will produce a quadrupole anisotropy in the temperature distribution. The currently available upperbound on the quadrupole anisotropy, $C_2^{1/2}$, will constrain the amplitude of the waves with $\lambda \simeq H_0^{-1}$ to be less than about 10^{-4}.

6.5 Anisotropies due to peculiar velocities

The first two terms in (6.39) represent the contribution to $(\delta T/T)$ due to the peculiar velocity of the observer and the source. Let us first consider the $(\mathbf{v}_{\rm obs} \cdot \mathbf{n})$ term. This gives

$$\frac{\delta T}{T} = v \cos\theta, \tag{6.57}$$

where θ is the angle between the direction of motion and the direction of observation. In the expansion of $(\delta T/T)$ in spherical harmonics, this contributes a dipole $(l = 1)$ term.

Such a dipole anisotropyhas been detected. From the observed dipole moment of the $(\delta T/T)$ we can determine both v and the direction of our motion with respect to the comoving frame. The best estimate[9] on dipole anisotropy currently available is from COBE, which gives $(v/c) = 0.00122 \pm 0.00006$ corresponding to a speed of $365 \pm 18\,\mathrm{km\,s^{-1}}$.

This motion is due to the addition of several velocity vectors: (i) The motion of the earth around the sun; (ii) The motion of the Sun in the galaxy; (iii) The motion of the Milky Way towards Andromeda in the Local Group and (iv) any motion of our Local Group with respect to the comoving frame. Subtracting the first three contributions we can determine the motion of our Local Group. Calculations lead to a speed of about $600\,\mathrm{km\,s^{-1}}$ for the Local Group towards the direction $l \simeq 270°, b \simeq 30°$. We will discuss this motion in detail in the next chapter.

In contrast to $\mathbf{v}_{\rm obs}$, the velocity $\mathbf{v}_{\rm em}$ at the surface of last scattering is caused by the velocity field associated with the density contrast. From (6.37), we find that

$$\mathbf{n} \cdot \mathbf{v}(t, \mathbf{x}) = -\frac{(\dot{\delta}/\delta)}{4\pi G\rho_b a} \mathbf{n} \cdot \nabla\phi = -\frac{\dot{a}}{4\pi G\rho_b a^2} (\nabla\phi) \cdot \mathbf{n}, \tag{6.58}$$

since $(\dot{\delta}/\delta) = (\dot{a}/a)$. In Fourier space, this relation becomes

$$\begin{aligned} \mathbf{n} \cdot \mathbf{v}(t, \mathbf{k}) &= \frac{\dot{a}}{(4\pi G\rho_b a^2)} (i\mathbf{k} \cdot \mathbf{n})\phi_{\mathbf{k}} = \frac{\dot{a}(i\mathbf{k} \cdot \mathbf{n})}{(4\pi G\rho_b a^2)} \frac{(4\pi G\rho_b a^2)}{k^2} \delta_k \\ &= \dot{a}(i\mathbf{k} \cdot \mathbf{n}) \left(\frac{\delta_k}{k^2}\right). \end{aligned} \tag{6.59}$$

The contribution of this term to $(\delta T/T)$ can now be analyzed exactly as in section 6.4, since (6.59) is identical in structure to (6.45) except for the extra $(\mathbf{k} \cdot \mathbf{n})$ term. Because of this additional factor, the angular dependence is now different from (6.53). We see from (6.59) that, for this contribution,

$$\left\langle \left(\frac{\delta T}{T}\right)^2 \right\rangle_k \propto k^2 \frac{|\delta_k|^2}{k^4} d^3k \propto |\delta_k|^2 k \left(\frac{dk}{k}\right), \tag{6.60}$$

so that each logarithmic interval provides an amount $|\delta_k|^2 k$. The angular dependence will, therefore, be

$$\left(\frac{\delta T}{T}\right)_k \propto k^{-1}[k^3|\delta_k|^2]^{1/2} \simeq \begin{cases} \lambda^{\alpha-1/2} \propto \theta^{\alpha-1/2} & (\theta < \theta_H) \\ \lambda^{\alpha-5/2} \propto \theta^{\alpha-5/2} & (\theta > \theta_H) \end{cases} . \quad (6.61)$$

For the scale invariant spectrum ($\alpha = 3/2$),

$$\left(\frac{\delta T}{T}\right) \simeq \begin{cases} \theta & (\theta < \theta_H) \\ \theta^{-1} & (\theta > \theta_H) \end{cases} . \quad (6.62)$$

Quite clearly, this effect is smaller than the one due to the gravitational potential for large angles: $\theta > \theta_H$. This is to be expected because peculiar velocities will have very little power on scales bigger than the Hubble radius. It dominates over the Sachs–Wolfe effect at small scales (by one power of θ) but neither effect is very significant at small θ.

6.6 Intrinsic anisotropies

The effects discussed in the last two sections will produce a $(\delta T/T)$ even if the MBR had no distortions in the last scattering surface. However, we do expect certain intrinsic temperature fluctuation in this surface because of the fluctuations δ_R in the energy density of radiation. Since $\rho_R \propto T^4$, it follows that $(\delta T/T) = (1/4)(\delta\rho_R/\rho_R) = (1/4)\delta_R$.

To estimate δ_R exactly, we have to solve the coupled differential equations governing the evolution of δ_B, $\delta_{\rm DM}$ and δ_R during the process of decoupling[10]. However, the essential features of the result can be obtained by the following argument.

At scales smaller than the Hubble radius, Thomson scattering keeps the photons and the baryons tightly coupled. In such a case, we find that (see exercise 4.9(g)) $\delta_R \approx (4/3)\delta_B$. Therefore, to within an order of magnitude, $(\delta T/T) \approx (1/3)\delta_B(t_{\rm dec})$ is determined essentially by the value of δ_B at the time of decoupling.

The value of δ_B on the last scattering surface depends on the nature of the dark matter. Consider, as an extreme example, a high density ($\Omega > 0.5$) universe in which there is no non-baryonic dark matter. Since δ_B in such a universe will grow only by a factor $(a_0/a_{\rm dec}) \simeq 10^3$ from the time of decoupling till today, and since $\delta_B > 1$ today, $\delta_B(t_{\rm dec})$ should be at least 10^{-3}. Therefore,

$$\left(\frac{\delta T}{T}\right) \geq \frac{1}{3} \times 10^{-3} \simeq 3 \times 10^{-4}. \quad (6.63)$$

If $\Omega < 0.5$, then the growth of perturbations effectively stops at $z_c \simeq (\Omega^{-1} - 2)$. In that case the fluctuations in δ_B could have only grown by the factor

$$\left(\frac{a_c}{a_{\rm dec}}\right) = \frac{1+z_{\rm dec}}{1+z_c} \simeq \frac{10^3\Omega}{(1-\Omega)}. \tag{6.64}$$

Therefore,

$$\left(\frac{\delta T}{T}\right) \geq \frac{1}{3}\left(\frac{1}{\Omega}-1\right)10^{-3} \simeq 3\times 10^{-3} \quad (\text{for } \Omega = 0.1). \tag{6.65}$$

Since $(\delta T/T) \propto \delta_B$, each Fourier mode with proper wavelength λ will contribute to $(\delta T/T)$ at the angular scale $\theta(\lambda) = 34''(\Omega h)\ (\lambda/1\,\text{Mpc})$. Galactic and larger scales correspond to the $(50''\text{–}5')$ range. These predicted anisotropies are much higher than the observational bounds in these angular scales. The current upper limit on the relevant angular scales is about 5×10^{-5}. This bound rules out a purely baryonic universe.

The above estimate, however, is somewhat oversimplified. We shall see in the next section that the finite duration of decoupling reduces the value of $(\delta T/T)$ at angular scales below about $5'$. The precise estimate of $(\delta T/T)$, therefore, requires a numerical integration of the perturbation equations. However, such detailed calculations still predict values in excess of the observed bounds.

It is important to stress that the above constraint is extremely powerful. We have seen earlier that the primordial nucleosynthesis rules out baryonic universes with Ω greater than about 0.2. However, there is no direct observational evidence demanding Ω to be larger than 0.2; therefore one had to take seriously the possibility of a purely baryonic universe with $\Omega \lesssim 0.2$. The calculation performed above rules out this possibility. The only way out of this constraint is by invoking re-ionization of the universe at some redshift $z < z_{\rm dec}$; we will discuss this process in the next section.

The situation is different in the presence of dark matter. We saw in chapter 4 that the perturbations in dark matter can grow from $t = t_{eq}$ onwards. Hence at $t = t_{\rm dec}$, $\delta_{\rm DM}$ will be higher than δ_B by the factor $(a_{\rm dec}/a_{eq}) \simeq 20\Omega h^2$. Therefore, in models with dark matter we get

$$\left(\frac{\delta T}{T}\right) = \frac{1}{3}\delta_B(t_{\rm dec}) = \frac{1}{60\Omega h^2}\delta_{\rm DM}(t_{\rm dec}). \tag{6.66}$$

Taking $\delta_{\rm DM}(t_{\rm dec}) \gtrsim 10^{-3}$ (so that $\delta_{\rm DM}(t_0) \gtrsim 1$), this predicts the anisotropy

$$\left(\frac{\delta T}{T}\right) \simeq 1.6(\Omega h^2)^{-1}\times 10^{-5}, \tag{6.67}$$

which is less than the observational limit for $\Omega = 1, h = 1$ and is marginally consistent with observations if $h \simeq 0.7$. A more detailed calculation, based on the numerical integration of the coupled equations gives a still lower value of about 6×10^{-6}.

The angular dependence of these anisotropies is complementary to that due to the Sachs–Wolfe effect. Since $(\delta T/T) \propto (\delta\rho/\rho)$, each logarithmic interval contributes the amount $k^3|\delta_k|^2$ to $< (\delta T/T)^2 >$. Hence

$$\left(\frac{\delta T}{T}\right) \simeq k^{3/2}|\delta_k| \simeq \begin{cases} \lambda^{\alpha-3/2} \propto \theta^{\alpha-3/2} & (\theta < \theta_H) \\ \lambda^{\alpha-5/2} \propto \theta^{\alpha-5/2} & (\theta > \theta_H) \end{cases} \tag{6.68}$$

which becomes, for $\alpha = 3/2$,

$$\left(\frac{\delta T}{T}\right) \simeq \begin{cases} \text{constant} & (\theta < \theta_H) \\ \theta^{-1} & (\theta > \theta_H). \end{cases} \tag{6.69}$$

This contribution will dominate over those due to peculiar velocity and the Sachs–Wolfe effect at small angles.

The comparison between theory and observation is somewhat complicated in the case of small angles. We saw earlier that the small angle observations are usually performed using two (or three) antennas. If the antennas respond sharply and pick up signals from a single, precise direction, then we can compare the observed values of $[C(0) - C(\alpha)]$ with theoretical predictions directly. Unfortunately, most antennas have a finite beam size; that is, angles below some minimum value σ will not be properly resolved. Because of this complication, the value of $(\Delta T/T)^2$ measured at some angular scale α will be a fairly complicated function of $|\delta_k|^2$. The precise form of this relation will depend on the response profile of the antenna. For a 2-beam system with Gaussian response, the relation between $(\Delta T/T)^2$ and $|\delta_k|^2$ turns out to be

$$\left(\frac{\Delta T}{T}\right)^2_\alpha = \frac{V^{-1}}{2\pi^2}\int_0^\infty k^2\, dk \int_{-1}^{+1} d\mu\, |\sigma_T(\mu, k)|^2 W_1(k, \mu; \sigma) W_2(k, \mu; \alpha), \tag{6.70}$$

where

$$\begin{aligned} W_1(k, \mu; \sigma) &= \exp\left[-4k^2 H_0^{-2}(1-\mu^2)\sigma^2\right], \\ W_2(k, \mu; \alpha) &= 1 - J_0\left[2kH_0^{-1}(1-\mu^2)^{1/2}\alpha\right] \end{aligned} \tag{6.71}$$

and $\sigma_T(\mu, k)$ is the actual temperature fluctuation on the $z = z_{\text{dec}}$ surface. The angle μ is $\cos^{-1}(\mathbf{k}\cdot\mathbf{n})$ and α is the angular separation between the antennas. The window function W_1 arises because the finite beam width σ of the antennas cuts off contribution at angles less than σ (that is, at wavelengths less than σH_0^{-1}). The $(1 - J_0)$ factor cuts off contributions at angles larger than α. Thus the observed values and predicted values cannot be compared in a simple manner for small angles.

Table 6.1. MBR anisotropies: comparison of models and observations[11]

Model	$(\Delta T/T)$ at $7.2'$	$(\Delta T/T)$ at $6°$	Comment
Observation	1.5×10^{-5}	4.8×10^{-5}	–
Baryon-dominated $\Omega = \Omega_B = 0.1$	10^{-3}	–	Ruled out
Baryon-dominated $\Omega = \Omega_B = 1.0$	5×10^{-5}	–	Ruled out; also violates nucleosynthesis bound
HDM; $z_{nl} = 1; b = 0.53$ $\Omega = 1, \Omega_B = 0.1$	2×10^{-5}	2×10^{-5}	Marginally ruled out
CDM; $\Omega = 0.2, \Omega_{\rm DM} = 0.17$ $\Omega_B = 0.03$	1.5×10^{-4}	4×10^{-5}	Ruled out
CDM; $\Omega = 0.2, \Omega_{\rm DM} = 0.1$ $\Omega_B = 0.1$	1.7×10^{-4}	5×10^{-5}	Ruled out
CDM; $b = 1.7; \Omega = 1$ $\Omega_B = 0.2$	6×10^{-6}	8×10^{-6}	Possible
CDM; $b = 1.7; \Omega = 1$ $\Omega_B = 0.1$	7×10^{-6}	7×10^{-6}	Possible

We shall now provide a detailed comparison of theory and observations[11]. The theoretical values are based on exact numerical integration of the perturbed equations. From among the observational bounds, we shall choose the limits of 1.5×10^{-5} at $7.2'$ and 4.8×10^{-5} at $6°$ since these are most restrictive. The results from various theoretical models are presented in Table 6.1 (see reference 11 for a more detailed table and discussion).

This table shows that the bounds on MBR anisotropies act as strong constraints on the models for structure formation. The baryon dominated models (models without non-baryonic dark matter) are clearly ruled out. (The results quoted were based on the initial perturbations being adiabatic. There are also strong constraints on models based on primordial isothermal fluctuations[12] though they can be made marginally viable by invoking re-ionization.) The hot dark matter models using a neutrino of mass about 25 eV are also ruled out, though only marginally. These models involve (anti) biasing value of b=0.53. (We shall discuss the need for antibiasing in chapter 8). The parameters chosen here are the ones which give least anisotropies. Cold dark matter models without any biasing *require* $\Omega = 0.2$. (This will also be discussed in chapter 8.) Such models are again ruled out for any sensible combination of $\Omega_{\rm DM}$ and Ω_B. Finally we have listed two cold dark matter models which are viable. All of them have $\Omega = 1$ and hence *require* biasing. The predictions are calculated for $b = 1.7$ and the result scales linearly with this parameter. For a range of Ω_B values the bounds are respected. However, notice that the theoretical predictions are only about a factor 2 lower than the observed values. If the observations improve by this factor, either these anisotropies will be seen or even these models will be ruled out.

6.7 Damping of the anisotropies

The discussion in sections 6.4 to 6.6 suggests that there will be some amount of net temperature anisotropy at all angular scales. At large angles the main contribution is from the gravitational potential while at small angles the dominant term is that due to the intrinsic anisotropy of the radiation field.

There are, however, other effects which can wipe out the $(\delta T/T)$ at small scales. Because of these effects, the value of $(\delta T/T)$ is significantly reduced at small angles. We shall now discuss these processes.

At scales less than $\lambda_s \simeq 10\,\text{Mpc}$, the viscous drag between photons and electrons wipes out baryonic perturbations (see chapter 4; we have taken $\Omega \simeq 10\Omega_B$). This length scale subtends an angle of about $5'$ in the sky. Thus fluctuations at smaller scales are severely damped.

Damping at these scales also occurs for another reason. We saw in chapter 3 that the probability $P(z)$ for the photon we receive to have been last scattered in a redshift interval dz around $z_{\rm dec}$ is given by $(d\tau/dz)\exp(-\tau)$, where τ is the optical depth. This function $P(z)$ is well-approximated by a Gaussian peaked at $z_{\rm dec} \simeq 1065$ with a width of about $\Delta z \simeq 80$. Thus there is significant probability for the photons to have originated anywhere in this band of $\Delta z \simeq 80$. This Δz corresponds to a proper length of $\Delta l \simeq 15\Omega^{-1/2}\,\text{Mpc}$. Modes with $\lambda < \Delta l$ will receive contribution from several layers of recombining plasma. Such a superposition of incoherent fluctuations will wipe out $(\delta T/T)$ at $\lambda < \Delta l$. The angle subtended by this length Δl is about $8'\Omega^{1/2}h$. Thus we expect very little temperature anisotropy to survive below, say, $5'$.

The second effect which can wipe out anisotropies at small scales is 're-ionization'. In the standard scenario described in chapter 3, baryonic matter becomes neutral around $z = 10^3$ and remains so until today. If this is the case, one would have expected to see a certain amount of neutral hydrogen in the intergalactic medium. (It will be quite difficult for galaxy formation to be 100 per cent efficient, thereby succeeding in making *all* the gas condense out as bound structures.) The light we receive from distant sources (like quasars) has to pass through this neutral hydrogen before reaching us. We should, therefore, be able to see the characteristic signature of neutral hydrogen in the absorption spectra of quasars. No such signature is seen. (This observation will be discussed in chapter 9.) The absence of any detectable neutral hydrogen in the galactic medium suggests that there could have been a re-ionization of the intergalactic medium at some redshift higher than about 4. Though no single, satisfactory, mechanism for such a re-ionization is known, this possibility cannot be completely ruled out.

Let us suppose that the re-ionization occurred at a redshift of $z_{\rm ion}$ and that the intergalactic medium remained fully ionized for all $z < z_{\rm ion}$. Then the optical depth due to Thomson scattering up to a redshift of z $(< z_{\rm ion})$ will be

$$\begin{aligned}\tau &= \int \sigma_T n_e \, dt = \left(\frac{\sigma_T \Omega_B \rho_c}{m_p}\right) \int_0^z H_0^{-1} \, dz \, (1+z)^3 \frac{1}{(1+z)^2 (1+\Omega z)^{1/2}} \\ &= \frac{\sigma_T H_0}{4\pi G m_p} \frac{\Omega_B}{\Omega} \left[2 - 3\Omega + (1+\Omega z)^{1/2} (\Omega z + 3\Omega - 2)\right].\end{aligned} \tag{6.72}$$

This expression has the limiting forms

$$\begin{aligned}\tau &= \frac{\sigma_T H_0}{4\pi G m_p} \frac{3}{2} (\Omega_B z) = 0.026 (\Omega_B z) \quad (\text{for } \Omega z \ll 1); \\ &= \frac{\sigma_T H_0}{4\pi G m_p} \left(\frac{\Omega_B}{\Omega^{1/2}}\right) z^{3/2} = 0.017 \left(\frac{\Omega_B}{\Omega^{1/2}}\right) z^{3/2} (\text{for } \Omega z \gg 1).\end{aligned} \tag{6.73}$$

If $\tau \simeq 1$ for some $z \leq z_{\rm ion}$, then the photons will be significantly scattered by the re-ionized plasma. This will partially wipe out information about the $z = z_{\rm dec}$ surface. For universes with $\Omega \simeq 1$, $\Omega_B \simeq 0.1$, $\tau = 1$ occurs near $z \simeq 70$ while for $\Omega \simeq \Omega_B \simeq 0.2$, $\tau = 1$ can occur at much lower redshift: $z \simeq 25$; thus in standard dark matter models re-ionization has to occur at fairly high redshifts (greater than 70) to produce any effect. If such re-ionization occurred at a redshift of $z_{\rm ion}$, its primary effect will be to wipe out the original $(\delta T/T)$ at scales smaller than the Hubble radius $d_H(z_{\rm ion})$ at $z_{\rm ion}$. From (6.2), we see that d_H subtends the angle

$$\theta_{\rm ion}(z_{\rm ion}) \cong 3^\circ \Omega^{1/2} \left(\frac{100}{z_{\rm ion}}\right)^{1/2} \tag{6.74}$$

in the sky. Since $\tau = 1$ corresponds to $z \cong 15.08 \Omega^{1/3} (\Omega_B h)^{-2/3}$, the maximum value of $\theta_{\rm ion}$ will be $\theta_m(z_m) \simeq 7.4^\circ (\Omega_B \Omega h)^{1/3}$. If such re-ionization occurs, the trace of primordial anisotropies will remain only at larger angles.

While re-ionization wipes out the original anisotropies, the peculiar velocities at $z \simeq z_{\rm ion}$ regenerate new anisotropies at small scales[13]. The actual value of these anisotropies is fairly model dependent. In the simplest models, these secondary anisotropies arise because of the coupling between the motion of the scatterers and the fluctuation in the electron density along the line of sight. This effect is dominant in the angular scales $1'$ to $10'$ and produces $(\delta T/T)$ of the order of 10^{-5}.

Though re-ionization is a possibility, it may not be a *likely* possibility. To provide a physical mechanism for re-ionization, it is necessary to form the first generation of structures at $z > z_{\rm ion}$. In standard scenarios with adiabatic initial perturbations, this can happen only at small $z_{\rm ion}$; in that case, the optical depth τ will be far less than unity and the effect will be insignificant. (It is comparatively easier to form structures at high redshifts in models with isocurvature perturbations.)

If the universe is baryon dominated with $\Omega = \Omega_B \simeq 0.2$ (which satisfies the constraint from nucleosynthesis), we can lower $z_{\rm ion}$ to about 25 and increase $\theta_{\rm ion} \simeq 6°$. The observational bound discussed in the last section, of course, is no longer relevant because it was based on the value of $(\delta\rho/\rho)$ at $z = z_{\rm dec}$. This is probably the only way of reconciling the baryon dominated universe with observations. However, for this model to work, some physical mechanism for re-ionization at $z \gtrsim 25$ needs to be introduced.

6.8 Spectral distortions due to ionized gas

There are two further situations in which the interaction between charged particles and the photons produces distortions in the MBR spectrum.

If the matter in the universe was reionized at some redshift in the interval $0 < z < 1100$, then significant amount of energy must have been supplied to matter by some astrophysical process. Once the matter is ionized, the coupling between photons and charged particles will transfer part of this energy into the MBR. Depending on the redshift at which this energy injection occurs, this process can distort the spectrum of the MBR.

Secondly, it must be noted that there are regions in the universe – like the clusters of galaxies – which contain hot, ionized gas. When the MBR photons pass through these regions, they will be scattered by the electrons (which are at much higher temperature) and gain energy. This will distort the MBR spectrum in the vicinity of a cluster of galaxies. We shall now study these two effects.

The spectral distortions which arise due to re-ionization depend crucially on the redshift at which the re-ionization occurs. Once the matter is ionized, Compton scattering, bremstrahlung and its inverse process, free–free absorption will come into operation and the plasma will evolve towards a new equilibrium configuration. Whether such an equilibrium can be reached will depend critically on the various timescales of the problem[14].

The free–free processes are effective at low frequencies while Compton scattering affects all frequencies. To distinguish various possibilities which can arise, it is convenient to define some characteristic redshifts and frequencies. The time scale ($t_{\rm ff}$) for the free–free processes at frequencies $\omega \simeq T$ and the time scale for Compton scattering (t_c), defined in chapter 3, can be expressed in terms of the redshift as

$$\begin{aligned} t_{\rm ff} &= \frac{3}{32\pi^3}\left(\frac{m\sqrt{6\pi mT}}{e^6 n_e^2}\right) T^3 \cong 0.4 \times 10^{24}\,{\rm s}\,(\Omega h^2)^{-2} z^{-5/2}, \\ t_c &= \frac{1}{\sigma_T n_e}\left(\frac{m}{T}\right) \cong 10^{28}\,{\rm s}\,z^{-4}(\Omega h^2)^{-1}, \end{aligned} \tag{6.75}$$

where we have assumed that (i) the matter is fully ionized; (ii) $z \gg 1$ and (iii) $(T_m - T_r) \ll T_r$ where T_r and T_m are radiation and matter temperatures. As described in chapter 3, the effective timescale for absorption is the geometric mean

$$\bar{t} = (t_{\rm ff} t_c)^{1/2} \cong 0.6 \times 10^{26}\,{\rm s}\,(\Omega h^2)^{-3/2} z^{-13/4}. \tag{6.76}$$

Free–free processes cease to be effective in neutralizing the distortions caused by scattering, when $\bar{t}$ becomes larger than H^{-1}. Using the form of $H(t)$ in the radiation dominated phase, we find that $\bar{t} = H^{-1}(t)$ corresponds to the redshift

$$z_{\rm ff} \simeq 10^5 (\Omega h^2)^{-6/5}. \tag{6.77}$$

If any energy is released into the plasma at $z > z_{\rm ff}$ it will be thermalized quickly. The depletion (or enhancement) of photons in any frequency range due to scattering will be balanced by free–free emission (or absorption). (Of course, energy release at $z > z_{\rm ff}$ is not 're-ionization' because neutral atoms have not yet formed.)

The situation is more complicated if energy is released at $z < z_{\rm ff}$. The free–free processes are not very effective but we have to check whether scattering is effective either. To do this, we can define a parameter y, usually called the Compton-y parameter, by the expression:

$$y = \int_t^{t_0} n_e \sigma_T \left(\frac{T_m}{m_e}\right) dt \cong \int_0^z n_e(z) \sigma_T \left(\frac{T_m}{m_e}\right) \left|\left(\frac{dt}{dz}\right)\right| dz. \tag{6.78}$$

To evaluate this expression we need to know the density of *free* electrons and *matter* temperature T_m as a function of redshift. In the absence of detailed information about the ionization history, we can only estimate y for $z \gg z_{\rm dec}$. Assuming that most of the contribution to y is from high

z, we get

$$y \cong \left(\frac{\sigma_T \Omega_B \rho_c}{m_p m_e}\right) \int H_0^{-1} \frac{dz}{(1+z)^2(1+\Omega z)^{1/2}} T_0 (1+z)(1+z)^3$$
$$\simeq 1.24 \times 10^{-11} \left(\frac{\Omega_B}{\Omega^{1/2}}\right) z^{5/2}. \tag{6.79}$$

Compton scatterings will be effective at redshifts for which $y \geq 1$. This gives us another characteristic redshift z_c, defined by $y(z_c) = 1$. From (6.79), we find

$$z_c \simeq 2 \times 10^4 (\Omega \Omega_B^{-2})^{1/5}. \tag{6.80}$$

(This is a rather crude estimate because the universe will be making a transition from radiation dominated to matter dominated phase around this redshift; using a more exact form of integrand in (6.79) reduces z_c slightly.)

In the range $z_{\rm ff} > z > z_c$, Compton scattering is effective but not free–free processes. The scattering will transfer the energy injected into the matter to the radiation; this will increase the energy of the radiation but cannot change the *number* of photons. So, clearly, the radiation cannot reach a new *Planck* spectrum because it has more energy (per photon) than a Planck spectrum has. The equilibrium distribution with the conservation of total number will be a Bose–Einstein distribution; this is the spectrum to which the radiation will evolve if energy is released in the interval $z_{\rm ff} > z > z_c$.

There is a further complication which needs to be tackled. The effectiveness of free–free processes was decided in the above analysis using $t_{\rm ff}$ which, in turn, was computed for frequencies $\nu \simeq T$. Since the $t_{\rm ff}$ has a frequency dependence of the form

$$t_{\rm ff} = t_{\rm ff}(\omega = T) \frac{(\omega/T)^3}{(1 - e^{-\omega/T})} \simeq t_{\rm ff}(\omega = T) \left(\frac{\omega}{T}\right)^2 \tag{6.81}$$

for $\omega \ll T$, the free–free processes can be quite effective at low frequencies ($\nu \ll T$), even after it has ceased to be effective at $\nu \simeq T$ (see exercise 3.13 for the frequency dependence of $t_{\rm ff}$). Thus at low frequencies, free–free processes can re-establish the Planck spectrum. The balance between Compton distortion and free–free thermalization occurs at a critical frequency ω_B at which $t_c = t_{\rm ff}$:

$$\frac{\omega_B}{T} \simeq 200 (\Omega h^2)^{1/2} z^{-3/4}. \tag{6.82}$$

For $\omega < \omega_B$, free–free processes can establish the Planck spectrum; for $\omega > \omega_B$, the previous analysis is applicable and the spectrum will be of the Bose–Einstein form.

We shall now turn to the case of energy injection at $z < z_c$. The most interesting situation occurs if z is also less than z_{dec} and some physical process re-ionizes matter. This will heat up the matter to a temperature much higher than that of radiation. So we now have to consider the case with $T_m \gg T_r$ (unlike the previous situation, in which $(T_m - T_r) \ll T_r$). Since $t_{\text{ff}} \gg H^{-1}$, we can ignore the free–free processes for most of the frequency range and study only the distortions in the radiation produced by the hot matter through Compton scattering.

Such a process is governed by the equation (called the Kompaneets equation):

$$\frac{\partial n}{\partial t} = \frac{\sigma n_e}{m} \frac{1}{\nu^2} \frac{\partial}{\partial \nu} \left\{ \nu^4 \left[n(n+1) + \frac{T_e}{n} \frac{\partial n}{\partial \nu} \right] \right\}, \tag{6.83}$$

where $n(\nu, t)$ is the photon distribution function and T_e is the electron temperature. (For a derivation, see exercise 6.3.) Introducing the dimensionless variables $x = (\nu / T_e)$ and y, as defined in (6.78), with

$$dy = \sigma n_e \left(\frac{T_e}{m} \right) dt = \left(\frac{T_e}{m} \right) d\tau, \tag{6.84}$$

this equation can be transformed to

$$\frac{\partial n}{\partial y} = \frac{1}{x^2} \frac{\partial}{\partial x} \left\{ x^4 \left[n(n+1) + \frac{\partial n}{\partial x} \right] \right\}. \tag{6.85}$$

Since this evolution is entirely due to Compton scattering it should conserve the total number of photons; this fact can be easily verified:

$$\frac{d}{dt} \int n \nu^2 \, d\nu = 0. \tag{6.86}$$

Hence the equilibrium solution to (6.85) corresponds to a Bose–Einstein distribution with $n = [\exp \beta(\nu - \mu) - 1]^{-1}$, rather than a Planck spectrum. This is the key reason for spectral distortion in the Compton dominated regime: Suppose the spectrum *was* originally Planckian. If energy is now injected into the system, it cannot evolve to another *Planck spectrum* with a new higher temperature. A Planck spectrum at a higher temperature will require more number of photons; but the evolution preserves the number of photons.

For the case we are interested in, viz. that of ionized matter interacting with MBR photons in the redshift range of $0 < z < z_{\text{dec}}$, it is safe to assume that the matter temperature T_e will be much higher than the radiation temperature T_r. In that case, we can ignore the $n(n+1)$ term in (6.85) and obtain

$$\frac{\partial n}{\partial y} \cong \frac{1}{x^2} \frac{\partial}{\partial x} \left\{ x^4 \frac{\partial n}{\partial x} \right\}. \tag{6.87}$$

This equation, being linear, can be easily solved [15] by transforming the independent variable from x to z where $\ln x = z - 3y$. This will lead to a diffusion-type equation in (y, z). Solving it and changing variables to (x, y) we get

$$n(x,y) = \frac{1}{(4\pi y)^{1/2}} \int_0^\infty \frac{d\zeta}{\zeta} n(\zeta, 0) \exp\left[-\frac{1}{4y}\left(3y + \ln\frac{x}{\zeta}\right)^2\right]. \tag{6.88}$$

All the properties of the system can be obtained from this solution. To begin with, notice that the spectrum at the present epoch $n(\nu, t_0)$ depends on t_0 *only* through the parameter

$$y = \int_{t_1}^{t_0} \frac{T_e(t)}{m} \sigma n_e(t)\, dt. \tag{6.89}$$

The information about various evolutionary histories, having different $T_e(t)$ and $n_e(t)$ is contained in this single number, called the Compton-y parameter. Further, by direct integration of (6.88), we can determine the total number and energy of the photons:

$$\begin{aligned} n_{\text{total}}(t_0) &= \int_0^\infty n(x,y)x^2\, dx = \int_0^\infty n(x,0)x^2\, dx = n_{\text{total}}(0); \\ \epsilon_{\text{total}}(t_0) &= \int_0^\infty n(x,y)x^3\, dx = e^{4y} \int_0^\infty n(x,o)x^3\, dx = \epsilon_0 e^{4y}. \end{aligned} \tag{6.90}$$

There is a change in the energy without any change in the photon number. Let us suppose that the initial spectrum was Planckian with $n(x,0) = (\exp x - 1)^{-1}$. Since the photons are now being redistributed at different frequencies, with an increase in the net energy, we expect an overall shift in the spectrum to *higher* frequencies. This would imply that the effective temperature of the spectrum at the Rayleigh–Jeans end will be *lowered.* To calculate the amount of lowering, we may proceed as follows: For small x, the Planck spectrum $n(x,0)$ becomes $(\exp x - 1)^{-1} \approx x^{-1}$. Substituting $n(x,y) = x^{-1}f(y)$ in our equation, we see that $f' = -2f$. So, in the $x \ll 1$ limit, the solution can be approximated by

$$n(x,y) \cong x^{-1}\mathrm{e}^{-2y}; \qquad (x \ll 1). \tag{6.91}$$

(The same result can be obtained from the general solution after more algebra.) In this long-wavelength limit, $n(x,y) \propto T$; the reduction from $n(x,0)$ to $n(x,y)$ by the factor e^{-2y}, therefore, corresponds to the decrease in temperature in the Rayleigh–Jeans limit

$$T_{\text{RJ}} = T_r \mathrm{e}^{-2y}. \tag{6.92}$$

To make concrete predictions, we need an estimate of the electron temperature. If most of the diffuse X-ray background observed in the universe is contributed by the ionized intergalactic medium, then the value of T_e at present will be[16] about 4.4×10^8 K; at an earlier redshift, the temperature would have been $T_e(z) = 4.4 \times 10^8(1+z)^2$ K. Using this one can compute the value of y. A calculation similar to that in (6.79) gives

$$y \simeq 5.3 \times 10^{-3}(\Omega h) \int_0^{z_{\rm ion}} \frac{(1+z)^3}{(1+\Omega z)^{1/2}} dz \simeq 1.5 \times 10^{-3}(\Omega h^2)^{1/2} z_{\rm ion}^{7/2}. \tag{6.93}$$

(We have assumed that the matter is fully ionized.) From the observations of MBR spectrum one can estimate the deviations from the Planck spectrum (if any). No such deviations have been seen. The results from COBE suggest that the MBR is a Planck spectrum with $T = 2.735 \pm 0.06$ K. This result can be also stated as an upper bound on the value of y: $y \lesssim 10^{-4}$. Comparison with the above expression shows that this bound rules out any re-ionization at $z_{\rm ion} \gtrsim 1$. The estimated value of y, of course, can be reduced by assuming a lower value of T_e (today); even then, it is not possible to respect the constraint of $y < 10^{-4}$ and still produce all the soft X-ray background from the ionized intergalactic medium.

The bound on y can also be stated as a bound on the amount of energy injected into the radiation field: From (6.90) we find that:

$$\frac{\Delta \epsilon}{\epsilon_r} = \mathrm{e}^{4y} - 1 \cong 4y. \tag{6.94}$$

Therefore $(\Delta\epsilon/\epsilon_r) \lesssim 4 \times 10^{-4}$.

The following point, however, should be noted: The analysis given above assumes that efficient transfer of energy from electrons to photons can take place in spite of the expansion of the universe. The cooling time for a plasma (through Compton scattering) is given by

$$t_{\rm cool} \cong \frac{1}{\sigma_T n_\gamma}\left(\frac{m_e}{T}\right). \tag{6.95}$$

This timescale exceeds the Hubble time $(\dot{a}/a)^{-1}$ for $z < 8$. Thus, for $z_{\rm ion} < 8$, only a small part of the energy injected into the matter will be transferred to radiation.

Finally, let us consider the interaction of the MBR photons with the hot gas in a cluster of galaxies[17]. If the optical depth is τ, then a fraction τ of the MBR photons, seen in the direction of the cluster may be assumed to have been scattered (at least) once by the hot electrons. This fraction gains in energy and is redistributed at the higher frequency region of the spectrum. Since the total number of photons is conserved, this results in a lateral shift of the spectrum. The net effect is similar to that in (6.92). The temperature in the Rayleigh–Jeans region of the spectrum decreases by

$$\frac{\delta T}{T_r} = -\frac{2T_e}{m_e}\tau. \tag{6.96}$$

This effect (called the Zeldovich–Sunyaev effect) can be quite large for some clusters and has been observed in several cases. For example, in the clusters 0016+16, A655 and A2218, the observed dip is $-1.4, -0.7$ and -0.7 mK respectively[18]. Since no independent measurements of T_e and τ are available, there is no clear-cut theoretical prediction which can be compared with the observation. However, the value is consistent with the parameters $T_e \simeq 5\,\mathrm{keV}$ and $\tau \simeq (1/80)$ for the gas in the cluster.

Exercises

6.1 The bounds on the distortions in the MBR allow one to put constraints on several astrophysical processes. Consider, as an example, the following scenario: A particular species of wimp with mass m is unstable and decays with a lifetime τ. The decay products are all very light (i.e., all of them have masses far smaller than m) and one of the decay products is a photon. What constraints can be placed on the life time τ based on the contribution of the photons (arising from the decay) to the background radiation in the universe? Consider the three cases (i) $H_0^{-1} \leq \tau$, (ii) $t_{\rm dec} < \tau < H_0^{-1}$ and (iii) $\tau < t_{\rm dec}$ separately. (Assume that the original freeze out of the wimp occurs as described in chapter 3.)

6.2 Consider a photon travelling through a large massive structure which is slowly collapsing. During the time taken for the photon to cross the structure, the gravitational potential well of the condensate changes by a non-negligible amount. Estimate the value of $(\delta T/T)$ induced by such a process.

6.3 Consider the evolution of the photon distribution function $n(\omega, t)$ through its interaction with non-relativistic electrons. Let the electrons be described by a thermal distribution $f(E)$ at temperature T. Assume that only Compton scattering operates and that the energy transfer Q in each collision

$$Q = \frac{\hbar(\omega_1 - \omega)}{kT}$$

is small. Derive the equation satisfied by $n(\omega, t)$ along the following lines:

(a) The exact Boltzmann equation describing the evolution of $n(\omega, t)$ is

$$\frac{\partial n}{\partial t} = \int d^3\mathbf{p} \int d\Omega \left(\frac{d\sigma}{d\Omega}\right)$$
$$\times \left\{ f_e(\mathbf{p}_1) n(\omega_1) \left[1 + n(\omega)\right] - f_e(\mathbf{p}) n(\omega) \left[1 + n(\omega_1)\right] \right\}$$

where σ is the cross section for the process $p + \omega \rightleftharpoons p_1 + \omega_1$. Explain the origin of the two terms on the right hand side, especially the $(1 + n)$ factors.

(b) Since energy transfer Q is small, expand $n(\omega_1)$ and $f_e(p_1)$ in a Taylor series in Q retaining terms up to the quadratic order. Show that, to this order,

$$\frac{\partial n}{\partial t} = [n' + n(1+n)] \int d^3p \int d\Omega \left(\frac{d\sigma}{d\Omega}\right) f_e Q$$
$$+ \left[\frac{1}{2}n'' + n'(n+1) + \frac{1}{2}n(n+1)\right] \int d^3p \int d\Omega \frac{d\sigma}{d\Omega} f_e Q^2$$

where $n'' = (\partial^2 n/\partial\omega^2)$ etc.

(c) Show that the second integral in the above equation is $[2xn_e\sigma_T(T/m) + \mathcal{O}(T/m)^2]$ where $x = (\omega/T)$.

(d) Argue that, since Compton scattering conserves number of photons, this equation must have the form

$$\frac{\partial n}{\partial t} = -\frac{1}{x^2}\frac{\partial}{\partial x}[x^2 j(x)]$$

Compare this form with $(\partial n/\partial t)$ calculated above to conclude that j has the form $g(x)[n' + h(n,x)]$.

(e) Use the fact that Bose–Einstein distribution must be a static solution to this equation to derive the form of h and j. Hence obtain the final result

$$\frac{\partial n}{\partial t} = \left(\frac{T}{m}\right)\frac{1}{x^2}\frac{\partial}{\partial x}\left[x^4\left\{n(n+1) + \frac{\partial n}{\partial x}\right\}\right].$$

6.4 In chapter 4, we decomposed the velocity field u_i of matter into a 'rotation' term (ω_{ij}), 'shear' term (σ_{ij}) and the Hubble flow $(H\delta_{ij})$. What bounds can be put on (ω/H) and (σ/H) from the observed isotropy of MBR?

6.5 A cosmic ray proton of energy 10^{20} eV will see – in its rest frame – microwave photons as having an energy of about 100 MeV. Such an energetic photon, hitting a proton at rest, is almost at the threshold for producing a pion.

(a) How would a terrestrial observer view this phenomenon?

(b) What is the mean free path for a proton with energy higher than 10^{20} eV?

6.6 It is possible to produce $e\bar{e}$ pairs in the energetic collisions of γ-rays. The cross section σ for this process is about 10^{-25} cm^2. Consider a cosmic ray photon of energy E (as measured by the terrestrial observer) interacting with microwave photons.

(a) Show that $e\bar{e}$ pairs should be produced by cosmic ray photons with $E \gtrsim 2.5 \times 10^{14}$ eV.

(b) Show that the mean-free-path for this process (i.e. the mean distance a cosmic ray photon will travel before producing $e\bar{e}$) is about 10^{22} cm. (This is smaller than the size of the galaxy.)

(c) What will be the fate of these $e\bar{e}$ pairs?

6.7 Consider the possibility that MBR originates from a collection of discrete sources and is not a primordial relic. (a) Suppose MBR is due to the integrated emission from N sources per steradian, each of flux density S which existed at some particular epoch in the past. Let the beam angle of the telescope be Ω. Show that the lack of statistical fluctuations in MBR will imply $N \gtrsim 10^{12}\,\mathrm{sr}^{-1}$. Is this an acceptable number? (b) Suppose MBR is due to the emission from a uniform population of sources with luminosity P and extending up to cH_0^{-1}, with $N(\geq S) \propto S^{-3/2}$ for $S \geq S_{min}$. Show that $N(\geq S_{\rm min}) \simeq (27\Omega)^{-1}(\Delta I/I)^{-3}$. Is this bound acceptable?

7
Velocity fields

7.1 Introduction

In the absence of inhomogeneities, comoving particles in the Friedmann universe will have velocities determined by Hubble's law. The inhomogeneities in the universe will perturb the Hubble flow and will induce 'peculiar' velocities to the galaxies. By measuring the peculiar velocity field one can obtain valuable information about the underlying mass distribution.

This short chapter discusses the large scale velocity field in the universe, which can serve as an important diagnostic device between various models. Section 7.2 describes the various attempts to measure this velocity field and compares the results. The constraints obtained by combining the velocity data with the bounds on MBR anisotropies are discussed in section 7.3.

7.2 Large scale velocity fields

There are two important predictions which emerge from the linear theory. The first one is the existence of the anisotropies in the MBR, which was discussed in the last chapter. The second prediction, which we will take up now, is related to the 'peculiar velocities' for galaxies. As we have seen in chapter 4, the linear theory relates the peculiar velocity $\mathbf{v}(\mathbf{x})$ (that is, the velocity over and above the Hubble velocity) and the density contrast $\delta(\mathbf{x})$ by

$$\mathbf{v}(\mathbf{x}) = H_0 \frac{\Omega^{0.6}}{4\pi} \int \delta(\mathbf{y}) \frac{(\mathbf{x}-\mathbf{y})}{|\mathbf{x}-\mathbf{y}|^3} d^3\mathbf{y}. \tag{7.1}$$

This equation shows that the density contrast $\delta(\mathbf{x})$ can be determined if the peculiar velocity field $\mathbf{v}(\mathbf{x})$ is known from the observation. On the other hand, if we have independent information on $\mathbf{v}(\mathbf{x})$ and $\delta(\mathbf{x})$, then this equation can be used to test the linear theory. We shall discuss both these approaches in this chapter.

The information about the velocity field is usually used in the following manner: Given the distribution of *luminous* matter in the universe, one can predict a velocity field $\mathbf{v}_{\text{lum}}(\mathbf{x})$ using the linear theory. Comparing this field with the observed $\mathbf{v}(\mathbf{x})$ one can hope to answer two different questions: (a) Is the luminous matter distributed in the same manner as the dark matter? (b) *If* luminous and dark matter are distributed in a similar manner, what is the value of Ω? These issues can be stated in a more transparent manner by rewriting (7.1) as

$$\nabla \cdot \mathbf{v} = -H_0 \Omega^{0.6} \delta(\mathbf{x}). \tag{7.2}$$

In other words, the divergence of the observed velocity field gives us the *actual* distribution of mass in the universe. Comparing it with the distribution of *luminous* matter, we can check whether the dark and luminous matter are distributed similarly. If they are, then the right hand side of (7.2) can be used to estimate Ω. To do this successfully, we need an estimate of the distances to the galaxies which is *not* based on their redshift.

The velocity field at large scales also provides a direct estimate of the primordial perturbation spectrum. Though the specific relation between $\mathbf{v}$ and δ will be model dependent, we certainly expect the deviations from homogeneity to be connected with $\mathbf{v}$ in some form. In particular, one should be able to determine the statistical properties of the density field from that of the velocity field if the latter is known with sufficient accuracy. Because of these reasons, it is of considerable interest to determine $\mathbf{v}(\mathbf{x})$; this task, however, turns out to be far from easy[1].

The only peculiar velocity which can be measured with reasonable certainty is our own. This velocity can be inferred from the dipole anisotropy of the MBR. To specify this velocity (and others) it is convenient to use two different coordinate systems, called 'galactic' and 'supergalactic' coordinates. The galactic coordinates are defined with the plane of equator set on the disc of our galaxy and the poles defined by the local perpendicular to this plane. The symbols b and l are used to denote the latitude and longitude with the latitude running from $-90°$ at south pole to $+90°$ at north pole. The axes are oriented such that the location of the sun is at $l = 0$, $b = 0$ and the sense of rotation of the sun is towards $l = 90°, b = 0$. The supergalactic coordinates are nearly perpendicular to the galactic coordinates. The supergalactic north pole is in the direction of $l = 227°$ and $b = -6°$. This makes the equatorial plane of supergalactic coordinates nearly perpendicular to the galactic equatorial plane. The coordinates in the supergalactic plane will be denoted by (B, L). The supergalactic plane is defined in this manner because a large concentration of galaxies exist in this plane (see figure 7.1). Most of the interesting large scale structures at distances less than $50h^{-1}$ Mpc from us lie in the supergalactic plane. These include the superclusters Hydra–Centaurus, Perseus–Pisces, Pavo–Indus–Telescopium etc.

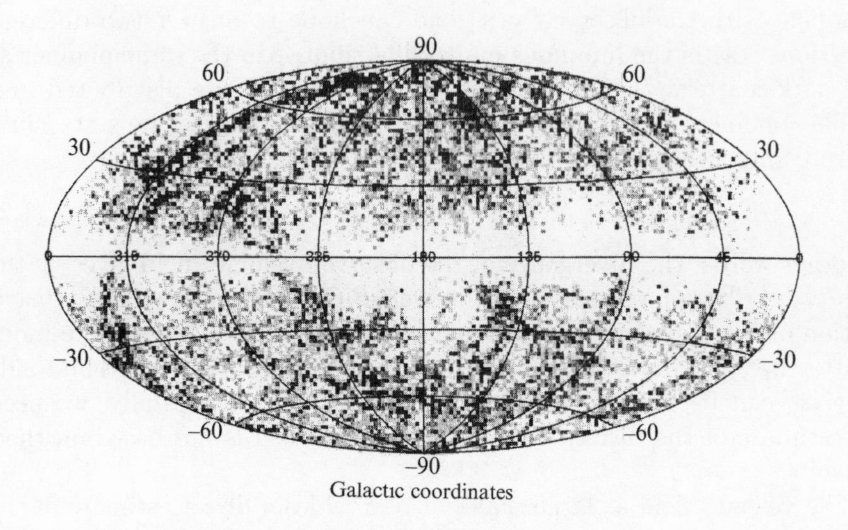

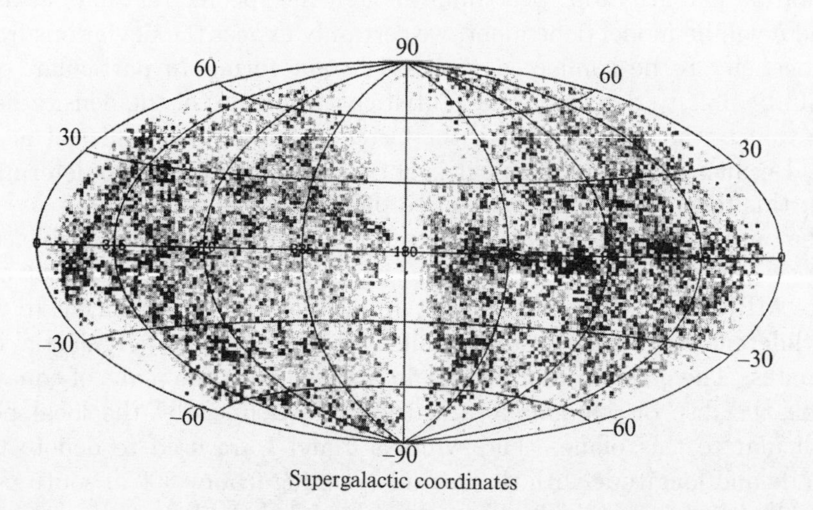

Fig. 7.1. These are two-dimensional plots of galactic density weighted by the amount of light produced per unit area. (above) The distribution in the galactic coordinates. The marked absence of galaxies near the equator is caused by the obscuration of dust in the plane of the disc of our galaxy. (below) The same distribution in the supergalactic coordinates showing the concentration of galaxies in the supergalactic plane. (This figure is reproduced, with permission, from the review by D. Burstein, Rep. Prog. Phys., **53**, 421 (1990); this figure was originally designed by Nigel Sharp.)

The dipole anisotropy of the MBR suggests that[2] *our Local Group* is moving with respect to MBR, at a speed of $614 \pm 50\,\mathrm{km\,s^{-1}}$ in the direction $l = 269 \pm 5^\circ, b = 28 \pm 5^\circ$. To avoid possible misunderstanding, it is worthwhile to state clearly what this velocity refers to. The quantity which is actually observed in an MBR experiment will be the velocity of earth with respect to MBR. If the rotation and revolution of earth – which vary on a very short time scale, but are known quite precisely – are subtracted out, we obtain the 'heliocentric' velocity which is about $360\,\mathrm{km\,s^{-1}}$ in the direction of $l = 265^\circ, b = 50^\circ$. This is the velocity which corresponds to the MBR anisotropy that a hypothetical observer located on the sun will measure. If we further correct for the motion of the sun (which is due to the sun's rotation in the Milky Way) and the motion of the Milky Way relative to galaxies in the Local Group, we obtain the velocity vector with respect to the centre of mass of the Local Group. The last two steps are achieved as follows:

A star in a *circular* orbit, at the location of the sun, will be moving with a speed of $220\,\mathrm{km\,s^{-1}}$ towards $(l = 90^\circ, b = 0)$. The motion of such a hypothetical star is used to define a frame of reference, usually called the 'local standard of rest', LSR. The actual motion of the sun deviates from the LSR somewhat; the velocity of the sun *with respect to* LSR is $20\,\mathrm{km\,s^{-1}}$ in the direction of $(l = 57^\circ, b = 23^\circ)$. Since the (v_x, v_y, v_z) components of a vector $\mathbf{v}$ in the direction (l, b) – which we will denote as $v(l, b)$ collectively – are $(v\cos b\cos l, v\,\cos b\sin l, v\sin b)$, we can easily compute[3] the velocity of sun with respect to the centre of the Milky Way, $v_{\odot-\mathrm{MW}}$. We get:

$$v_{\odot-\mathrm{MW}} = (10, 235.4, 7.8)\,\mathrm{km\,s^{-1}}. \tag{7.3}$$

To proceed further, we need to know the motion of the centre of our galaxy with respect to the centre of mass of the Local Group. This value, however, is somewhat uncertain[4]. As a first approximation, we may assume that the centre of the Milky Way is moving towards Andromeda which is in the direction specified by the unit vector $(-0.476, 0.79, -0.41)$. The relative velocity of Andromeda towards the Milky Way is about $119\,\mathrm{km\,s^{-1}}$. If we assume that $m_{\mathrm{MW}}\mathbf{v}_{\mathrm{MW}} = -m_A\mathbf{v}_A$ and that $m_A \approx 2m_{\mathrm{MW}}$, then we find that the centre of the Milky Way must be moving with the velocity $(-38, 62.7, -29)\,\mathrm{km\,s^{-1}}$ with respect to the centre of mass of the Local Group. Combining this value with the previous result, we find that the velocity of the sun with respect to the centre of mass of the Local Group is $(-28, 298, -21)\,\mathrm{km\,s^{-1}}$. Different measurements of the motion in the Local Group have led to different values for this number. Detailed analysis, taking into account the velocities of all the members of the Local Group, for example, gives[4] the value:

$$v_{\odot-\mathrm{LG}} = (-89.3, 292.6, -36.96)\,\mathrm{km\,s^{-1}}. \tag{7.4}$$

To facilitate easy comparison of the results, one usually uses the 'standard' result[4] of $(0, 300, 0)\,\mathrm{km\,s^{-1}}$. Given some specified value for $\mathbf{v}_{\odot-\mathrm{LG}}$ and the velocity of the sun with respect to MBR, we can compute the velocity of the centre of mass of the Local Group with respect to MBR. Taking $\mathbf{v}_{\odot-\mathrm{LG}} = (0, 300, 0)\,\mathrm{km\,s^{-1}}$ gives

$$\mathbf{v}_{\mathrm{LG-MBR}} = 614 \pm 50\,\mathrm{km\,s^{-1}}(l = 269^\circ \pm 5^\circ, b = 28^\circ \pm 5^\circ). \qquad (7.5)$$

On the other hand, if we take $\mathbf{v}_{\odot-\mathrm{LG}} = (-89.3, 292.6, -37)\,\mathrm{km\,s^{-1}}$ then we find that $\mathbf{v}_{\mathrm{LG-MBR}} = 622 \pm 20\,\mathrm{km\,s^{-1}}$ ($l = 277^\circ \pm 2^\circ$, $b = 30^\circ \pm 2^\circ$). Notice that the final values which we obtain for this velocity are not too different.

The two immediate questions which arise from the above observation are the following: (i) What is the physical origin of this, rather large, velocity of the Local Group and (ii) how far does it extend? In other words, what is the size of the region within which the motion of the Local Group is shared by other galaxies?

To answer the first question we may proceed as follows: Since peculiar velocities are generated by the mass concentration around us, it is natural to see whether our velocity points towards any of the known mass concentration. One of the largest clusters near us is the Virgo cluster which is in the direction (284°, 74°). The v_{MBR} can be resolved into three components in the following convenient manner. We first resolve the velocity vector into a component perpendicular to the supergalactic plane (v_z) and another in the supergalactic plane. The component in the supergalactic plane is further resolved into one along the direction of the Virgo cluster (v_v) and one in perpendicular direction (v_s). These components have the values $v_z = -355 \pm 25\,\mathrm{km\,s^{-1}}$, $v_v = 418 \pm 25\,\mathrm{km\,s^{-1}}$ and $v_s = 277 \pm 25\,\mathrm{km\,s^{-1}}$. It is obvious that the Virgo cluster cannot be the sole source of this peculiar velocity; it accounts, at best, only for about 46 per cent of the total amplitude. Two other nearby clusters are Hydra and Centaurus and it is certainly conceivable that some of the gravitational pull is exerted by these systems. To settle this issue, it is necessary to measure the effects of a deeper and more extensive set of galaxies.

In the linear theory, peculiar velocities are in the same direction as the peculiar acceleration, which can be computed by adding the contributions (M_i/r_i^2) from each galaxy vectorially. Assuming that galaxies trace mass and that they have a constant (M/L) ratio one can replace this sum by the addition of (L_i/r_i^2) for each galaxy. Such calculations have been performed using optical and infrared catalogues[5]. The 'light dipoles' are in reasonable alignment (7° to 15°) with the MBR dipole. These results suggest that the mass distribution causing the acceleration lies outside the Local Group and extends to a sphere of $40h^{-1}$ Mpc or so.

If most of the accelerating mass is far away, then all the galaxies in our local region will feel typically the same acceleration and will move together as a unit. In other words, there could exist a 'bulk flow' in our neighbourhood. This leads us to the second question. which is an intrinsically more difficult one. While it is rather easy to measure *our* peculiar velocity with respect to MBR, it is not so easy to estimate the peculiar velocities of other galaxies with respect to MBR. What is available observationally are the redshifts (z) of the various galaxies around us. Redshift data will give an equivalent *radial* velocity (v) for each galaxy. But this radial velocity will be partly due to Hubble flow and partly due to peculiar motions. To isolate the latter, we need an estimate of the distance (r) to the galaxy (which should be, of course, not based on the redshift). If this information is available then one can compute the radial peculiar velocity, $v_{\rm pec}$, by the relation

$$v_{\rm pec} = v - Hr, \tag{7.6}$$

thereby obtaining a detailed map of the radial velocity field of the universe. There have been several attempts to map the peculiar velocity field around us by using different types of astronomical objects (spirals, ellipticals ...) and by using different estimators for the distance. Since these distance estimators are rather crucial to our discussion, we shall first examine some of their properties.

There are two distance estimators which have been used extensively in these studies. These are based on the infrared Tully–Fisher (IRTF) relation for spirals ($L \propto v^4$ where L is the absolute luminosity and v is the circular velocity) and the diameter–velocity dispersion (D_n–σ) relation for ellipticals. (This relation is $D_n^3 \propto \sigma^4$ where D_n is the angular diameter and σ is the central velocity dispersion[6]. The diameter D_n is defined using the isophote within which the mean surface brightness is 20.75 mag arc sec^{-2}. Normally, the 'diameters' are defined either using a faint isophote at 25 mag arc sec^{-2} or using central brightness in the photographic plate. Both these extremes are avoided in the above definition for D_n.) These relations represent empirically determined correlations between a distance dependent quantity (like flux or angular diameter) and another distance-independent quantity (circular velocity or central velocity dispersion). These relations can be tested and calibrated using galaxies in a cluster. Since the cluster members are all at almost the same distance from us, the apparent luminosity can be used in these calibrations. The observed scatter in these relations is quite small: about 20 per cent for the D_n–σ relation and 25 per cent for the IRTF relation[6]. Notice, however, that to study a 500 km s^{-1} peculiar velocity at 5000 km s^{-1} distance – as we would like to do – one needs *ten* per cent accuracy in distance measurement. (Here, and in what follows, it is convenient to specify distances in terms of equivalent Hubble velocities. In this notation, a velocity of $5000\,\mathrm{km\,s^{-1}}$ will correspond to a distance $(5000\,\mathrm{km\,s^{-1}}/100h\,\mathrm{km\,s^{-1}\,Mpc^{-1}}) = 50h^{-1}\,\mathrm{Mpc}$, etc.) This is impossible to achieve for single galaxies and we must concentrate on the clusters, in which statistical errors can be reduced because of the existence of a large number of galaxies.

The *theoretical* status of these relations is as follows: The Tully–Fisher relation can be justified heuristically in models which have a power spectrum with an effective index $n \lesssim 0$ near galactic scales (We will say more about this result in the next chapter). The velocities are determined by the depth of the dark matter potential well; if there exists some correlation between the amount of *baryonic* matter trapped by a well and the depth of the well (as is quite likely to exist), we do expect a correlation between luminosity and the velocity dispersion. What seems surprising is the fact that observations show very little scatter in this relation; one would have expected such a dynamical relation to be influenced strongly by the environment.

The D_n–σ relation, on the other hand, can be derived from the following two facts: (i) The brightness profiles of all bright ellipticals follow the de Vaucouleurs law written in the form

$$\frac{B(r)}{B_0} = \exp[-7.67(r/R)^{1/4}] \tag{7.7}$$

(see chapter 1; the notation has been changed slightly). (ii) The 'mass-to-light' ratio (which we will discuss in detail in chapter 11) in the inner parts of a galaxy is a gently rising function of the luminosity; $(M/L) \propto L^m$ with $m \simeq 0.25$. From the second fact and the virial theorem it follows that we can write,

$$\begin{aligned}\sigma^2 &\propto (M/R) \propto (L/R)(M/L) \propto L^{m+1}R^{-1} \propto (B_0R^2)^{m+1}R^{-1} \\ &\propto R^{1+2m}B_0^{m+1}\end{aligned} \tag{7.8}$$

where we have used the fact that, for systems governed by the de Vaucouleurs profile, the total luminosity, L, scales as B_0R^2. Further, note that the diameter D_n is defined as the diameter within which the *mean* surface brightness has some fixed value B_c (usually 20.75 mag arc $\sec^{-2}$ in blue). For the ellipticals obeying de Vaucouleurs law, it is easy to see that the ratio (D_n/R) will only be a function of (B_0/B_c). For the case we are interested in $B_0 \gg B_c$, and so we can approximate this function as a power law; numerical calculation shows that $(D_n/R) \propto (B_0/B_c)^k \propto B_0^k$ with $k \simeq 0.8$. Using this relation in the expression for σ^2, we get

$$\sigma^2 \propto R^{1+2m}B_0^{m+1} \propto D_n^{1+2m}B_0^{[1+m-k(1+2m)]}. \tag{7.9}$$

For $m \simeq 0.25$, $k \simeq 0.8$ this gives $D_n \propto \sigma^{4/3}\, B_0^{-1/30}$; ignoring the weak dependence on B_0, we get a result which is quite close to the observed value.

Though the scatter in these relations is not too much, there are several other sources of error which make the distance estimation using these methods a difficult task. They include an unknown, systematic, dependence of the 'zero-point' of D_n–σ and IRTF relation on the environment[7] and the selection effects which arise due to the fact that samples are flux limited[8] (called the Malmquist bias; see exercise 7.1).

We shall now discuss the results of peculiar velocity measurements based on these distance estimators. A study of nearly 385 elliptical galaxies using the D_n–σ relation suggested[9] that the galaxies in the Centaurus cluster themselves are moving in the same direction as us. Detailed modelling of these galaxies within a velocity sphere of 6000 km s^{-1} shows that there is a bulk motion of about 600 km s^{-1} in the direction of $l = 312°, b = 6°$. This direction is still about 40° away from the MBR dipole direction. We shall now examine this result in greater detail.

The first question to settle will be the estimate of the size of the region in which all the galaxies exhibit a common, shared flow. This can be decided by plotting the peculiar velocities of nearby galaxies, measured in the Local Group velocity frame, against the observed distance. If a galaxy completely shares the MBR motion of the Local Group, its peculiar velocity in the Local Group frame should be nearly zero. Such an analysis shows[10] that galaxies within a distance of about 500 km s^{-1} (which is equivalent to a distance of $5h^{-1}$ Mpc) share the MBR motion of the Local Group. The motion of the galaxies begin to deviate in a systemic manner around (600–750) km s^{-1}. This motion seems to be shared by a part of the Coma–Sculptor cloud. This cloud is in the form of a narrow, thin distribution fully contained in the plane of the Local Supercluster. The major axis of this cloud extends nearly 750 km s^{-1} in the direction $l = 134°, b = 62°$ and about 400 km s^{-1} in the opposite direction. The minor axis is along $l = 315°, b = 28°$. The v_z velocity of the Local Group is shared at least in projection by the whole cloud. There is some evidence that the galaxies at the edge of the cloud are moving towards the centre of the cloud.

To present the peculiar velocities at larger distances it is best to use a velocity frame in which the MBR is isotropic. Further, it is convenient to project the velocities on to the supergalactic plane. Such a projection of peculiar velocities is given in figure 7.2.

The overall impression from this figure is that there exists a net flow towards the Centaurus region (upper left side of the figure) along with a net inward motion on either side of the Supergalactic plane from the Local

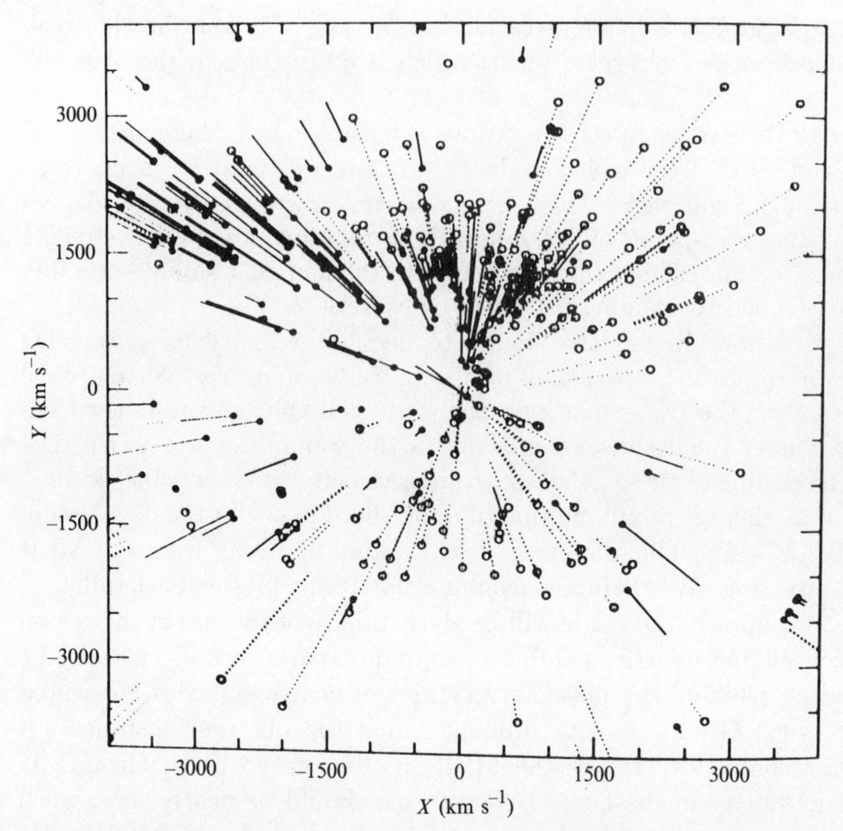

Fig. 7.2. The projected velocities of all galaxies which are within the supergalactic latitudes of $\pm 22.5°$ and within the 'velocity distance' of $5000\,\mathrm{km\,s^{-1}}$ are shown. All the velocities are measured in the MBR frame. The lengths of the lines are proportional to the total (*not* projected) velocity. Velocities away from the Local Group are plotted with filled dots at the predicted distance while the velocities directed towards the Local Group are shown with open circles and dotted lines. Notice that positive peculiar velocities are predominantly in the upper left quadrant. (This figure is reproduced, with permission, from the review by D. Burstein, Rep. Prog. Phys., **53**, 421 (1990).)

Group. Analyzing this data one concludes that there exist significant dipole and quadrupole components in the flow. The net dipole motion has an amplitude of $500 \pm 50\,\mathrm{km\,s^{-1}}$ in the direction $l = 314 \pm 15°, b = 14 \pm 5°$. It must be emphasized that the peculiar velocities shown in figure 7.2 use the D_n–σ relation for estimating distances; any systematic error in this relation could affect the results drastically even though such a possibility seems unlikely.

A simple way of verifying such a dipole motion is to plot the peculiar velocity against the cosine of the angle between the direction of a galaxy and the predicted direction of flow. Figure 7.3 shows such a cosine diagram which confirms the interpretation given above.

One possible way of modelling this velocity field is in terms of a large overdensity ('Great Attractor' or 'GA') situated in the general direction of the dipole flow. Such a modelling suggests that a large spherical mass (of about $5 \times 10^{16}\,\mathrm{M}_{\odot}$) lying at a distance of $4500\,\mathrm{km\,s}^{-1}$ with a $\rho \propto r^{-2}$ density profile will induce the observed velocity field. (The precise parameters describing GA have changed in subsequent analysis with different data sets; but the typical values are as suggested above. The term 'Great Attractor' has also been interpreted differently in different contexts. It essentially identifies a *region* in the universe, centred 4200–4500 km s^{-1} away with a density profile of r^{-2}, and not a single, specific *object* like, say, the Virgo supercluster.) Following the original work there have been several detailed investigations along similar lines. Additional support for

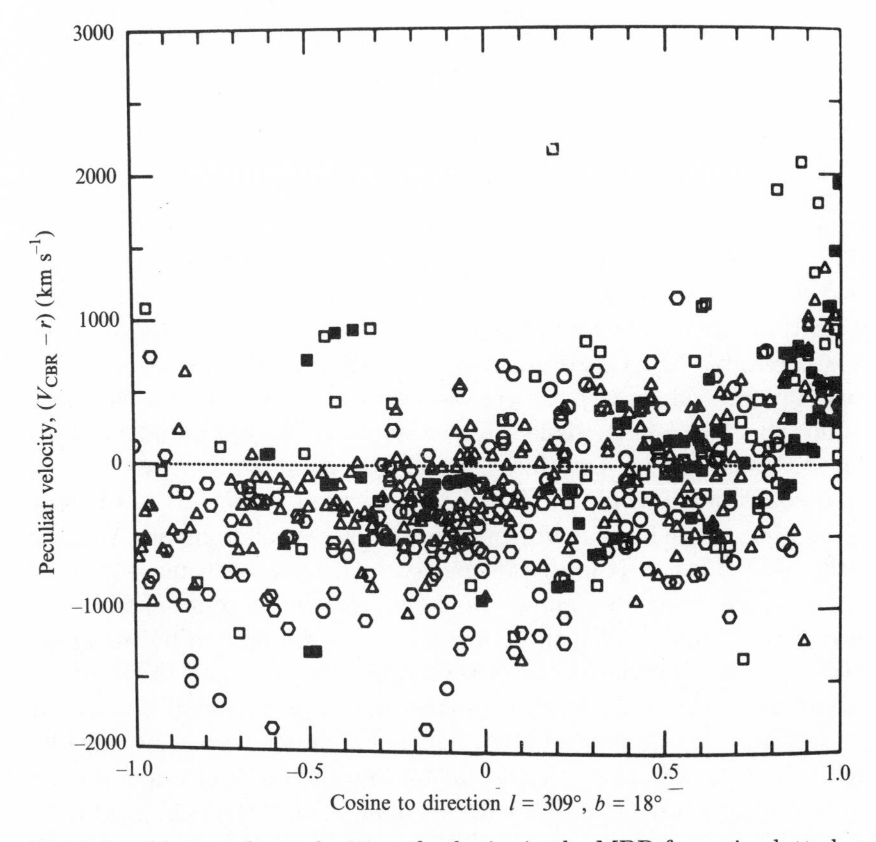

Fig. 7.3. The peculiar velocities of galaxies in the MBR frame is plotted against the cosine of the angle between the direction of that galaxy and direction of the dipole. (This figure is reproduced, with permission, from the review by D. Burstein, Rep. Prog. Phys., **523**, 421 (1990).)

the GA interpretation came from the study of spirals using the infrared Tully–Fischer relation for the distance estimate, and from detailed investigation of the Hydra–Centaurus supercluster[11]. A new redshift survey covering one steradian around GA and subsequent measurements of distances to more than about 1000 of these galaxies reinforces[12] the picture of GA. Some of the studies also seem to indicate that the galaxies on the other side of the GA are falling towards GA, though this result is yet to be firmly established. It must be mentioned that an enormous supercluster (now called the 'Shapley supercluster') has been detected at a distance of $14,000\,\mathrm{km\,s^{-1}}$ beyond the GA. The effect of this supercluster on the conclusions drawn above is still uncertain.

The study described above – and other similar attempts – provides us with the radial component of the peculiar velocities of galaxies. The information contained in these velocities can be used to generate the density distribution in the following manner[13]: we have seen in chapter 4 that the velocity field in the Friedmann universe is irrotational at sufficiently late times. It is, therefore, possible to represent the velocity as $\mathbf{v} = -\nabla\psi$ where ψ is a 'velocity potential'. Given the radial velocity field $v_r(r, \theta, \phi)$ we can construct ψ by the integral

$$\psi(r, \theta, \phi) = -\int_0^r v_r(r', \theta, \phi)\, dr'. \tag{7.10}$$

Once ψ is known everywhere, tangential components can be constructed by taking the gradient. This result arises from the fact that the gravitational field is completely determined by a single scalar function and hence, to the extent that the velocity is determined by gravity, it must be also expressible in terms of a single scalar function of space.

To use this procedure in practice, we need to smooth the velocity field over some suitable length scale. Given such a smooth radial velocity field, it is straightforward to integrate (7.10) numerically on a grid in spherical coordinates. This method has been used extensively in recent years to obtain the density field from the velocity measurement. A study of nearly 1000 galaxies (the distances to which have been measured by D_n–σ and IRTF relations) shows that the velocity potential, within a distance of $6000\,\mathrm{km\,s^{-1}}$ from the Local Group, is dominated by the great attractor[13]. The difference in the potential between GA and LG is about $2\times 10^6\,\mathrm{km^2\,s^{-2}}$. The peak density contrast in GA, when smoothed over a size of $1400\,\mathrm{km\,s^{-1}}$, is about 1.2 ± 0.4. The Local Supercluster appears like a ridge on the side of GA. The Virgo cluster and the Local Group are both seen to be falling towards the GA with peculiar velocities $658\pm121\,\mathrm{km\,s^{-1}}$ and $565 \pm 125\,\mathrm{km\,s^{-1}}$ respectively. The mass distribution derived from the velocity field agrees by and large with the known galaxy distribution except near the Perseus–Pisces supercluster. This discrepancy is possibly due to lack of adequate sampling of this area.

This analysis can also be used to predict the root-mean-square velocity at various scales. Within a sphere of radius $4000\,\mathrm{km\,s^{-1}}$ and $6000\,\mathrm{km\,s^{-1}}$ (centred on the Local Group) the average velocities are $388 \pm 67\,\mathrm{km\,s^{-1}}$ towards $L = 177°$, $B = -15°$ and $327 \pm 82\,\mathrm{km\,s^{-1}}$ towards $L = 194°$, $B = 5°$ respectively. (As we shall see later, the *predicted* root-mean-square velocities in unbiased cold dark matter models are somewhat lower than these values.)

To test the theoretical predictions we have to compare the density distribution obtained from the velocity field with the actual density distribution in the universe. There are, however, two serious difficulties with this project: Firstly, note that the velocity field probes the distribution of the total mass while observations are confined to the distribution of the luminous matter. One usually tries to get around this difficulty by introducing a 'bias factor', b, so that $(\delta N/N)_{\mathrm{gal}} = b(\delta\rho/\rho)_{\mathrm{true}}$. *If* this assumption is valid then we only need to scale the observed value by b^{-1}; however, it is not clear how correct this assumption is.

Secondly, even obtaining reliable information about the distribution of galaxies around the GA is not easy. The direction of the Great Attractor is towards very low galactic latitude. Even if we correct for the scattering by the Milky Way, we still have to face the problem of incompleteness of the galaxy surveys in this region.

Probably the best information we have about the density field comes from the spectroscopic surveys of the IRAS galaxies[14]. This sample of IRAS point sources, covering more than 88 per cent of the sky, has the redshifts for more than 2500 galaxies with flux greater than 1.9 Jy at 60 μ. Almost all of them are spirals. The advantage of using a sample selected in the infrared band is that it provides a much better sky coverage due to negligible galactic extinction. Optical surveys are seriously incomplete at $|b| < 20°$ while the IRAS survey is more complete up to $|b| \gtrsim 5°$. From the number counts of the galaxies, the density field is constructed by a fairly complicated statistical technique. The results from this analysis[14], however, do not seem to quite match with the GA picture discussed above.

Figure 7.4 compares the mass distribution traced by IRAS galaxies with the mass distribution calculated from the peculiar velocities measured by the D_n–σ method. The disagreement is striking. The IRAS map shows two peaks of high density; one corresponding to the Great Attractor region (on the left) and one corresponding to the Perseus–Pisces supercluster. In contrast the map derived from peculiar velocities has only one high density region dominated by the Great Attractor. The disparity can also be seen in the predicted acceleration of the Local Group. In the IRAS data most of this acceleration comes from within $3000\,\mathrm{km\,s^{-1}}$ while

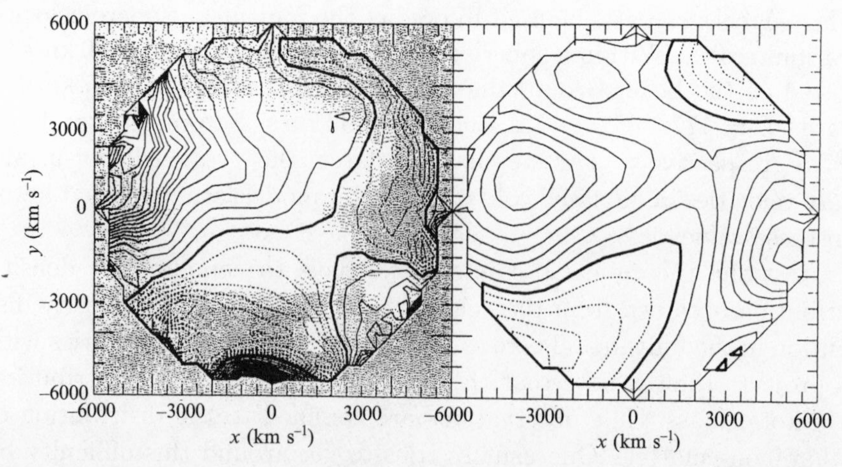

Fig. 7.4. The mass distribution inferred from the IRAS survey is quite different from that calculated from the measurement of the peculiar velocities. Contours corresponding to greater-than-average densities are indicated by solid lines while those corresponding to less-than-average are shown by dotted lines. The IRAS map shows two peaks while the peculiar velocities seem to be dominated by a single one. (Reprinted, by permission, from Nature, **350**, 395; copyright (C) 1991, Macmillan Magazines Ltd.)

in the GA model very little of the acceleration comes from such a close region.

A more recent (QDOT) redshift survey of a sample of 2163 IRAS galaxies also leads to results[15] which are not quite in agreement with the GA model. In this survey, redshifts were measured systematically for over 2000 IRAS galaxies. Given the redshifts of the galaxies, the peculiar velocities are calculated by a complicated iterative procedure. Roughly speaking, this procedure is as follows: An initial distance is attributed to the galaxies based on the redshift after correcting for several known effects (like membership or proximity of a cluster of known distance, the motion in the local region, the dipole velocity with respect to the MBR etc.). These positions of the galaxies are used to predict peculiar velocities of all the galaxies. These peculiar velocities are now taken into account to calculate revised positions of the galaxies and the iteration is continued until convergence is obtained. The resulting velocity field can be compared with the one discussed before. It turns out that this peculiar velocity field can be generated entirely by a small number of clusters with large halos around them. (Halos of typically $30h^{-1}$ Mpc in radius are required). The combined effect of these clusters is to produce at the location of the Local Group the peculiar velocity of magnitude $511\,\mathrm{km\,s^{-1}}$ towards $(304^\circ, 11^\circ)$; which is to be compared with the previous result of $535 \pm 60\,\mathrm{km\,s^{-1}}$ towards $(309^\circ, 18^\circ)$. Notice that the former model does not require any large concentration of mass in any one particular direction, though it does require large halos around the clusters. The motion of the Local Group is generated due to the attraction of masses within a distance of about $100h^{-1}$ Mpc. The direction of acceleration agrees reasonably well with the MBR dipole direction.

The reasons for the discrepancy between IRAS results and the D_n–σ survey are not clear. In comparing the theoretical distribution with the one based on IRAS galaxies, the following point should be kept in mind. The galaxies which emit strongly in the infrared represent late-type systems which are under-represented in the high-density regions. It is not clear how serious this effect is; but it could certainly make large mass concentrations less conspicuous in the IRAS maps.

It must be stressed that these conclusions (including the existence or otherwise of a GA in the IRAS-QDOT sample, the disparity between the two surveys etc.) are still very tentative. The statistical analysis of these data is plagued by several complications and it will take some time for a clear picture to emerge. On the other hand, *if* it eventually turns out that the IRAS results are correct, then the contradiction could force us to one of the following conclusions: (1) There is a serious, systematic, error in the distance measurements using the D_n–σ relation. It could even be that the measurement to some major cluster – like Centaurus – is in error, producing the effect of a spurious bulk flow. (2) The biasing hypothesis which we used is too simple-minded. The actual distribution of matter is related to the distribution of galaxies in a much more complicated manner. Only more detailed observations and study can tell us the real reason.

7.3 Theoretical constraints on peculiar velocities

In the linear perturbation theory which we are using, there exists a one-to-one correspondence between the density contrast $(\delta\rho/\rho)_\lambda$ at some scale λ and the velocity perturbation at that scale. This is most easily seen by writing (7.2) in the Fourier space:

$$v_\lambda \simeq \frac{\lambda}{H_0^{-1}} \Omega^{0.6} \left(\frac{\delta\rho}{\rho}\right)_\lambda . \tag{7.11}$$

In other words, $(v_\lambda/H_0\lambda) \simeq \Omega^{0.6}(\delta\rho/\rho)_\lambda$. This relation shows that, if $(\delta\rho/\rho)$ falls faster than λ^{-1} – which is equivalent to $(\delta M/M)$ falling faster than $M^{-1/3}$, a condition that is satisfied by the Harrison–Zeldovich spectrum – then one can ignore contributions from scales $\lambda \gg L$ in computing v_λ at $\lambda = L$. It is, therefore, meaningful to average the velocity field over some radius R_v and determine the root-mean-square power in the velocity field as a function of R_v. To do this formally, we will introduce a window function $W_v(R_v;\mathbf{x})$ and write

$$\begin{aligned} v^2(R_v) &= \left\langle \left(\int v(\mathbf{r}+\mathbf{x}) W_v(R_v;\mathbf{x})\, d^3\mathbf{x} \right)^2 \right\rangle \\ &\cong \frac{H_0^2 \Omega^{1.2}}{2\pi^2} \int_0^\infty |W_v(k)|^2 P(k)\, dk \end{aligned} \tag{7.12}$$

where $W_v(k)$ is the Fourier transform of the window function and $P(k) = |\delta_k|^2$ is the power spectrum of the density field. If $P(k) \propto k^n$, it is clear that $v^2(R) \propto R^{-(n+1)}$. At large scales, $n \simeq 1$ in the cold dark matter models giving $v(R) \propto R^{-1}$. The proportionality constant depends on the form of the window function, normalization convention and, of course, on the power spectrum. If we take $(\delta M/M) \simeq 1$ at $R = 10h^{-1}\,\mathrm{Mpc}$, it is equivalent to assuming that $v_R \simeq HR$ at $R = 10h^{-1}\,\mathrm{Mpc}$. This leads to the normalization: $v_{10} = 1000\,\mathrm{km\,s^{-1}}$. Since $v(R) \propto R^{-1}$, we find that v at $R \simeq 60h^{-1}\,\mathrm{Mpc}$ (corresponding to a velocity distance of $6000\,\mathrm{km\,s^{-1}}$) is $v_{60} \simeq (1000\,\mathrm{km\,s^{-1}}/6) \simeq 166\,\mathrm{km\,s^{-1}}$. This value is clearly much smaller than the observed streaming velocities of $350\,\mathrm{km\,s^{-1}}$ or so at these scales. A more careful calculation gives about $220\,\mathrm{km\,s^{-1}}$ at $R \simeq 60h^{-1}\,\mathrm{Mpc}$ which is still on the smaller side.

One can also compute the two-point correlation function of the velocity field $c(\mathbf{r}) =< \mathbf{v}(\mathbf{x}) \cdot \mathbf{v}(\mathbf{x}+\mathbf{r}) >$ (see exercise 7.6). It has been argued[16] that both IRAS and GA data show a correlation length which is higher than the correlation length which is predicted by the CDM models; but these results depend on the detailed assumptions which are made regarding the bias parameter as well as on the way the data is analyzed to obtain the correlation lengths.

The probability for the existence of a mass concentration like GA can also be estimated using CDM mass functions. In an $\Omega = 1$ model with the bias parameter $b = 2.5$, this probability is negligible; it is a 7.3σ event. Even with $b = 1.5$ and much more relaxed assumptions about the GA configuration, it is a 2.4σ event. These results can be understood as follows: We know that the density contrast at any given location obeys a Gaussian probability distribution with the variance $\sigma^2 =< \delta^2 > \propto M^{-2\beta}$ with $\beta = (1/2)(1+n/3)$. The typical fluctuation expected in the density over a scale l will be of the order of $\sigma(l) \propto M^{-\beta}(l) \propto l^{-3\beta}$. We can, therefore, write: δ (expected) $\equiv \delta_{\mathrm{ex}} \simeq (l/R_c)^{-3\beta}$ where R_c is the normalization scale ($R_c \simeq 8h^{-1}\,\mathrm{Mpc}$). If we take $l_{\mathrm{GA}} \simeq 50h^{-1}\,\mathrm{Mpc}$ then $\delta_{\mathrm{ex}} \ll 1$. But the observed density contrast is much higher. We have seen that $(v/v_H) \simeq (\Omega^{0.6}/3)\delta$; at the scale of GA, $(v/v_H) \simeq 0.1$ giving δ (observed) $= \delta_{\mathrm{GA}} \simeq (0.3\,/\Omega^{0.6})$. The probability for such a region to have formed by now is

$$P_{\rm GA} = \frac{1}{\sqrt{2\pi}\delta_{\rm ex}} \int_{\delta_{\rm GA}}^{\infty} \exp\left(-\frac{\delta^2}{2\delta_{\rm ex}^2}\right) d\delta \simeq \left(\frac{\sqrt{2\pi}\delta_{\rm GA}}{\delta_{\rm ex}}\right) \exp\left(-\frac{\delta_{\rm GA}^2}{2\delta_{\rm ex}^2}\right) \tag{7.13}$$

for $\delta_{\rm ex} \ll \delta_{\rm GA}$. For $\Omega = 1$, $P_{\rm GA} \simeq 4 \times 10^{-2}$, 5×10^{-6} and 1.7×10^{-27} for $n = -1, 0$ and $+1$ respectively. In the large scales we are considering, the effective index, $n_{\rm eff}$, of the power spectrum is nearly unity. So such regions are very improbable in the CDM models.

There has been considerable discussion in the literature about these issues and whether cold dark matter can correctly explain the large velocities which are observed[16]. It seems that, though one may be able to push the model to the limit and just manage to account for the large scale power in the velocity field, the most natural scenarios have difficulty in explaining the observations.

One can obtain much more stringent – and interesting – constraints by combining the bounds on MBR anisotropy with the results on large scale streaming. Since the same density perturbations lead to temperature anisotropy $(\Delta T/T)$ and the peculiar velocity v, it is clear that a large value for one will imply a large value for the other, *irrespective of the detailed form of the power spectrum.* Since we have an *upper* bound on $(\Delta T/T)$ and a large value for v, there is a potential for contradiction which needs to be checked.

An order of magnitude comparison between the two effects is easy to make[17]. The contribution to temperature anisotropy $(\Delta T/T)_\lambda$ from δ_λ at scale λ is $(\Delta T/T)_\lambda \simeq \phi_\lambda \simeq (\lambda/H^{-1})^2\, \delta_\lambda$. Using $v_\lambda \simeq (H\lambda)\delta_\lambda$ we can write this as

$$\left(\frac{\Delta T}{T}\right)_\lambda \simeq \left(\frac{\lambda v_\lambda}{H^{-1}}\right) \simeq 3 \times 10^{-5} \left(\frac{v}{500\,{\rm km\,s^{-1}}}\right) \left(\frac{\lambda}{50h^{-1}\,{\rm Mpc}}\right). \tag{7.14}$$

It is clear that the observed velocity field is (at best) only marginally consistent with the bounds on MBR anisotropy.

A more rigorous bound can be obtained as follows: For a given power spectrum $P(k)$, the root-mean-square peculiar velocity is

$$v^2(R) = \frac{\Omega^{1.2} H_0^2}{2\pi^2} \int_0^\infty P(k) {\rm e}^{-k^2 R^2}\, dk \tag{7.15}$$

where we have used a Gaussian window function. The root-mean-square temperature fluctuation due to the Sachs–Wolfe effect is

$$\left(\frac{\Delta T}{T}\right)^2 = \frac{F^2(\Omega)}{8\pi^2 H^{-4}} \int_0^\infty \frac{P(k)}{k^2} dk. \tag{7.16}$$

(This expression is identical to the one derived in chapter 6 except for the factor $F(\Omega)$. In chapter 6, we considered only the $\Omega = 1$ model for which

F is unity; in general, $F(\Omega) \simeq \Omega^{0.3}$, to a good approximation.) We now ask: Which $P(k)$ will minimize the value of $(\Delta T/T)$ for a constant value of v^2 at some fixed R? This is equivalent to minimizing the functional

$$K[P(k)] = \frac{\int_0^\infty P(k)k^{-2}\,dk}{\int_0^\infty P(k)\exp(-k^2R^2)\,dk}. \tag{7.17}$$

The minimum for this expression is easy to find. The integrand in the numerator can be written as $f(k)P(k)\exp(-k^2R^2)$ where $f(k) = k^{-2}\exp(k^2R^2)$. Clearly $f(k)$ has the minimum value of (eR^2) at $k = R^{-1}$; so the numerator is always greater than or equal to eR^2 times the denominator. Hence $K[P(k)] \geq eR^2$ with the minimum being reached for the delta function power spectrum $P(k) \propto \delta_{\rm Dirac}(k - R^{-1})$. Fixing the normalization using (7.15), we find

$$P_{\rm min}(k) = \left(\frac{2\pi^2 ev^2}{H_0^2\Omega^{1.2}}\right)\delta_{\rm Dirac}(k - R^{-1}). \tag{7.18}$$

The temperature anisotropy caused by this spectrum is

$$\begin{aligned}\left(\frac{\Delta T}{T}\right)_{\rm min} &= \sqrt{\frac{e}{2}}\Omega^{-0.3}\left(\frac{vR}{H_0^{-1}}\right) \\ &\simeq 3.6\times 10^{-5}\,\Omega^{-0.3}\left(\frac{v}{500\,{\rm km\,s^{-1}}}\right)\left(\frac{R}{50h^{-1}\,{\rm Mpc}}\right)\end{aligned} \tag{7.19}$$

which is slightly higher than our order of magnitude estimate. This anisotropy will be seen at the angular scale of $\theta \simeq (\Omega/2)(R/H_0^{-1}) \simeq (\Omega/2)^\circ(R/50h^{-1}\,{\rm Mpc})$. It should be noted that the bound is applicable only at the large angles in which the Sachs–Wolfe effect dominates. Within the domain of applicability, the bound is *independent* of the form of the power spectrum $P(k)$.

One can do better[18], if we use the fact that the root-mean-square fluctuations in the mass $(\Delta M/M)_R$ are also generated by the same power spectrum $P(k)$. We have seen in chapter 4 that

$$\left(\frac{\delta M}{M}\right)_L^2 = \frac{\nu^2}{2\pi^2}\int_0^\infty k^2\,dk\,P(k){\rm e}^{-k^2L^2} \tag{7.20}$$

where we have used a Gaussian window function of length scale L. We have also multiplied the expression for $(\delta M/M)$ by an overall constant ν which has the following interpretation: We saw in chapter 5 that we can treat $\delta(\mathbf{x})$ as a Gaussian random variable. The density contrast at any location will have a dispersion $\sigma^2 =< \delta^2 >$. If we now consider a region which arose from a fluctuation which was $\nu\sigma$ higher than the mean, then the corresponding mass fluctuation $(\delta M/M)$ will be ν times higher than the usual value.

Comparing the expressions for $v^2, (\Delta T/T)^2$ and $(\Delta M/M)^2$ we can again derive bounds as in the previous case. We now get

$$
\begin{aligned}
v^2(R) &\geq \frac{e\Omega^{1.2}H^2}{\nu^2}(L^2 - R^2)\left(\frac{\delta M}{M}\right)_L^2, \\
\left(\frac{\Delta T}{T}\right)^2 &\geq \frac{e^2\Omega^{0.6}H^4L^4}{16\nu^2}\left(\frac{\delta M}{M}\right)_L^2, \\
\left(\frac{\Delta T}{T}\right)^2 &\geq \frac{eH^2R^2}{4\Omega^{0.6}}v^2(R)
\end{aligned}
\tag{7.21}
$$

where we have assumed that $L > R$. When L and R are comparable, these constraints are not of much value. But one can also obtain, using the Schwarz inequality, the bound:

$$
\left(\frac{\delta M}{M}\right)_L^2 \left(\frac{\Delta T}{T}\right)^2 \geq \frac{\nu^2\Omega^{-1.8}}{4}v^4(l); \quad l = (L/2) \tag{7.22}
$$

which is useful even when $R \simeq L$ (see exercise 7.2). It must be mentioned that all the bounds obtained above ignore the fact that the finite thickness of the decoupling surface smears $(\Delta T/T)$ over a small size, D, say. The bounds obtained above are valid for $L \gg D, R \gg D$. (It is, however, possible to take into account the effect of D by smearing $(\Delta T/T)$ with a Gaussian window function.) Substituting some characteristic numbers, we get,

$$
\begin{aligned}
\left(\frac{\delta M}{M}\right) &\lesssim 4 \times 10^{-2}\nu\Omega^{-0.6}\left(\frac{L}{100h^{-1}\,\mathrm{Mpc}}\right)^{-1}\left(\frac{v}{600\,\mathrm{km\,s}^{-1}}\right) \text{ (if } L \gg R), \\
\left(\frac{\delta M}{M}\right) &\lesssim 1.3 \times 10^{-2}\nu\Omega^{-0.3}\left(\frac{L}{100h^{-1}\,\mathrm{Mpc}}\right)^{-2}\left(\frac{\Delta T/T}{10^{-5}}\right), \\
v &\lesssim 200\,\mathrm{km\,s}^{-1}\,\Omega^{0.3}\left(\frac{R}{60h^{-1}\,\mathrm{Mpc}}\right)^{-1}\left(\frac{\Delta T/T}{10^{-5}}\right), \\
v &\lesssim 1000\,\mathrm{km\,s}^{-1}\nu^{-1}\Omega^{0.9}\left(\frac{\delta M}{M}\right)\left(\frac{\Delta T/T}{6 \times 10^{-6}}\right) \text{ (at } R = L/2).
\end{aligned}
\tag{7.23}
$$

It should be stressed again that these constraints are independent of the form of $P(k)$ and hence must be respected by a wide class of models in which the observed structures, MBR anisotropy and large scale peculiar velocities all arise due to gravity.

The bounds may be compared with observations if we can determine $(\Delta M/M)$. One way of estimating this quantity at large scales is to use the galaxy–galaxy, cluster–cluster correlation functions, extrapolated to large scales. From the definition of correlation functions, we have

$$
\left(\frac{\delta N}{N}\right)_r^2 \simeq \xi(r) \tag{7.24}
$$

where the left hand side refers to the mean square fluctuation in the number in a region of size r. If we further take $(\delta N/N) = (\delta M/M)$ we can compare the bounds with observations. The relevant curves are shown in figure 7.5. It is clear from the figure that the galaxies can be easily accommodated with $\Omega = 1, \nu = 1$ but clusters probably require a higher value of ν. Though the conclusions from this figure should still be considered tentative, it does illustrate the power of the method. In particular, any detection of MBR anisotropy will help us improve the bound considerably.

Finally, we discuss the connection between MBR anisotropies and the existence of mass concentrations like the 'Great Attractor'[19]. From the relation $\nabla \cdot \mathbf{v} = -H_0\Omega^{0.6}\,(\delta\rho/\rho)$ and $\mathbf{v} = -\nabla\psi$ it follows that $\nabla^2\psi = H_0\Omega^{0.6}(\delta\rho/\rho)$. On the other hand, $\nabla^2\phi = 4\pi G(\delta\rho) = (3/2)H_0^2(\delta\rho/\rho)$. Therefore, we find that $\phi = (3/2)H_0\Omega^{0.4}\psi$. The construction of the local velocity potential from the observation is thus equivalent to determining

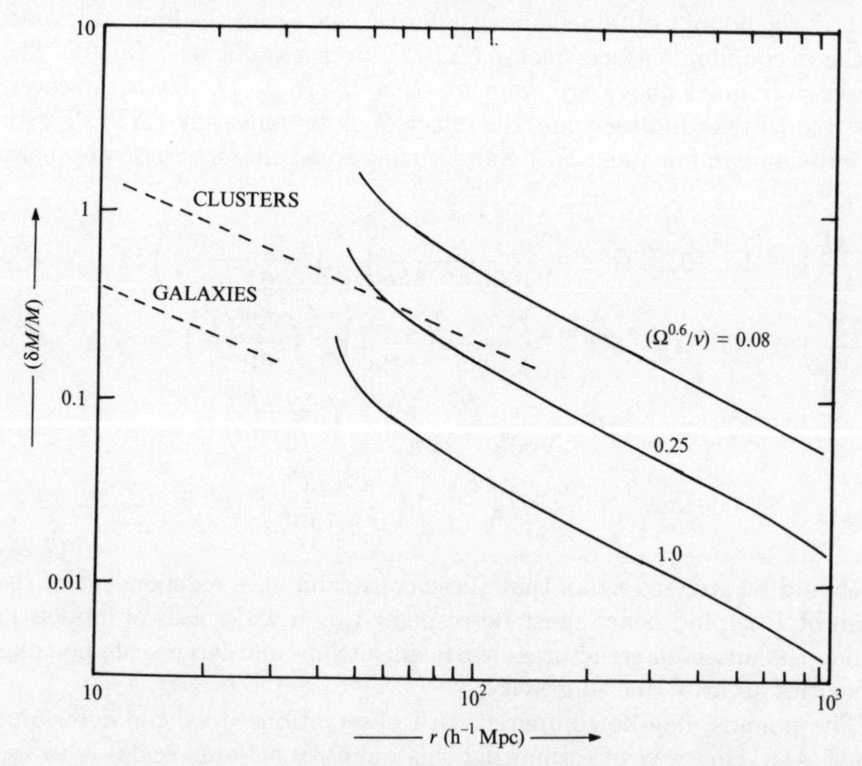

Fig. 7.5. The upper limit on $(\delta M/M)$ derived in the text is plotted against r for various values of $(\Omega^{0.6}/\nu)$. The broken lines are the expected amplitudes of $(\delta N/N)$ based on the galaxy–galaxy and cluster–cluster correlation functions. (Reprinted, by permission, from Nature, **332**, 328; copyright (C) 1988, Macmillan Magazines Ltd.)

the gravitational potential ϕ in our local region at the present epoch. Once $\phi(\mathbf{x}, t)$ is known at present $(t = t_0)$, we can easily extrapolate back the form of this potential to the epoch of decoupling. In the $\Omega = 1$ universe, the gravitational potential is independent of time (see chapter 6) and hence no change is required; if $\Omega \neq 1$, $\phi \propto [D(t)/a(t)]$ where $D(t)$ is the growing solution to the linear perturbation equation. Since the explicit form of this expression is known, the form of ϕ at $t = t_{\rm dec}$ can be easily determined.

Consider now a distant, extragalactic observer who is looking in our direction. The gravitational potential ϕ and the peculiar velocity field $\mathbf{v}(\mathbf{x})$ will induce a $(\Delta T/T)$ towards our direction for this extragalactic observer. Since we know both $\phi(\mathbf{x})$ and $\mathbf{v}(\mathbf{x})$ accurately, we can predict the pattern of the temperature variation which the extragalactic observer will see. Working out the details numerically, one finds that $(\Delta T/T)_{\rm SW} \simeq 1.7 \times 10^{-5}$ at about 1° scale due to the Sachs–Wolfe effect and $(\Delta T/T)_D \simeq 4.9 \times 10^{-5}$ due to the Doppler effect, in the best possible direction. These are somewhat higher than the 'typical' value 5×10^{-6} which should arise more frequently.

Assuming our location is not special in the universe, there should exist other 'Great Attractor' like regions elsewhere. If that is the case, *we* should be able to see the temperature anisotropy derived above. Note that the value of $(\Delta T/T)$ quoted above is *larger* than the bound which is currently available. The absence of their detection is probably due to 'seeing conditions' not being ideal. As the coverage and sensitivity of MBR measurements improve, temperature anisotropies due to large mass concentrations, like 'Great Attractors' – if they exist – should surface.

Exercises

7.1 (Devised by D. Narasimha) Consider a method for estimating distances to groups of galaxies far away in the following manner: We first estimate the probability distribution of the bolometric magnitude of the brightest star in each galaxy (or brightest galaxy in a cluster) in the nearby region. Having determined the mean absolute magnitude and the variance of the distribution we carry out a similar procedure for a distant group. From the mean *apparent* magnitude of the brightest member of the distant population, we can estimate the distance. The variance in the apparent magnitude will give us an idea as to whether the distant population is similar to the nearby one.

(a) What are the possible sources of error in this procedure?

(b) In the sample used as a standard, we have taken the volume average of the sample to estimate the mean luminosity. Since the samples obtained by the usual survey are limited by the detectability of the instruments which are used, we will always be faced with a sample in which the low-brightness members are under-represented. The devia-

tion of the observed distribution (in a flux limited sample) from the true distribution of a volume sample is called 'Malmquist bias.' Derive an expression for the observed mean and variance of a Gaussian probability distribution – which in reality has zero mean and a variance of σ – if the observations are incapable of detecting those members of the sample which are $N\sigma$ lower than the mean in the luminosity.
(c) If we know the spatial integrated probability distribution as a function of absolute magnitude, can we use this information to correct for Malmquist bias?
(d) Suppose the probability density is proportional to $\exp(M - M_0)$ for $M < M_0$ and zero for $M > M_0$. Calculate the mean and variance of this sample. Show that Malmquist bias will change the mean but not the variance.

7.2 Derive the bounds on $(\Delta M/M)$ etc. discussed in section 7.3 of the text.

7.3 An interesting statistic[20] to describe large scale velocity fields is the 'cosmic Mach number'. This quantity $\mathcal{M}(R)$ at a scale R is defined to be the ratio $[\overline{v}(R)/\sigma(R)]$ where $\overline{v}(R)$ is the mean velocity of a patch of size R and $\sigma(R)$ is the velocity dispersion of objects relative to this patch. More precisely, we define it as follows: Let $v_s(\mathbf{r}; a)$ be the velocity field, smoothed with a window function of resolution a. The mean velocity of a patch of size R, centred at $\mathbf{r}_0$ is defined to be

$$\overline{\mathbf{v}}(R; a, \mathbf{r}_0) = \int \mathbf{v}_s(\mathbf{r}, a) W(\mathbf{r} - \mathbf{r}_0; R)\, d^3\mathbf{r},$$

where W is a Gaussian window function of size R. Further,

$$\sigma^2(R; a; \mathbf{r}_0) = \int |\mathbf{v}_s(\mathbf{r}, a)|^2 W(\mathbf{r} - \mathbf{r}_0; R)\, d^3\mathbf{r} - |\overline{\mathbf{v}}(R; a; \mathbf{r}_0)|^2.$$

The Mach number is defined to be

$$\mathcal{M}^2(R; a) \equiv \frac{< |\overline{\mathbf{v}}(R; a; \mathbf{r}_0)|^2 >}{< \sigma^2(R; a; \mathbf{r}_0) >},$$

where the average is over all locations $\mathbf{r}_0$.
(a) What physical meaning can be attributed to $\mathcal{M}$? In a universe which has large scale homogeneity and small scale fluctuations, how will $\mathcal{M}(R)$ change with R?
(b) Relate $\mathcal{M}(R)$ to the power spectrum of fluctuations. Show that $\mathcal{M}(R)$ is independent of the normalization used for $P(k)$. Estimate $\mathcal{M}(R)$ for a power law form of $P(k)$.

7.4 Given the density contrast $\delta(\mathbf{x})$ at all space, one can compute the peculiar velocity $\mathbf{v}(0)$ at origin to be

$$\mathbf{v}(0) = \frac{H_0 \Omega_0^{0.6}}{4\pi} \int \delta(\mathbf{x}) \frac{\mathbf{x}}{|\mathbf{x}|^3}\, d^3\mathbf{x}.$$

Often, we will know $\delta(\mathbf{x})$ only up to some depth R. So the quantity which is actually observed will be

$$\mathbf{v}_R = \frac{H_0\Omega_0^{0.6}}{4\pi}\int \delta(\mathbf{x})\frac{\mathbf{x}}{x^3}W(x/R)\,d^3\mathbf{x},$$

where $W(z)$ is a window function. For a sharp cutoff in the measurements, $W(z) = 1$ for $z < 1$ and zero otherwise.
(a) Show that, with such a sharp cutoff,

$$\mathbf{v} = \mathbf{v}_R - \frac{1}{3}H_0\Omega_0^{0.6}\mathbf{x}_{\rm c.m} + \mathbf{v}_{\rm bulk}$$

where

$$\mathbf{x}_{\rm c.m} \equiv \frac{3}{4\pi R^3}\int \delta(\mathbf{x})W(x/R)\,d^3x; \quad \mathbf{v}_{\rm bulk} = \frac{3}{4\pi R^3}\int \mathbf{v}(\mathbf{x})W(x/R)\,d^3\mathbf{x}.$$

Interpret this result (see reference 21).
(b) Assume that $\delta(\mathbf{x})$ is described by a Gaussian random process with power spectrum $P(k)$. Show that $(\mathbf{v}, \mathbf{v}_R)$ is a six-dimensional, zero mean, Gaussian variate with the covariance matrix:

$$< v_R^\alpha v_R^\beta >= \delta^{\alpha\beta}\sigma_R^2; \quad < v^\alpha v^\beta >= \delta^{\alpha\beta}\sigma^2; \quad < v_R^\alpha v^\beta >= \gamma\sigma\sigma_R\delta^{\alpha\beta}$$

with

$$\sigma^2 = \frac{H_0^2\Omega_0^{1.2}}{6\pi^2}\int_0^\infty dk\ P(k);$$

$$\sigma_R^2 = \frac{H_0^2\Omega_0^{1.6}}{6\pi^2}\int_0^\infty dk\ P(k)\tilde{W}^2(kR);$$

$$\gamma = \frac{H_0^2\Omega_0^{1.2}}{6\pi^2\sigma\sigma_R}\int_0^\infty dk\ P(k)\tilde{W}(kR)$$

where

$$\tilde{W}(kR) = kR\int_0^\infty dy\ W(y)j_1(kRy).$$

The $j_1(r)$ is the first spherical Bessel function.
(c) The 'true' velocity field is $\mathbf{v}$ while the observed one is $\mathbf{v}_R$. When will this difference be important?

7.5 The bulk flow results discussed in the text assume that MBR provides the cosmic rest frame. Suppose this is not the case and that the MBR has a systematic, large scale deviation from the fundamental rest frame of comoving observers. What constraints can be imposed on such a deviation? How should the deviation be arranged if the bulk flow should disappear? What observations can decide this issue?

7.6 Consider the velocity correlation function $C(\mathbf{r}) =< \mathbf{v}(\mathbf{x})\cdot\mathbf{v}(\mathbf{x}+\mathbf{r}) >$. Show that, in the linear approximation C satisfies the equation $\nabla^2 C \cong -H_0^2\Omega_0^{1.2}\ \xi_{\rm gg}$. If $P(k) \propto k^n$, then $\xi_{\rm gg} \propto r^{-(n+3)}$. Solve the equation for $C(r)$ in this case to obtain $C(r) = \alpha r^{-1} + \beta n r^{-(n+1)}$ for $n > -1$. How does one fix α and β? What is the physical interpretation of the two terms?

Part three

Towards a more complete picture

8

The nonlinear evolution

8.1 Introduction

The linear perturbation theory developed in chapters 4 and 5 fails when the density contrast becomes nearly unity. Since most of the observed structures in the universe – like galaxies, clusters etc. – have density contrasts far in excess of unity, their structure can be understood only by a fully nonlinear theory. This chapter discusses the nonlinear evolution of perturbations, starting from where we left off in chapters 4 and 5. Nonlinear evolution can be studied analytically if some simplifying assumptions are made. Such simplified analytic models are studied in sections 8.2, 8.5 and 8.6. The results of these calculations are used to understand the properties of the galaxies in sections 8.3, 8.4 and 8.7. Sections 8.8 and 8.9 attempt to provide a more detailed modelling of the properties of spiral and elliptical galaxies and discuss the difficulties encountered in such attempts. Finally, section 8.10 reviews the results of N-body simulations used in studying nonlinear evolution.

8.2 Spherical model for the nonlinear collapse

The evolution of density perturbations in the linear regime was analyzed in chapters 4 and 5. The final result of this analysis was an expression for the processed power spectrum $P(k)$ at $t \gtrsim t_{\rm dec}$. The observed isotropy of the MBR guarantees that the density contrast δ_k must have been quite small ($\delta_k \lesssim 10^{-4}$ or so) at this epoch, implying that the evolution of the density contrast can be studied using linear theory at $t \gtrsim t_{\rm dec}$ and that δ_k grows in proportion to the scale factor $a(t)$. At some later time, $t_{\rm nl}(\lambda)$, the density contrast at a wavelength λ will become comparable to unity. For $t > t_{\rm nl}(\lambda)$, the linear perturbation theory fails at this wavelength and we have to study the evolution using some other techniques.

The Fourier transform $\delta_k(t)$ of the density contrast $\delta(t, \mathbf{x})$ was useful in the linear regime because each mode was evolving independently. Since this is no longer true in the nonlinear limit, there is no specific advantage in using the Fourier components; it is better to study the evolution of $\delta(\mathbf{x}, t)$ directly, in the $\mathbf{x}$ space.

This may be done as follows: Consider the density contrast $\delta(\mathbf{x}, t_i)$ in the universe at some time t_i. This density contrast will divide the universe into several overdense ($\delta > 0$) and under dense ($\delta < 0$) regions. It is reasonable to expect that regions which are significantly overdense will collapse and (eventually) form gravitationally bound objects. In these overdense regions, the self-gravity of the local mass concentration will work *against* the expansion of the universe; i.e., this region will expand at a progressively *slower* rate compared to the background universe. Such a slowing down will increase the density contrast between the overdense region and the background universe and – consequently – make the gravitational potential of the local mass concentration (in that region) more and more dominant. Eventually, such a region will collapse under its own self-gravity and will form a bound system.

The details of the above process will depend on the initial density profile. The simplest model which one can study analytically[1] is based on the assumption that the overdense region is spherically symmetric (about some point). Let us suppose that the overdense region we are interested in has an initial density distribution

$$\rho(r, t_i) = \rho_b(t_i) + \delta\rho(r, t_i) = \rho_b(t_i)[1 + \delta_i(r)] \tag{8.1}$$

where $\delta_i(r) = \delta(r, t_i)$ is the initial density contrast which is some specified, *non-increasing*, function of r. Since we are now interested in perturbations with $\lambda \ll d_H$, the size R of the overdense region (which may be taken to be the scale over which δ_i is significant) can be taken to be much smaller than the Hubble radius. In this case, we can study the dynamics of this region using the Newtonian approximation developed in chapter 4. In the Newtonian limit, it is convenient to use the proper radial coordinate $r = a(t)|\mathbf{x}|$ where $\mathbf{x}$ is the comoving Friedmann coordinate. The dynamics of the overdense region is determined by the gravitational potential

$$\begin{aligned}\phi_{\text{total}}(r, t) = \phi_b(r, t) + \delta\phi(r, t) &= -\frac{1}{2}\left(\frac{\ddot{a}}{a}\right) r^2 + \delta\phi(r, t) \\ &= \frac{2\pi}{3} G\rho_b r^2 + \delta\phi(r, t)\end{aligned} \tag{8.2}$$

where ϕ_b is the equivalent Newtonian potential of the Friedmann metric (see chapter 4) and $\delta\phi$ is the potential generated due to the excess density $\delta\rho(r, t)$. The motion of a thin shell of particles located at a distance r is governed by the equation

$$\frac{d^2\mathbf{r}}{dt^2} = -\nabla\phi_{\text{total}} = -\frac{4\pi G\rho_b(t)}{3}\mathbf{r} - \nabla(\delta\phi) = -\frac{GM_b}{r^3}\mathbf{r} - \frac{G\delta M(r, t)}{r^3}\mathbf{r}. \tag{8.3}$$

In writing the second term, we have used the fact that, for a spherically symmetric density distribution, the gravitational force only depends on the mass δM contained inside the shell. Here M_b and $\delta M(r,t)$ stand for

$$M_b = \frac{4\pi}{3}\rho_b(t)r^3 = \frac{4\pi}{3}\rho_b(t)a^3(t)x^3 = \text{constant}; \tag{8.4}$$

$$\delta M(r,t) = 4\pi \int_0^r \delta\rho(q,t)q^2\, dq = 4\pi\rho_b(t)\int_0^r q^2\delta(q,t)\, dq. \tag{8.5}$$

To simplify the analysis of the problem, we will assume that the spherical shells do not cross each other during the evolution. That is, if we initially label the shells as $1, 2, \cdots$ etc. with the radii $r_1 < r_2 < r_3 \cdots$ etc., then the subsequent evolution is assumed to preserve the ordering $r_1 < r_2 \cdots$. In such a case the mass contained within a shell of radius r does not change with time: $\delta M(r,t) = \delta M(r,t_i) = \text{constant}$. We can now combine the two terms in (8.3) to write

$$d^2r/dt^2 = -GM/r^2, \tag{8.6}$$

where

$$M = \rho_b\left(\frac{4\pi}{3}r_i^3\right)(1+\overline{\delta}_i), \quad \overline{\delta}_i = \left(\frac{3}{4\pi r_i^3}\right)\int_0^{r_i}\delta_i(r)4\pi r^2\, dr. \tag{8.7}$$

Here r_i is the initial radius of the shell with mass M and $\overline{\delta}_i$ is the average value of δ within r_i at time t_i. The first integral of equation (8.6) is

$$\frac{1}{2}\left(\frac{dr}{dt}\right)^2 - \frac{GM}{r} = E \tag{8.8}$$

where E is a constant of integration. The sign of E determines whether a given mass shell will expand forever or eventually decouple from the expansion and collapse. If $E > 0$, it follows from (8.8) that $\dot{r}^2$ will never become zero; the shell will expand for ever. On the other hand, if $E < 0$ then as r increases $\dot{r}$ will eventually become zero and later negative, implying a contraction and collapse.

This condition for the collapse of an overdense region can be expressed in a more convenient form. To do this, let us consider the terms in (8.8) at the initial instant $t = t_i$. It is convenient to choose t_i to be the time at which δ is quite small so that the overdense region was expanding along with the background. That is, we shall assume that the peculiar velocities v_i are negligible at $t = t_i$ (a more general case is studied in exercise 8.1). Then, $\dot{r}_i = (\dot{a}/a)r_i = H(t_i)r_i \equiv H_i r_i$ at time t_i, and the initial kinetic energy will be

$$K_i \equiv \left(\frac{\dot{r}^2}{2}\right)_{t=t_i} = \frac{H_i^2 r_i^2}{2}. \tag{8.9}$$

The potential energy at $t = t_i$ is $U = -|U|$ where

$$|U| = \left(\frac{GM}{r}\right)_{t=t_i} = G\frac{4\pi}{3}\rho_b(t_i)r_i^2(1+\overline{\delta}_i) = \frac{1}{2}H_i^2 r_i^2 \Omega_i(1+\overline{\delta}_i)$$
$$= K_i\Omega_i(1+\overline{\delta}_i) \tag{8.10}$$

with $\Omega_i = (\rho_b(t_i)/\rho_c(t_i))$ denoting the *initial* value of the density parameter Ω of the smooth background universe. The total energy of the shell is, therefore,

$$E = K_i - K_i\Omega_i(1+\overline{\delta}_i) = K_i\Omega_i[\Omega_i^{-1} - (1+\overline{\delta}_i)]. \tag{8.11}$$

The condition $E < 0$ for the shell to collapse (eventually), becomes $(1+\overline{\delta}_i) > \Omega_i^{-1}$, or

$$\overline{\delta}_i > [\Omega_i^{-1} - 1]. \tag{8.12}$$

In a closed or flat universe (with $\Omega_i^{-1} \leq 1$), this condition is satisfied by any overdense region with $\overline{\delta} > 0$. In this case, the overdense regions will always collapse although (as we will see) smaller overdensities will take longer times to turn-around and collapse. In an open universe with $\Omega_i < 1$, the overdensity has to be above a critical value for collapse to occur. For a general density distribution $\delta_i(r)$, only shells within a critical initial radius r_{cr}, such that $\overline{\delta}_i(r_{cr}) = \Omega_i^{-1} - 1$, will be able to collapse.

Let us now consider a shell with $E < 0$, which expands to a maximum radius r_m and then collapses. The maximum radius r_m which such a shell attains can be easily derived. To do this, note that at the instant of maximum expansion, we have $\dot{r} = 0$ giving

$$E = -GM/r_m = -(r_i/r_m)K_i\Omega_i(1+\overline{\delta}_i). \tag{8.13}$$

Equating this expression for E with the one in (8.11), we get

$$\frac{r_m}{r_i} = \frac{(1+\overline{\delta}_i)}{\overline{\delta}_i - (\Omega_i^{-1} - 1)}. \tag{8.14}$$

Clearly, $r_m \gg r_i$ if $\overline{\delta}_i \gtrsim (\Omega_i^{-1} - 1)$; shells which are only slightly overdense, compared to the critical value $(\Omega_i^{-1} - 1)$, will expand much further and can take a long time to collapse.

The time evolution of the shell can be found by integrating the equations of motion. The solution to equation (8.8), for $E < 0$, is given in a parametric form by

$$r = A(1-\cos\theta), \quad t + T = B(\theta - \sin\theta); \quad A^3 = GMB^2 \tag{8.15}$$

where A and B are constants related to each other as shown. The parameter θ increases with increasing t, while r increases to a maximum value before decreasing to zero. The constant T allows us to set the initial condition that at $t = t_i$, $r = r_i$. A shell enclosing mass M and initially expanding with the background universe will progressively slow down, reach a maximum radius at $\theta = \pi$, 'turn-around' and collapse. The epoch of maximum radius is also referred to as the epoch of 'turn-around'. At the 'turn-around', $dr/dt = 0$ and $r = r_m$.

The constants A and B can be determined by using (8.14). At $\theta = \pi$, $r(\pi) = r_m = 2A$; comparing with (8.14), we get

$$A = \frac{r_i}{2}\frac{(1+\overline{\delta}_i)}{[\overline{\delta}_i - (\Omega_i^{-1} - 1)]}. \tag{8.16}$$

Using $A^3 = GMB^2$, and the expression for M from (8.7) we find B to be:

$$B = \frac{1+\overline{\delta}_i}{2H_i\Omega_i^{1/2}[\overline{\delta}_i - (\Omega_i^{-1} - 1)]^{3/2}}. \tag{8.17}$$

The value of T can be fixed by setting $r = r_i$ at $t = t_i$. As an example, consider the case in which the background universe is flat ($\Omega_i = 1$). Then

$$A = \frac{r_i}{2}\left(\frac{1+\overline{\delta}_i}{\overline{\delta}_i}\right); \quad B = \frac{1}{2H_i}\frac{(1+\overline{\delta}_i)}{\overline{\delta}_i^{3/2}}. \tag{8.18}$$

At $t = t_i$ we have to satisfy the conditions

$$r_i = \frac{r_i}{2}\left(\frac{1+\overline{\delta}_i}{\overline{\delta}_i}\right)(1 - \cos\theta_i) \tag{8.19}$$

$$t_i + T = \frac{1}{2H_i}\left(\frac{1+\overline{\delta}_i}{\overline{\delta}_i^{3/2}}\right)(\theta_i - \sin\theta_i). \tag{8.20}$$

From (8.19), we get, $\cos\theta_i = (1-\overline{\delta}_i)\,(1+\overline{\delta}_i)^{-1}$. Since δ_i is expected to be quite small, we can approximate this relation as $\cos\theta_i \simeq 1 - 2\delta_i$, obtaining $\theta_i^2 = 4\delta_i$. Substituting in (8.20), we get

$$H_i(t_i + T) = \frac{2}{3}(1+\delta_i). \tag{8.21}$$

Or, since $H_i t_i = (2/3)$ for the $\Omega = 1$ universe, $H_i T = (2/3)\delta_i$. This shows that $(T/t_i) = \delta_i \ll 1$. Hence, we will ignore T in what follows. (Similar conclusions hold for models with $\Omega_i \neq 1$, as long as $\delta_i \ll 1$.) The equation (8.15), with the constants A and B fixed by (8.16) and (8.17), give the complete information about how each perturbed mass shell evolves. These equations can be used to work out all the characteristics of a spherical perturbation.

Consider, for example, the evolution of mean density within each mass shell. Since M is constant for each mass shell, the mean density within a shell is

$$\overline{\rho}(t) = (3M/4\pi r^3) = \frac{3M}{4\pi A^3(1-\cos\theta)^3}. \tag{8.22}$$

In the special case in which the initial density enhancement is homogeneous, the average density calculated above is also the actual density. The density profile of such a constant density sphere is often referred to as the 'top-hat' profile. To work out the time evolution of the density contrast $\overline{\delta}(r,t)$, one also needs to know how the background density evolves. In the simplest case of a flat universe with $k=0$, the expansion factor $a(t)$ and density $\rho_b(t)$ of the background are given by:

$$a \propto t^{2/3}; \quad \rho_b(t) = \frac{1}{6\pi G t^2}. \tag{8.23}$$

Dividing the mean density $\overline{\rho}(r,t)$ in equation (8.22) by the background density, we get the mean density contrast:

$$\frac{\overline{\rho}(r,t)}{\rho_b(t)} = 1 + \overline{\delta}(r,t) = \frac{3M}{4\pi A^3}\frac{6\pi G B^2(\theta-\sin\theta)^2}{(1-\cos\theta)^2}, \tag{8.24}$$

where we have used the relation between t and θ given in equation (8.15) and set $T=0$. Since $A^3 = GMB^2$ it follows that

$$\overline{\delta} = \frac{9}{2}\frac{(\theta-\sin\theta)^2}{(1-\cos\theta)^3} - 1. \tag{8.25}$$

The linear evolution for the average density contrast is recovered in the limit of small t. In this limit, we have

$$\overline{\delta} \approx \frac{3\theta^2}{20}; \quad t \approx \frac{B\theta^3}{6} \tag{8.26}$$

so that

$$\overline{\delta} = \frac{3}{20}\left(\frac{6t}{B}\right)^{2/3}. \tag{8.27}$$

For a flat universe with $\Omega_i = 1$ and $H_i = 2/(3t_i)$

$$B = \frac{3}{4}\frac{t_i}{\overline{\delta}_i^{3/2}}(1+\overline{\delta}_i). \tag{8.28}$$

Using this value for B in (8.27) we find, to the leading order,

$$\overline{\delta} = \frac{3}{5}\overline{\delta}_i\left(\frac{t}{t_i}\right)^{2/3} \propto a(t). \tag{8.29}$$

This is the correct growth law ($\delta \propto t^{2/3}$) for the purely growing mode in the linear regime if the initial peculiar velocity is zero; the origin of the factor (3/5) is discussed in exercise 8.1.

For the $\Omega_i = 1$ model, A and B are given by (8.18). Assuming that $\overline{\delta}_i$ is small compared with unity, and retaining only the leading terms of $\overline{\delta}_i$ in A and B, we can write:

$$A \cong \frac{r_i}{2\overline{\delta}_i}; \quad B \cong \frac{3t_i}{4\overline{\delta}_i^{3/2}}. \tag{8.30}$$

For further discussion, it is convenient to use two other variables x and $\overline{\delta}_0$ in place of r_i and δ_i. The quantity x is the comoving radius: $x = r_i[a(t_0)/a(t_i)]$ corresponding to r_i; the parameter $\overline{\delta}_0$ is defined as: $\overline{\delta}_0 = (a(t_0)/a(t_i))\ (3\delta_i/5) = (3/5)\delta_i(1+z_i)$. This is the present value of the density contrast, as predicted by the linear theory, if the density contrast was δ_i at the redshift z_i. In terms of x and δ_0, we have:

$$A = \frac{3x}{10\delta_0}; \quad B = \left(\frac{3}{5}\right)^{3/2} \frac{3t_0}{4\delta_0^{3/2}}. \tag{8.31}$$

Hereafter we will omit the overbar on $\overline{\delta}$ when no confusion can arise. Collecting all our results together, the evolution of a spherical overdense region can be summarized by the following equations:

$$r(t) = \frac{r_i}{2\delta_i}(1-\cos\theta) = \frac{3x}{10\delta_0}(1-\cos\theta), \tag{8.32}$$

$$t = \frac{3t_i}{4\delta_i^{3/2}}(\theta - \sin\theta) = \left(\frac{3}{5}\right)^{3/2} \frac{3t_0}{4\delta_0^{3/2}}(\theta-\sin\theta), \tag{8.33}$$

$$\overline{\rho}(t) = \rho_b(t)\frac{9(\theta-\sin\theta)^2}{2(1-\cos\theta)^3}, \tag{8.34}$$

The density can be expressed in terms of the redshift by using the relation $(t/t_i)^{2/3} = (1+z_i)(1+z)^{-1}$. This gives

$$(1+z) = \left(\frac{4}{3}\right)^{2/3} \frac{\delta_i(1+z_i)}{(\theta-\sin\theta)^{2/3}} = \left(\frac{5}{3}\right)\left(\frac{4}{3}\right)^{2/3} \frac{\delta_0}{(\theta-\sin\theta)^{2/3}}; \tag{8.35}$$

$$\delta = \frac{9}{2}\frac{(\theta-\sin\theta)^2}{(1-\cos\theta)^3} - 1. \tag{8.36}$$

Given an initial density contrast δ_i at redshift z_i, these equations define (implicitly) the function $\delta(z)$ for $z > z_i$. Equation (8.35) defines θ in terms of z (implicitly); equation (8.36) gives the density contrast at that $\theta(z)$. For comparison note that linear evolution gives the density contrast δ_L where

$$\delta_L = \frac{\overline{\rho}_L}{\rho_b} - 1 = \frac{3}{5}\frac{\delta_i(1+z_i)}{1+z} = \frac{3}{5}\left(\frac{3}{4}\right)^{2/3}(\theta-\sin\theta)^{2/3}. \tag{8.37}$$

We can estimate the accuracy of the linear theory by comparing $\delta(z)$ and $\delta_L(z)$. To begin with, for $z \gg 1$, we have $\theta \ll 1$ and we get $\delta(z) \simeq \delta_L(z)$. When $\theta = (\pi/2)$, $\delta_L = (3/5)(3/4)^{2/3}(\pi/2 - 1)^{2/3} = 0.341$ while $\delta = (9/2)(\pi/2 - 1)^2 - 1 = 0.466$; thus the actual density contrast is about 40 per cent higher. When $\theta = (2\pi/3), \delta_L = 0.568$ and $\delta = 1.01 \simeq 1$. If we interpret $\delta = 1$ as the transition point to nonlinearity, then such a transition occurs at $\theta = (2\pi/3)$, $\delta_L \simeq 0.57$. From (8.35), we see that this occurs at the redshift $(1 + z_{\rm nl}) = 1.06\delta_i(1 + z_i) = (\delta_0/0.57)$. The next important stage occurs at $\theta = \pi$ when the spherical region reaches the maximum radius of expansion. From our equations, we find that the redshift z_m, the proper radius of the shell r_m and the average density contrast δ_m at 'turn-around' are:

$$
\begin{aligned}
(1 + z_m) &= \frac{\delta_i(1 + z_i)}{\pi^{2/3}(3/4)^{2/3}} = 0.57(1 + z_i)\delta_i = \frac{5}{3}\frac{\delta_0}{(3\pi/4)^{2/3}} \cong \frac{\delta_0}{1.062}, \\
r_m &= \frac{3x}{5\delta_0}, \\
\left(\frac{\overline{\rho}}{\rho_b}\right)_m &= 1 + \overline{\delta}_m = \frac{9\pi^2}{16} \approx 5.6.
\end{aligned}
\tag{8.38}
$$

The first equation gives the redshift at turn-around for a region, parametrized by the (hypothetical) linear density contrast δ_0 at the present epoch. If, for example, $\delta_i \simeq 10^{-3}$ at $z_i \simeq 10^4$, such a perturbation would have turned around at $(1 + z_m) \simeq 5.7$ or when $z_m \simeq 4.7$. The second equation gives the maximum radius reached by the perturbation. The third equation shows that the region under consideration is nearly 6 times denser than the background universe, at turn-around. This corresponds to a density contrast of $\delta_m \approx 4.6$ which is definitely in the nonlinear regime. The linear evolution gives $\delta_L = 1.063$ at $\theta = \pi$.

After the spherical overdense region turns around it will continue to contract. Equation (8.34) suggests that at $\theta = 2\pi$ all the mass will collapse to a point. However, long before this happens, the approximation that matter is distributed in spherical shells and that random velocities of the particles are small, will break down. The collisionless component of density, viz. the dark matter, will reach virial equilibrium by a process known as 'violent relaxation'. This process arises as follows[2]: During the collapse there will be large fluctuations in the gravitational potential, in a time scale of the order of the free-fall collapse time, $t_{\rm dyn} \simeq (G\rho)^{-1/2}$. Since the potential is changing with time, individual particles do not follow orbits which conserve the energy. Clearly, the change in the energy of a particle depends in a complex way on its initial position and velocity, but the net effect will be to widen the range of energies available to the particles. Thus, a potential varying in time can provide a relaxation mechanism for the particles which operates in a timescale $t_{\rm dyn}$ which is much smaller than the two-body-relaxation time t_R. This process has been termed 'violent relaxation' (for more details, see exercise 8.2).

The above process will relax the collisionless (dark matter) component to a configuration with radius $r_{\rm vir}$, velocity dispersion v and density $\rho_{\rm coll}$. (The behaviour of the baryonic component is a little more complicated and we will discuss it separately later.) Such a virialized system can be used to model the structures which we see in the universe. We shall now estimate the physical parameters of such a system.

After virialization of the collapsed shell, the potential energy U and the kinetic energy K will be related by $|U| = 2K$ so that the total energy $\mathcal{E} = U + K = -K$. At $t = t_m$ all the energy was in the form of potential energy. For a spherically symmetric system with constant density, $\mathcal{E} \approx -3GM^2/5r_m$. The 'virial velocity' v and the 'virial radius' $r_{\rm vir}$ for the collapsing mass can be estimated by the equations:

$$K \equiv \frac{Mv^2}{2} = -\mathcal{E} = \frac{3GM^2}{5r_m}; \quad |U| = \frac{3GM^2}{5r_{\rm vir}} = 2K = Mv^2. \tag{8.39}$$

We get:

$$v = (6GM/5r_m)^{1/2}; \quad r_{\rm vir} = r_m/2. \tag{8.40}$$

The time taken for the fluctuation to reach virial equilibrium, $t_{\rm coll}$, is essentially the time corresponding to $\theta = 2\pi$. From equation (8.35), we find that the redshift at collapse, $z_{\rm coll}$, is

$$(1 + z_{\rm coll}) = \frac{\delta_i(1 + z_i)}{(2\pi)^{2/3}(3/4)^{2/3}} = 0.36\delta_i(1 + z_i) = 0.63(1 + z_m) = \frac{\delta_0}{1.686}. \tag{8.41}$$

The density of the collapsed object can also be determined fairly easily. Since $r_{\rm vir} = (r_m/2)$, the mean density of the collapsed object is $\rho_{\rm coll} = 8\rho_m$ where ρ_m is the density of the object at turn-around. Further, $\rho_m \cong 5.6\rho_b(t_m)$ and $\rho_b(t_m) = (1 + z_m)^3 \, (1 + z_{\rm coll})^{-3} \rho_b(t_{\rm coll})$. Combining these relations, we get

$$\rho_{\rm coll} \simeq 2^3 \rho_m \simeq 44.8\rho_b(t_m) \simeq 170\rho_b(t_{\rm coll}) \simeq 170\rho_0(1 + z_{\rm coll})^3 \tag{8.42}$$

where ρ_0 is the present cosmological density. This result determines $\rho_{\rm coll}$ in terms of the redshift of formation of a bound object (see figure 8.1). (For comparison, it may be noted that linear theory predicts $\delta_L = 1.686$ at $\theta = 2\pi$.) Once the system has virialized, its density and size do not change. Since $\rho_b \propto a^{-3}$, the density contrast δ increases as a^3 for $t > t_{\rm coll}$.

Let us now consider the collapse of the baryonic component, for which a similar result holds. During the collapse, the gaseous mixture of hydrogen and helium develops shocks and gets reheated to a temperature

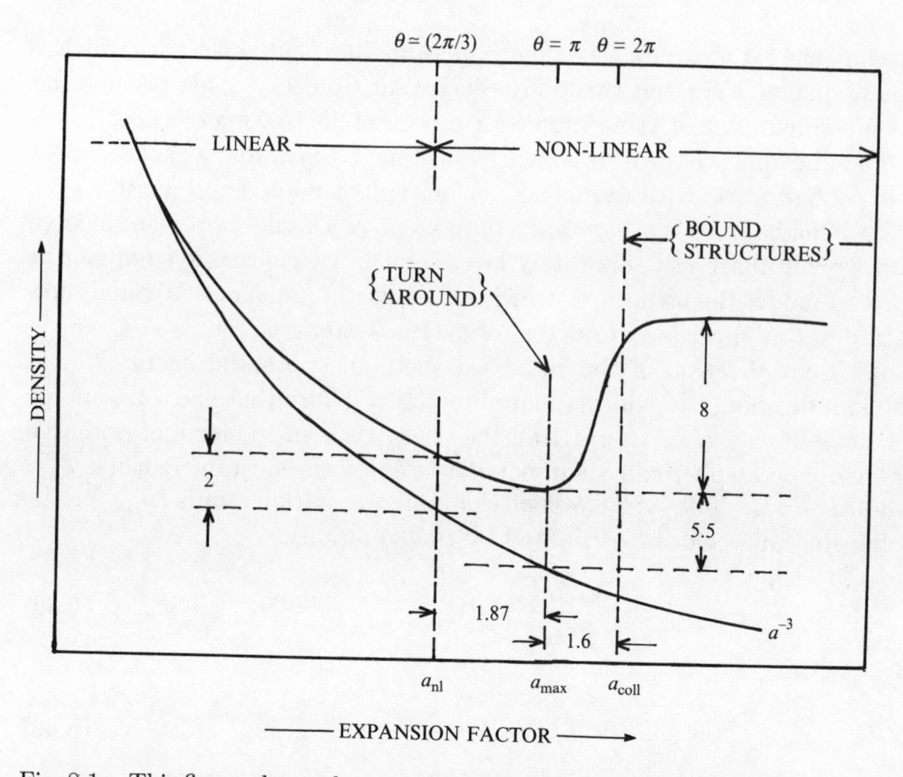

Fig. 8.1. This figure shows the growth of density in a spherical over-dense region. The lower curve shows the evolution of the background density in a matter-dominated case. The upper curve is the density of the spherical inhomogeneity. In the linear region, the contrast grows as a. Once the nonlinear stage is reached, the spherical region collapses faster, virializes and forms a bound structure. The density of the bound structure remains constant thereafter.

at which pressure balance can prevent further collapse. At this stage the thermal energy will be comparable to the gravitational potential energy. The temperature of the gas, $T_{\rm vir}$, is related to the velocity dispersion v^2 by $3\rho_{\rm gas}T_{\rm vir}/2\mu = \rho_{\rm gas}v^2/2$, where $\rho_{\rm gas}$ is the gas density and μ is its mean molecular weight. This gives $T_{\rm vir} = \mu v^2/3$. It is useful to express the above results with typical numbers for the various quantities shown explicitly. If the He fraction is Y by weight and the gas is fully ionized, then

$$\mu = \frac{(m_{\rm H}n_{\rm H} + m_{\rm He}n_{\rm He})}{(2n_{\rm H} + 3n_{\rm He})} = \frac{m_{\rm H}}{2}\left(\frac{1+Y}{1+0.375Y}\right) \cong 0.57m_{\rm H}, \qquad (8.43)$$

if $Y = 0.25$. Apart from the cosmological parameters, two parameters need to be specified. These may be chosen to be the mass M of the over-dense region and the redshift of formation $z_{\rm coll}$. Using the cosmological parameters

$$\begin{aligned}\rho_0 &= 1.88 \times 10^{-29}\Omega h^2 \,\mathrm{g\,cm^{-3}},\\ x &= 0.92(\Omega h^2)^{-1/3}(M/10^{12}\,\mathrm{M_\odot})^{1/3}\,\mathrm{Mpc},\\ t_0 &= 0.65 \times 10^{10} h^{-1}\,\mathrm{yr}.\end{aligned} \tag{8.44}$$

and $\delta_0 = 1.686(1 + z_{\mathrm{coll}})$, we find

$$\begin{aligned} r_{\mathrm{vir}} &= 258(1+z_{\mathrm{coll}})^{-1}\left(\frac{M}{10^{12}\,\mathrm{M_\odot}}\right)^{1/3} h_{0.5}^{-2/3}\,\mathrm{kpc}\\ &= 434\delta_0^{-1}\, h_{0.5}^{-2/3}\, M_{12}^{1/3}\,\mathrm{kpc},\\ v &= 100(1+z_{\mathrm{coll}})^{1/2}\left(\frac{M}{10^{12}\,\mathrm{M_\odot}}\right)^{1/3} h_{0.5}^{1/3}\,\mathrm{km\,s^{-1}}\\ &= 77\delta_0^{1/2}\, M_{12}^{1/3}\, h_{0.5}^{1/3}\,\mathrm{km\,s^{-1}},\\ T_{\mathrm{vir}} &= 2.32\times 10^5(1+z_{\mathrm{coll}})\left(\frac{M}{10^{12}\,\mathrm{M_\odot}}\right)^{2/3} h_{0.5}^{2/3}\,\mathrm{K}\\ &= 1.36\times 10^5 \delta_0 M_{12}^{2/3}\, h_{0.5}^{2/3}\,\mathrm{K}.\end{aligned} \tag{8.45}$$

Also note that

$$t_{\mathrm{coll}} = t_0(1+z_{\mathrm{coll}})^{-3/2}; \quad (1+z_m) = 1.59(1+z_{\mathrm{coll}}). \tag{8.46}$$

These expressions use $h_{0.5}$, the Hubble constant in units of $50\,\mathrm{km\,s^{-1}\,M}$ pc^{-1}; we have also set $\Omega = 1$. The above results can be used to estimate the typical parameters of collapsed objects once we are given M and the collapse redshift. For example, if objects with $M = 10^{12}\,\mathrm{M_\odot}$ (which is typical of galaxies) collapse at a redshift of, say 2, then one gets $r_{\mathrm{vir}} \approx 86\,\mathrm{kpc}$, $t_{\mathrm{coll}} \approx 1.2\times 10^9\,\mathrm{yr}$, $v \approx 173\,\mathrm{km\,s^{-1}}$, $T_{\mathrm{vir}} \approx 7\times 10^5\,\mathrm{K}$. The density contrast of the galaxy at present will be $(\rho_{\mathrm{coll}}/\rho_0) \approx 170(1+z_{\mathrm{coll}})^3 \approx 4.6\times 10^3$.

These values are broadly in agreement with the parameters which one associates with a galactic halo. The linear evolution, studied in chapter 4, combined with the spherical collapse model discussed above, seems to be capable of producing structures of the correct magnitude. The virial radius of the baryonic content of the galaxy will be much smaller because baryons can cool by radiative processes and contract further. This will be discussed in section 8.3.

The equation (8.33) also provides a relation we needed in chapter 5. The time of formation of a bound structure (t_{coll}) is related to the density contrast δ_i at an earlier time t_i by $t_{\mathrm{coll}} \propto t_i\delta_i^{-3/2}$. That is, the minimum density contrast needed at time t_i for a bound structure to form at $t \le t_{\mathrm{coll}}$ scales as $(\delta_i)_{\mathrm{min}} \propto (t_i/t_{\mathrm{coll}})^{2/3}$. This result was used in chapter 5.

The spherical top-hat model can be used to estimate nonlinear density contrast in the following way: We start with some density contrast δ_i at z_i, and compute the density contrast δ_0 at present using the linear theory to be $\delta_0 = (3/5)\delta_i(1+z_i)$. The *actual* density contrast, of course, will be higher and can be calculated as follows: (1) If $\delta_0 < 1.063$, then we can find a $\theta(\delta_0)$ in the range $0 < \theta < \pi$ by inverting the relation (8.37): $\delta_0 = (3/5)(3/4)^{2/3}(\theta - \sin\theta)^{2/3}$. The correct density contrast can now be obtained from (8.36) using this value of $\theta(\delta_0)$. (2) If $\delta_0 > 1.686$, then our analysis shows that a bound structure would have already formed at $(1+z_{\rm coll}) = (\delta_0/1.686)$ with the density $\rho_{\rm coll} \cong 170\rho_0(1+z_{\rm coll})^3 = (170/1.686^3)\rho_0\delta_0^3 \simeq 35.5\rho_0\delta_0^3$. The correct density contrast is, therefore, $\delta = (\rho_{\rm coll}/\rho_0) - 1 = 35.5\delta_0^3 - 1$. (3) For $1.063 < \delta_0 < 1.686$, the spherical collapse model is a bad approximation and cannot be used to make reliable predictions. The actual density contrast increases by two orders of magnitude during this interval.

We end this section by mentioning an extremely simple *general relativistic* solution which describes the evolution of a spherical inhomogeneity in the Friedmann universe (see exercise 8.3). It turns out[3] that such a situation can be described by a metric of the form

$$ds^2 = dt^2 - a^2(x,t)\left[\frac{dx^2}{1-k(x)x^2}\left[\frac{(ax)'}{a}\right]^2 - x^2(d\theta^2 + \sin\theta\, d\phi^2)\right] \quad (8.47)$$

where $a(x,t)$ is a *space dependent* 'expansion factor', and $k(x)$ is a space dependent curvature constant. The metric can be written in the above form as long as mass shells at different values of x do not cross ; a condition which will be satisfied by density distributions in which ρ decreases monotonically with x. The Einstein equations determining the time evolution of the expansion factor $a(x,t)$ and the matter density $\rho(x,t)$ turn out to be:

$$\frac{\dot{a}^2 + k(x)}{a^2} = \frac{C}{a^3} = \frac{8\pi G\rho(x,t)}{3}\frac{(ax)'}{a} \quad (8.48)$$

where C is a constant. Here (8.48) is actually two equations, one giving the evolution of a and the other that of ρ. If ρ, and hence a and k, are independent of x these equations reduce to the standard equations for a Friedmann universe.

This simple generalization of the homogeneous universe model offers considerable insight into the way a spherical overdense (or underdense) region behaves. Equation (8.48) shows that the behaviour of a mass shell at a comoving radius x is completely specified by the *local* value of the curvature constant k. If at some x, $k(x) \le 0$, the corresponding mass shell will expand for ever, while if $k(x) > 0$, it will turn around at some stage and collapse.

We are now in a position to understand the evolution of different types of spherical density perturbations that may arise in the Friedmann universe. Consider the case when $k(x)$ is positive for $x < x_0$, is zero at x_0 and tends to a constant negative value, say -1, far away from the origin. One way of realizing such a situation is to embed a density hill centred around the origin in an open Friedmann universe and start off the universe expanding uniformly. From the evolution equation (8.48) we can infer that the region $x < x_0$ will eventually collapse, while the region $x \geq x_0$ will expand for ever. Here we see quite clearly that condensation in a local part of the universe does not alter the global behaviour of an open Friedmann universe. Similarly one can construct expanding voids in a closed universe. In this case, one demands that $k(x) < 0$ for $x < x_0$ (say) and positive elsewhere. This situation can be realized if there is a deep enough density valley in a closed universe. The region within x_0 will keep on expanding, whereas the region outside will initially expand at a slower rate and eventually recollapse.

8.3 Scaling laws

The analysis in the previous section dealt with the nonlinear collapse of a *single* overdense region in an otherwise smooth universe. To model the structure formation correctly, we need to find the full *power spectrum* of bound objects which are formed due to the nonlinear collapse. This is a considerably more difficult task in which only limited success has been achieved.

The mass function of the bound objects can be calculated in a fairly straightforward manner, once a choice is made for the filtering function. (This was discussed in chapter 5.) While this is adequate for some purposes, it does not provide a dynamical picture of the collapse. Somewhat more detailed modelling is possible if the form of the power spectrum at $t = t_{\rm dec}$ is known. The further evolution depends crucially on whether the dark matter is cold or hot. Different kinds of analytic approximations are needed for the two cases. We shall first consider the case of cold dark matter. The approximate analysis of the hot dark matter will be taken up in sections 8.5 and 8.6.

We saw in chapter 5 that the density inhomogeneity can be characterized by a Gaussian distribution with some variance, σ, which is related to the power spectrum of the fluctuations. Labelling the fluctuations by the mass $M \propto \lambda^3 \propto k^{-3}$, we can relate the mean square fluctuation in the mass to the variance of the Gaussian as,

$$\sigma^2(M) =< (\delta M/M)^2 >= CM^{-(3+n)/3} \tag{8.49}$$

where n is the index of the power spectrum (with $P(k) \propto k^n$) and C is the normalization constant which should be fixed by comparison with the observations. Since $P(k)$ is not a strict power law, n should be thought of as an approximate local value $d(\ln P)/d(\ln k)$ in the relevant range.

The quantity $\sigma(M)$ was interpreted in chapter 5 as the typical (excess) mass contrast at some scale $R \propto M^{1/3}$. Since $(\delta M/M) \propto [(\delta\rho)R^3/\rho_b R^3]$ $\propto (\delta\rho/\rho_b)$, we can take the quantity $\sigma(M)$ to be proportional to the average density contrast $\overline{\delta}$ inside a region of radius R which was the parameter used in the last section. More generally, we can set $\delta = \nu\sigma$ to describe a spherical region with density contrast which is ν times the standard deviation. Using this expression in (8.49) we can express all the physical quantities in terms of the mass of the overdense region. We then find the following scalings: $t_{\text{coll}} \propto \nu^{-3/2} M^{(n+3)/4}$; $\rho \propto \nu^3 M^{-(n+3)/2}$; $r_{\text{vir}} \propto r_m \propto \nu^{-1} M^{(n+5)/6}$; $v \propto \nu^{1/2} M^{(1-n)/12}$; $T_{\text{vir}} \propto \nu M^{(1-n)/6}$. The same scalings have been obtained from simpler dimensional arguments and the linear theory in chapter 5. But only a detailed model can provide us with the constants of proportionality appearing in these relations.

For $n > -3$, the variance σ decreases with increasing M; then the scaling $t_{\text{coll}} \propto M^{(n+3)/4}$ shows that, on the average, smaller masses turn around and collapse earlier than larger masses. Structures grow by the gradual separation and recollapse of progressively larger units. As each unit condenses out, it will in general be made up of a number of smaller condensations which had collapsed earlier. This leads to a heirarchical pattern of clustering.

When a larger mass collapses, its substructure is likely to be erased[4] rapidly by the mergers and tidal disruption of its subunits, provided the specific binding energy $(GM/r_{\text{vir}}) \propto v^2$ increases with M. Since $v \propto M^{(1-n)/12}$, this happens for $n < 1$. In this case, the evolution of structure will be self-similar in time with a characteristic mass $M_c(t)$ which grows with time as $M_c(t) \propto t^{4/(n+3)}$. For masses much larger than $M_c(t)$, the fluctuations will still be in the linear regime; on scales comparable to $M_c(t)$ structure will be turning around and collapsing and will show a heirarchical pattern; while on mass scales much smaller than $M_c(t)$, the structure would have been smoothed out by nonlinear relaxation effects.

It should be stressed that the processed spectrum at $t = t_{\text{dec}}$ is not a pure power law. So the scaling laws derived above can only be applied piecewise, over mass intervals in which $P(k)$ can be approximated (locally) as a power law. In the cold dark matter models $n \gtrsim -3$ at small M, increases with increasing M and reaches the asymptotic value of $n = 1$ for $M \gtrsim 10^{15}\,\text{M}_\odot$. The power spectrum on galactic scales can be approximated by $n \approx -2$. In this case, one sees from the relation $v \propto M^{(1-n)/12}$ that $M \propto v^4$. This relation connects the *total* mass of the system with the velocity dispersion in the gravitational potential produced by this mass. If we assume that the total mass is proportional to the luminosity L of the system and that the velocity dispersion is of the same order as the rotational velocity σ of visible objects (stars, gas, ...etc.) in this potential, it follows that $L \propto \sigma^4$. This was one of the relations used in chapter 7 to estimate the distances to the galaxies.

8.4 The masses of galaxies

Galaxies have typical masses of about $10^{11}\,\mathrm{M}_\odot$. Theories for galaxy formation, based purely on the gravitational instability of density fluctuations do not provide any natural explanation for this characteristic mass. It is, therefore, necessary to understand the extra physical considerations which lead to this characteristic mass scale.

To begin with, it should be noted that the part of a galaxy which is directly accessible to observations is the baryonic part, though the gravitationally dominant part may be the dark matter. The dynamics of the baryonic part can be properly described[5] only if the cooling mechanisms in the gas is also taken into account. Consider a gas cloud of mass M and radius R, which is supported against gravitational collapse by gas pressure. To provide this support, the gas should have a temperature T where $T = (\mu v^2/3) = (\mu/3)(6GM/10r_v) \simeq (GM\mu/5R)$ if we identify the virial radius r_v with R. Because of this rather high temperature, the gas will be radiating energy and cooling. Once the temperature changes due to cooling, the delicate balance between gravity and pressure support can be affected. The evolution of such a cloud will depend crucially on the relative values of the cooling timescale,

$$t_{\mathrm{cool}} = \frac{E}{\dot{E}} \approx \frac{3\rho kT}{2\mu\Lambda(T)} \tag{8.50}$$

and the dynamical timescale

$$t_{\mathrm{dyn}} \approx \frac{\pi}{2}\left[\frac{2GM}{R^3}\right]^{-1/2} = 5\times 10^7\,\mathrm{yr}\left(\frac{n}{1\,\mathrm{cm}^{-3}}\right)^{-1/2}. \tag{8.51}$$

Here ρ is the average *baryonic* density and $\Lambda(T)$ gives the cooling rate of the gas at temperature T. Note that we have taken t_{dyn} to be the freefall time of a uniform density sphere of radius R.

There are three possibilities which should be distinguished as regards the evolution of such a cloud. Firstly, if t_{cool} is greater than the Hubble time, H^{-1}, then the cloud could not have evolved much since its formation. On the other hand, if $H^{-1} > t_{\mathrm{cool}} > t_{\mathrm{dyn}}$, the gas can cool; but as it cools the cloud can retain the pressure support by adjusting its pressure distribution. In this case the collapse of the cloud will be quasi-static on a timescale of order t_{cool}. Finally there is the possibility that $t_{\mathrm{cool}} < t_{\mathrm{dyn}}$. In this case the cloud will cool rapidly (relative to its dynamical timescale) to a minimum temperature. This will lead to the loss of pressure support and the gas will undergo an almost freefall collapse. Fragmentation into smaller units can now occur because, as the collapse proceeds isothermally, smaller and smaller mass scales will become gravitationally unstable.

The criterion $t_{\rm cool} < t_{\rm dyn}$ can determine the masses of galaxies. Only when this condition is satisfied can a gravitating gas cloud collapse appreciably and fragment into stars. Further, in any heirarchical theory of galaxy formation, unless a gas cloud cools within a dynamical timescale and becomes appreciably bound, collapse on a larger scale will disrupt it. In these theories, galaxies are the first structures which have resisted such disruption by being able to satisfy the above criterion.

Let us first examine this model without introducing any dark matter. The cooling of primordial gas is mainly due to three processes: bremsstrahlung, recombinations in the hydrogen–helium plasma and Compton scattering of hot electrons by the colder cosmic background photons. As discussed in chapter 6, Compton cooling is important only at redshifts higher than $z \simeq 8$ or so. Since galaxy scales become nonlinear only at $z \lesssim 10$ we can ignore the Compton cooling. The cooling rate of the gas due to bremsstrahlung and recombination can then be written as[6]

$$\Lambda(T) = (A_B T^{1/2} + A_R T^{-1/2})\rho^2 \tag{8.52}$$

where the $A_B \propto (e^6 n^2 T^{1/2}/m_e^{3/2})$ term represents the cooling due to bremsstrahlung and the $A_R \simeq e^4 m A_B$ term arises from the cooling due to recombination. (The temperature dependence of both these processes was discussed in chapter 1.) This expression is valid for temperatures above 10^4 K; for lower temperatures, the cooling rate drops drastically since hydrogen can no longer be significantly ionized by collisions. Introducing the numerical values appropriate for a hydrogen–helium plasma (with a helium abundance $Y = 0.25$ and some admixture of metals) the expression for $t_{\rm cool}$ becomes

$$t_{\rm cool} = 8\times 10^6\,{\rm yr}\left(\frac{n}{1\,{\rm cm}^{-3}}\right)^{-1}\left[\left(\frac{T}{10^6\,{\rm K}}\right)^{-1/2} + 1.5 f_m \left(\frac{T}{10^6\,{\rm K}}\right)^{-3/2}\right]^{-1}. \tag{8.53}$$

Here n is the number density of gas particles and the factor f_m takes into account the possibility that the gas may be enriched with metals : $f_m \approx 1$ when there are no metals and $f_m \approx 30$ for solar abundance of metals. For gas with primordial abundance ($f_m \approx 1$), one can see from (8.53) that there is a transition temperature $T^* \approx 10^6$ K. For $T > T^*$ bremsstrahlung dominates while for $T < T^*$, the line cooling dominates.

Let us now consider the ratio $\tau = (t_{\rm cool}/t_{\rm dyn})$. The condition $\tau = 1$ defines a curve on the ρ–T space, which demarcates the region of parameter space in which cooling occurs rapidly within a dynamical time, from the region of weak cooling (see figure 8.2).

For $T < T^*$, when line cooling is dominant, we have $t_{\rm cool} \propto (T^{3/2}/\rho)$ and $t_{\rm dyn} \propto \rho^{-1/2}$ giving $\tau \propto (T^{3/2}/\rho^{1/2}) \propto M$; hence the $\tau = 1$ curve will be parallel to the lines of constant mass in the $\rho - T$ plane. Substituting the numbers and using the expression for the cooling time from (8.53) we find that $\tau = 1$ implies $f(T/10^6\,{\rm K})^{3/2}(n/{\rm cm}^{-3})^{-1/2} = 4.28$. Expressing the mass of the cloud as $M = (5RT/G\mu) = 2.1 \times 10^{11}\,{\rm M}_\odot(T/10^6\,{\rm K})^{3/2}(n/{\rm cm}^{-3})^{-1/2}$ we can write

$$\tau = \frac{t_{\rm cool}}{t_{\rm dyn}} \approx \frac{M}{9 \times 10^{11}\,{\rm M}_\odot}, \tag{8.54}$$

if $\mu = 0.57$ and $f = 1$. Thus the criterion for efficient cooling can be satisfied for masses below a critical mass of about $10^{12}\,{\rm M}_\odot$, provided $T < 10^6\,{\rm K}$.

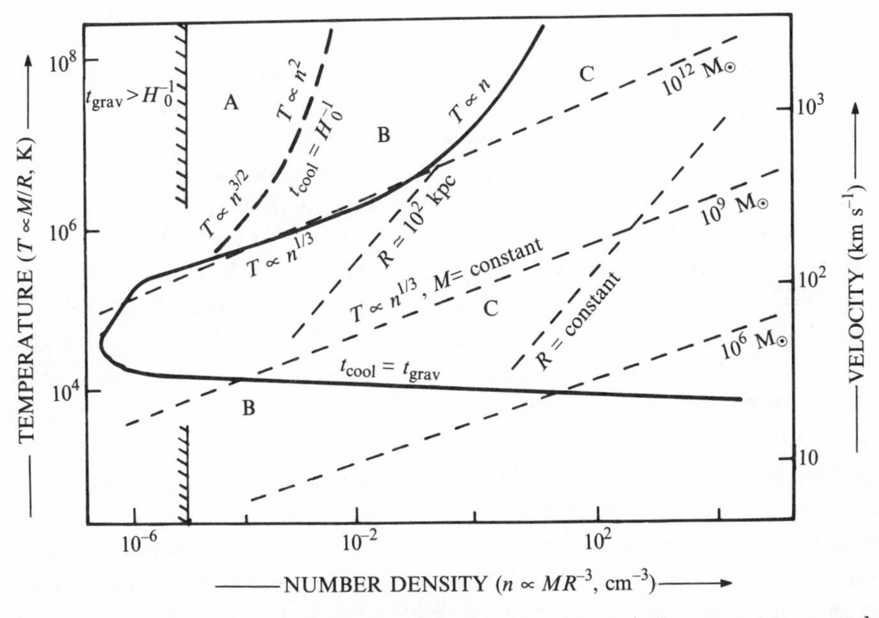

Fig. 8.2. Diagram showing the regions in which various cooling processes can be efficient. The thick line is obtained by equating the timescale for cooling with the free-fall timescale. The top part of the curve is dominated by the bremsstrahlung process while the left end of the curve is contributed by the line cooling of the ionized gas. (The exact shape of the curve depends on the composition of the gas and the amount of metals present.) The thick broken line on the top portion is obtained by equating cooling time to the Hubble time. Similarly the vertical shaded line is obtained by equating free-fall time and with Hubble time. Also marked for reference are the lines of constant mass and constant radius.

On the other hand, for $T > T^*$, when bremsstralung dominates the cooling process, $t_{\rm cool} \propto (T^{1/2}/\rho)$ and $t_{\rm dyn} \propto \rho^{-1/2}$. So $\tau \propto (T^{1/2}/\rho^{1/2}) \propto R$, and the curve $\tau = 1$ will be parallel to the lines of constant radius in the ρ–T space. We now find that $\tau = 1$ implies $(T/10^6\,{\rm K})^{1/2}\,(n/{\rm cm}^{-3})^{-1/2} = 6.43$. Expressing the radius of the cloud as $R = (GM\mu/5T) = 13\,{\rm kpc}\ (T/10^6\,{\rm K})^{1/2}(n/{\rm cm}^{-3})^{-1/2}$ we get

$$\tau = \frac{t_{\rm cool}}{t_{\rm dyn}} \cong \frac{R}{80\,{\rm kpc}}. \tag{8.55}$$

Therefore, clouds with high temperature ($T > T^*$) have to shrink below a critical radius of about 10^2 kpc before being able to cool efficiently to form galaxies.

These features are illustrated schematically in figure 8.2 which is usually called a 'cooling diagram'. The ρ–T space is divided into three regimes A, B and C. A gas cloud with constant mass evolves roughly along lines of constant M_J, with $T \propto \rho^{1/3}$, if it is pressure supported. Gas clouds in region A have $t_{\rm cool} > t_{\rm Hubble}$ and cool very little. Those in region B cool slowly and undergo quasi-static collapse, with the pressure balancing gravity at each instant, until they enter the region C where $\tau < 1$. Gas clouds in C can cool efficiently to form galaxies because they have masses below $10^{12}\,{\rm M}_\odot$ or radius below 10^2 kpc. These masses and radii compare well with the scales characteristic of galaxies.

Let us now consider the effects of including the dark matter component. The dynamical timescale is now determined by the total density of dark matter and baryons, whereas the cooling time still depends only on the density of the baryonic gas. In this case, the gas will not be at the virial temperature initially. It is only during collapse that the gas gets heated up by shocks produced when different bits of gas run into each other. If the cooling timescale of the shocked gas is larger than the dynamical timescale in which the cloud settles down to an equilibrium, then the gas will eventually get heated up to the virial temperature. On the other hand, if the cooling time was shorter, the gas may *never* reach such a pressure supported equilibrium. Efficient cooling will result in the gas sinking to the centre of the dark matter potential well which is being formed, until halted by rotation or fragmentation into stars.

Clearly, it is again the ratio of the cooling time to the dynamical time of the object which governs the evolution. Further, notice that smaller mass clumps are disrupted as larger masses turn around and collapse. However, if the gas component can cool efficiently enough, it may shrink sufficiently close to the centre of the dark matter potential and thus resist further disruption. This process will break the hierarchy. Galaxies could be, again, thought of as the first structures that have survived the disruption due to heirarchical clustering.

The spherical model can be used to estimate the relevant dynamical timescale. We assume $t_{\rm dyn}$ to be comparable to $(t_{\rm coll}/2)$, the time taken for a spherical top-hat fluctuation to collapse after turning around. This expression is the same as the $t_{\rm dyn}$ given in (8.51) above, provided we identify R in (8.51) with the radius of turn-around r_m. Then

$$t_{\rm dyn} \approx \frac{t_{\rm coll}}{2} \approx 1.5 \times 10^9 \left(\frac{M}{10^{12}\,{\rm M}_\odot}\right)^{-1/2} \left(\frac{r_m}{200\,{\rm kpc}}\right)^{3/2} {\rm yr}. \qquad (8.56)$$

For estimating the cooling timescale, we use (8.53) and assume that the gas makes up a fraction F of the total mass and is uniformly distributed within a radius $r_m/2$. The gas temperature is taken to be of order the virial temperature obtained in the spherical model; that is, $T_{\rm vir} \simeq (\mu v^2/3)$, where $v^2 \simeq (6GM/5r_m)$. This corresponds to the temperature achieved by heating by shocks which have a velocity of order of the virial velocity. In that case,

$$t_{\rm cool} \approx 2.4 \times 10^9 f_m^{-1} \left(\frac{F}{0.1}\right)^{-1} \left(\frac{M}{10^{12}\,{\rm M}_\odot}\right)^{1/2} \left(\frac{r_m}{200\,{\rm kpc}}\right)^{3/2} {\rm yr}. \quad (8.57)$$

We have assumed that the line cooling dominates at the temperature $T = T_{\rm vir}$ relevant to the galaxies, and adopted a typical value of $F \simeq 0.1$. Note that the collapse, in general, is likely to be highly inhomogeneous and the above estimates are only supposed to give a rough idea of the numbers involved. From the last two equations we get

$$\tau = \left(\frac{t_{\rm cool}}{t_{\rm dyn}}\right) \approx 1.6 f_m^{-1} \left(\frac{F}{0.1}\right)^{-1} \left(\frac{M}{10^{12}\,{\rm M}_\odot}\right) \qquad (8.58)$$

so that efficient cooling (with $\tau < 1$) requires

$$M < M_{\rm crit} \approx 6.4 \times 10^{11}\,{\rm M}_\odot\, f_m \left(\frac{F}{0.1}\right). \qquad (8.59)$$

It is clear that masses of order of galactic masses are again picked out preferentially even when the dark matter is included.

The procedure outlined above can be used to analyze any particular theory of structure formation involving heirarchical clustering. The starting point will be the cooling diagram, in which the $\tau = 1$ curve is plotted. Given the power spectrum of density fluctuations, one can work out density contrast at various scales $\delta_0 = \nu\sigma(M)$. Then the various properties, like ρ and T of the collapsed objects which are formed, can be estimated using the spherical model. We saw that these properties depend only on one parameter M, once the density contrast δ_0 is fixed. Thus, for each value of ν one gets a curve on the ρ–T plane, giving the properties of collapsed objects. These curves assume that the proto-condensations have virialized, but that the gas has not cooled and condensed. Cooling moves points on these curves to higher densities. In the same diagram one can also plot, for comparison, the observed positions of galaxies, groups and clusters of galaxies.

A simplified form of such a cooling diagram, for a particular version of cold dark matter theory[7], is given in figure 8.3. This figure suggests that, while galaxies show evidence for having cooled and condensed within their dark halos, groups and clusters of galaxies have too long a cooling time to have dissipated much of their energy. From the diagram one can also see that gas clouds with mass in the range of $10^8\,\mathrm{M}_\odot < M < 10^{12}\,\mathrm{M}_\odot$

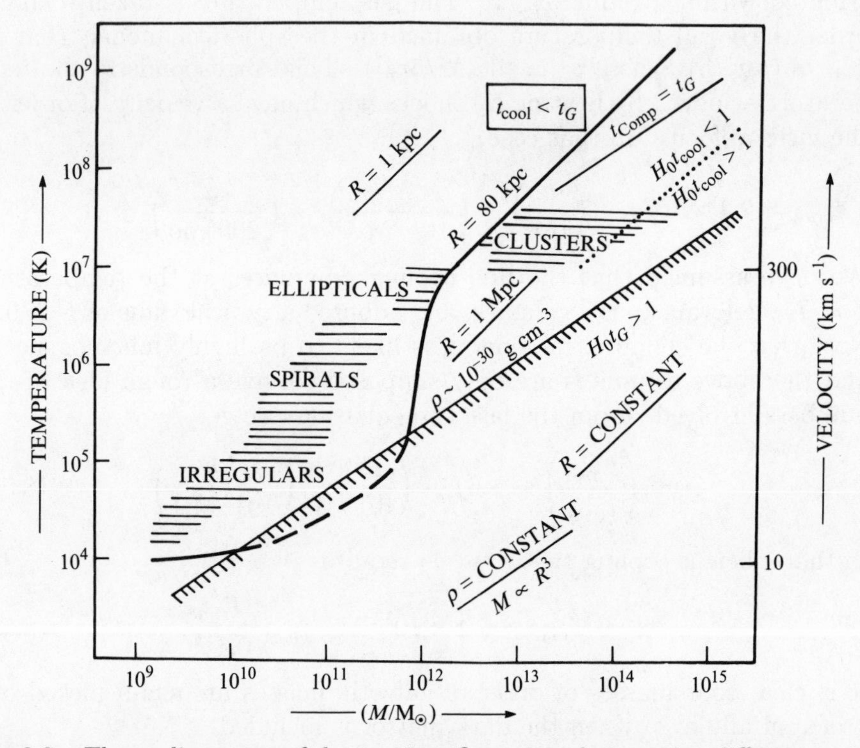

Fig. 8.3. The cooling curve of the previous figure is redrawn using different variables. The y-axis corresponds to the gravitational potential energy per particle and the x-axis, the mass. (These were the variables used in chapter 1). The slanted, shaded, line corresponds to the line $H_0 t_G = 1$. Objects to the right of (and below) this line cannot collapse within the age of the universe. The thick line in the middle is the cooling curve obtained by $t_{\rm cool} = t_G$. Objects to the left of (and above) this curve can cool efficiently while the objects to the right of this curve cannot. The dotted curve on the right delineates the regions which can cool efficiently within the Hubble time. Also shown is a thin straight line obtained by equating the Compton cooling time and the free-fall time. Typical locations of clusters and different kinds of galaxies are indicated. Also shown are the lines of constant radii and constant density.

can cool within the dynamical timescale. The lower limit comes from the fact that the cooling rate drops drastically below about 10^4 K, when hydrogen can no longer be significantly ionized by collisions.

Some complicating features which affect the above simple ideas deserve mention. Firstly, note that we have ignored the formation of stars and the feedback of this process on the gas. If the star formation is very efficient, the supernovae from the massive stars may provide an important heat input. It may even drive out the gas if the potential well is shallow enough[8]. In fact, such effects may be crucial in preventing all the baryons from being locked up in small objects, before typical galaxies form. Also, we see from (8.58) that if the gas were enriched with metals, much larger masses can cool within a dynamical time because of the increased cooling rate. So the chemical history of the gas could be important.

Finally, we discuss the effect of Compton cooling, which has been ignored so far. The cooling rate of a gas with electron density n_e and temperature T embedded in a blackbody radiation field of density ρ_R and temperature T_r is given by (see chapter 6)

$$\Lambda_{\text{Comp}} = \frac{4\sigma_T n_e \rho_R (T - T_r)}{m_e}. \tag{8.60}$$

The cooling time for matter, due to inverse Compton scattering off the cosmic background photons, will be therefore,

$$t_{\text{Comp}} = \frac{3 m_p m_e (1+z)^{-4}}{8\mu\sigma_T \Omega_R \rho_c} \approx 2.1 \times 10^{12} (1+z)^{-4}\,\text{yr}. \tag{8.61}$$

Here we have assumed that $T \gg T_r$ and used $\rho_R(z) = \Omega_R \rho_c (1+z)^4$ to take into account the expansion of the universe. Comparing t_{Comp} with the dynamical time in (8.56) we get

$$\tau_{\text{Comp}} = \frac{t_{\text{Comp}}}{t_{\text{dyn}}} \approx 2 \times 10^2 (1 + z_{\text{coll}})^{-5/2}. \tag{8.62}$$

This ratio is less than unity for $z_{\text{coll}} > 7$, independent of the mass of the collapsing object. So Compton cooling can efficiently cool an object only if it collapses at a redshift higher than $z \simeq 10$, whatever its mass. It is not clear whether galaxies can collapse that early; but if they do, then galaxy scales cannot be preferentially picked out through the cooling processes outlined above. An interesting feature emerges if one plots the the line $\tau_{\text{Comp}} = 1$ in the cooling diagram (see figure 8.3). Note that this line is parallel to the lines of constant density. Galaxies and clusters are neatly separated by the Compton cooling line, suggesting that galaxy formation ceased when Compton cooling became inefficient. If galaxies could form earlier than a redshift of $z \approx 10$, then the above fact could provide a reason for their characteristic masses, which is quite different from the one discussed earlier.

8.5 Zeldovich approximation

The analysis so far was based on the assumption of spherical symmetry and used the power spectrum of the cold dark matter models in which smaller masses become nonlinear first. The evolution, however, will be different if the power spectrum is that produced in a universe dominated by hot dark matter. It was shown in chapter 4 that the power spectrum for hot dark matter is peaked at a mass scale of about $M_{\rm FS} \approx 10^{14}\,{\rm M}_\odot$. Therefore, the first structures which form due to nonlinearity will *all* have masses around this value. There will be very little power on small scales. To analyze this scenario, it is preferable to use a different kind of approximation.

Such an approximation, proposed by Zeldovich[9], is possible for scales which are much smaller than d_H where Newtonian analysis is possible. The starting point of the Zeldovich approximation is the result from the linear theory for the growth of small perturbations, expressed as a relation between the Eulerian and Lagrangian co-ordinates of the particles. In a smooth universe with uniform density $\rho_b(t)$, the actual position of any particle $\mathbf{r}(t)$, is related to its initial (Lagrangian) location $\mathbf{q}$ by

$$\mathbf{r}(t) = a(t)\mathbf{q}. \tag{8.63}$$

This result, of course, is altered in the presence of growing density perturbations. In the linear regime, the only modification needed is the addition of a separable function of t and $\mathbf{q}$ of the form, say, $f(t)\mathbf{p}(\mathbf{q}) = a(t)b(t)\mathbf{p}(\mathbf{q})$. That is, we can take

$$\mathbf{r}(t) \equiv a(t)\mathbf{x}(t) = a(t)[\mathbf{q} + b(t)\mathbf{p}(\mathbf{q})] \tag{8.64}$$

where $\mathbf{x}(t)$ is the comoving Eulerian coordinate. This equation gives the comoving position ($\mathbf{x}$) and proper position ($\mathbf{r}$) of a particle at time t, given that at some time in the past it had the comoving position $\mathbf{q}$. To demonstrate that this equation correctly describes the linear evolution, let us calculate how the perturbed density evolves when the individual particles move according to (8.64). If the initial, unperturbed, density is $\overline{\rho}$ (which is independent of $\mathbf{q}$), then the conservation of mass implies that the perturbed density will be

$$\rho(\mathbf{r}, t)\, d^3\mathbf{r} = \overline{\rho}\, d^3\mathbf{q}. \tag{8.65}$$

Therefore

$$\rho(\mathbf{r}, t) = \overline{\rho}\det(\partial q_i/\partial r_j) = \frac{\overline{\rho}/a^3}{\det(\partial x_j/\partial q_i)} = \frac{\rho_b(t)}{\det(\delta_{ij} + b(t)(\partial p_j/\partial q_i))} \tag{8.66}$$

where we have set $\rho_b(t) = [\bar{\rho}/a^3(t)]$. Expanding the Jacobian to the first order in the perturbation $b(t)\mathbf{p}(\mathbf{q})$, we get

$$\frac{\delta\rho}{\rho} = \frac{(\rho - \rho_b)}{\rho_b} = -b(t)\nabla_{\mathbf{q}} \cdot \mathbf{p}. \tag{8.67}$$

On the other hand, the linear theory predicts that

$$\frac{\delta\rho}{\rho}(\mathbf{x}, t) = g(t)\delta_i(\mathbf{x}) = g(t)\sum_{\mathbf{k}} A_{\mathbf{k}} \exp\{i\mathbf{k} \cdot [\mathbf{q} + b(t_i)\mathbf{p}(\mathbf{q})]\} \tag{8.68}$$

where $g(t)$ is the function describing the time evolution of the growing mode of the density contrast and $A_{\mathbf{k}}$ is the Fourier transform of the initial density contrast, δ_i. For the $\Omega = 1$ universe, for example, $g(t) = (3/5)(t/t_i)^{2/3}$. Let us choose the initial moment t_i such that the term $b(t_i)\mathbf{p}$ is negligible compared to $\mathbf{q}$. Then, if we identify $b(t)$ with $g(t)$ and set

$$\mathbf{p}(\mathbf{q}) = \sum_{\mathbf{k}} \frac{i\mathbf{k}}{k^2} A_{\mathbf{k}} \exp(i\mathbf{k} \cdot \mathbf{q}), \tag{8.69}$$

we see that the approximation given by the relation (8.64) correctly reproduces the result of the linear theory for the growth of small density perturbations. Thus, the relation (8.64) is definitely correct in the linear approximation.

From the definition of $\mathbf{p}(\mathbf{q})$ given in (8.69), one has

$$\mathbf{p}(\mathbf{q}) = \nabla_{\mathbf{q}}\Phi_0(\mathbf{q}) \tag{8.70}$$

where

$$\Phi_0(\mathbf{q}) = \sum_{\mathbf{k}} \frac{A_{\mathbf{k}} \exp(i\mathbf{k} \cdot \mathbf{q})}{k^2}. \tag{8.71}$$

This relation allows one to obtain a simple physical interpretation for $\Phi_0(\mathbf{q})$ and $\mathbf{p}(\mathbf{q})$. Note that

$$\nabla_{\mathbf{q}} \cdot \mathbf{p} = \nabla^2\Phi_0 = \sum_{\mathbf{k}} A_{\mathbf{k}} \exp(i\mathbf{k} \cdot \mathbf{q}) = -\frac{(\rho - \rho_b)}{b\rho_b}. \tag{8.72}$$

Using the Einstein equation $\ddot{a} = -(4\pi G\rho_b a)/3$, we can write this equation as

$$\nabla_{\mathbf{q}}^2\Phi_0 = \frac{4\pi Ga^2(\rho - \rho_b)}{(3ab\ddot{a})}. \tag{8.73}$$

The equation for the gravitational potential ϕ in the perturbed universe is

$$\nabla_{\mathbf{x}}^2\phi = 4\pi Ga^2(\rho - \rho_b). \tag{8.74}$$

Comparing these two expressions at an early epoch (say, at $t = t_i$) when $\mathbf{x}$ is very nearly equal to $\mathbf{q}$, we get:

$$\phi = 3ab\ddot{a}\Phi_0. \tag{8.75}$$

Thus Φ_0 is proportional to the gravitational potential of the linear theory and $\mathbf{p}(\mathbf{q})$ is proportional to the peculiar velocity field of the linear theory.

Zeldovich suggested that while (8.64) is in accordance with the linear theory, it may also provide a good approximate description of the evolution of density perturbations in the *nonlinear regime* where $\delta\rho/\rho$ greatly exceeds unity. To study the consequences of this hypothesis, it is best to proceed as follows: Since $\mathbf{p}(\mathbf{q})$ is a gradient of a scalar function, the Jacobian in (8.66) is a real symmetric matrix. This matrix can be diagonalized at every point $\mathbf{q}$, to yield a set of eigenvalues and principal axes as a function of $\mathbf{q}$. If the eigenvalues of $(\partial p_j/\partial q_i)$ are $[-\lambda_1(\mathbf{q}), -\lambda_2(\mathbf{q}), -\lambda_3(\mathbf{q})]$ then the perturbed density is given by

$$\rho(\mathbf{r}, t) = \frac{\rho_b(t)}{(1 - b(t)\lambda_1(\mathbf{q}))(1 - b(t)\lambda_2(\mathbf{q}))(1 - b(t)\lambda_3(\mathbf{q}))} \tag{8.76}$$

where $\mathbf{q}$ can be expressed as a function of $\mathbf{r}$ by solving (8.64). This expression describes the effect of deformation of an infinitesimal, cubical volume (with the faces of the cube determined by the eigenvectors corresponding to λ_n) and the consequent change in the density. For a growing perturbation, $b(t)$ increases with time; therefore, a positive λ denotes collapse and negative λ signals expansion. In an overdense region the density will become infinite if one of the terms in brackets in the denominator of (8.76) becomes zero. In the generic case, these eigenvalues will be different from each other; let $\lambda_1 \geq \lambda_2 \geq \lambda_3$. At any particular value of $\mathbf{q}$ one of them, say λ_1, will be maximum. Then the density will diverge for the first time when $(1 - b(t)\lambda_1) = 0$; at this instant the material contained in a cube in the $\mathbf{q}$ space gets compressed to a sheet in the $\mathbf{r}$ space, along the principal axis corresponding to λ_1. Thus sheetlike structures, or 'pancakes', will be the first nonlinear structures to form when gravitational instability amplifies density perturbations.

Notice that the description uses trajectories which are built out of the linear theory. Such a description, of course, cannot be exact. To understand the nature of the approximation, we may proceed as follows: Given the acceleration field $\ddot{\mathbf{x}}$, we can compute the density distribution ρ as

$$\nabla_x \cdot \ddot{\mathbf{x}} = -4\pi G\rho, \tag{8.77}$$

provided the potential producing the acceleration ($\ddot{\mathbf{x}} = -\nabla\phi_{acc}$) is the same as the potential generated by the density ($\nabla^2\phi_N = 4\pi G\rho$). For the *exact* trajectories, equation (8.77) will be an identity; for the approximate ansatz we are using for $\mathbf{x}(t)$, we can explicitly compute the quantity $Q = (\nabla\cdot\ddot{\mathbf{x}} + 4\pi G\rho)$. The smallness of this quantity will provide a measure of validity for the approximation.

Since $\mathbf{x}(t)$ is known, we can explicitly evaluate this quantity Q. The calculation is most easily done in the locally diagonal coordinates (see exercise 8.4) and expressed in terms of the 'invariants' of the tensor $D_{ij} = (\partial p_j/\partial q_i)$:

$$I_1 = \lambda_1 + \lambda_2 + \lambda_3; \quad I_2 = \lambda_1\lambda_2 + \lambda_2\lambda_3 + \lambda_3\lambda_1; \quad I_3 = \lambda_1\lambda_2\lambda_3. \tag{8.78}$$

We find that

$$hQ = 3\left[\frac{\ddot{a}}{a} + \frac{4\pi G}{3}\frac{\rho_0}{a^3}\right] - \left[\ddot{b} + \frac{2\dot{a}}{a}\dot{b} + 3\frac{\ddot{a}}{a}b\right] f + 3\frac{\ddot{a}}{a}g, \tag{8.79}$$

with

$$f = I_1 - 2bI_2 + 3b^2 I_3; \quad g = -b^2 I_2 + 2b^3 I_3; \quad h = (1 - bI_1 + b^2 I_2 - b^3 I_3). \tag{8.80}$$

The term in the first square bracket vanishes identically; the second term will vanish when $b(t)$ satisfies the linear perturbation equation

$$\ddot{b} + \frac{2\dot{a}}{a}\dot{b} + 3\frac{\ddot{a}}{a}b = 0. \tag{8.81}$$

Thus the inaccuracy Q is of the order of the non-vanishing third term $3(\ddot{a}/a)g$. The fractional inaccuracy is of the order of

$$\left(\frac{g}{h}\right) \cong \frac{(b^2 I_2 - 2b^3 I_3)}{(1 - bI_1 + b^2 I_2 - b^3 I_3)} = \frac{b^2 I_2 - 2b^3 I_3}{(1 - b\lambda_1)(1 - b\lambda_2)(1 - b\lambda_3)}. \tag{8.82}$$

In a planar collapse with $\lambda_2 \approx \lambda_3 \approx 0$, the approximation is excellent. If the collapse first occurs when $b\lambda_1 = 1$, the approximation remains quite good till that event.

Notice that, in spite of the formal similarity of (8.64) to the inertial motion of the comoving co-ordinate of the particles, it actually describes gravitational instability in an expanding universe. If there were no gravity and only expansion, $b(t)$ would in fact decrease with time. The interesting difference between the Zeldovich approximation and the linear theory is that the former predicts the origin of the nonlinear objects from the high peaks of $\lambda_1(\mathbf{q})$, while the latter predicts the formation of structure from the peaks of $\delta(\mathbf{q}) \propto (\lambda_1 + \lambda_2 + \lambda_3)$.

The distribution of λ can also be used to obtain a qualitative picture of the collapse which occurs in the Zeldovich approximation. Let us suppose that the eigenvalues λ_i are uncorrelated. (This assumption, of course, is not strictly true but appears to be a fair approximation.) Then the probability of any given eigenvalue being positive (or negative) is (1/2). A set of points which expand along *all* the three directions will end up as a void; this will happen to a fraction $(1/2)^3 = (1/8)$ of the mass. The remaining fraction (7/8) will collapse at least along one of the axes. Of these, a fraction (3/8) of mass will collapse into pancakes, another (3/8) into filaments and (1/8) will eventually collapse in all directions. Even in the latter two cases, the *initial* collapse will be along *one* direction, producing pancakes (also see exercise 8.5).

When the trajectories of the Zeldovich approximation intersect, density increases considerably on the caustics. The baryons which fall into the potential well near caustics are compressed to high densities; the velocity of the collapsing gas can exceed the local speed of sound forming a shock wave. The gas is shock-heated to a temperature of about $T \simeq 1\,\text{keV}$ and cools by radiation. The cooled gas should, eventually, form galaxies.

Numerical simulations have been employed[10] to test how well Zeldovich approximation works. It is found that at the beginning of the nonlinear stage it reproduces the density distribution quite well and also reproduces in an excellent fashion the formation, appearance and location of the pancakes. At later times, however, Zeldovich approximation predicts the caustics to increasingly blur out and the pancake to thicken, while the N-body simulations show that pancakes remain relatively thin. It turns out that the thickness of the pancakes stabilizes quickly even in a collisonless medium due to the action of gravity. Particles falling into pancakes oscillate about the central region rather than move out progressively along the direction of their initial velocity as predicted by (8.64). The N-body simulations also show that particles flow along pancakes to form filaments at the intersection of pancakes and finally clumps at the intersection of filaments, whose sharpness is not well reproduced by the Zeldovich approximation.

Some of these problems can be overcome by an extension of the Zeldovich approximation which we will discuss in the next section. But before doing that it is worth asking why the Zeldovich approximation works *at all.* At least until the formation of caustics, equation (8.64) describes the evolution of density perturbations very well, even though the density contrasts are already highly nonlinear. Why should this be so?

There are several ways of answering this question. First of all, note that even when the term $b\mathbf{p}$ is small relative to $\mathbf{q}$, the *density perturbation* need not be small since $\delta\rho/\rho$ depends on the derivatives of $b\mathbf{p}$ and not on its magnitude. So, even if particle positions are *not* significantly perturbed, the density contrasts can be nonlinear. The Zeldovich approximation exploits this feature by describing the perturbations in terms of the *perturbed trajectories* of particles extrapolated from linear theory, instead of trying to extrapolate the evolution of *density contrasts.* Another way of understanding this result is by realizing that the basic quantity which is used in the Zeldovich approximation is the perturbed Newtonian potential extrapolated from linear theory. The *actual* potential may not deviate from linear theory too much even when the density contrasts become highly nonlinear, because potentials are much smoother functions than densities.

Lastly, the following analogy from the physics of gravitational lensing may be helpful in visualizing the Zeldovich approximation. Consider a plane wavefront which crosses a perturbing mass. Rays of light (or, equivalently, trajectories of photons) which were originally parallel to one another will now get deviated. These rays can cross later, producing caustics at which the intensity will be infinitely high, even if the deviations of the rays are only slight. We can describe these deviations using linearized equations and still work out the properties of caustics. On the other hand, we could not have studied caustics using equations which are linearized with respect to the intensity. In our present case, the density contrast and particle trajectories are analogous to the intensity and rays in optics.

8.6 The adhesion model

It is possible to handle some of the difficulties mentioned in the last section by an improved approximation scheme[11]. The new procedure allows one to describe the formation of pancakes, filaments and clumps in a relatively simple manner though the internal structure of pancakes and other objects cannot be inferred from this model.

The starting point of this model is the Zeldovich approximation (8.64) expressed in a slightly different form. Let us consider the peculiar velocity $\mathbf{V}(t) \equiv a(t)\, d\mathbf{x}/dt$, predicted by the Zeldovich approximation. Using (8.64), we find that

$$\mathbf{V} = a(t)\dot{b}\mathbf{p}(\mathbf{q}). \tag{8.83}$$

We now define a new velocity variable $\mathbf{v} \equiv \mathbf{V}/(a\dot{b})$, and use $b(t)$ as a new time variable instead of t. Then, from (8.64) we get

$$\mathbf{v} \equiv \left(\frac{\mathbf{V}}{a\dot{b}}\right) = \frac{1}{\dot{b}}\frac{d\mathbf{x}}{dt} = \frac{d\mathbf{x}}{db} = \mathbf{p}(\mathbf{q}). \tag{8.84}$$

Since $\mathbf{p}(\mathbf{q})$ is constant in time, the derivative of $\mathbf{v}$ with respect to b vanishes:

$$\frac{d\mathbf{v}}{db} = \frac{\partial \mathbf{v}}{\partial b} + \mathbf{v}\cdot\nabla\mathbf{v} = 0. \tag{8.85}$$

This is similar to the Euler equation for the velocity field in fluid mechanics. We can also recast the equation of continuity in the expanding universe

$$\frac{\partial \rho}{\partial t} + \frac{3}{a}\frac{da}{dt} + \frac{1}{a}\nabla\cdot(\rho\mathbf{V}) = 0, \tag{8.86}$$

in terms of the new variables by defining a new density variable $\eta \equiv a^3\rho(\mathbf{x}, t)$. We then get

$$\frac{\partial \eta}{\partial b} + \nabla \cdot (\eta \mathbf{v}) = 0. \tag{8.87}$$

These equations describe the evolution of a force free fluid with b as the time coordinate. Assuming that at small b the density is homogeneous, one can immediately write down its solution in Lagrangian form

$$\mathbf{x} = \mathbf{q} + b\mathbf{p}(\mathbf{q}); \quad \eta = \frac{\eta_0}{\det(\partial x^j / \partial q^i)}. \tag{8.88}$$

As is to be expected, this is essentially the Zeldovich solution (8.64) written in terms of comoving coordinates, where $\mathbf{p}$ is the initial velocity field.

Since equations (8.85) and (8.86) are equivalent to the Zeldovich approximation, they also fail to describe the evolution of density inhomogeneities after the formation of pancakes. This model can be improved by modifying (8.85) by adding a 'viscous' term to its right hand side. The modified equations are intended to describe the location of the structures and not the evolution of the internal structure of pancakes or other clumps. So the form chosen for this additional viscous term should be irrelevant. A particularly simple form for this term is $\nu\nabla^2\mathbf{v}$, which leads to

$$\frac{\partial \mathbf{v}}{\partial b} + \mathbf{v} \cdot \nabla \mathbf{v} = \nu \nabla^2 \mathbf{v} \tag{8.89}$$

in place of (8.85). Note that as $\nu \to 0$, the viscosity term does not influence the motion, except in regions where there are rapid variations in velocity, that is, at caustics. In this limit the evolution outside caustics follows the Zeldovich approximation while the multistream flows at the caustics are prevented by the viscosity. The main advantage of using this particular additional term is that (8.89), called Burger's equation[12], has an analytic solution for irrotational velocity fields.

It was shown in chapter 4 that the rotational part of the velocity decays away in the expanding universe. We can, therefore, express the velocity field at late times as the gradient of a velocity potential Φ; that is $\mathbf{v} = \nabla\Phi$. Then, from (8.89), we see that Φ obeys the equation

$$\frac{\partial \Phi}{\partial b} + \frac{1}{2}(\nabla\Phi)^2 = \nu\nabla^2\Phi. \tag{8.90}$$

Writing

$$\Phi = -2\nu \ln U, \tag{8.91}$$

this equation is transformed to the linear diffusion equation

$$\frac{\partial U}{\partial b} = \nu \nabla^2 U. \tag{8.92}$$

The general solution to the diffusion equation is well-known:

$$U(\mathbf{x}, b) = \left(\frac{1}{4\pi\nu b}\right)^{3/2} \int U(\mathbf{q}, 0) \exp\left[-\frac{(\mathbf{x}-\mathbf{q})^2}{4\nu b}\right] d^3\mathbf{q}. \tag{8.93}$$

From (8.91) we find that $U(\mathbf{q}, 0) = \exp(-\Phi_0(\mathbf{q})/2\nu)$, where $\Phi_0(\mathbf{q})$ gives the initial value of the velocity potential and is the same function defined previously in section 8.5. Expressing U in terms of Φ, and taking the gradient, we find the velocity to be

$$\mathbf{v}(\mathbf{x}, b) = \frac{\int [(\mathbf{x}-\mathbf{q})/b] \exp(-G(\mathbf{x}, \mathbf{q}, b)/2\nu)\, d^3\mathbf{q}}{\int \exp(-G(\mathbf{x}, \mathbf{q}, b)/2\nu)\, d^3\mathbf{q}} \tag{8.94}$$

where

$$G(\mathbf{x}, \mathbf{q}, b) = \Phi_0(\mathbf{q}) + \frac{(\mathbf{x}-\mathbf{q})^2}{2b}. \tag{8.95}$$

The solution (8.94) in the limit $\nu \to 0$ is of particular interest. As ν tends to zero, the main contribution to the integral in (8.94) comes from the vicinity of the *absolute minimum* of $G(\mathbf{x}, \mathbf{q}, b)$ treated as a function of $\mathbf{q}$. This point $\mathbf{q}$ is the solution of the equation

$$\frac{(\mathbf{q}-\mathbf{x})}{b} + \nabla\Phi_0 = 0 \tag{8.96}$$

which corresponds to the Zeldovich solution, expressed in comoving coordinates. If $\mathbf{q}_{min}$ is a solution of (8.96) at which G is also an absolute minimum, then the 'velocity' $\mathbf{v}(\mathbf{x}, b)$ is determined by the steepest descent approximation to the integral in (8.94):

$$\mathbf{v}(\mathbf{x}, b) = \frac{\mathbf{x} - \mathbf{q}_{min}(\mathbf{x}, b)}{b}. \tag{8.97}$$

At early times when the density contrasts are small, one expects a unique solution, say $\mathbf{q}_1$, to (8.96); at every Eulerian point $\mathbf{x}$ there is only one particle which has come from $\mathbf{q}_1$. But later on for some $\mathbf{x}$ there may be several roots $\mathbf{q}_1, \mathbf{q}_2, \mathbf{q}_3, \cdots$ say, all of which satisfy (8.96). This means that several particles from different $\mathbf{q}$ would have all come to the same $\mathbf{x}$ under the original Zeldovich approximation. But as a solution to (8.94), one still gets a *unique* $\mathbf{q}_{min}$ and $\mathbf{v}$, viz. the one for which G is an absolute minimum. This is because the other particles – which could reach $\mathbf{x}$ if the medium was collisionless – got 'stuck' in the pancakes earlier due to the additional term $\nu\nabla^2\mathbf{v}$. Of course there will be $\mathbf{x}$ for which G will have an absolute minimum simultaneously at several points $\mathbf{q}$. This can happen if fluid elements at these $\mathbf{q}$ have just met at this Eulerian point $\mathbf{x}$. The set of all such $\mathbf{x}$ will then trace out the caustics in Eulerian space in the limit $\nu \to 0$.

A simple geometrical method can be used to study the solution (8.94) in the limit of $\nu \to 0$. Given an Eulerian coordinate $\mathbf{x}$ and a time b we construct the paraboloid

$$P(\mathbf{x}, \mathbf{q}, b) = -\frac{(\mathbf{x} - \mathbf{q})^2}{2b} + h. \tag{8.98}$$

The discussion above shows that the coordinate of the absolute minimum of $G(\mathbf{x}, \mathbf{q}, b)$ is given by the point $\mathbf{q}$ at which this paraboloid is tangential to the hypersurface $\Phi_0(\mathbf{q})$, for the first time, as one increases h from $(-\infty)$ upwards. The fact that the paraboloid is tangential to Φ_0 is equivalent to (8.96) and the property that the $\mathbf{q}$ is an absolute minimum is guaranteed by demanding that the paraboloid is tangential for the *first* time as h is increased. The Eulerian coordinate $\mathbf{x}$ of the particle, which has the Lagrangian coordinate $\mathbf{q}$, is by construction the coordinate of the top of the paraboloid. At early times, when b is small, the paraboloid has large curvature (proportional to b^{-1}); it is very narrow and is tangential to the hypersurface at only one point. But as time goes on, b increases and the paraboloid becomes shallower. It may then be tangential to Φ_0 at two points as one keeps increasing h (see figure 8.4).

In this case these two points have just run into a pancake. The case in which the paraboloid is tangential at two points is degenerate and signals the formation of sheets in the Eulerian space. At later stages, the

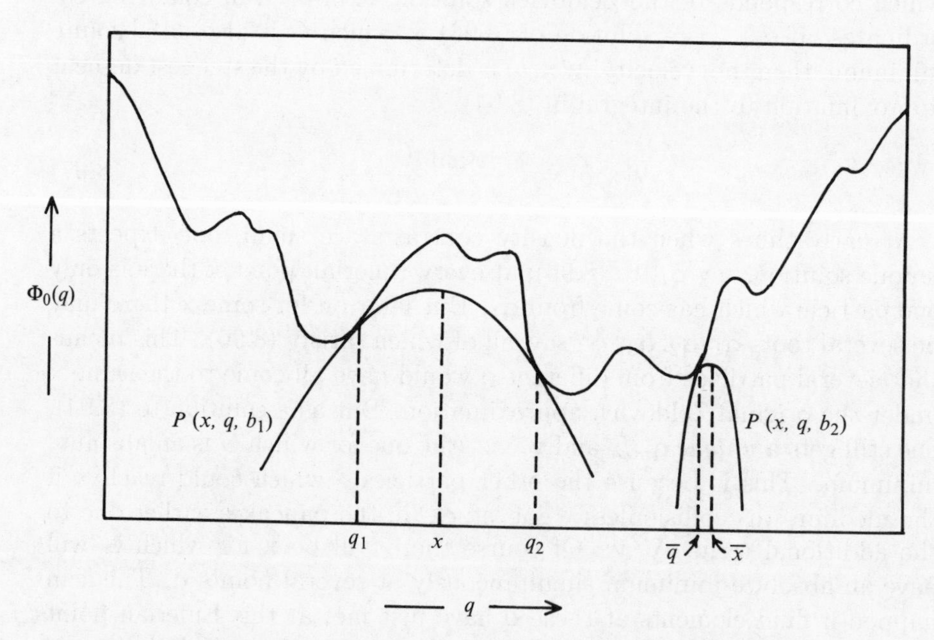

Fig. 8.4. Geometrical construction illustrating the procedure for obtaining the saddle points in the adhesion approximation (one dimensional case).

paraboloids may touch the surface Φ_0 at three or even four points. The apexes of these paraboloids then indicate the positions of filaments and clumps formed.

Numerical simulations have been carried out using the adhesion model and the results have been compared with those from the N-body simulations[13]. Both the numerical solution of (8.94) and simulations using the geometrical technique have been used to evolve the initial density field. Such comparisons show that the adhesion model reproduces the results of fully nonlinear calculations very well. In fact, because (8.94) has an exact solution, we can find the density field at some time without following the evolution for intermediate times, unlike in an N-body simulation. Efficient algorithms using much shorter computing time than N-body simulations can be developed to implement the adhesion model. Thus one can simulate larger volumes of space provided an approximate treatment of small scale structure is acceptable.

Why does the adhesion model work? In other words, why do the particles 'stick' along the pancake rather than flow in the perpendicular direction, causing pancakes to thicken? Since the real system is collisionless, the only relaxation processes which are effective are the two-body relaxation and violent relaxation. In the high-density regions – like near a caustic – these processes can operate within a dynamical timescale[14]. Further, the action of the relaxation process will be highly anisotropic because the velocity field near a pancake will be anisotropic. These features lead to an efficient gravitational relaxation process near the caustics which prevents the pancakes from thickening.

Another aspect of the adhesion model is worth pointing out. In this framework, the evolution of the density field is completely determined if the initial gravitational potential perturbation ϕ ($\propto \Phi_0$) is specified. For Gaussian random fluctuations, the power spectrum of ϕ is proportional to $P(k)k^{-4}$, where $P(k)$ is the power spectrum of the density fluctuation. So the ϕ field has more large wavelength (small k) power than the density fluctuation field. This may have interesting consequences for the way the characteristic mass M_c of collapsed objects grows with time. In section 8.3 we saw that if $-3 \leq n \leq 1$ we have the scaling law

$$M_c \propto t^{4/(n+3)} \propto a^{6/(n+3)}. \tag{8.99}$$

It was generally believed that the same scaling law would hold even for $1 \leq n \leq 4$. For spectra steeper than $n = 4$, nonlinear generation of the long wavelength part of the spectrum due to mode-coupling terms could dominate any intrinsic large scale power, thereby producing an effective $n = 4$ power spectrum even if initially $n \geq 4$ (see chapter 5). So a limiting growth law for M_c had been proposed for $n \geq 4$ in the form

$$M_c \propto t^{4/7} \propto a^{6/7}. \tag{8.100}$$

On the other hand, the adhesion model leads to a very different scaling law for spectra with $n > 1$. Note that, as t becomes large, $b(t)$ becomes large and the top of the paraboloid becomes very flat compared to the peaks and troughs of the initial potential Φ_0. In this limit, for any x, the paraboloid is tangential to the hypersurface $\Phi_0(q)$ practically at its local minima and finally at the deepest minima. Thus the asymptotic behaviour of how the structure evolves is governed by the statistics and spatial distribution of the deepest minima of the initial potential Φ_0. As b increases, the characteristic scale at which an absolute minimum can exist will increase depending on the statistics of the minima. (This will also make the masses of the collapsed structures increase). In particular, it turns[15] out that (8.99) holds only if the variance of the gravitational potential diverges as k tends to zero. For a pure power law spectrum with a small wavelength cut-off $k_{\rm max}$ the variance of the potential is given by

$$\sigma_\phi^2 \propto \int_0^{k_{\rm max}} [P(k)k^{-4}]k^2\,dk \propto \int_0^{k_{\rm max}} k^{n-2}\,dk \tag{8.101}$$

which diverges at small k only for $n \leq 1$. For $n > 1$, when σ_ϕ is finite, it is found that M_c grows according to the limiting form

$$M_c \propto a^{3/2}, \tag{8.102}$$

corresponding to the growth law for $n = 1$ in (8.99). Thus the adhesion model predicts a more rapid growth of structure for the steep power spectra compared to the results based on an extrapolation of the linear theory, even when the effects of mode-coupling are taken into account in a simple manner. There is some supportive evidence, from one dimensional numerical simulations[15], that the adhesion model indeed predicts the correct limiting growth law of M_c; that is, simulations seem to support the form (8.102) rather than (8.99) for the steep spectra. Whether this result has any implication for the theories of structure formation is yet to be explored.

One may ask whether the cooling processes discussed earlier (in section 8.4) are relevant in setting the galactic scales in the scenarios involving pancake formation, in which large condensates are formed first and smaller masses arise due to fragmentation. Such considerations are still relevant if the fragments are subgalactic, as indeed seems to be the case in the pancake models. The coagulation of subgalactic clouds to form galaxies leads to a picture not very different from inhomogeneous protogalactic collapse. The condition for efficient cooling of the shocks, which are formed when two clouds collide, leads to a condition which is essentially equivalent to $\tau < 1$.

8.7 The angular momentum of galaxies

Another key parameter characterizing galaxies, apart from their masses, is their angular momentum. One important class of galaxies, the disc galaxies, owe their equilibrium to rotational support. The origin of galactic angular momentum is, therefore, an important question in the theory of galaxy formation. At present the most popular idea[16] is that galaxies acquire their angular momentum through the tidal torques due to their neighbours. We will now examine this mechanism.

The growth of the spin of a protogalaxy, due to tidal torques, is most easily analyzed using the Zeldovich approximation. If the density field has lot of power on scales much smaller than galaxies, these scales will become nonlinear much before galactic scales do; if this happens the Zeldovich approximation will cease to describe the evolution of galaxies. To avoid this formal difficulty, one should apply Zeldovich's equation (8.64) to a density field which is smoothed (by a window function) on protogalactic scales rather than to the actual density field. (This construction assumes that the small scale nonlinear structures have negligible influence on the gravitational evolution of larger, quasilinear scales; this assumption seems reasonable but further work needs to be done to justify it.)

The angular momentum of the mass which will eventually form a galaxy is given by

$$\mathbf{L}(t) = \int_V [\mathbf{r}(\mathbf{q},t) - \bar{\mathbf{r}}(t)] \times [\mathbf{V}(\mathbf{q},t) - \bar{\mathbf{V}}]\rho(\mathbf{r},t)\, d^3\mathbf{r}, \tag{8.103}$$

where $\mathbf{r}$ and $\mathbf{V}$ describe the proper position and the peculiar velocity of a mass element. The integral is over the region which will eventually end up forming the galaxy, $\bar{\mathbf{r}}$ is the centre of mass and $\bar{\mathbf{V}}$ the peculiar velocity of the centre of mass of the system at time t. Using equation (8.65) for the conservation of mass, we can convert the integral to one over the Lagrangian volume, say V_L, which initially contained the galaxy mass. Substituting the Zeldovich solution (8.64) for $\mathbf{r}$, we get

$$\mathbf{L}(t) = a^2 \int_{V_L} [([\mathbf{q}-\bar{\mathbf{q}}]+b(t)[\mathbf{p}(\mathbf{q})-\mathbf{p}(\bar{\mathbf{q}})]) \times \dot{b}(\mathbf{p}(\mathbf{q})-\mathbf{p}(\bar{\mathbf{q}}))]\bar{\rho}\, d^3\mathbf{q}. \tag{8.104}$$

The cross product of the b and the $\dot{b}$ terms vanishes because they are parallel. Since $\mathbf{p}(\mathbf{q}) = \nabla\Phi_0(\mathbf{q})$, we can write this expression as

$$\mathbf{L}(t) = \bar{\rho}a^2\dot{b} \int_{V_L} (\mathbf{q} - \bar{\mathbf{q}}) \times [\nabla\Phi_0(\mathbf{q}) - \nabla\Phi_0(\bar{\mathbf{q}})]\, d^3\mathbf{q}. \tag{8.105}$$

Thus the angular momentum acquired by the protogalaxy is of first order in the perturbation. Its time evolution is determined by the behaviour of $a^2\dot{b}$, since the integral and $\bar{\rho}$ are constant in time. For a spatially flat universe with $b(t) \propto a(t) \propto t^{2/3}$, we find that $a^2\dot{b} \propto t$ and so the angular momentum grows linearly with time.

Further insight into the meaning of (8.105) can be obtained if we assume that Φ_0 within V_L can be approximated by the first three terms of its Taylor series about $\bar{\mathbf{q}}$

$$\Phi_0(\mathbf{q}) = \Phi_0(\bar{\mathbf{q}}) + (q_i - \bar{q}_i)\left(\frac{\partial \Phi_0}{\partial q_i}\right)_{\bar{\mathbf{q}}} + \frac{1}{2}(q_i - \bar{q}_i)\left(\frac{\partial^2 \Phi_0}{\partial q_i \partial q_j}\right)_{\bar{\mathbf{q}}}(q_j - \bar{q}_j). \tag{8.106}$$

Substituting this result in equation (8.105) we find that

$$L_i(t) = a^2 \dot{b} \epsilon_{ijk} I_{jl} T_{lk}, \tag{8.107}$$

where

$$I_{jl} = \int_{V_L} (q_j - \bar{q}_j)(q_l - \bar{q}_l)\bar{\rho}\, d^3\mathbf{q} \tag{8.108}$$

is the moment of inertia tensor of the mass in V_L, and

$$T_{lk} = \frac{\partial^2 \Phi_0}{\partial q_l \partial q_k} \tag{8.109}$$

is proportional to the tidal gravitational field at $\bar{\mathbf{q}}$.

The tensor product in (8.107) gives the torque on an extended body in a tidal field; $\mathbf{L}$ vanishes if and only if the tensors T_{ij} and I_{ij} have the same principal axes. Since the inertia tensor depends only on the shape of the protogalaxy, while the tensor T_{ij} depends in addition on the distribution of neighbouring protoclumps, this will not happen in the general case. Equation (8.107) then shows that the angular momentum of a galaxy can arise due to the coupling of the first order tidal field with the zeroth order quadrupole moment of the irregular mass distribution.

Although the above calculation, deriving (8.107), provides some insight into how $\mathbf{L}$ originates due to tidal torquing, it cannot be used in a clear-cut way to make actual estimates. The problem is basically twofold : Firstly, it is difficult to decide how V_L should be specified so that it encompasses all the matter that will end up in a collapsing protogalaxy. Secondly, one has to take into account the evolution of $\mathbf{L}$ during the fully nonlinear stages of protogalactic evolution. At present the best way of addressing both these problems seems to be via N-body simulations[17].

In all such studies, the angular momentum is usually quoted in terms of the parameter

$$\lambda \equiv \frac{L E^{1/2}}{G M^{5/2}}, \tag{8.110}$$

which was discussed in chapter 1. This parameter is the ratio between the *actual* angular frequency ω of the system and the *hypothetical* angular frequency ω_0 that is needed to support the system purely by rotation. It follows that a self-gravitating system with appreciable rotational support has a λ comparable to unity.

The N-body simulations[17] indicate a broad distribution of λ values, in the range of 0.1 to 0.01, with a median value of $\lambda \approx 0.05$ for the collapsed objects. In these simulations, both white noise initial conditions ($n = 0$) and initial conditions which have more power on large scales (as predicted in the models of galaxy formation with cold dark matter), have been studied. It is found that $L(t) \propto t$ when the density contrasts are small relative to unity, in accordance with the prediction of the linear theory. As clumps become nonlinear, L grows more slowly and can even decrease at later times depending upon the existence of substructures at smaller scales. For example, if an object has a significant amount of substructure, it initially acquires more angular momentum because of a stronger coupling to the tidal field. But, subsequently, as the subclumps in the object sink to the centre and merge, they lose their orbital angular momentum to the outer parts, resulting in a high density, low-L core.

It is also found that the median value of λ for the collapsed objects is quite insensitive to the shape of the initial power spectrum of density fluctuation or the magnitude of its initial overdensity. This result can be understood as follows: Higher peaks have a shorter collapse time; so one may have thought that they will have less time to get spun up. But, we have seen in chapter 5 that higher peaks are also more clustered and hence experience stronger torques. Which of these two effects is dominant depends on the power spectrum; the results from the simulations confirm that, for cold dark matter spectra, these two effects nearly cancel.

The net result of these simulations can be summarized as follows: Tidal torques can give angular momentum to a protogalaxy. But both analytical estimates and N-body simulations indicate that the resulting λ distribution has a large spread with a median value of about 0.05 which is more or less independent of the shape of the initial spectrum of density fluctuations and the peak height. Tidal torques are therefore able to give, at best, only five to ten per cent of the angular momentum needed for the rotational support. We shall now try to understand how more angular momentum can be acquired by a protogalaxy.

8.8 Formation of disc galaxies

The previous sections have been devoted to general ideas on whether it is possible to account for two of the basic properties of galaxies, their characteristic mass and angular momentum. In this and the next section we consider more specific details of the formation of different types of galaxies. For our purposes, we will ignore these fine distinctions which exist in galactic morphology and only consider two basic types of galactic systems – the discs (or spirals) and ellipticals. This section is devoted to the disc galaxies and the next one to the formation of ellipticals.

The disc galaxies are supported against gravity by rotation and have $\lambda \simeq 0.4 - 0.5$. The rotation velocity of the disc remains approximately constant except near the centre, where it decreases to zero. Typical rotation velocities range from $(200 - 300)\,\mathrm{km\,s^{-1}}$. In the heirarchical clustering models, with a dominant component of dark matter, these galaxies can be thought of as structures with a luminous core and extended halo. The dark halo forms when a galactic scale fluctuation in the dark matter component grows, turns around, collapses, virializes and settles into an equilibrium. Meanwhile, the gas in the halo also collapses with the halo and can attain a pressure supported equilibrium if the shocks heat up the gas to the virial temperature of the system. However, the gas can cool radiatively and thus collapse *further* into the core of the dark halo. Since the gas makes up only a small fraction (about 0.1) of the total mass, the gravity of the halo dominates that of the gas in the initial stages. At this stage, star formation is expected to be suppressed by the tidal forces of the halo, except in regions of large gas density[18]. But as the gas sinks to the halo core, it becomes dominated by its self-gravity and can fragment to form stars. This results in a luminous galaxy in the core of an extended dark halo.

Dissipative collapse in a dark halo seems to be crucial to understand the rotation of disc galaxies as well. We noted in the last section that tidal torques can supply only 5 to 10 per cent of the angular momentum required for the rotational support, leading to $\lambda_i \simeq 0.05$. This is far below the observed value of the spin parameter $\lambda_d \simeq 0.4 - 0.5$ of the disc galaxies. One can attempt to reconcile the theory with observation by noting that, as the gas collapses due to cooling, the binding energy increases, while mass and angular momentum remain the same. Since the spin parameter $\lambda \propto |E|^{1/2}$, it will increase as the binding energy increases. This idea, however, works only if massive dark halos exist.

To see this, let us first ignore the presence of massive dark halos around galaxies, and assume that a protogalaxy was just a self-gravitating cloud of baryonic gas. The binding energy of the protogalaxy will be $|E| \simeq GM^2/R$, where R is its characteristic radius. Since M is constant during collapse, $|E| \propto R^{-1}$ and so $\lambda \propto R^{-1/2}$. The gas cloud has to collapse by a factor of about $(\lambda_d/\lambda_i)^2 \approx (0.5/0.05)^2 \simeq 100$, before it can spin up sufficiently to form a rotationally supported system, where λ_i is the initial value of λ produced by tidal torques. Therefore, to form a rotationally supported galactic disc of mass $10^{11}\,\mathrm{M}_\odot$ and radius $10\,\mathrm{kpc}$, matter needs to collapse from an initial radius of $1\,\mathrm{Mpc}$. This process would take an inordinately long time of about $t_{\rm coll} = (\pi/2)(R^3/2GM)^{1/2} \simeq 5.3 \times 10^{10}\,\mathrm{yr}$, much longer than the age of the universe. Note that even the material in the core, with a scale length of $r_c \simeq 3\,\mathrm{kpc}$, would have to collapse from a distance of $300\,\mathrm{kpc}$ and will take about $10^{10}\,\mathrm{yr}$.

This difficulty is easily avoided if a massive halo exists. In the presence of a massive dark halo, the initial spin parameter of the system, before collapse of the gas, can be written as $\lambda_i = (L|E|^{1/2}/GM^{5/2})$ where the various quantities, L, E and M refer to the *combined* dark matter–gas system, although the contribution from the gas is negligible compared to that of the dark matter. After the collapse, the gas becomes self-gravitating and the spin parameter of the resulting disc galaxy will be $\lambda_d = (L_d|E_d|^{1/2}/GM_d^{5/2})$, where the parameters now refer to the disc. So we find that

$$\frac{\lambda_d}{\lambda_i} = \left(\frac{L_d}{L}\right)\left(\frac{|E_d|}{|E|}\right)^{1/2}\left(\frac{M_d}{M}\right)^{-5/2}. \tag{8.111}$$

The energy of the virialized dark matter–gas system, assuming that the gas has not yet collapsed, can be written as $|E| = k_1(GM^2/R_c)$, while that of the disc is given by $|E_d| = k_2(GM_d^2/r_c)$. Here R_c and r_c are the characteristic radii associated with the combined system and the disc respectively, while k_1, k_2 are constants of order unity which depend on the precise density profile and geometry of the two systems. The ratio of the binding energy of the collapsed disc to that of the combined system is then

$$\frac{|E_d|}{|E|} = \frac{k_2}{k_1}\left(\frac{M_d}{M}\right)^2\left(\frac{r_c}{R_c}\right)^{-1}. \tag{8.112}$$

Further, the total angular momentum (per unit mass) acquired by the gas, destined to form the disc, should be the same as that of the dark matter. This is because all the material in the system experiences the same external torques before the gas separates out due to cooling. Assuming that the gas conserves its angular momentum during the collapse, we have $(L_d/M_d) = (L/M)$. Hence,

$$\begin{aligned}\frac{\lambda_d}{\lambda_i} &= \left(\frac{M_d}{M}\right)\left(\frac{k_2}{k_1}\right)^{1/2}\left(\frac{M_d}{M}\right)\left(\frac{R_c}{r_c}\right)^{1/2}\left(\frac{M}{M_d}\right)^{5/2}\\ &= \left(\frac{k_2}{k_1}\right)^{1/2}\left(\frac{R_c}{r_c}\right)^{1/2}\left(\frac{M}{M_d}\right)^{1/2}\end{aligned} \tag{8.113}$$

where we have used (8.112) to simplify (8.111). The gas originally occupied the same region as the halo before collapsing and so had a precollapse radius of R_c. Hence the collapse factor for the gas is

$$\frac{R_c}{r_c} = \left(\frac{k_1}{k_2}\right)\left(\frac{M_d}{M}\right)\left(\frac{\lambda_d}{\lambda_i}\right)^2. \tag{8.114}$$

We see that the required collapse factor for the gas to attain rotational support has been reduced by a factor (M_d/M), from what was required in the absence of a dominant dark halo. For a typical galaxy with a halo which is ten times as massive as the disc, one needs a collapse by only a factor of about 10 or so before the gas can spin up sufficiently to attain rotational support.

The arguments given above can be improved if one assumes realistic models for the halo and the disc. Such computations, incorporating a more realistic rotation curve and including the self-gravity of the disc, agree substantially with the above rough estimate[19].

8.9 The formation of elliptical galaxies

We now turn to consider the formation of the second basic type of galaxy, the ellipticals. Elliptical galaxies are systems of stars which are supported against gravity by their random motions. Observations suggest that ellipticals are triaxial and owe their shape to anisotropic velocity dispersions[20] (see exercise 8.6). In terms of the spin parameter, the slow rotation of ellipticals implies a value of $\lambda \simeq 0.05$.

The simplest model for the formation of the ellipticals can be built along the following lines: Ellipticals were formed at high redshifts by *dissipationless* collapse and formation of stars, while spirals formed later on with considerable dissipation. Stars can form rapidly in dense protogalaxies even before the maximum radius is reached, and then relax violently to form the ellipticals. The secondary infall can reproduce the observed profiles. In contrast, the disc galaxies result when star formation is inefficient and the left-over gas gathers as a disc perpendicular to the direction of the angular momentum vector[21]. The major problem with this model is that the ellipticity of the galaxy has to be explained as due to flattening by rotation. However, as we discussed above, this is inconsistent with the observations. Hence this model needs to be modified.

Thoughthis model may not be completely correct, it must be noted that the rate of star formation is likely to be a crucial parameter in any scheme to explain the appearance of the different types of galaxies. This is because, to form a thin disc, it is necessary that star formation does not exhaust the gas supply; on the other hand, in ellipticals, star formation must have been efficient enough that the stars formed before the gas could settle into a disc. It is also likely that there are other parameters which are equally crucial. One such parameter will be the angular momentum.

We saw in section 8.8 that tidal torques can provide an angular momentum corresponding to the spin parameter $\lambda \simeq 0.05$. This value is ideal to explain the observed rotation of elliptical galaxies provided further enhancement is avoided. It is, therefore, necessary to assume that the galactic scale fluctuations collapse without much dissipation. To explain both the typical mass and the size of ellipticals, this collapse must have occurred quite early. For example, using the spherical top-hat model we can easily show that to obtain $r_{\rm vir} \simeq 10\,{\rm kpc}$ and $M = 10^{11}\,{\rm M}_\odot$, one needs $\delta_0 = \delta_i(1+z_i) \simeq 20$ and that the collapse should occur at redshift $z \simeq 11$. It is difficult to reconcile such an early epoch of galaxy formation with the isotropy of microwave radiation. It is also difficult to understand the existence of disc-like systems (which have λ comparable to unity) in this picture.

On the other hand, if one argues (as we did in the last section) that galaxies form by dissipative collapse in the potential well of a massive dark halo, then λ will increase to a value of about unity. Although this is needed for explaining the discs, we do not want λ to be so high for the ellipticals. So this model, which explains many facets of disc galaxies, fails to explain why rotation is not dynamically important for ellipticals. We discuss below some of the ideas which have been put forward in the attempts to solve this problem.

One widely discussed idea is that all galaxies were originally formed as discs and ellipticals arise from the merger of disc systems[22]. There are some obvious positive features in this hypothesis. Firstly, galaxy mergers are actually seen to occur in the universe today and probably were more frequent in the past. Secondly, a number of ellipticals which have smooth light profile, do show signs that they have experienced mergers. These signs include tidal tails, shells and in some cases, gas discs inclined to the principal planes. Furthermore, numerical simulations of mergers between galaxies show that the resultant systems usually resemble the ellipticals. Their density profiles closely match the density profile observed in ellipticals. Since discs which are merging will have their spins randomly oriented with respect to each other, the resultant spin angular momentum can be considerably smaller than that of the original discs; if several disc galaxies are involved in the merger, then the final orbital angular momentum will be negligible .

The merger hypothesis also has several potential difficulties: (1) Ellipticals are more abundant in rich clusters. In such clusters the galaxies have typical velocities of $1000\,{\rm km\,s^{-1}}$, which makes mergers rather improbable. (2) In a dissipationless merger, if the energy and mass are nearly conserved, the energy per unit mass (which is a measure of the depth of the potential well) of the remnant will be similar to that of the discs. But observations show that ellipticals have much deeper potential wells than typical discs. (3) Some ellipticals have higher phase space densities in the core than found in any disc galaxy. A dissipationless merger cannot increase the maximum value of the phase space density.

Some of these problems may be tackled by refining and modifying the original merger hypothesis, by including the effects of heirarchical clustering, dark matter and dissipation. The first problem, for example, can be resolved in the theories of galaxy formation involving heirarchical clustering. We saw earlier that, in such theories, the velocity dispersion increases with mass as long as the index of the power spectrum n satisfies the condition $n < 1$. This implies that, in small subclusters with low mass, the random bulk velocities of the galaxies will be smaller than that of a rich cluster. In such an environment mergers can take place more easily to form ellipticals. These subclusters later combine hierarchically to form rich clusters, thereby providing the rich clusters with ellipticals.

Some of the other problems can be resolved if one takes into account the presence of extended dark halos around disc galaxies. The presence of the dark halos leads to significant dynamical friction on the galaxies as they move through each other's halo, leading to the merger of the galaxies even if they were initially on a parabolic orbit. Numerical simulations[23] suggest that a compact group of galaxies evolves through a sequence of mergers on a timescale which is only of the order of a few crossing times. The dark halo also acts as a sink for energy. As the discs merge they can become more strongly bound by transferring the energy to the halo. The dynamical friction also helps the galaxies to lose their orbital angular momentum before merging, thereby leading to a merger remnant with low angular momentum.

Several of the other difficulties can be satisfactorily tackled only if the presence of gas in the merging galaxies is properly taken into account. During the merger, this gas can dissipate energy and sink to the centre, leading to cores with high phase space densities. This may, in fact, be essential at least for the low-luminosity ellipticals. For example, the phase space density for M32 is about $0.08\,\mathrm{M}_\odot\,\mathrm{pc}^{-3}(\mathrm{km\,s}^{-1})^{-3}$ which is a factor 10^7 higher than that of typical discs[24]. The formation of the elliptical galaxy M32 must have involved a considerable amount of dissipation. But it is not clear whether some other observations, like the excess[25] of globular clusters seen in the ellipticals compared with spirals, can be explained in these models.

Another idea, which has been suggested to explain the origin of elliptical galaxies, is worth mentioning in this context. It was noticed[26] that when a protogalaxy undergoes a clumpy collapse, significant amount of angular momentum can be transferred outwards as the clumps sink to the centre and merge. This leads to slowly rotating cores embedded in extended halos. If stars could form in these clumps before the protogalaxy collapses, then the resultant stellar system in the core will also be slowly rotating and will be an ideal candidate for an elliptical. On the other hand, it was found that quiescent protogalactic collapses did not lead to such a transfer of angular momentum. These halos could then be ideal sites for disc formation. This picture of the formation of the elliptical galaxy is very similar to the picture based on mergers except that the progenitors are subgalactic clumps and not full-blown discs. It is not yet clear whether this idea will work in detail. The formation of ellipticals is still an open question.

8.10 Nonlinear evolution using N-body simulations

A direct method of following the nonlinear evolution of the density perturbations is by cosmological N-body simulations. This is especially so, if the universe is dominated by collisionless particles, since the evolution of large scale structure is then driven by gravity alone. We will now discuss some of the results obtained in such simulations, concentrating specifically on two of the well-explored theories of galaxy formation, the hot and the cold dark matter models.

In an N-body simulation, one describes the matter distribution as a collection of N particles interacting via gravity. The state of the system at any time t is given by the positions and the velocities of the particles which are evolved through a sequence of small time-steps. From the positions of the particles, one can compute the force on any specific particle due to all other $(N-1)$ particles. This result is used to update the velocities which can then be used to update particle positions. In order to get a good representation of a smooth density distribution one has to make N as large as possible. The main limitation on the value of N arises from the computer time required to calculate the forces.

Three different schemes for the computation of forces have been extensively explored in the literature. The method which is simplest in concept is the direct summation method (also known as the 'particle–particle' or PP scheme), where the force on a particle is calculated by directly summing the inverse square law forces due to all other particles. In general one does not use the strict $(1/r^2)$ law all the way up to $r = 0$; instead, the force law is smoothed at small distances by modifying r^{-2} to, say, $(r^2+a^2)^{-1}$. Although such a softening of the force results in some loss of spatial resolution, it proves useful in reducing the computing time. This is because under a strictly $1/r^2$ force law, the velocities of the particles change very rapidly at very small separations. One then has to use correspondingly small time-steps to follow their trajectories accurately, which is costly in terms of computer time. The PP method gives fairly accurate description of the forces but is not suitable for a very large number of particles. About N^2 operations are required to evaluate the forces on N particles due to the other $N-1$ particles, and hence the computing time increases very rapidly with N. Clever algorithms using individual particle time-steps and a temporal hierarchy of force evaluations can reduce the number of operations to about $N^{1.6}$; but still, in practice, it turns out that the PP method is really useful only for $N \lesssim 10^4$.

A significant gain in N can be obtained if one calculates the forces on particles, not directly, but from a potential obtained by solving the Poisson equation on a mesh. The particle–mesh or PM scheme adopts such an approach. All the variables such as the density or the potential, which are functions of the space coordinates, are approximated by their values on a regular array of mesh points. Differential operators are replaced by finite difference approximations on the mesh. The potentials and forces at the location of particles are derived by interpolation of the values defined on the mesh. The densities at the mesh points are obtained by the opposite process of assigning, in a well-defined manner, the particle mass to a number of nearby mesh points.

Once the densities are specified on the mesh, say as $\rho(l,m,n)$, where (l,m,n) are integers giving the position of a mesh point, one can calculate the potential on the mesh $\phi(l,m,n)$ as the sum

$$\phi(l,m,n) = \sum_{l'm',n'} G(l-l', m-m', n-n')\rho(l',m',n') \tag{8.115}$$

where $G(l,m,n)$ is the Green function of the Laplacian operator defined on the mesh. Using the convolution theorem, one can calculate ϕ by multiplying the finite Fourier transforms of G and ρ, and then taking the Fourier transform of the product. The main advantage of the PM method lies in the fact that an efficient algorithm, the Fast Fourier Transform (FFT), can then be employed to work out the potential. Also notice that the Green function has to be worked out only once. The assigning of the mass of the particles to the mesh points, computing the forces on the particles, updating the velocities and positions all require about $10N$ operations. Finding the potential on the mesh takes about $5M^3 \log_2 M^3$ operations for a $(M \times M \times M)$ mesh[27]. The total number of operations, therefore, is about $(10N + 5M^3 \log_2 M^3)$. Thus, for a fixed M and sufficiently large N, the PM method requires less computer time than the PP method. For example, if $M = 32$ and $N = 10^5$ the number of operations required in a time-step of the PM simulation is about 4×10^6, compared with 10^{11} in the PP method. The enormous gain in speed using the PM method is at the cost of resolution in the evaluation of the force field. This is acceptable while studying problems in which the potential varies at scales which are at least as large as a few mesh lengths. Such a situation arises while studying the nonlinear evolution of a HDM universe, basically because the density fluctuations in such models have no power on small scales. Indeed, the PM method has been extensively used to simulate the evolution of HDM models.

On the other hand, if the nonlinearity is reached at about the same time over a wide range of length scales, as in a CDM universe, the PM method has to be modified. This brings us to the third scheme for evaluating the force, called the P^{3}M technique. The P^{3}M scheme tries to combine the advantages of both the PP and PM methods. The basic idea is to split up the interparticle forces into two parts: a short range, rapidly varying, part due to nearby particles and a slowly varying part due to more widely separated particles. The PP method is used to find the total short range part of the force on each particle and the PM method is used for the slowly varying force contribution. The P^3M scheme has the advantage that it can evaluate the short range force *accurately* and the long range force *rapidly*. Its main disadvantage, compared to the PM method, is the extra time taken in calculating the short range force by direct summation. This adds a number of about $N_n N$ to the total number of operations where N_n is the typical number of neighbours which contribute to the short range force. The P^3M method has been extensively used in simulations of the CDM models. We now turn to discuss the results obtained for two particular theories, the CDM and HDM models.

In the 'standard' cold dark matter model, one assumes the dark matter to be made of cold, collisionless particles with negligible random velocities. One also assumes the primordial density fluctuation field to be a random Gaussian field with a power spectrum of the Harrison–Zeldovich form. The post-recombination power spectrum of density fluctuations, is of course, different from the primordial $n = 1$ power law due to the processes discussed in chapter 4. It bends gently from the $n = -3$ power law on the subgalactic scales to an $n = 1$ power law on scales larger than λ_{eq}. Since $(\delta M/M)$ does not vary very much from the subgalactic to the cluster scales, there is substantial 'cross-talk' between different scales.

The first series of N-body simulations to examine the nonlinear evolution of the CDM universe were mainly carried out by Davis, Frenk, Efstathiou and White[28]. In the first paper of this series, Davis and collaborators followed the evolution of 32, 768 particles using a P^3M code with periodic boundary conditions. They considered an ensemble of models and studied both the $\Omega = 1$ and the $\Omega < 1$ models. The main result of this study came from an examination of the two-point correlation function ξ of the mass distribution. Recall that ξ for galaxies is well approximated by a power law of the form: $\xi(r) = (r/r_0)^{-1.8}$, where $r_0 \simeq 5h^{-1}\,\text{Mpc}$. Any model of galaxy formation should be able to reproduce this behaviour. In the simulations with the CDM power spectrum it was found that the nonlinear evolution leads to a progressive steepening of the two-point correlation function. As a result, there is only one instant of time when the auto-correlation function has approximately the same slope as that observed for the galaxies. However, at this instant of time, which occurs very 'early' in the simulations, the amplitude of ξ is smaller than the observed value unless: (i) $h \lesssim 0.22$ in a $\Omega = 1$ universe or (ii) the universe is open with $\Omega < 1$. The first possibility is ruled out since the Hubble constant is bounded by $0.4 < h < 1$. The second possibility has two difficulties: Firstly, models with $\Omega < 1$ tend to produce too much MBR anisotropy. Secondly, there is a strong theoretical prejudice favouring the value $\Omega = 1$ (we will discuss this aspect in detail in chapter 10).

It seems, therefore, that we need to do some further work to get around this difficulty.

One way of retaining $\Omega = 1$ in a CDM model is to keep $\Omega_{\rm CDM} \simeq 0.2$ and invoke a smooth contribution $\Omega_{\rm smooth} = 1 - \Omega_{\rm CDM} \simeq 0.8$ due to relativistic particles[29] or the cosmological constant. These models, however, require severe fine tuning of parameters. Another way of retaining the value $\Omega = 1$ in the CDM picture will be the following: We may relax the assumption that galaxies trace the mass distribution. It is possible that galaxies form more readily in regions of higher density; that is, around the high peaks in a suitably smoothed version of the linear density fluctuation distribution. We saw in chapter 5 that such high peaks are more strongly correlated than the underlying mass distribution. If one postulates, for example, that galaxies form only in peaks with excess fractional density contrasts above, say, 2.5σ (where σ is the root mean square mass fluctuation), then one can account for both the amplitude and the slope of the observed ξ, with $h \simeq 0.5$ and $\Omega = 1$. In this model the galaxies form a 'biased' subset of the mass distribution.

Considerable effort has been spent on how to realize such 'biased galaxy formation' in reality. The first attempts in this direction concentrated on radiative and hydrodynamic processes which could suppress the galaxy formation in halos which collapse late, from the low peaks in the density field[30]. Later on, it has been pointed out that a purely gravitational process may itself lead to biasing. For example, it was found from the N-body simulations that the strength of the galaxy clustering is larger for galaxies with a deeper potential well. This enhancement ranges from a factor of about 1.8 for galaxies with circular velocities $v_c > 100\,{\rm km\,s}^{-1}$ to about 5 for $v_c > 250\,{\rm km\,s}^{-1}$. This bias arises because clumps in higher density regions behave as though they are embedded in a (local) Friedmann universe with higher density. Since perturbations grow faster in a higher density background, these objects will collapse earlier and accrete more matter than similar objects in the low density regions. So the typical velocity dispersion and mass of clumps will be larger in the protoclusters than in the protovoids. It is possible that this enhanced clustering of the bright galaxies relative to the underlying mass distribution may be sufficient to reconcile a flat CDM universe with observed galaxy correlation function.

A somewhat different approach is based on a feature called 'velocity bias'. In this approach[31] a bias in the velocities of galaxies arises due to the dynamical friction of the galaxies moving through the dark matter. Such dynamical friction leads to a reduction in galaxy velocities on scales up to the cluster scales. This reduction implies that any mass estimate on cluster scales, using galaxy velocities, is likely to be an underestimate. The resulting concentration of the galaxies, with respect to the mass, results in an enhanced amplitude of the galaxy auto-correlation function. One may then be able to reconcile a flat CDM universe with observations of galaxy clustering.

Adopting the high peak model of biasing, N-body simulations have been used to examine how well the CDM model can explain a number of other features of galaxies. On the galactic scales the CDM model agrees very well with the observations; the dark galactic halos are predicted to have flat rotation curves as observed and the correct abundance of dark halos as a function of their potential well depth is obtained[32]. However it has been a somewhat controversial issue whether the CDM model works equally well at large scales[33] and explains the observations. The sheets, filaments and voids seen in the CfA survey, the abundance of rich clusters and their correlation function, the bulk flows of $600\,\mathrm{km\,s^{-1}}$ on scales of about 50 Mpc are some of the large scale features which are difficult to understand in CDM models. The sympathizers of the CDM model have maintained that the filaments, large sheets and voids do arise in the simulations of the CDM model; that the abundance of rich clusters can be understood if one takes proper account of projection effects; that bulk velocities can also be explained.

The standard CDM model also seems to be in disagreement with some of the recent galaxy surveys. The first one[34] – the APM Galaxy Survey – has been compiled from the scans of 185 copies of UK Schmidt J Survey photographs using the Automated Plate Measuring System in Cambridge. Each plate covers a region of $5.8^\circ \times 5.8^\circ$ and the total network of plates covers 4300 square degrees in the region $\delta < -20^\circ$ and $b \lesssim -40^\circ$. The final catalogue has about 2×10^6 galaxies reaching up to a magnitude of 20.5, corresponding to a depth of $600h^{-1}$ Mpc. The data is used to produce the two-dimensional angular correlation function of the galaxies in the sky. This result can be compared with that of CDM numerical simulations. The results show that the CDM model predicts significantly lower power at large scales (at scales above about $25h^{-1}$ Mpc) compared with the observations. The situation gets worse as h is increased; even for $h = 0.4$ the fit is quite bad.

Similar discrepancies have arisen in a sparse sample redshift survey of IRAS galaxies[35]. The IRAS point source catalogue is complete to a flux limit of 0.6 Jy at 60 μm flux. Over 2000 galaxies are selected out of this catalogue at the rate of 1 in 6, covering the entire sky. The median redshift of the sample is about 8800 km $\mathrm{s^{-1}}$. The clustering properties of these galaxies were studied by calculating the fluctuations in the galaxy count in roughly cubical cells of volume $V = l^3$. The fluctuations in the number density can then be compared with the results obtained in a CDM model. The theory predicts significantly lower power compared to the observations, at scales larger than about 20 Mpc.

At present these two observations constitute the greatest challenge to the CDM model. It appears that these observations rule out a $\Omega = 1$ CDM model with an acceptable value of h if the bias parameter is constant throughout the universe. Since the value of the Hubble constant is fairly well constrained (the fitting anyway requires a h pretty close to the acceptable lower bound), the resolution of the difficulty should lie in changing the other two assumptions. Changing a constant bias parameter to a scale dependent *function* will, of course, trivially solve the problem. But such a step, unless supplemented by a detailed physical mechanism for the biasing scheme, will take away the predictive power of the theory. The alternative is to consider models with $\Omega < 1$, or to construct mechanisms which enhance the large scale power. The former idea can be realized, for example, by invoking a cosmological constant ($\Omega_{\rm vac} \simeq 0.8$) and keeping $\Omega_{DM} \simeq 0.2$. The latter idea can be implemented by tampering with the inflationary potential or by invoking an unstable dark matter candidate of correct mass and lifetime. All these alternatives require very special parameter values.

Let us now turn to the second major model for the structure formation, the hot dark matter model. The original popularity of this model was due to the fact that there is a natural candidate for hot dark matter, the neutrino. As we described in chapters 3 and 4, massive neutrinos with a mass of about 30 eV were still relativistic when scales up to the cluster scales entered the horizon. As a result, the free streaming of the neutrinos wipes out any primordial fluctuations on scales less than $\lambda_{\rm FS} = 41(m_\nu/30\,{\rm eV})^{-1}\,{\rm Mpc}$. Due to this cut-off the first structures to form in the neutrino dominated universe are cluster mass objects. The cut-off in the small scale power also makes the initial density and potential field smooth on scales of the order of $\lambda_{\rm FS}$. The nonlinear evolution then proceeds initially according to the Zeldovich approximation, through the formation of pancakes, filaments and finally massive clumps. The N-body simulations[36] of this evolution have been performed by a number of authors using PM codes. These simulations brought out several potential problems with the hot dark matter model, which led to a decline in its popularity.

The basic problem is due to the large value of $\lambda_{\rm FS}$. In the HDM model, galaxies can only form after the collapse of cluster-sized pancakes, say, by fragmentation. Suppose one would like the galaxies to start forming sufficiently early, say at redshifts $z_{\rm form}$, which should be comparable to the redshifts of the farthest quasars. Then some fraction of the matter must have gone through the pancakes by this redshift. For example, one may define the epoch of galaxy formation as the epoch when one per cent of the particles in a simulation have passed through the caustics. It turns out that after the formation of the first pancakes, the clustering in a hot dark matter universe proceeds rapidly until most of the mass is in very massive clumps. And, unless $z_{\rm form} \lesssim 0.5$, the autocorrelation function of particles which have gone through a caustic (and identified as potential galaxies) has a much larger amplitude than that of the observed correlation function of the galaxies.* The clustering of real galaxies is also significantly weaker than that of the total mass distribution (of neutrinos) for any acceptable redshift of galaxy formation.

White and others also pointed out another difficulty with the hot dark matter scenario. The simulations produce very massive clusters if $z_{\rm form} \simeq (2\text{–}3)$. These clusters are unlike any known object in the universe and the accretion of gas into their potential wells would produce very large X-ray luminosities (about $10^{46}\,{\rm erg\,s^{-1}}$) and large (about 45 keV) gas temperatures. The existing X-ray observations do not reveal any such source although simulations predict that a large number of them should have been detected[37].

One possible weakness in these arguments against the HDM model is the considerable uncertainty which exists in deciding how the galaxies are related to the mass distribution. If, for example, galaxies do not form in the dense filaments and clumps, then the galaxy distribution may be much smoother than that of the dark matter. Such antibiasing offers one way of reconciling[38] the hot dark mater model with the observed galaxy clustering. In this picture, one would also have to argue that the gas was somehow prevented from falling into the deep potential wells of the massive clusters. Another possibility[39], still within the framework of the galaxies tracing the neutrino mass distribution, has been explored by Centrella and collaborators. They identified the present epoch as the time when the two-particle correlation function in their simulation matched the observed ξ. This choice means that their model is less evolved than the model due to White and collaborators; the mass is not completely concentrated in giant clusters at the present epoch. They argue that to form quasars at high redshift, it is not essential that one per cent of the mass goes through the caustics, like White and others assumed; rather, it is sufficient if a small fraction of the mass has become nonlinear with $\delta\rho/\rho \gtrsim 1$. Their model has the first nonlinear structures developing at a redshift of about 7 with most of the galaxies forming between $z = 7$ and $z = 1$. Also, in their way of normalizing the hot dark matter spectrum, very massive clusters do not form and the problems with the X-ray observations may be avoided. These authors conclude that the present state of ignorance about how galaxies form warrants caution in rejecting the hot dark matter model. However, it is not clear whether

---

* The COBE results also create difficulties for HDM scenarios; see Appendix C.

simple nonlinearity in the density contrast is enough to form quasars and galaxies.

We see from the above discussions that, in their present form, neither the cold dark matter nor the hot dark matter model is able to accommodate all the observed features of galaxies.

Exercises

8.1 The purpose of this exercise is to work out the details of the spherical collapse model when the initial (peculiar) velocity is non-zero. Assume that, at $t = t_i$, $\dot{r}_i = H_i r_i + v_i(r_i)$. The total energy will now be

$$E = -\frac{H_i^2 r_i^2 \Omega_i}{2} D; \quad D = (\overline{\delta}_i + 1) - \Omega_i^{-1}\left[1 + \left(\frac{v_i}{H_i r_i}\right)\right]^2.$$

(a) Show that the shell will recollapse provided $\overline{\delta}_i > \Omega_i^{-1}\,(1 + v_i/H_i r_i)^2 - 1$. In this case, also show that the maximum radius reached by the shell is $r_m = r_i D^{-1}\,(1 + \overline{\delta}_i)$.

(b) Solve the equations of motion to obtain the solution of the form

$$r = A(1 - \cos\theta); \quad t = B(\theta - \sin\theta) + T; \quad A^3 = GMB^2.$$

(c) Determine the constants A and B using the expression for maximum radius:

$$A = \frac{r_i}{2D}(1 + \overline{\delta}_i); \quad B = \frac{1 + \overline{\delta}_i}{2H_i \Omega_i^{1/2} D^{3/2}}.$$

The constant T is fixed by the conditions

$$T = t_i - B(\theta_i - \sin\theta_i); \quad r_i = A(1 - \cos\theta_i).$$

(d) Consider now a $k = 0$ universe with $a \propto t^{2/3}$, $\rho_b = (6\pi G t^2)^{-1}$. Show that the density contrast is

$$\overline{\delta} = \frac{9}{2}\frac{[(\theta - \sin\theta) + T/B]^2}{(1 - \cos\theta)^3} - 1.$$

Expand this quantity in powers of $(t - T)$ and obtain, for small $(t - T)$

$$\overline{\delta} = \frac{3}{20}\left[\frac{6(t - T)}{B}\right]^{2/3} + \frac{2T}{(t - T)}.$$

Further, since $\Omega_i = 1$, $H_i = 2/(3t_i)$, show that to leading order

$$B = \frac{3}{4}\frac{t_i}{(\overline{\delta}_i - 3t_i v_i/r_i)^{3/2}}.$$

(e) Hence show

$$\overline{\delta} = \frac{3}{5}\left(\frac{t}{t_i}\right)^{2/3}\left[\overline{\delta}_i - \frac{3t_i v_i}{r_i}\right] + \frac{2}{5}\left(\frac{t_i}{t}\right)\left[\overline{\delta}_i + \frac{9t_i v_i}{2r_i}\right].$$

(f) Solve the linear perturbation equations in the matter dominated, Newtonian case assuming arbitrary initial conditions at $t = t_i$: $\delta_{\mathbf{k}} = \delta_k(t_i)$, $\mathbf{v}_{\mathbf{k}} = \mathbf{v}_{\mathbf{k}}(t_i)$ at $t = t_i$. Show that

$$\delta_{\mathbf{k}}(t) = A_{\mathbf{k}} t^{2/3} + B_{\mathbf{k}} t^{-1}$$

$$\mathbf{v}_{\mathbf{k}}(t) = -\frac{2}{3}\frac{i\mathbf{k}}{k^2}\left[A_{\mathbf{k}} t^{-1/3} - \frac{3}{2} B_{\mathbf{k}} t^{-2}\right] + \left(\frac{t_i}{t}\right)^{4/3}\left[\mathbf{v}_{\mathbf{k}}(t_i) + \frac{\mathbf{k}}{k^2}\left[\mathbf{k}\cdot\mathbf{v}_{\mathbf{k}}(t_i)\right]\right],$$

with

$$A_{\mathbf{k}} = \frac{3}{5t_i^{2/3}}\left[\delta_{\mathbf{k}}(t_i) + it_i\mathbf{k}\cdot\mathbf{v}_{\mathbf{k}}(t_i)\right]$$

$$B_{\mathbf{k}} = \frac{2}{5} t_i \left[\delta_{\mathbf{k}}(t_i) - \frac{3}{2} t_i i\mathbf{k}(\mathbf{k}\cdot\mathbf{v}_{\mathbf{k}}(t_i))\right].$$

Compare the solutions in (e) and (f) and show that they are consistent. When $v_i = 0$, (3/5)th of the amplitude is in the growing mode.

8.2 Consider a system evolving through the collisionless Boltzmann equation. Let it be far away from a steady state configuration at some initial moment.

(a) Argue that each particle is acted upon by a mean gravitational field which is varying 'rapidly' in time. Estimate the timescale in which the energy of a particle will change significantly.

(b) Consider a collisionless system of particles evolving under the influence of an *external*, static gravitational field $\phi(r)$. Assume that $\phi(r) < 0$, is smooth and increases monotonically to $\phi = 0$ sufficiently fast as $r \to \infty$. At $t = 0$, the particles occupy a small region of volume V in phase space. Sketch how the shape of this region will change with time in a generic situation. Suppose the 'resolution limit' in phase space is (V/K) with some fixed K. Estimate the timescale after which all the allowed region of phase space will *appear* to be occupied.

(c) Argue that the two (different) processes described above can lead to a relaxation of collisionless systems at a much shorter timescale than the two-body collisions.

(d) Consider now the final equilibrium state reached by such a collisionless evolution. Define a coarse-grained distribution function f_c by averaging f over cells of some small size μ in the phase space. Let the initial distribution function be f_i. The evolution proceeds such that: (i) The volume of a region with a given phase density is conserved (ii) Total energy is conserved. Also assume that each phase element undergoes 'complete relaxation'. What will be the most probable form of final f_c?

8.3 This exercise provides a derivation of the general relativistic, inhomogeneous model and assumes familiarity with general relativity. Consider a metric of the form

$$ds^2 = dt^2 - e^{2\alpha}\,dx^2 - e^{2\beta}(d\theta^2 + \sin^2\theta\,d\phi^2)$$

with $\alpha = \alpha(x,t)$, $\beta = \beta(x,t)$. The components of the Einstein tensor $G^\mu_\nu = R^\mu_\nu - (1/2)\delta^\mu_\nu R$ for this metric are given by

$$G^0_0 = \dot\beta^2 + 2\dot\alpha\dot\beta + \mathrm{e}^{-2\beta} - \mathrm{e}^{-2\alpha}(2\beta'' + 3\beta'^2 - 2\alpha'\beta'),$$
$$G^1_1 = 2\ddot\beta + 3\dot\beta^2 + \mathrm{e}^{-2\beta} - (\beta')^2\mathrm{e}^{-2\alpha},$$
$$G^2_2 = \ddot\alpha + \dot\alpha^2 + \ddot\beta + \dot\beta^2 + \dot\alpha\dot\beta - \mathrm{e}^{-2\alpha}(\beta'' + \beta'^2 - \alpha'\beta'),$$
$$G^1_0 = \mathrm{e}^{-2\alpha}\left[2\dot\beta' + 2\dot\beta\beta' - 2\dot\alpha\beta'\right].$$

The source is taken to be of the form $T^\mu_\nu = \mathrm{dia}[\rho(x,t), 0, 0, 0]$.
(a) Show that the G^1_0 equation implies $\mathrm{e}^\beta\beta' = \mathrm{e}^\alpha g(x)$ with some $g(x)$. Put $\exp[\beta(x,t)] = xa(x,t)$ and show that

$$\exp[\alpha(x,t)] = \frac{(ax)'}{[1 - x^2/R^2(x)]^{1/2}},$$

where $R(x)$ is arbitrary.
(b) Recast the G^0_0 and G^1_1 equations into the form

$$\frac{8\pi G\rho}{3}\frac{\partial}{\partial x}(ax)^3 = \frac{\partial}{\partial x}\left(\dot a^2 ax^3 + \frac{ax^3}{R^2}\right),$$
$$2\frac{\ddot a}{a} + \frac{\dot a^2}{a^2} + \frac{1}{a^2R^2} = 0.$$

(c) Show that the solution to these can be expressed in the form

$$\dot a^2 a + aR^{-2} = F(x); \qquad \frac{8\pi G}{3}\frac{\partial}{\partial x}(ax)^3 = \frac{\partial}{\partial x}(x^3F),$$

where $F(x)$ is arbitrary. Convert these equations into the form given in the text.

8.4 The accuracy of the Zeldovich approximation can be estimated as follows: From the relation $x_\alpha = a(t)\ [q_\alpha - b(t)p_\alpha(\mathbf{q})]$ where $p_\alpha(\mathbf{q}) = (\partial\Phi/\partial q_\alpha)$ it follows that $\dot x_\alpha = (\dot a/a)x_\alpha - a\dot b p_\alpha$.
(a) Differentiate this relation once again and rewrite the expression in the form

$$\ddot{\mathbf{x}} = \frac{\ddot a}{a}\mathbf{x} - (2\dot a\dot b + a\ddot b)\mathbf{p}.$$

(b) We need to calculate the divergence of this expression with respect to $\mathbf{x}$. The first term gives $3(\ddot a/a)$; to evaluate $\nabla_x \cdot \mathbf{p}$ proceed as follows: Show that

$$\nabla_x \cdot \mathbf{p} = \frac{\partial p_\alpha}{\partial x^\beta} = \frac{\partial p_\alpha}{\partial q^\beta}.\frac{\partial q^\beta}{\partial x^\alpha} = \left(\frac{\partial^2\Phi}{\partial q^\alpha\partial q^\beta}\right)\left[\frac{\partial x^\alpha}{\partial q^\beta}\right]^{-1}$$
$$= d_{\alpha\beta}\left[a(\delta_{\alpha\beta} - bd_{\alpha\beta})\right]^{-1},$$

where we have defined a matrix $d_{\alpha\beta} \equiv (\partial^2\Phi/\partial q^\alpha\partial q^\beta)$.
(c) In the locally diagonal system $d_{ik} = \mathrm{dia}(\lambda_1, \lambda_2, \lambda_3)$ and $[a(\delta_{ik} - bd_{ik})]^{-1} = a^{-1}\mathrm{dia}\ [(1-b\lambda_1)^{-1}, (1-b\lambda_2)^{-1}, (1-b\lambda_3)^{-1}]$. In this frame

show that

$$\nabla_x \cdot \mathbf{p} = \frac{1}{a}\left[\frac{\lambda_1}{1-b\lambda_1} + \frac{\lambda_2}{1-b\lambda_2} + \frac{\lambda_3}{1-b\lambda_3}\right]$$
$$= \frac{1}{a}\frac{(I_1 - 2bI_2 + 3b^2 I_3)}{(1 - bI_1 + b^2 I_2 - b^3 I_3)}$$

where I_1, I_2 and I_3 are the invariants defined in the text.
(d) Using $4\pi G\rho = 4\pi G\rho_0 a^{-3}(1 - bI_1 + b^2 I_2 - b^3 I_3)^{-1}, = 4\pi G\rho_0 a^{-3} h^{-1}$, compute $Q = \nabla \cdot \ddot{\mathbf{x}} + 4\pi G\rho$. Show that

$$hQ = 3\frac{\ddot{a}}{a}(1 - bI_1 + b^2 I_2 - b^3 I_3) - \left(2\frac{\dot{a}}{a}\dot{b} + \ddot{b}\right)(I_1 - 2bI_2 + 3b^2 I_3) + \frac{4\pi G\rho_0}{a^3}.$$

Rewrite this expression as

$$hQ = \left(3\frac{\ddot{a}}{a} + \frac{4\pi G\rho_0}{a^3}\right)$$
$$- \left(\ddot{b} + 2\frac{\dot{a}}{a}\dot{b} + 3\frac{\ddot{a}}{a}b\right)(I_1 - 2bI_2 + 3b^2 I_3) + 3\frac{\ddot{a}}{a}\left[-b^2 I_2 + 2b^3 I_3\right].$$

This is the form used in the text.

8.5 Galaxy redshifts are often used as one of the 'spatial coordinates' in denoting the position of a galaxy. However, since galaxies have peculiar velocities, the redshift space will not match properly with coordinate space. Consider, for example, the formation of caustics in the redshift space.
(a) The net velocity at any location, in Zeldovich approximation, is given by $\mathbf{u} = (\dot{a}/a)\mathbf{x} - a\dot{b}\mathbf{V}\ (\mathbf{q})$ where $\mathbf{V} = \nabla_q \mathbf{\Phi}(\mathbf{q})$ in an obvious notation. Show that the redshift of a galaxy at Lagrangian coordinate $\mathbf{q}$ will be

$$z = aq_3 - (1+f)abV_3$$

where we have taken the line-of-sight to be along the third axis and $f = (\dot{b}/b)/(\dot{a}/a) \simeq \Omega^{0.6}$.
(b) Let the redshift space coordinates be $(x = x_1, y = x_2, z = x_3 + v/H)$. On transforming from a Lagrangian system to the redshift coordinates, the entries in the third row of the Jacobian matrix will be multiplied by $(1+f)$ factors. Show that, in the simplest case, this will give the density

$$\rho(x,t,z,t) = \frac{\rho_0}{a^3}\frac{1}{(1-b\lambda_1)(1-b\lambda_2)(1-b(1+f)\lambda_3)}.$$

Hence show that the redshift-space caustic occurs at $b\lambda_3 = (1+f)^{-1}$ $= (1/2)$ for $\Omega = 1$. In general, show that the 'redshift caustics' will precede the 'density caustics' if a region is collapsing.

8.6 Consider a system described by collisionless Boltzmann equations. Let

$$\rho = \int f\, d^3\mathbf{v}; \quad \sigma_{ij} = \frac{1}{\rho}\int f v_i v_j\, d^3\mathbf{v}; \quad u_i = \frac{1}{\rho}\int f v_i\, d^3\mathbf{v}.$$

(a) Prove the tensor virial equation:

$$\frac{1}{2}\frac{d^2}{dt^2}I_{jk} = 2T_{jk} + \Pi_{jk} + W_{jk}$$

where

$$I_{jk} = \int \rho x_j x_k \, d^3\mathbf{x}; \quad T_{jk} = \frac{1}{2}\int \rho u_j u_k \, d^3\mathbf{x};$$

$$\Pi_{jk} = \int \rho \left[\sigma_{jk} - u_j u_k\right] d^3\mathbf{x}$$

and

$$W_{jk} = -\frac{1}{2}G \int \rho(\mathbf{x})\rho(\mathbf{x}')\frac{(x_j' - x_j)(x_k' - x_k)}{|\mathbf{x}' - \mathbf{x}|^3} d^3\mathbf{x}' \, d^3\mathbf{x}.$$

(b) Consider an axisymmetric elliptical galaxy with symmetry axis along z and line-of-sight along the x-axis. In steady state, show that tensor virial equation implies

$$\frac{2T_{xx} + \Pi_{xx}}{2T_{zz} + \Pi_{zz}} = \frac{W_{xx}}{W_{zz}}.$$

If the only streaming motion is a rotation about the z-axis, $T_{zz} = 0$. Define

$$\frac{1}{2}Mv_0^2 \equiv 2T_{xx} = \frac{1}{2}\int \rho u_\phi^2 \, d^3\mathbf{x}; \quad \Pi_{xx} = M\sigma_0^2; \quad \Pi_{zz} = (1-\delta)M\sigma_0^2$$

and write the above relation as

$$\frac{v_0^2}{\sigma_0^2} = 2(1-\delta)\frac{W_{xx}}{W_{zz}} - 2.$$

(c) Argue that, if the isodensity surfaces are concentric, similar ellipsoids, then $(W_{xx}/W_{zz}) = f(\epsilon)$ is a function of the ellipticities alone. Thus the above equation determines the ellipticity in terms of (v_0/σ_0) and δ. Suppose (v_0/σ_0) is plotted against ϵ for various values of δ. How should these curves look? (Observed (v_0/σ_0) and ϵ of the ellipticals suggest that a given ϵ is maintained due to large δ and small (v_0/σ_0) rather than by small δ and large (v_0/σ_0); hence the flattening is not due to rotation.)

9
High redshift objects

9.1 Introduction

The most direct probe of the models for nonlinear evolution will be structures seen at high redshifts, some of which are shown in figure 9.1. This chapter discusses the constraints that can be imposed on the models for galaxy formation from the study of such objects. Section 9.2 covers the search for 'primeval' galaxies – which are galaxies in their formation stages – and discusses some of the candidates. Quasars and their absorption spectra are studied in the next two sections. The last section describes the models for the radio galaxies and the constraints which arise from them.

9.2 Primeval galaxies

The isotropy of the MBR indicates that density perturbations were in the linear regime (with $\delta \lesssim 10^{-4}$) at $z \simeq 10^3$. It is, therefore, clear that the nonlinear phase of the galaxy formation occurs very much after the decoupling, probably at $z < 50$ or so. Hence the most direct way of probing the nonlinear structure will be to study objects with large redshifts. As we shall see in the next few sections, such a study will be useful in constraining different models for the structure formation. There also exists the possibility that we may actually be able to detect a galaxy during its formation stage, just as we observe star-forming regions in the Milky Way. We shall now consider several issues related to the detection of such a 'primeval' galaxy.

To begin with, it is possible[1] to impose some constraints on the epoch of galaxy formation using the spherical top-hat model. We saw in the last chapter that the density of a spherical overdense region is $(9\pi^2/16)$ times higher than the background density ρ_b when the region has expanded to the maximum size (i.e., at turn-around). If the material contracts further by a factor f_c to form the galaxy, then the density will increase by another factor f_c^3. Therefore, the observed density of galaxies today, $\rho_{\rm obs}$, can be expressed as

$$\rho_{\rm obs} \simeq f_c^3 \frac{9\pi^2}{16} \rho_b(t_m) \simeq 5.6 \Omega \rho_c (1+z_g)^3 f_c^3, \tag{9.1}$$

where z_g is the redshift at turn-around. For a luminous galaxy, with a of mass of about 10^{11} M$_\odot$, contained within, say, a radius $r \simeq 10\,$kpc, we have $(\rho_{\rm obs}/\rho_c) \simeq 10^5$; so from (9.1) we find that

$$z_g \simeq \frac{1}{f_c} \left(\frac{\rho_{\rm obs}}{5.6\Omega\rho_c} \right)^{1/3} \simeq \frac{30}{f_c \Omega^{1/3}}. \tag{9.2}$$

If the collapse factor was much greater than unity, one also has to allow for the change in ρ_b during the time taken for the collapse.

Since $t_{\rm coll} \simeq 2t_m$ and $\rho_b \propto t^{-2}$, the density contrast increases by another factor of 2^2 and the z_g obtained above has to be divided by a further factor of order $2^{2/3}$. In the case of dissipationless collapse, we estimated that $f_c \cong 2$. On the other hand, for the gas which forms the disc, we need larger collapse factors of about 10, if the disc has to attain the necessary rotational support (see chapter 8). So we find that $z_g \lesssim 15\Omega^{-1/3}$ for the formation of the halo cores and $z_g \lesssim 3\Omega^{-1/3}$ for the formation of the discs.

The above estimate suggests that galaxy formation could be a fairly recent event and is probably accessible to direct observation. There has been considerable effort in recent years to detect a galaxy in its formation. To understand these efforts (and results) it is first necessary to determine

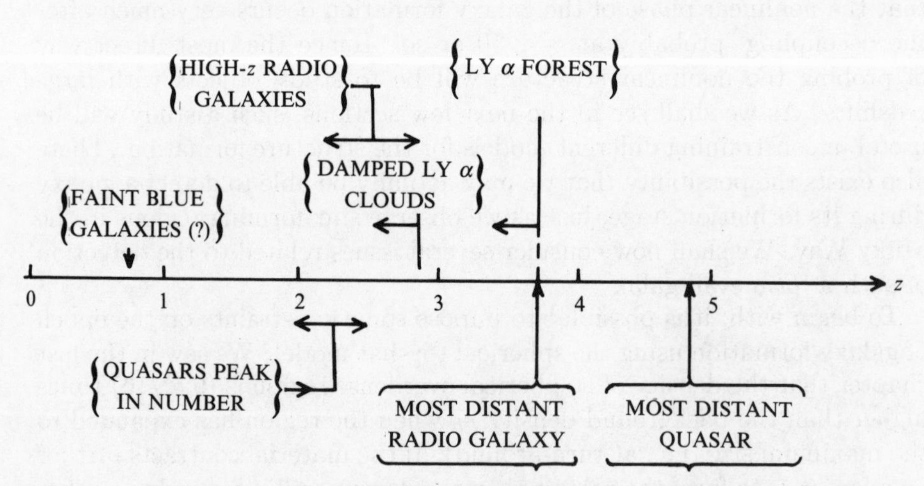

Fig. 9.1. Diagram showing the various high redshift objects discussed in this chapter.

how a young galaxy will look. Hence we shall begin with a review of some basic facts about the galactic evolution.

Broadly speaking, one may think of elliptical galaxies (in which not much star formation is occurring today) as galaxies which had undergone an early, rapid phase of star formation while discs (in which star formation goes on today) are systems in which star formation has been almost constant since their inception. The 'colour' (which corresponds to the average wavelength at which the light is emitted) and luminosity of the galaxy evolve as the stellar populations contained in it change. Using our knowledge of stellar evolution, we can thus hope to understand some aspects of galactic evolution[2].

Let us consider the ellipticals first. We saw in chapter 1 that more massive stars have shorter lifetimes; for example, stars more massive than $1.25\,\mathrm{M}_\odot$ live for less than 5 Gyr. If the star formation in ellipticals took place as a 'burst' long ago, we should expect them to contain only stars with $M \lesssim 1\,\mathrm{M}_\odot$. These stars are most luminous for the first 1 Gyr or so and then evolve over to the 'giant' phase. Thus the light from the ellipticals should be that produced by these giants and the colour should be predominantly red. Observations show that this is indeed the case. Since the temperatures of the low mass stars do not depend sensitively on the mass of the star, the colour of the elliptical does not change much after the first 1.5 Gyr or so, by which time heavier stars would have disappeared.

On the other hand, the luminosity of the ellipticals will continue to evolve. We can write the luminosity of an elliptical as

$$L \simeq \left(E(M)\frac{dN}{dM}\right)_{M_G} \left|\left(\frac{dM_G}{dt}\right)\right|, \tag{9.3}$$

where $E(M)$ is the energy emitted by a giant-branch star of mass M, (dN/dM) is the number of stars in the mass interval dM (called the 'initial mass function', IMF) around the giant branch and $|dM_G/dt|$ is the rate at which stars enter the giant branch. (In other words, $M_G(t)$ is the mass of the star that leaves the main sequence at an age t.) The last two functions can be obtained from stellar evolution theory and are well approximated by

$$\left(\frac{dN}{dM}\right) \simeq K\left(\frac{M}{1\,\mathrm{M}_\odot}\right)^{-(1+x)} ; \quad \frac{M_G}{1\,\mathrm{M}_\odot} \simeq \left(\frac{t}{10\,\mathrm{Gyr}}\right)^{-y} \tag{9.4}$$

with $x \simeq 1.5$, $y \simeq 0.4$. Substituting these values, we get

$$\left(\frac{dM_G}{dt}\right) \propto [-yt^{-(1+y)}] \propto [-y(M_G)^{1+(1/y)}]; \quad L \propto EM^{1/y-x}. \tag{9.5}$$

Hence

$$\frac{d\ln L}{d\ln t} = \left[\frac{d\ln E}{d\ln M} + (y^{-1} - x)\right]\frac{d\ln M}{d\ln t} = xy - \left(1 + y\frac{d\ln E}{d\ln M}\right). \qquad (9.6)$$

Since E depends only very weakly on M and $xy \simeq 0.6$, we find that $L \propto t^{-0.4}$. Thus the luminosity of the ellipticals decreases with time and they must have been brighter in the past. At the redshift of $z \simeq 1$, $t = 2^{-3/2}t_0$ in an $\Omega = 1$ universe; an elliptical will be $2^{0.6} \simeq 1.5$ times brighter at $z \simeq 1$.

The evolution of spirals can be worked out along similar principles. Here we need to know both the IMF *and* the star formation rate $\dot{M}_{\rm star}(t)$. The results will depend on the specific forms assumed for these functions. Suppose, for example, that the star formation rate is constant. This would mean that the number of young, blue stars would remain constant while that of the older stars increases, making the system more luminous and redder.

The above analysis can also be extended to cover the initial star formation phase of a protogalaxy. Calculations show that a star formation rate of $1\,{\rm M}_\odot$ per year results in a luminosity of about $2.2 \times 10^9\,{\rm L}_\odot$ if the IMF is of the 'standard' form[3]. So a galaxy of mass M is expected to have an initial luminosity of

$$L = 2.2 \times 10^{12}\,{\rm L}_\odot \left(\frac{M}{10^{11}\,{\rm M}_\odot}\right)\left(\frac{t_{\rm form}}{10^8\,{\rm yr}}\right)^{-1}, \qquad (9.7)$$

where $t_{\rm form}$ is the time during which the initial burst of star formation takes place. This equation shows that, if the star formation was in a rapid burst (with $t_{\rm form} \simeq 10^8$ yr), then a typical galaxy will be quite luminous compared to the characteristic luminosity $L_* = 1.6 \times 10^{10}h^{-2}\,{\rm L}_\odot$ which occurs in the galaxy luminosity function. Such a scenario *may* be applicable in the hot dark matter models. But in the cold dark matter models, galaxies form over a much more extended period of time. Note that the expansion time H^{-1} at $z \simeq 1$ will be about $3.5 \times 10^9 h^{-1}$ yr; if galaxies are assembled over such timescales, then $L \simeq 6 \times 10^{10}h^{-1}\,{\rm L}_\odot \simeq L_*$. It will be difficult to detect these objects at $z \gtrsim 1$.

In these models, a primeval galaxy will look like an enlarged version of the star-forming region in our own galaxy, with appropriate redshift. The spectrum will be flat with an abrupt cut-off at the 'Lyman limit'. This is the wavelength of 912 Å, below which almost all photons emitted by the system will be absorbed through the ionization of hydrogen. There could also be strong, narrow emission lines associated with the ambient gas which are excited by the newly formed young stars.

Suppose we wish to search for a primeval galaxy, say, at its most luminous phase. If this phase occurs at redshift $z \gtrsim 5$, then the Lyman limit absorption will shift to the wavelength of $912(1+z)$ Å $\gtrsim 5472$ Å which is in the optical band. Hence the dip in their spectral intensity will occur in the optical band thereby making them invisible at optical wavelengths. On the other hand, the emission from young stars in the ultraviolet and visible wavelengths will be redshifted to the infrared band for such z. Hence the infrared band is the obvious choice for searching for such objects. However, these searches in the infrared have not led to any candidate for a primeval galaxy at large z so far.

If primeval galaxies exist at much lower redshifts, then optical searches would be more suitable. The null results of such searches[4] were once taken to imply that galaxies cannot form at redshifts below about 6. But these conclusions were based on models of galaxy formation which assumed rapid star formation rates, viz. that a galaxy of stars formed within a dynamical timescale of the system. In many galaxy formation theories, stars could form over an extended timescale in small subunits, which then merge to form the 'final galaxy'. As we saw above, such a primeval galaxy will have a low enough surface brightness to have escaped detection in the searches mentioned above.

An apparent positive signal in the primeval galaxy search came recently from the counts of faint galaxies. Such counts are now available up to $B = 27$ magnitude (that is up to 7×10^{-31} erg cm^{-2} Hz^{-1} at 4500 Å; blue band) and up to $K = 23$ magnitude (that is up to 4×10^{-30} erg cm^{-2} Hz^{-1} at 22000 Å; infrared)[5]. All the galaxy images up to a limiting flux are included in these counts. So the counts do not refer to 'standard candles' but cover objects with a wide range of intrinsic properties. A galaxy with a given magnitude can be at a wide range of redshifts; fainter galaxies can be thought of as more distant only in a statistical sense. In the blue band, the galaxy counts up to $B = 21$ can be modelled fairly easily (without invoking any galactic evolution or esoteric cosmological models); beyond $B = 21$, the counts increase much more rapidly compared to what is expected in an $\Omega = 1$ model with no galactic evolution. For example, there are about 4.5×10^5 galaxies per square degree which are brighter than 27th magnitude[5]. The proper volume of the Friedmann universe up to the redshift of $z = 4$ is

$$\mathcal{V} = 4\pi a_0^3 \int_0^{r(z)} \frac{r^2\, dr}{\sqrt{1 - kr^2}} = \begin{cases} 3.7 \times 10^6 h^{-3}\,\mathrm{Mpc}^3; & (\Omega = 1) \\ 1.4 \times 10^7 h^{-3}\,\mathrm{Mpc}^3; & (\Omega = 0.1). \end{cases} \tag{9.8}$$

(Half of this volume is at $z = 2.2$ for $\Omega = 1$ and at $z = 2.7$ for $\Omega = 0.1$.) The galaxy luminosity function, determined from local observations is (see chapter 1)

$$\Phi(L)\, dL\, dV = \Phi_\star (L/L_\star)^{-\alpha} \exp(-L/L_\star)(dL/L_\star)\, dV, \tag{9.9}$$

with $\Phi_\star \simeq 0.012 h^3\,\mathrm{Mpc}^{-3}$, $L_\star \simeq 10^{10} h^{-2}\,\mathrm{L}_\odot$ and $\alpha \simeq$ (1.1–1.2). If we set the total number of galaxies with luminosities higher than some luminosity x as equal to the count 4.5×10^5, then we get the equation

$$\int_x^\infty \Phi(L)V\, dL = 4.5 \times 10^5, \tag{9.10}$$

which can be solved for x. We find that $x = (L_\star/400)$ for $\Omega = 1$ and $x = (L_\star/10)$ for $\Omega = 0.1$. That is, to account for the observed count, we have to include galaxies which are *considerably* fainter than $L_\star$ today. But the fact that we actually observe them implies that these galaxies must have been *much* brighter at $z \simeq (3\text{–}4)$.

It is, however, possible to account for these counts by invoking galactic evolution[6] and a formation redshift of $z_f \simeq 3$. If there is evolution in the luminosity, then one can see galaxies at much farther distances and the excess counts are interpreted as due to galaxies at high redshifts in the rapid star formation phase. This idea, however, runs into some trouble because spectroscopic data (which is available up to $B = 24$ where the galaxy count is $10^4\,\text{deg}^{-2}$) now shows that most of these galaxies[7] are actually at low redshifts of $z \simeq (0.3\text{–}0.6)$.

Though this rules out the blue galaxies as high redshift objects, it raises several cosmologically interesting questions. Lower redshifts imply that less proper volume is available, thereby making matters worse. We are now left with two alternative explanations for the faint galaxies: (i) The universe actually has a lot more volume at moderate redshifts compared with the $\Omega = 1$ model. For example, a model with a cosmological constant and $\Omega = 0.1$ will provide sufficient extra volume to explain the excess counts. (ii) There existed at $z \simeq (0.3\text{–}0.5)$ a major population of blue, dwarf galaxies which is not counted in the present day luminosity functions. (That is, they are unusually dim today.) It seems that observations in the K-band are *not* consistent with the first alternative; these counts strongly favour a $\Omega = 1$ universe with zero cosmological constant[8].

Thus we are left with the second alternative. These galaxies, which have about a tenth of the luminosity of a normal galaxy in the blue band, must have somehow faded below detectability by $z \simeq 0$. This is really puzzling because these galaxies contain as many baryons as normal galaxies. The baryonic content of these galaxies can be estimated in an indirect way as follows: During stellar evolution, hydrogen and helium are converted into heavier elements (usually called 'metals') with the emission of radiation. The massive stars which dominate the blue light of these galaxies are also the prime source of metallicity in them. The amount of this star light – which is mostly in the blue and ultraviolet bands – is well correlated with the amount of metals produced. Fortunately, this correlation does not have any strong dependence on the redshift. Even if we do not resolve the individual galaxies below some luminosity, we do know that the starlight of these galaxies must be contributing to the extragalactic sky background today. Most of this sky background at $B = (22\text{–}24)$ comes from the 'dead' galaxies at $z \simeq (0.3\text{–}0.4)$, and is about $10^{-24}\,\mathrm{erg\,s^{-1}\,cm^{-2}\,Hz^{-1}\,deg^{-2}}$. Theoretical calculations[9], on the other hand, lead to the estimate

$$S = 3.6 \times 10^{-25} \left(\frac{\rho Z}{10^{-34}\,\mathrm{g\,cm^{-3}}} \right) \mathrm{erg\,cm^{-2}\,s^{-1}\,Hz^{-1}\,deg^{-2}}, \qquad (9.11)$$

where ρZ is the density of metals. Comparison shows that $\rho Z \simeq 3 \times 10^{-34}\,\mathrm{g\,cm^{-3}}$. So, even if Z was as high as 0.02, ρ must be about $10^{-32}\,\mathrm{g\,cm^{-3}}$; in fact, Z is likely to be much lower; in which case, the baryonic content of these galaxies is higher than that in local populations.

The large abundance of these low redshift galaxies (with a surface density of about $10^4\,\mathrm{deg^{-2}}$ up to 24th B-magnitude), compared to what is expected in the models of counts without any evolution, implies that significant changes are taking place in the galaxy population even at such low z. In particular, the galaxy luminosity function itself may be altering in shape. Roughly speaking, the data suggests a luminosity function

$$\phi(L) \propto L^{-\alpha(z)} \exp(-L/L_*); \quad \alpha(z) \simeq 1 + 2z, \qquad (9.12)$$

for $L < L_*$ (and *not* at higher L). Such a change in shape can arise due to progressive merging of smaller galaxies into bigger ones[10]. Another possibility[11] is that the slope at the faint end of the galaxy luminosity function steepens with the redshift, possibly due to the increase in the starburst activity of the low-luminosity galaxies at larger z. The galaxy number counts in the K-band may help to distinguish between these possibilities. It appears that the K-band counts are consistent with a model having no evolution. Since the K-band light is a more robust tracer of the mass of the galaxies, this makes the models involving large scale merging less promising[12]. If the blue galaxies are indeed low mass galaxies undergoing sporadic bursts of star formation, then it is not clear that they represent a major episode of galaxy formation, though they seem to be producing a large fraction of present day metals.

We shall now turn to objects which are known to be at higher redshifts and discuss what they imply for models of galaxy formation.

9.3 Quasars and galaxy formation

Since high redshift objects are also at larger distance from us, they have to be intrinsically quite bright to be seen by us. For example, a giant elliptical galaxy in the nearby ($z \simeq 0$) region will fade below the limits of detectability (say, 26 magnitude using CCDs) if placed at a redshift of about $z \simeq 1$ in an $\Omega = 1$ universe. The same limit also applies to clusters of galaxies. Spiral galaxies like ours are difficult to detect even at $z \simeq 0.5$.

The objects with the highest redshifts discovered so far are the quasi-stellar objects – or, quasars, for short – and the radio galaxies. The quasar PC1247 + 3406 with $z = 4.897$, holds[13] the present record for the highest redshift; more than 20 quasars with $z > 4$ are known. Several galaxies at high redshift have been discovered recently using their radio properties; the one with highest redshift ($z = 3.8$) being[14] the galaxy 4C 41.17.

Systematic observations of quasars have by now produced the following picture regarding the abundance of quasars at different redshifts. The quasar luminosity function $n(L, z)$, defined as the number of quasars per unit comoving volume per unit luminosity at a given redshift, shows a systematic trend. Integrating n over L in some range (L_{min}, ∞) we obtain the redshift dependence of the number density of quasars above a fixed luminosity. There is good evidence[15] to suggest that this integrated comoving number density of quasars increases with redshift up to a redshift of $z \simeq 2$. This increase occurs both for radio-quiet quasars and for radio-loud quasars. At the same time, with the exception of the brightest quasars, the number density of quasars and radio galaxies shows a *decline* between redshifts of 2 and 4. This decline is gradual, and not a sharp cut-off. (It should, however, be noted that this decline does occur at a timescale which is short compared with the Hubble time at that redshift. This indicates a remarkable synchronization in quasar activity all over the space around this time – a feature which has not been adequately explained.) Such an epoch of peak quasar density at $z \simeq 2$ could indicate that this epoch is in some way special; perhaps because it was the epoch when galaxies which could host quasars formed in abundance.

The high redshift objects, especially the quasars, constrain the models for galaxy formation in three important ways. First of all, the very existence of these objects demands that fully nonlinear structures must have formed by such high redshifts. As we shall see, this is not an easy task. Secondly, the high redshift radio galaxies are different from their low z counterparts, which might indicate the vastly different physical conditions at these redshifts. Any comprehensive model for structure formation must explain these differences. Lastly, the quasars allow us to study the properties of several types of intervening objects, which cause absorption lines in the spectra of the quasar. A study of the absorption line systems can, therefore, be used to probe the contents of the universe at redshifts lower than the redshift of the quasar. We shall now discuss the first constraint in detail; the second and the third constraints will be taken up in the subsequent sections.

The very existence of quasars at high redshifts, say at $z > 4$, imposes crucial constraints on galaxy formation. Quasars are believed to be powered by the accretion of matter onto a massive black hole, located at the centre of the galaxy. So, before any quasar activity can begin, some galaxies must have evolved at least to the stage of developing a compact and massive black hole. The overdensity involved in a quasar is quite high compared to objects which have just virialized. Thus these galaxies should have formed and virialized at redshifts higher than that of the quasar allowing sometime for the evolution. However, in most of the theoretical models, galaxies form at a relatively late epoch of about $z \simeq 2$. In this case, it will be difficult to account for the large number of quasars observed at $z \gtrsim 2$.

The luminosity of the quasars which are powered by accretion is constrained by a maximum (limiting) luminosity known as the Eddington luminosity. This is the luminosity, L_E, above which the radiation pressure on an accreted fluid element exceeds the gravitational attraction of the black hole, thereby preventing the accretion. We can estimate L_E as follows: The number density $n(r)$ of photons crossing a sphere of radius r, centred at the quasar of luminosity L is $(L/4\pi r^2)\,(\hbar\omega)^{-1}$ where ω is some average frequency. The rate of collisions between the photons and ionized matter will be $n\sigma_T$. Assuming that the momentum transfer per collision is $(\hbar\omega/c)$, we can estimate the net force exerted by the photons on the surrounding (ionized) matter to be

$$f_{\rm rad} = (n\sigma_T).\left(\frac{\hbar\omega}{c}\right) = \left(\frac{L}{4\pi r^2}\right)\left(\frac{1}{\hbar\omega}\right)\sigma_T\left(\frac{\hbar\omega}{c}\right) = \left(\frac{L\sigma_T}{4\pi c r^2}\right). \quad (9.13)$$

This force will be larger than the gravitational force $f_g = (GM_{\rm BH}m_p/r^2)$ on the fluid element at r, if $L > L_E$ where,

$$L_E = \frac{4\pi G m_p c M_{\rm BH}}{\sigma_T} \approx 1.3\times 10^{47}\left(\frac{M_{\rm BH}}{10^9\,{\rm M}_\odot}\right)\,{\rm erg\,s^{-1}}. \quad (9.14)$$

Super-Eddington luminosities are possible, but need special models involving, for example, the electromagnetic extraction of the rotational energy of a spinning black hole. Unless such special models are involved, one can infer a characteristic black hole mass from (9.14) using the observed luminosity of the quasar. The quasars with $z > 4$ have typical luminosities of $10^{47}\,{\rm erg\,s^{-1}}$, in a universe with $\Omega = 1$ and $h = 0.5$. This requires black holes of mass $M_{\rm BH} \simeq 10^9\,{\rm M}_\odot$.

From the luminosity one can also estimate the amount of fuel that must be present to power the quasar for a lifetime t_Q. If ϵ is the efficiency with which the rest mass energy of the fuel is converted into radiation, then the fuel mass is

$$M_f = \frac{Lt_Q}{\epsilon} \approx 2\times 10^9\,{\rm M}_\odot\,(L/10^{47}\,{\rm erg\,s^{-1}})(t_Q/10^8\,{\rm yr})(\epsilon/0.1)^{-1}. \quad (9.15)$$

So if the lifetime is about 10^8 yr, and the efficiency is about ten per cent, the required fuel mass is comparable to the mass of the central black hole.

The mass estimated above corresponds to that involved in the central engine of the quasar. This mass will, in general, be a small fraction F, of the mass of the host galaxy. We can write F as a product of three factors:[16] a fraction f_b of matter in the universe which is baryonic; some fraction $f_{\rm ret}$ of the baryons originally associated with the galaxy, which was retained when the galaxy was formed (the remaining mass could be expelled via a supernova-driven wind); a fraction $f_{\rm hole}$ of the baryons retained, which participates in the collapse to form the compact central object. In standard cold dark matter models, $f_b \simeq 0.1$ for an $\Omega = 1$ universe while $f_{\rm ret}$ will depend crucially on the depth of the potential well of the galaxy[17] or, equivalently, on the circular velocity v_c; $f_{\rm ret} \simeq 0.1$ for $v_c \lesssim 100\,{\rm km\,s^{-1}}$. The quantity $f_{\rm hole}$ depends on the way the central mass accumulates and how efficiently the system can lose angular momentum and sink to the centre, and is difficult to estimate reliably. A very optimistic estimate for $F = f_b t_{\rm ret} f_{\rm hole}$ will be $F \simeq 0.01$; $F \simeq 10^{-3}$ is more likely. We then get, for the mass of the host galaxy

$$M_G = 2 \times 10^{11}\,{\rm M_\odot} \left(\frac{L}{10^{47}\,{\rm erg\,s^{-1}}}\right)\left(\frac{t_Q}{10^8\,{\rm yr}}\right)\left(\frac{\epsilon}{0.1}\right)^{-1}\left(\frac{F}{0.01}\right)^{-1}. \tag{9.16}$$

For $(F/0.01) \simeq (0.1\text{–}1)$, with the earlier value being more likely, (9.16) implies a mass for the host galaxy in the range of $(10^{11}\text{–}10^{12})\,{\rm M_\odot}$, if other dimensionless parameters in (9.15) are of order unity. Therefore the existence of quasars at $z > 4$ suggests that a reasonable number of objects with galactic masses should have formed before this redshift. We shall now examine whether this is possible.

It is straightforward to estimate the typical mass of a collapsed object at any z in the hierarchical models[18]. Consider a sphere of radius R, containing an excess density which is ν times the root-mean-square value $\sigma(M)$ where M is the average mass in a sphere of radius R. We saw in chapter 8 that $\sigma(M) \propto M^{-(3+n)/6}$. Let $\sigma(M) = (M/M_0)^{-(3+n)/6}$ where M_0 is a constant which needs to be determined. We can estimate M_0 by using the fact that (see chapter 5), the root-mean-square fluctuation in the number density of galaxies, $(\delta N/N)$, is about 0.9 at $10h^{-1}\,{\rm Mpc}$. A sphere of radius $10h^{-1}\,{\rm Mpc}$ contains – on average – the mass $M = 1.15 \times 10^{15}(h^{-1}\Omega)\,{\rm M_\odot}$. Assuming that $(\delta N/N) = b(\delta M/M) = b\sigma(M)$ where b is the biasing factor, we get,

$$\left(\frac{\delta N}{N}\right) = b\left(\frac{M}{M_0}\right)^{-(3+n)/6}. \tag{9.17}$$

Setting $(\delta N/N) = 0.9$ at $M = 1.15 \times 10^{15}(\Omega h^{-1})\,\mathrm{M}_\odot$ we can determine the value of M_0. It is easy to see that this normalization leads to

$$\begin{aligned} \sigma(M) &= \left(\frac{0.9}{b}\right)\left(\frac{M}{1.15 \times 10^{15}(\Omega h^{-1})\,\mathrm{M}_\odot}\right)^{-(3+n)/6} \\ &= \left(\frac{0.9}{b}\right)\left(\frac{M}{2.3 \times 10^{15}\Omega h_{0.5}^{-1}\,\mathrm{M}_\odot}\right)^{-(3+n)/6} . \end{aligned} \qquad (9.18)$$

In the last expression, we have scaled the mass for $h = 0.5$ which is a more realistic value. A relative mass fluctuation of $\nu\sigma(M)$ will correspond to the density contrast

$$\delta_0(M) = \nu\sigma(M) = 0.9\left(\frac{\nu}{b}\right)\left(\frac{M}{2.3 \times 10^{15}\Omega h_{0.5}^{-1}\,\mathrm{M}_\odot}\right)^{-(3+n)/6} . \qquad (9.19)$$

In a $\Omega = 1$ universe, the collapse redshift $z_{\rm coll}$ is related to δ_0 by

$$(1 + z_{\rm coll}) = \frac{\delta_0}{1.686}. \qquad (9.20)$$

Combining (9.19) and (9.20) we get the characteristic mass which collapses at a redshift z to be

$$M_c(z) = 2.3 \times 10^{15}\,\mathrm{M}_\odot \left(\frac{1.686b(1+z)}{0.9\nu}\right)^{-(6/(n+3))} h_{0.5}. \qquad (9.21)$$

From (9.21) we see that M_c drops steeply with the redshift for negative n. For example, $M_c \propto (1+z)^{-6}$ for $n = -2$, which is the relevant exponent for galactic scales. Since $b > 1$, M_c is further decreased by a factor $b^{6/(n+3)}$. We see that, for $b = 1, n = -1$, the characteristic mass is comparable to the galactic masses for $z \simeq (4$–$5)$ while for $n = -2$, this happens only at redshifts $z \simeq (1$–$2)$. So the theories in which galactic scales correspond to $n = -2$, predict very few galactic mass objects at $z > 4$. In the cold dark matter theories, the effective value of n varies with mass scale, from $n \gtrsim -3$ at small masses through $n \simeq -2$ at galactic scales to $n \simeq (-1$ to $0)$ at cluster mass scales. These models also need a biasing factor $b \simeq 2$, which further reduces M_c. Thus, in these models, galactic scale objects are rare at redshifts $z \simeq 4$; in fact, galaxies form in abundance only at a $z \lesssim 2$.

The procedure used for normalizing the spectrum plays an important role in the above considerations and deserves closer scrutiny. Let us suppose that, at the *present* epoch, the fractional excess density contrast in galaxies $\delta N/N$ is unity at some mass scale M_0. Since $\sigma(M) = (\delta M/M) = (1/b)(\delta N/N)$ (by definition of biasing parameter), one would like to set $\sigma(M_0) = (1/b)$. If $\sigma(M)$ calculated from the theory is expressed as $(1/b)(M/M_c)^{-n}$, then we will take $M_c = M_0$. This procedure which was adopted above, however, is not quite correct. The scale at which $(\delta N/N)$ is unity is *already* nonlinear and we are not quite justified in using the linear theory at this scale. It would be preferable to find some scale M_1 at which $(\delta N/N)$ is small, say 0.1, and use $\sigma(M_1) = (1/b)(\delta N/N)_{M_1} = (0.1/b)$. Unfortunately, we cannot implement this procedure because observations are quite unreliable at scales for which $(\delta N/N) < 1$.

There is, however, another way of tackling this normalization problem by using the results of nonlinear evolution based on the spherical top-hat model. To obtain $(\delta\rho/\rho) \simeq 1$ at the present epoch (at some scale) one would require a *smaller* initial density contrast in the nonlinear theory than in the linear theory. In the spherical model $\delta = [\rho/\rho_b - 1]$ is unity when $\theta \approx 2\pi/3$ (see chapter 8). For this value of θ we get the fractional excess density contrast, linearly extrapolated to the present epoch, to be $\delta_0 = 0.57$. In other words, we should actually normalize our spectrum by demanding $\sigma(M_0) = 0.57/b$ rather than by $\sigma(M_0) = (1/b)$. Another way of stating this result is as follows: Suppose we start with some density contrast δ_i at some time t_i and evolve it till the present, using *linear* theory. Suppose further that δ_i is chosen at t_i in such a way that the density contrast now is $\delta_0 = 0.57$. Then, the *same* initial contrast δ_i would have actually evolved to $\delta_0 = 1$, because of nonlinear effects. With this correct normalization the characteristic mass which collapses at any z becomes

$$M_{\rm c,nl}(z) = 1.2 \times 10^{15} h_{0.5}\, \mathrm{M}_\odot \left(\frac{1.686b(1+z)}{0.57\nu}\right)^{-(6/(n+3))} . \qquad (9.22)$$

where we have used the fact that $\delta N/N = 1$ on a scale of $8h^{-1}$ Mpc (see figure 9.2).

We see that the effect of this more precise normalization is equivalent to introducing an extra bias factor of $0.57^{-1} \simeq 2$ in the expression for M_c in (9.21). Thus $M_{\rm c,nl}(z)$ is in general smaller than M_c. For example, $M_{\rm c,nl}(z=4) = 3.7 \times 10^{11}\, \mathrm{M}_\odot$ for $n = -1$ compared with $M_c = 1.6 \times 10^{12}\, \mathrm{M}_\odot$. Thus, incorporating the effect of nonlinear evolution leads to an even smaller abundance of quasars at high z.

The low probability of existence of the host galaxies of quasars in the cold dark matter theory can be quantified[16] by using the mass function discussed in chapter 5. The mass function gives the comoving number density $f(M,z)\,dM$ of collapsed objects in a mass range dM as a function of redshift. Assuming that every halo of mass M_G (or greater) forms a quasar with lifetime t_Q, the expected number density of quasars will be

$$N_Q(z) \approx f_Q \int_{M_G}^{\infty} f(M,z)\, dM. \qquad (9.23)$$

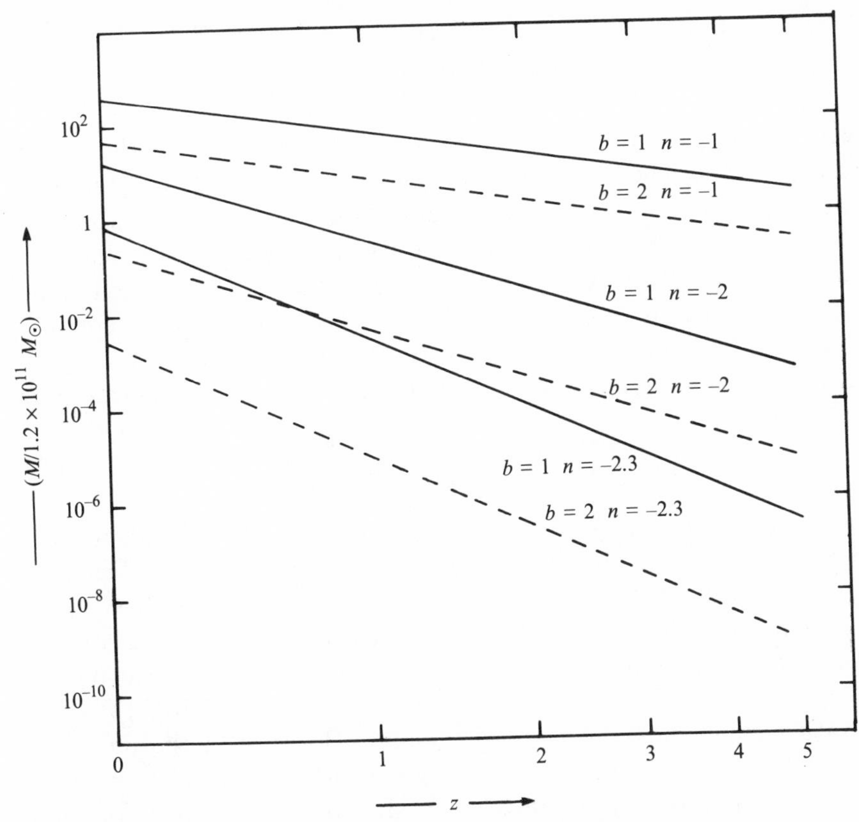

Fig. 9.2. The typical mass which becomes nonlinear at various redshifts is shown. The calculation is based on the spherical collapse model and is parametrized by the index of the power spectrum (n) and the biasing parameter (b). We have also set $\Omega = 1$ and $h = 0.5$. In the standard cold dark matter model, galactic scales correspond to $n \simeq -2, b \simeq 2$.

Here $f_Q = \min(1, t_Q/t(z))$; this factor f_Q takes into account the fact that only a fraction of order $[(t_Q/t(z)]$ of halos will display quasar activity if the quasar lifetime t_Q is smaller than the age of the universe $t(z)$ at redshift z. Using the form of $f(M,z)$ derived in chapter 5 and evaluating this integral, one can show that

$$N_Q(z) = 10^{-3} A(1+z)^{5/2} \exp[-B(1+z)^2] \text{ Mpc}^{-3}, \tag{9.24}$$

where $B \propto M_Q^{(3+n)/3}$ and $A \propto M_Q^{(n-3)/6}$. In the relevant range, the power spectrum has the index $n \simeq -2.3$. Substituting the numbers and expressing the quasar mass M_Q in terms of the luminosity, we get

$$B \simeq 0.21 \left[\left(\frac{L}{10^{47}\,\text{erg}\,\text{s}^{-1}} \right) \left(\frac{t_Q}{10^8\,\text{yr}} \right) \left(\frac{\epsilon}{0.1} \right)^{-1} \left(\frac{F}{0.01} \right)^{-1} \right]^{0.226} ; \tag{9.25}$$

$$A \simeq \left(\frac{t_Q}{10^8\,\mathrm{yr}}\right)\left[\left(\frac{L}{10^{47}\,\mathrm{erg\,s^{-1}}}\right)\left(\frac{t_Q}{10^8\,\mathrm{yr}}\right)\left(\frac{\epsilon}{0.1}\right)^{-1}\left(\frac{F}{0.01}\right)^{-1}\right]^{-0.866}. \tag{9.26}$$

Clearly, the exponential factor leads to a sharp decline in the quasar density at $z \gtrsim 2$. The observed (comoving) number density of luminous quasars with $L \gtrsim 10^{47}\,\mathrm{erg\,s^{-1}}$ at $z \simeq 2$ is about $1.2 \times 10^{-7} h_{0.5}^{-3}\,\mathrm{Mpc^{-3}}$ in a $\Omega = 1$ universe. One can estimate a critical redshift z_{crit} at which the quasar density in (9.24) will drop below this value. We mentioned earlier that the comoving number density of the brightest quasars does not show any decrease between $z \sim 2$ to $z \sim 4$. So it may be desirable for z_{crit} to be at least greater than about 4. This implies from (9.24) the condition

$$\left(\frac{t_Q}{10^8\,\mathrm{yr}}\right) \lesssim 10\left(\frac{\epsilon}{0.1}\right)\left(\frac{F}{0.01}\right). \tag{9.27}$$

The value of F depends on uncertain astrophysics but is unlikely to be greater than 10^{-3}, so that $(F/0.01) \simeq 0.1$. The radiative efficiency ϵ cannot substantially exceed 0.1. So the right hand side is unlikely to be larger than unity. The inequality (9.27) then implies that high redshift ($z \simeq 4$) quasars can only exist in sufficient numbers if the quasar lifetime t_Q is shorter than about 10^8 yr. Because of equation (9.16), this leads to a rather small mass for the host galaxy. However, if the lifetime is so short that $M_f < M_{\mathrm{BH}}$, we should not use the *fuel* mass to estimate the mass of the host galaxy; instead we should use the black hole mass M_{BH} and write $M_G \approx F M_{\mathrm{BH}}$. But note that the host galaxy will take a minimum time t_s to create a black hole of mass M, even if it accretes mass to a seed black hole at the Eddington limit. This timescale t_s is determined by setting $M_f = M_{\mathrm{BH}}$ and $L = L_E$ in our earlier expressions:

$$t_S = \epsilon t_E = \frac{\epsilon \sigma_T c}{(4\pi G m_p)} \approx 4 \times 10^7 \left(\frac{\epsilon}{0.1}\right)\mathrm{yr}. \tag{9.28}$$

(This is also the timescale in which the mass of a black hole, accreting at the Eddington limit, doubles due to the accretion.) This equation imposes a *lower* limit to t_Q, viz. that $t_Q \gtrsim t_s$. With this constraint, equation (9.27) can still be satisfied but not by a wide margin. Further if $z_{\mathrm{crit}} \sim 4$, the abundance of high luminosity quasars must decline exponentially for $z > 4$; which does not seem to be supported by recent observations[19].

The conclusions reached above also depend directly on the estimate for number of quasars, $N(L > 10^{47})$, with luminosity greater than $10^{47}\mathrm{erg\,s^{-1}}$ or, more directly, on the number of host galaxies with masses greater than about $10^{12}\,\mathrm{M_\odot}$. This value can go *up* by a factor 10 if we estimate it using low redshift samples of massive galaxies. In that case, the troubles will begin at a lower redshift of about $z \simeq 2$ or so.

Finally we should note that the above analysis assumes that the quasars can start radiating immediately after the collapse of their host galaxy. More realistically, it will take some time for the hosts to develop the compact nuclei which will power the quasars. In this case, one has to account for still higher redshifts which makes the problem worse.

It must also be emphasized that the existence of high redshift quasars is a potential problem for any theory in which galaxies form at low redshifts. For example, in the hot dark matter models superclusters collapse into pancake-like structures and subsequently fragment to form galaxies. Studies of nonlinear clustering on scales less than about 10 Mpc show that the supercluster collapse must have occurred quite recently at $z \lesssim 2$, so as to avoid excessive clustering. The existence of quasars with $z \gtrsim 4$ is hard to understand in these models as well.

The fact that quasars have high intrinsic luminosity may cause one to wonder how representative quasars will be of the general population of galaxies. But note that, if the quasar phase lasts only for a short time (compared to the Hubble time $H^{-1}(z)$ corresponding to the quasar redshift), then the number of galaxies which go through a quasar-like phase could be much larger than the number of quasars seen at that z. In that case, the quasars will be good tracers of the galaxy population at high z.

9.4 Absorption spectra of quasars

Radiation from a quasar at redshift z has to pass through the intervening matter before reaching us. The material on the path can absorb some of this radiation, thereby leaving a characteristic imprint on the (absorption) spectrum of the quasar. More interestingly, the absence of certain absorption lines can be used to constrain the density of intergalactic matter. This technique, for example, is used to put a bound on the neutral hydrogen which can exist in the intergalactic medium. This bound is obtained as follows:

We have seen in chapter 3 that neutral matter and radiation decoupled at $z \simeq 10^3$. Since it is quite unlikely that the formation of structures at lower redshifts could have been 100 per cent efficient, we would expect at least a small fraction of the neutral hydrogen to remain in the intergalactic medium with nearly uniform density. This neutral hydrogen could, in principle, be detected by examining the spectrum of a distant quasar. Neutral hydrogen absorbs Lyman-α photons, which are photons of wavelength 1216 Å whose energy corresponds to the energy difference between the ground state and the first exited state of the hydrogen atom. Because of the cosmological redshift, the photons which are absorbed will have a shorter wavelength at the source and the signature of the absorption will be seen at longer wavelengths at the observer. So the spectrum of the source should show a dip at wavelengths on the blue side (shortwards) of the Lyman-α emission line if neutral hydrogen is present between the source and the observer. The magnitude of this dip depends on the neutral hydrogen density and can be calculated using the optical depth, τ, for such absorption.[20]

The general procedure for calculating the optical depth τ has been described in section 2.5. The absorption cross section for Lyman-α photon is

$$\sigma(\nu) = \frac{\pi e^2}{m_e} f \delta(\nu - \nu_\alpha), \tag{9.29}$$

where $f = 0.416$ is a numerical constant called the 'oscillator strength' and ν_α is the frequency corresponding to the Lyman-α photon. An analysis similar to that carried out in section 2.5 will now give

$$\tau = \frac{n_H(\bar{z})}{(1+\bar{z})(1+\Omega\bar{z})^{1/2}} \frac{\pi e^2}{m_e \nu_\alpha} f \frac{1}{H_0} \approx 4.14 \times 10^{10} h^{-1} \frac{n_H(\bar{z})}{(1+\bar{z})(1+\Omega\bar{z})^{1/2}} \tag{9.30}$$

where $(1+\bar{z}) = (\nu_\alpha/\nu)$. The absorption will lead to a dip in the intensity of the quasar spectrum at wavelengths $\lambda < \lambda_\alpha$. Spectra of several quasars have been examined for such a dip. (Note that a redshift of about 2 would have brought the Lyman-α line into the visible part of the spectrum.) No such dip was seen, which led to an upper limit on the density of neutral hydrogen in the intergalactic medium. The limit on the optical depth[21] from the observations is $\tau \lesssim 0.05$ for quasars with a mean redshift of $\bar{z} \simeq 2.6$. This leads to the upper limit

$$n_H(z = 2.64) \lesssim 8.4 \times 10^{-12} h \,\mathrm{cm}^{-3}, \tag{9.31}$$

for a universe with $\Omega = 1$. This value should be compared with the baryon number density at this redshift:

$$n_b = 1.1 \times 10^{-5} (1+z)^3 \Omega_b h^2 \,\mathrm{cm}^{-3} \simeq 2 \times 10^{-4} (\Omega_b h^2) \,\mathrm{cm}^{-3}. \tag{9.32}$$

So we see that n_H is much smaller than n_b. Note that the optical depth is $\tau \simeq 10^4$ for this value of n_b; it is this large optical depth which allows us to impose such a strong bound.

The bound on n_H (called the 'Gunn–Peterson bound') implies that there is very little neutral hydrogen in the intergalactic medium. Either the process of structure formation is so efficient that the intergalactic medium is virtually empty or the hydrogen in the intergalactic medium must be fully ionized. Calculations show that 20 to 45 per cent of hydrogen remain uncollapsed at $z \simeq 4$ in the absence of reheating[22]; the uncollapsed component is actually *higher* if there is reheating. Hence the first situation is extremely unlikely and it is generally assumed that the hydrogen in the intergalactic medium must be almost completely ionized.

The Lyman-α absorption is absent[21] even in the spectra of quasars at higher redshifts of $z \simeq 4$. This raises the question of what could have caused the ionization of the intergalactic medium at these redshifts. This ionization requires significant amount of energy; to heat the gas to 10^6 K requires about 10^{60} erg per galaxy if $\Omega_B \approx 0.1$. At present there is no clear indication as to what sources can provide this much energy at $z \simeq 4$. Either the intergalactic medium has been collisionally heated to temperatures higher than about 10^6 K or a sufficient number of energetic photons have been produced by sources which have formed at higher redshifts ($z \gtrsim 4$) to completely photo-ionize the intergalactic medium. In either case, a strong constraint on the epoch of galaxy formation is implied. The heating of the intergalactic medium may occur in some scenarios of galaxy formation involving explosions. The latter possibility of photo-ionization has been extensively explored by several authors[22]. It turns out that quasars *cannot* provide enough emissivity in the form of ultraviolet photons. Other possible sources of photons, like young galaxies, primordial stars and decaying particles, also do not seem to be entirely satisfactory.

Another possibility, which arises in hierarchical models, is ionization of the intergalactic medium by high mass stars, which form in the subgalactic clumps which go nonlinear at higher redshifts[23]. From our analysis in the last section, it is clear that clumps with masses of $(10^7\text{–}10^8)\,\mathrm{M}_\odot$ can form in abundant quantities even as early as $z \simeq 10$. Even if a small fraction f (about 10^{-3}) of the baryons in these systems turned into massive stars, sufficient ultraviolet radiation would be emitted to photo-ionize the intergalactic medium. However, it is very difficult to estimate the actual value of f in a reliable manner.

The lack of Lyman-α absorption is, therefore, a mystery at present; it may contain some crucial aspect of the physics related to galaxy formation.

We shall next consider some of the absorption lines *actually* seen in the quasar spectra. The quasar absorption lines have been generally divided into several distinct categories. Of these the 'Lyman forest' lines, the 'damped Lyman-α systems' and the 'metal lines' are the most interesting in the context of galaxy formation and evolution. We begin our discussion with the Lyman forest lines.

Quasars show many narrow absorption lines at wavelengths shortward of the Lyman-α emission line. These are thought to be Lyman-α absorption lines arising in clumps of gas which exist in the path of the photon with each line being contributed by absorption in one cloud. Because these lines are so numerous they are usually referred to as the Lyman-α 'forest'. In fact, the line density in many objects is so high that it becomes difficult to find the continuum level to be used to determine any of the absorption line parameters. The parameters which can be measured in the spectra are the wavelengths, the equivalent widths and – in those cases in which the spectrum has sufficient resolution – the Doppler widths of the lines. Detailed studies have revealed several important properties of these lines.[24]

To begin with, the mean number density of lines per unit redshift, dN/dz, appears to evolve with epoch. For Lyman-α lines with 'rest' equivalent widths (that is, equivalent widths which are obtained after correcting for the redshift) higher than 0.32 Å, one finds that

$$\frac{dN}{dz} \approx k(1+z)^{\gamma}, \tag{9.33}$$

where $\gamma = 2.3 \pm 0.42$ for $1.5 \lesssim z \lesssim 3.8$. The constant k is more uncertain and is in the range (3–5). For comparison, we can calculate the expected number of absorption lines per unit redshift, assuming that the lines are due to absorption by clouds, of radius r_c and space density $n_c(z) = n_0(1+z)^3$, which exist along the line of sight to a quasar. We get

$$\begin{aligned} \frac{dN_c}{dz} &\approx \pi r_c^2 \times n_0(1+z)^3 \times \frac{c\,dt}{dz} \\ &\approx 0.02 \left(\frac{r_c}{10\,\mathrm{kpc}}\right)^2 \left(\frac{n_0}{10^{-2}\,\mathrm{Mpc}^{-3}}\right) \frac{(1+z)}{(1+\Omega z)^{1/2}}, \end{aligned} \tag{9.34}$$

where we have substituted typical numbers for r_c and n_0 corresponding to nearby galaxies. Comparing (9.33) and (9.34) we see that either the absorbing clouds have radii which are larger than 10 kpc or the clouds are more abundant than nearby galaxies. Moreover, it can be seen that the number density of the observed lines increases much more rapidly with z than our estimate, based on the assumption that the comoving number density of the absorbers is conserved. For example, if $\Omega = 1$, one expects $dN/dz \propto (1+z)^{1/2}$ with constant comoving number density of clouds, whereas observations indicate $dN/dz \propto (1+z)^{2.3}$. Any model for the Lyman-α forest has to explain this evolution.

The equivalent width and the profile of the Lyman-α absorption line can be used to derive the column density of neutral hydrogen (HI) and the Doppler parameter of the absorbing cloud. The HI column densities for the Lyman-α forest systems range from $10^{13}\,\mathrm{cm}^{-2}$ to $10^{15}\,\mathrm{cm}^{-2}$ and the column density distribution in this range is well approximated by a power law of the type

$$f(\bar{N})\,d\bar{N} \propto \bar{N}^{-\beta}\,d\bar{N}, \tag{9.35}$$

where f is the fraction of clouds with column densities between $\bar{N}$ and $\bar{N} + d\bar{N}$, and $\beta \simeq 1.75$. The lower limit of $\overline{N}_m \simeq 10^{13}\,\mathrm{cm}^{-2}$ in the applicability of the above distribution may not reflect the absence of systems with $\overline{N} < \overline{N}_m$; but may rather arise because they are below the detection threshold of the present data. The exponent β does not seem to depend strongly on the redshift.

The Doppler widths are usually quoted in terms of a parameter $b = \sqrt{2}\sigma$, where σ is the one dimensional velocity dispersion of the absorber. The Doppler parameter shows a broad peak around $b = (30\text{–}35)\,\mathrm{km\,s^{-1}}$ and the mean value shows no strong dependence on the redshift.

The clustering properties of the Lyman-α system have also been studied. If the absorbers were associated with galaxies, then one would have expected some degree of clustering. Interestingly, there is evidence for weak clustering in the velocity scales of about $150\,\mathrm{km\,s^{-1}}$ at the intermediate redshift of $z \simeq 2.5$, but at higher redshifts (about $z \sim 3.4$) the Lyman-α clouds appear to be distributed uniformly. So, there is a suggestion that clustering in the Lyman-α systems increases as the redshift decreases.

One key observation which has helped our understanding of these clouds is that the two images of the quasar Q2345+007, formed by gravitational lensing[25], contain a number of common Lyman-α absorption lines. In other words, the same cloud is intercepting the two rays forming the images. If that is the case, one can obtain a lower limit to the size of the absorbers: $r_c \simeq (5\text{–}25)\,\mathrm{kpc}$. Taking this value to be a rough estimate of the actual size and assuming a spherical shape for the absorbing clouds, we can estimate the HI density to be

$$n_{\mathrm{HI}} \simeq \frac{\bar{N}}{r_c} \simeq 3 \times 10^{-8} \left(\frac{\bar{N}}{10^{15}\,\mathrm{cm^{-2}}} \right) \left(\frac{r_c}{10\,\mathrm{kpc}} \right)^{-1} \mathrm{cm^{-3}}. \tag{9.36}$$

To make further progress one has to know the fraction of neutral hydrogen present, since clouds with $\sigma \gtrsim 10\,\mathrm{km\,s^{-1}}$ or $T \gtrsim 10^4\,\mathrm{K}$ are likely to be highly ionized. A popular assumption is that the clouds are photo-ionized, in which case the neutral hydrogen fraction depends on the ionizing flux of background photons present in the vicinity (called the metagalactic flux). This, in turn, can be estimated by noting how the number of Lyman-α lines changes as the redshift approaches that of the quasar. There is some indication that the number density of the lines above a fixed equivalent width decreases as one approaches the quasar redshift[26]. This effect, called the 'proximity effect', may be due to the increase in ionizing flux of photons from the quasar and the resulting decrease in HI column density, as one approaches the neighbourhood of the quasar. This change in the line number density as one approaches the quasar will be governed by the change in the ratio between the quasar flux and the flux of ionizing photons and can be used to estimate the background ionizing flux. Detailed study shows that[26] the background flux at the Lyman limit, $J(\nu_T)$, (the subscript 'T' is for 'threshold') is roughly constant in the redshift range $1.7 < z < 3.8$ and has a value of about $10^{-21.0\pm0.5}\,\mathrm{erg\,cm^{-2}\,s^{-1}\,Hz^{-1}\,sr^{-1}}$. The neutral hydrogen fraction can then be estimated from the equation governing the ionization equilibrium, $\alpha_H n^2 = \Gamma_H n_{\mathrm{HI}}$, where n is the total number density of the gas (assumed to be mostly hydrogen), Γ_H is the ionization rate and α_H is the recombination coefficient. Since the quasar spectrum in the relevant region falls as ν^{-1} we may take the ionizing flux $J(\nu) \propto \nu^{-1}$. Then we can write Γ_H as

$$\Gamma_H = \int J(\nu)\sigma_H(\nu)\,d\nu = J(\nu_T)\int_{\nu_T}^{\nu_{\max}} \left(\frac{\nu}{\nu_T}\right)^{-1} \sigma_H(\nu)\,d\nu \equiv J(\nu_T)G_H, \tag{9.37}$$

where σ_H is the cross section for ionization. Using[27] the values $\alpha_H = 4.36\times10^{-10}\,T^{-3/4}\,\mathrm{cm^3\,s^{-1}}$, $G_H = 3.2\times10^9\,\mathrm{erg^{-1}\,cm^2\,Hz\,sr}$, equation (9.37) and equation (9.36), we can estimate n to be

$$n \approx 4.9\times10^{-4} J_{21}^{1/2}\left(\frac{\overline{N}}{10^{15}\,\mathrm{cm^{-2}}}\right)^{1/2}\left(\frac{T}{10^4\,\mathrm{K}}\right)^{3/8}\left(\frac{r_c}{10\,\mathrm{kpc}}\right)^{-1/2}\mathrm{cm^{-3}}. \tag{9.38}$$

where $J_{21}^{1/2}$ stands for J measured in units of $10^{-21}\,\mathrm{erg\,cm^{-2}\,s^{-1}\,Hz^{-1}}$ $\mathrm{sr^{-1}}$. One can also estimate the mass of the cloud to be

$$M_c \simeq \frac{4\pi}{3} n m_p r_c^3 \approx 4.6\times10^7\,\mathrm{M_\odot}\,J_{21}^{1/2}\bar{N}_{15}^{1/2} r_{10}^{5/2} T_4^{3/8}. \tag{9.39}$$

in an obvious notation. Thus clouds with masses in the range of $(10^7\text{–}10^8)\,\mathrm{M_\odot}$ are typical.

It is important to understand how these clouds could have formed. For a cloud with the parameters determined above, the thermal energy exceeds the gravitational energy of the gas. In that case, the gas cannot be held together by the self-gravity but must be confined in some other way. Two possible confining agents have been suggested ; the pressure of a hotter intergalactic medium or the gravity of the dark matter.

The former possibility arises naturally if the clouds form through the thermal instabilities in shocks caused by either gravitational or hydrodynamical processes[28]. The shock could be either due to the explosions associated with the galaxy formation in some special scenarios or due to the pancaking which arises in the hot dark matter models. In these models we identify the Lyman-α clouds with those which are not massive enough to be gravitationally unstable but are, at the same time, of sufficient mass so that they are not evaporated by the surrounding hot intergalactic medium. The observed evolution of the number density of lines given in (9.33) is explained by using the decrease in the density and temperature of the intergalactic medium as the universe expands. The range of column densities at a fixed z is explained by invoking a range of masses for the clouds or variations in the pressure of the hot confining medium. The observed weak clustering of the lines, however, is not easy to interpret in this model. Also of potential importance is the unsuccessful search for voids in the Lyman-α forest, which indicates that the pressure of any confining medium is the same (within a factor 2) in the incipient voids as well as in clusters.

This model also constrains the properties of the intergalactic medium. For example, the temperature of the intergalactic medium can now be determined using the other parameters: The condition for pressure support $n_{\rm cloud}T_{\rm cloud} \simeq n_{\rm IGM}T_{\rm IGM}$ will imply a temperature $T_{\rm IGM} \simeq 4 \times 10^6\,{\rm K}$ if $n_{\rm IGM} \simeq 2 \times 10^{-7}(1+z)^3(\Omega_B/0.1)\,{\rm cm}^{-3} \simeq 1.3 \times 10^{-5}(\Omega_B/0.1)\,{\rm cm}^{-3}$, $n_{\rm cloud} \simeq 10^{-3}\,{\rm cm}^{-3}$ and $T_{\rm cloud} \simeq 10^4\,{\rm K}$, where we have set $z \simeq 3$.

An alternative model is based on the assumption that the clouds are gravitationally confined by 'mini' halos of cold dark matter[29]. In the models with the cold dark matter, smaller mass objects form first, cluster together and merge to form larger structures like galaxies. For the smallest masses, the potential well may be too shallow to capture any of the gas photo-ionized by the background flux, while for masses comparable to the galactic mass the gas may cool efficiently, satisfying the $\tau < 1$ criterion derived in chapter 8, and sink to the centre of the dark matter halo. There can exist dark matter halos with intermediate mass in which the captured gas may be stably confined being neither hot enough to escape nor cool enough to collapse; the heating by photo-ionization balances the cooling by radiative recombination in these clouds. Such minihalos have a virial velocity of about $30\,{\rm km\,s}^{-1}$ and characteristic masses of about $10^9\,{\rm M}_\odot$ with a ten per cent contribution from the baryons.

In this model the observed evolution of the Lyman-α forest may arise due to several factors. Radiative cooling of the gas, a decrease in the background ionizing flux with time and accretion of gas can all cause the gas to sink to the core and eventually fragment into stars. The minihalos could also get progressively destroyed if they get incorporated into larger clumps as galaxy formation proceeds. In the minihalo model, the distribution of column densities suggested by (9.35) may arise quite naturally from the fact that the lines-of-sight to the sources will have a range of impact parameters. Suppose the baryon density follows an isothermal dark halo density law with $n \propto r^{-2}$, then the condition $\alpha_H n^2 = \Gamma_H n_{\rm HI}$ gives $n_{\rm HI} \propto r^{-4}$; further, $\bar{N} \propto n_{\rm HI} r_p \propto r_p^{-3}$ where r_p is the impact parameter of the light rays. Assuming that r_p is randomly distributed we may estimate the probability for the line-of-sight to have an impact parameter in the range $(r, r+dr)$ to be proportional to $r\,dr \propto \bar{N}^{-1/3}.\bar{N}^{-4/3}\,d\bar{N}$; then we get $f(\bar{N})\,d\bar{N} \propto \bar{N}^{-5/3}\,d\bar{N}$, which is not too different from (9.35).

We now turn to another type of absorbers called the damped Lyman-α systems. These systems are potentially a very important probe of galaxies at large z, since they may contain a significant fraction of the baryons at large redshifts.

The damped Lyman-α systems were discovered in observations which were originally intended to discover the counterparts of the spiral galaxies at large redshifts. Since these discs are expected to have a large column density in neutral hydrogen, it was hoped that they would produce strong Lyman-α absorption lines which have been broadened by 'radiation damping'. (Note that, for large enough optical depth, the width of a line in determined by the *natural* line-width. Natural broadening dominates over the thermal broadening because the Lorentzian profile of the natural width decreases much more slowly than the Gaussian profile of the thermal broadening.) In a systematic study of the spectra of 68 quasars, 18 damped Lyman-α systems were discovered. These systems have the following properties[30].

The redshifts of the detected systems lie between 1.8 and 2.8. The absorbers have an average HI column density $< \bar{N} > \simeq 10^{21}\,\mathrm{cm}^{-2}$. By comparison the H_2 content is very low; studies show that the H_2 mass fraction is about 10^{-5}. In contrast, note that the H_2 mass fraction in the disc of our own galaxy is greater than about ten per cent, for lines-of-sight encountering comparable column densities of neutral hydrogen. There is also an upper limit to the dust content of these systems. A comparison of the optical continua of quasars located behind the damped systems, with a control sample, does not reveal any evidence of reddening. This suggests that the dust-to-gas ratio in the damped systems is less than half the value of that of the Milky Way. Though these absorbers lack H_2 and dust, they have metals. Low ionization states of carbon, silicon and iron (CII, SiII and FeII) are always detected while high ionization states, CIV and SiIV, are less common. The overall metal abundance is less than a tenth of the solar abundance. The velocity dispersion revealed by the metal lines, associated with the damped systems, ranges from (10 to 100) $\mathrm{km\,s^{-1}}$. On the other hand, the associated 21 cm absorption lines, in 7 of the damped systems, show the HI to be much more quiescent with a velocity dispersion which is less than $17\,\mathrm{km\,s^{-1}}$. This may be indicating that both a quiescent component (producing HI absorption) and a turbulent component (producing the metal lines) may be present in the absorber.

One of the most intriguing properties of these systems is their abundance. Unlike in the case of the Lyman-α forest, there is no evidence for redshift evolution of the damped Lyman-α systems; their redshift distribution is consistent with the absorbers having a constant comoving number density and cross section. However, the number of damped Lyman-α systems per unit redshift interval, with $\bar{N} \gtrsim 10^{20.3}\,\mathrm{cm}^{-2}$ is given by

$$\frac{dN_{\rm damp}}{dz} = 0.29 \pm 0.08; \quad < z >= 2.4, \tag{9.40}$$

where $< z >$ refers to the average redshift of these systems. In the redshift range $(0, 3)$, the average number of lines (per unit redshift) expected from the spiral galaxies is about 0.05 ± 0.03 (see, for example, equation (9.34)). So the number density of the damped systems is higher than that expected from the discs by a factor of about six. This implies that either the number density or the cross section of the absorbers is larger than that of a typical disc galaxy. The latter conclusion would suggest that the disc galaxies at high z have radii which are 3 times larger than the radii of the present day discs.

One can estimate the density contributed by these absorbers from the observed value of $dN_{\rm damp}/dz$ and using the fact they have a mean column density $< \bar{N} >= 10^{21}\,{\rm cm}^{-2}$. The mean mass density contributed by the damped Lyman-α systems at their mean redshift $< z >$ is

$$\begin{aligned} \rho_{\rm damp}(< z >) &= \mu m_p < \bar{N} > \frac{dN}{dt} \\ &= \mu m_p < \bar{N} > \frac{dN}{dz} \times H_0(1+ < z >)^2(1 + \Omega < z >)^{1/2}. \end{aligned} \tag{9.41}$$

Here $\mu = 1.4$ is the mean molecular weight of the gas and m_p is the proton mass. As the universe expands, this average mass density would have decreased by a factor $(1+ < z >)^3$. Comparing the resulting density with the critical density today, one gets the current density parameter of the HI making up the damped Lyman-α absorbers:

$$\begin{aligned} \Omega_{\rm damp} &= \mu m_p < \bar{N} > H_0 \frac{dN}{dz} \frac{(1 + \Omega < z >)^{1/2}}{\rho_c(1+ < z >)} \\ &\approx \begin{cases} 1.2h^{-1} \times 10^{-3} & \text{for } \Omega = 0.1 \\ 2.3h^{-1} \times 10^{-3} & \text{for } \Omega = 1 \end{cases}, \end{aligned} \tag{9.42}$$

where we have put in the appropriate numerical values from (9.40). For comparison, note that the mass density of the stars in disc galaxies gives an $\Omega \simeq 2h^{-1} \times 10^{-3}$. So we see that the damped Lyman-α systems may contain, at the redshift at which they are detected, a significant fraction of the baryons in the universe, almost comparable to that of luminous matter in galaxies. It is this fact which makes these absorbers such a crucial probe of the nature of protogalaxies at high z.

One possibility is that the damped Lyman-α systems arise in rotationally supported discs which are the progenitors of present day disc galaxies. However, in order to explain the number density of such systems which are observed, it is necessary to assume that these protodiscs were several times larger in the past. The main difficulty is then in understanding how one can form rotationally supported disc galaxies with radii several times the radius of present day discs. We saw in chapter 8 that, in hierarchical theories, tidal torques provide only about (5–10) per cent of the angular momentum needed for rotational support. To acquire rotational support the gas has to collapse by a factor of 10 in radius. Now, if the Lyman-α discs are rotationally supported with a radius of about 30 kpc, then this gas has to collapse from a distance of about 300 kpc. Even if the radius of the gas sphere at turn-around has this value, it would take at least 5.5 Gyr for the cloud to collapse (if $M \sim 10^{12}\,\mathrm{M}_{\odot}$). This age corresponds to the redshift of $z \simeq 1$ in a flat universe, which is too late to explain the damped Lyman-α population. One way out of this difficulty is to decrease the collapse factor; but whether this can be done in the context of rotationally supported discs is not clear.

On the other hand, if one gives up the demand for rotational support, it is much easier to produce objects with large sizes. One possibility is that the damped Lyman-α absorbers are caustics or pancakes of cool gas which have been produced from the collapse in an asymmetric protogalactic potential[31]. One can have the large sizes for these systems with modest collapse factors and reasonable timescales, since collapse has to occur only along the shortest axis of the system.

Another class of absorption lines which may probe the evolution of halos of galaxies at high z arise from the so called 'heavy element systems'. These lines are seen as narrow absorption lines at wavelengths longer than that of Lyman-α emission. The commonly encountered lines are those corresponding to magnesium (MgII), carbon (CII, CIV), silicon (SiIV) and iron (FeII). The typical neutral hydrogen column density is inferred to be about $3 \times 10^{18}\,\mathrm{cm}^{-2}$ and the Doppler parameter is in the range (5–25) $\mathrm{km\,s}^{-1}$. A number of reasons[32] seem to favour the hypothesis that the heavy element redshifts (with $z_{\mathrm{abs}} < z_{\mathrm{em}}$) arise in the intervening galaxies. The observed frequency, per unit redshift of these absorption systems, can be used to infer the required mean cross sections of galactic halos, and how they evolve, via (9.34). For example, it is found that[33], for CIV absorption systems, $< dN/dz > \simeq 2.5$ at a mean $< z > \simeq 2$ and that $dN/dz \propto (1+z)^{-1.2\pm0.7}$ in the redshift interval $1.3 < z < 3.4$. Since (9.34) gives $dN/dz \simeq 0.035$ from the galaxies with a cross section of 10 kpc and at $< z >= 2$, one deduces that a galactic halo cross section of about 85 kpc is required to explain the observed value of dN/dz. The observations and the models for the evolution of metal lines are perhaps still in a preliminary stage. But as these improve one expects the metal line systems to become a very important probe of the evolving gaseous halos of forming galaxies.

Finally the following point is worth noting: The existence of heavy element absorption systems shows that even randomly picked, low density regions of the intergalactic medium contain heavy elements (formed in stars) at quite high redshifts. This fact again suggests that a considerable amount of activity should have taken place all over the universe at these redshifts.

9.5 High redshift radio galaxies

The optical study of a large sample of strong, steep-spectrum radio sources has led to the discovery of several high redshift radio galaxies, including 4C41.17, the galaxy with the highest redshift of 3.8. At present more than 20 galaxies are known with $z > 2$, of which several have redshifts greater than 3.

The high z radio galaxies are relevant to galaxy formation in three ways: Firstly their optical properties seem to be very different from their lower z counterparts. So their study may help one to understand the physical conditions and the evolution of the high redshift galaxies, which host the radio sources. Secondly the very existence of the high z radio galaxies will constrain theories of galaxy formation, just as in the case of quasars discussed in section 9.3. Lastly, one may be able to make an independent estimate of the age and the minimum mass of these galaxies, from the flux they emit at various wavebands. This will provide additional constraints on the epoch of structure formation. We shall now discuss these features.

The high z radio galaxies are different from their counterparts at low redshifts in a number of ways. To begin with, high z radio galaxies exhibit optical emission – both lines and continuum – extended[34] over large regions, from several tens of kpc to over 200 kpc. The gas in several of these objects displays large velocity gradients (up to about $2000\,\mathrm{km\,s^{-1}}$). Finally, the region emitting in the optical band is elongated, and its major axis is preferentially aligned with the axis defined by the radio source. This is in complete contrast to what is seen in the case of radio galaxies at low redshift, where (if at all) the alignment is between the radio axis and the minor axis of the optical galaxy.

The most popular explanation[35] for the alignment effect is that the passage of radio beams has triggered a burst of star formation around the radio jet; the optical emission from these stars will be naturally aligned with the radio axis. There exists a nearby radio source, called 'Minkowski's object' which perhaps shows the best example of a starburst triggered by a radio source. Such a process may also explain the radio–optical alignment in high redshift galaxies.

There is also evidence for the infrared - radio alignments in some high z radio galaxies, including 4C41.17, the radio galaxy with the largest z. The obvious strategy will be to extrapolate the above picture and say that the infrared emission is also produced by the stars formed due to the passage of the radio jets. Since these stars have to be younger than the radio source, which itself is believed to have a lifetime $t \lesssim 10^8$ yr, the observed infrared emission has to arise from a relatively young stellar population. However this hypothesis leads to difficulties with two other sets of observations, which also bring to the forefront the question of the age of these high redshift radio galaxies. We shall now discuss these observations.

The first set of observations, which could be inconsistent with the above explanation of the infrared emission, is the observed colours of these galaxies. This can be seen by plotting the flux density against the wavelength at the epoch of emission (i.e. the wavelength corrected for the redshift). Note that the *infrared* observations of a high z galaxy give information about the the *red* to *near infrared* part of the spectrum in the 'rest frame' of the galaxy. This part of the spectrum is generally dominated by emission from the old red-giant stars. On the other hand, young massive stars generally contribute to the blue end of the spectrum. For an old elliptical galaxy in which there is very little ongoing star formation, the old stars dominate and the flux density is largest in the red part of the spectrum. So the red wavelength (in the rest frame) traces out more robustly the underlying stellar population of the galaxy. The spectral energy distribution for a number of high z radio galaxies has been determined. It turns out that these curves are generally flat in the ultraviolet and show a rise in flux in the red. If this 'red bump' is due to the existence of an old stellar population then those stars could not have been formed by the radio source. Hence one should hardly see any elongation of the infrared (which corresponds to red wavelength in the rest frame) images along the radio axis.

If the red bump is fitted with an old stellar population, then, in order to fit the flat ultraviolet part of the spectral energy distribution, one has to invoke a second burst of star formation. Detailed model building shows[36] that only about four per cent of the mass is associated with the radio activity which lasts for about 10^8 yr. In order to fit the red bump of 4C41.17, it turns out that the older stellar population must at least be 1.3 Gyr old. Since the age of the universe at $z = 3.8$ is only 1.2 Gyr (in a universe with $\Omega = 1$ and $h = 1/2$), these models have important implications for both galaxy formation and cosmology.

Unfortunately, these models are not unique and depend crucially on the evolutionary models used for the stars in the post main sequence branch. The latter is quite uncertain. For example, it is possible to invoke stellar evolution models which can explain the flat ultraviolet and the 'red bump' with young stellar populations[37]. These models use normal initial mass functions and produce the observed spectrum in less than 0.3 Gyr; the spectrum persists for about 0.6 Gyr if most of the stars are formed in less than 0.1 Gyr. If this is the case, then 4C41.17 has an age of about 0.33 Gyr and was formed at $z \simeq 4.9$ if $\Omega = 1, h = 0.5$.

There is another important observational constraint on any model which attempts to explain the infrared emission from the high z radio galaxies, viz., the infrared Hubble diagram[38]. A plot of the infrared K magnitude against the redshift, for complete samples of radio galaxies, shows a remarkably tight correlation between the magnitude and z. The K–z relation remains tight right up to the highest redshifts sampled, with a dispersion of only about 0.4 in magnitude, which remains constant over the redshift range $0 < z < 2$. The few systems known at $z > 3$ also fall roughly on this line. Any model for the high z radio galaxies must be consistent with the small scatter in this diagram and the fact that high z galaxies lie on the same line defined by galaxies at lower redshift. Note that the continuity and the small scatter in the infrared Hubble diagram exist in spite of the difference in the radio luminosity between the two samples by a factor six or so and a wide variation in the spectral energy distribution at shorter wavelengths. These features can be easily understood if the K band light arises dominantly from a uniform 'old' population of stars. In models involving young stellar populations, the contribution to the light in the K band is expected to evolve on rather short timescales, say in about 10^8 yr. It is then not a priori clear whether the small scatter and the continuity in the infrared Hubble diagram would be maintained, unless different components of stellar populations conspire together in a special manner.

The models which explain the spectral energy distributions of high z radio galaxies also predict the total luminous mass and the redshift, z_f, at which the galaxy was formed. For a cosmological model with $\Omega = 1$ and $h = 1/2$, the characteristic luminous mass[39] of these galaxies is about $3 \times 10^{11}\, \mathrm{M}_\odot$. If one takes into account the possible contribution of dark matter, the total mass of the galaxy could be an order of magnitude higher than the above value. The redshift of formation z_f depends on both the observed redshift and an estimate of the age of the galaxy from the model to account for its spectral energy distribution. For 4C41.17 at a redshift of 3.8, the luminous mass is about $5.7 \times 10^{11}\, \mathrm{M}_\odot$ and $z_f \simeq 4.9$. For lower values of Ω the z_f is lower but the masses are significantly larger. There are other models which imply that most of these radio galaxies have formed[40] well before their observed redshifts, with $z_f > (5\text{–}10)$, where the lower value corresponds to lower Ω. We can see from the above estimates that the very existence of high z radio galaxies, just as in the case of quasars, can set interesting constraints on galaxy formation theories. In particular, if many more radio galaxies are discovered at $z \gtrsim 3$, theories predicting galaxy formation at low z will be in serious trouble.

Exercises

9.1 It is possible to relate the mass, M, of a galaxy to the redshift of its formation, z_f, if we make a few assumptions regarding the detailed dynamics. This may be done in the following way: Numerical simulations suggest that $(\sigma/H_f R) \simeq 1.67$ where σ is the central velocity dispersion, R is the total radius of an isothermal halo and H_f is the Hubble constant at the time of formation. Also assume that: (i) the luminosity L and velocity dispersion σ are connected by the Faber–Jackson relation $L = L_0(\sigma/\sigma_0)^4$ where $L_0 = 1.5 \times 10^{10} h^{-2}\,\mathrm{L}_\odot$; $\sigma_0 = 225\,\mathrm{km\,s^{-1}}$ and (ii) the (M/L) ratio for the condensations is about $150h$ in solar units. Show that these relations imply $H_f^2 M^{1/2} =$ constant, where $H_f^2 = H_0^2(1+z_f)^2(1+\Omega z_f)$. Substitute the numerical values to obtain

$$\left(\frac{M}{10^{12}\,\mathrm{M}_\odot}\right)^{1/2} = \frac{315(2h)^3}{(1+z_f)^2(1+\Omega z_f)}.$$

How does this result compare with the value for the mass scale which is turning nonlinear at z_f?

10

The origin of perturbations

10.1 Introduction

The models for structure formation discussed so far are incomplete in one important aspect: they do not explain the origin of the initial inhomogeneity. This chapter discusses two different mechanisms by which the initial inhomogeneities in the universe can be generated. The first one is based on the inflationary models for the universe, in which the primordial quantum fluctuations get amplified and evolve to become classical seed perturbations. Sections 10.2 and 10.3 introduce the inflationary model and section 10.4 discusses how the perturbations are generated. The second method of producing the initial inhomogeneity is through some form of 'seeds' which could be present in the early universe. These seeds accrete matter gravitationally, leading to the formation of the structures. A prototype model which uses such a procedure is based on the cosmic strings and hot dark matter. The dynamics of the cosmic strings and the mechanism by which they produce the structures are described in section 10.5.

10.2 The concept of inflation

The models for the structure formation discussed so far were based on the growth of an initial density perturbation through gravitational instability. Such models should be considered incomplete until a physical mechanism capable of generating the initial perturbation is provided. However, there arises a basic difficulty in generating an initial perturbation in the standard Friedmann model, which has the expansion law of the form $a(t) \propto t^{1/2}$ for small t.

To understand this difficulty, let us recall some of the results of chapters 2 and 4. A perturbation with a physical scale λ_0 today would correspond to a proper length of $\lambda_0(a(t)/a_0) \propto t^n$ in the past, if we take $a(t) \propto t^n$. The characteristic expansion scale of the universe, on the other hand, is given by the Hubble radius $H^{-1}(t) = (\dot{a}/a)^{-1} = n^{-1}t$. In realistic cosmological models, $n < 1$ and hence the ratio $[\lambda(t)/H^{-1}(t)]$ increases as we go to the earlier epochs. In other words $\lambda(t)$ would have been *larger* than the Hubble radius at sufficiently high redshifts. We saw in chapter 2 that we can associate with every wavelength λ_0 the mass scale :

$$M(\lambda) = \frac{4\pi}{3}\bar{\rho}(t)\left(\frac{\lambda(t)}{2}\right)^3 = 1.5 \times 10^{11}\,\mathrm{M}_\odot\;\Omega h^2\left(\frac{\lambda}{1\,\mathrm{Mpc}}\right)^3, \qquad (10.1)$$

which remains constant during expansion of the universe (since $\bar{\rho} \propto a^{-3}$ and $\lambda \propto a$). The proper wavelength of a perturbation containing this mass will be bigger than the Hubble radius at all redshifts $z > z_{\mathrm{enter}}(M)$ where,

$$z_{\mathrm{en}}(M) \approx \begin{cases} 1.41 \times 10^5 (\Omega h^2)^{1/3} (M/10^{12}\,\mathrm{M}_\odot)^{-1/3}; \\ \qquad\qquad M < M_{\mathrm{eq}} \approx 3.2 \times 10^{14}\,\mathrm{M}_\odot\,(\Omega h^2)^{-2} \\ 1.10 \times 10^6 (\Omega h^2)^{-1/3} (M/10^{12}\,\mathrm{M}_\odot)^{-2/3}; \\ \qquad\qquad M > M_{\mathrm{eq}} \approx 3.2 \times 10^{14}\,\mathrm{M}_\odot\,(\Omega h^2)^{-2} \end{cases}. \qquad (10.2)$$

Notice that a galactic mass perturbation was bigger than the Hubble radius for redshifts larger than a moderate value of about 10^6.

This result leads to a major difficulty in conventional cosmology. Normal physical processes can act coherently only over sizes smaller than the Hubble radius. Thus any physical process leading to density perturbations at some early epoch, $t = t_i$, could only have operated at scales smaller than $H^{-1}(t_i)$. But most of the relevant astrophysical scales (corresponding to clusters, groups, galaxies, etc.) were much bigger than $H^{-1}(t)$ for reasonably early epochs. Thus, if we want the seed perturbations to have originated in the early universe, then it is difficult to understand how any physical process could have contributed to it.

To tackle this difficulty, we must arrange matters such that at sufficiently small t, $\lambda(t) < H^{-1}(t)$. If this can be done, then the physical processes can lead to an initial density perturbation. It is possible to do this if we make $a(t)$ increase rapidly (e.g. exponentially) with t for a brief period of time. Such a rapid growth is called 'inflation'.[1]

Consider a model for the universe in which the universe was radiation dominated up to, say, $t = t_i$, but expanded exponentially in the interval $t_i < t < t_f$:

$$a(t) = a_i \exp[H(t - t_i)], \qquad t_i \le t \le t_f. \qquad (10.3)$$

For $t > t_f$, the evolution is again radiation dominated $[a(t) \propto t^{1/2}]$ until $t = t_{\mathrm{eq}} \cong 5.8 \times 10^{10} (\Omega h^2)^{-2}\,\mathrm{s}$. The evolution becomes matter dominated for $t_{\mathrm{eq}} < t < t_{\mathrm{now}} = t_0$. Typical values for t_i, t_f, and H, may be taken to be

$$t_0 \approx 10^{-35}\,\mathrm{s}; \quad H \approx 10^{10}\,\mathrm{GeV}; \quad t_f \approx 70H^{-1}, \tag{10.4}$$

which give an overall 'inflation' by a factor of about $A \equiv \exp N \cong \exp(70) \approx 2.5 \times 10^{30}$ to the scale factor during the period $t_i < t < t_f$. (In this chapter we shall use the symbol H to denote the Hubble constant *during* the inflationary phase.) At $t = t_i$, the temperature of the universe is about 10^{14} GeV. During this exponential inflation, the temperature drops drastically; however the matter is expected to be reheated to the initial temperature of about 10^{14} GeV at $t \approx t_f$ by the physical processes which we will discuss below. Thus, inflation effectively changes the value of $S = T(t)a(t)$ by a factor $A = \exp(70) \approx 10^{30}$. Note that this quantity S is conserved during the non- inflationary phases of the expansion.

The most attractive feature of the inflationary model is the possibility of providing the seed perturbations which can grow to form the large scale structures. This is realized in the following manner:

During inflation, physical wavelengths grow exponentially ($\lambda \propto a \propto \exp(Ht)$) while the Hubble radius remains constant. Therefore, a given length scale has the possibility of crossing the Hubble radius *twice* in the inflationary models. Consider, for example, a wave length $\lambda_0 \simeq 2$ Mpc today which contains a mass of a typical galaxy, $1.2 \times 10^{12}(\Omega h^2)\,\mathrm{M}_\odot$. At the end of inflation, $t = t_f$, this scale would have been

$$\lambda(t_f) = \lambda_0 \frac{a(t_f)}{a(t_0)} = 2\,\mathrm{Mpc}\left(\frac{T_0}{T(t_f)}\right) \simeq 1.9 \times 10^{-2}\,\mathrm{cm}. \tag{10.5}$$

This value, of course, is much *larger* than the typical Hubble radius at that epoch, $H^{-1} \approx 2 \times 10^{-24}$ cm. But at the beginning of the inflation, this wavelength would correspond to a proper length of

$$\lambda(t_i) = \lambda(t_f)\frac{a(t_i)}{a(t_f)} = A^{-1}\lambda(t_i) = 1.8 \times 10^{-32}\,\mathrm{cm}. \tag{10.6}$$

This is much *smaller* than the Hubble radius. This is possible because the Hubble radius remains constant throughout the inflation while λ increases exponentially. In a time interval of about $\Delta t = t - t_i \simeq 18H^{-1}$, λ will grow as big as the Hubble radius during the inflationary phase. The situation is summarized[2] in figure 10.1.

We see that the scales which are astrophysically relevant today were much smaller than the Hubble radius at the onset of inflation. Hence, causal processes could have operated at these scales. During the inflation, the proper wavelength grows, and becomes equal to the Hubble radius H^{-1} at some time $t = t_{\mathrm{exit}}$. For a mode labelled by a wave vector $\mathbf{k}$, this happens at $t_{\mathrm{exit}}(k)$ where $(2\pi/k)a(t_{\mathrm{exit}}) = H^{-1}$; that is, when $(k/aH) =$

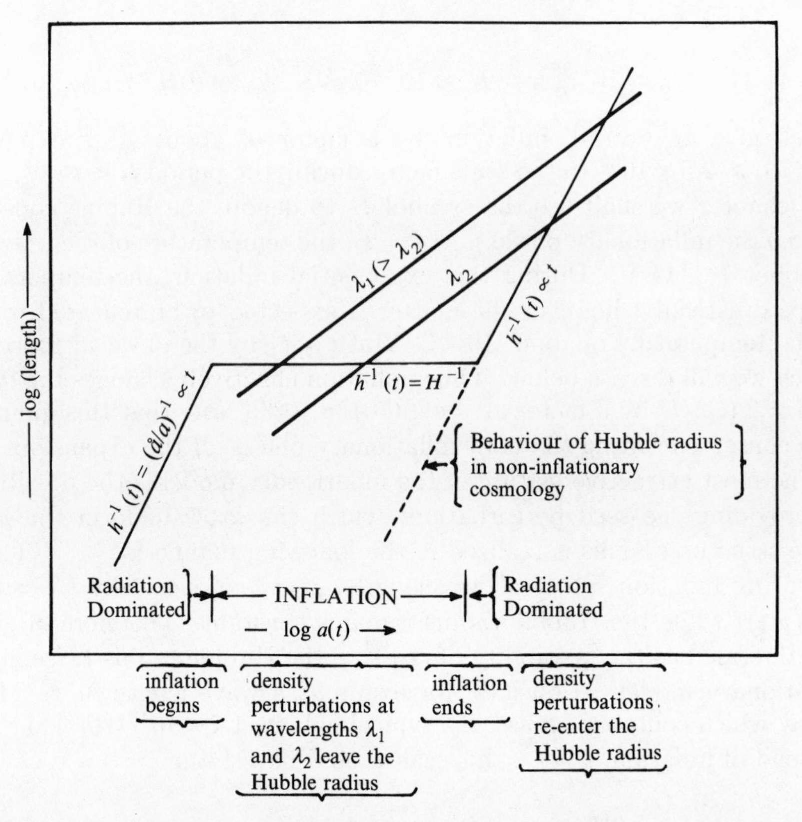

Fig. 10.1. The behaviour of the Hubble radius in an inflationary model is shown. Notice that it is possible for the wavelength of the perturbation to be smaller than the Hubble radius at two different epochs.

2π. In the radiation dominated era which occurs after the inflation, the proper length grows only as $t^{1/2}$ while the Hubble radius grows as t; Thus the Hubble radius 'catches up' with the proper wavelength at some later time $t = t_{\rm enter}(k)$. For $t > t_{\rm enter}$, this wavelength will be completely within the Hubble radius.

Thus inflationary models allow λ to be less than H^{-1} at *two* different epochs: during an early phase, $t < t_{\rm exit}(k)$ and during a late phase, $t > t_{\rm enter}(k)$. Any perturbation generated by physical processes at $t < t_{\rm exit}$ can be preserved intact during $t_{\rm exit} < t < t_{\rm enter}$ and can lead to formation of structures at $t > t_{\rm enter}$.

Since the physical processes taking place in this early epoch ($t < t_{\rm exit}$) are quantum mechanical by nature, quantum fluctuations in the matter fields can act as the seed perturbations. Therefore, in principle, we can now generate (and compute) the density inhomogeneities in the universe. Such computations will be discussed in sections 10.3 and 10.4.

The inflationary phase may also help us to understand certain other puzzling features of the Friedmann universe. We shall discuss two of them, conventionally called the 'horizon' and 'flatness' problems:

Consider the radiation dominated phase of the universe with the expansion law $a(t) \propto t^{1/2}$. The proper distance which a light signal could have travelled in the time interval $(0, t)$ is

$$R_H(t) \equiv a(t) \int_0^t \frac{dx}{a(x)} = 2t. \tag{10.7}$$

Therefore, causal communication between two observers O and O' can exist only if they are within a distance $2R_H(t) = 4t$. This boundary is called the 'particle horizon'. Two observers O and O' separated by a proper distance larger than $2R_H(t)$ at an epoch t could have never influenced each other. Hence there is no a priori reason to expect points O and O' to have similar physical environments.

If the present features of the universe were essentially determined at some early epoch – say, at $t = t_i$ when the temperature of the universe was $T \simeq 10^{14}$ GeV – then we would expect a sphere, which had a radius $2R_H\,(t_i)$ at that epoch, to have expanded to encompass the presently observed universe. This will provide a natural explanation for the observed homogeneity of our universe. From the initial epoch $T \simeq 10^{14}$ GeV to the present epoch with $T_0 = 2.75\,\mathrm{K} \approx 2.4 \times 10^{-4}$ eV the universe has expanded by a factor $(T/T_0) \simeq 4 \times 10^{26}$. However, when $T \simeq 10^{14}$ GeV the time was $t \simeq 10^{-35}$ s, and $2R_H \simeq 6 \times 10^{-25}$ cm. Thus the primordial sphere of homogeneity would have expanded only to a size of about 2.4×10^2 cm by today, a value far short of the size of the present universe. So the observed large scale isotropy of the MBR and the very large scale homogeneity of the universe cannot be explained in a natural fashion in the standard model.

The second difficulty (called the 'flatness problem') has to do with the approximate equality between the actual matter density of the present day universe and the critical density. The ratio between these two quantities $\Omega(t) = [\rho(t)/\rho_c(t)]$ could in principle have any positive value at the present epoch ($t = t_0$). Since the universe, treated as a dynamical system, is governed by a second order differential equation, one can independently specify a and $\dot{a}$ at some initial time $t = t_i$. This is equivalent to specifying the value of ρ and H at $t = t_i$, thereby fixing the initial value of Ω_i. Such a choice could, in general, lead to *widely different values* for $\rho \propto a^{-3}$ and $\rho_c \propto (\dot{a}/a)^2$ at any later epoch $t > t_i$. In other words, Ω will differ widely from unity at $t \gg t_i$. More quantitatively, we can write, at any time t, $\Omega(t) - 1 = (k/\dot{a}^2)$. Assuming $k \neq 0$, a comparison with the present epoch, $t = t_0$, gives

$$\Omega(t) - 1 = \frac{\dot{a}_o^2}{\dot{a}^2}\,(\Omega_0 - 1). \tag{10.8}$$

Expressed as a function of temperature, in the radiation dominated era, this relation becomes

$$\Omega(T) - 1 \cong 4 \times 10^{-15}(\Omega_0 - 1)\left(\frac{T}{1\,\mathrm{MeV}}\right)^{-2}. \tag{10.9}$$

The present astronomical observations imply that Ω_0 is between 0.1 to 5 (say) with liberal allowance given for systematic errors. Thus $|\Omega_0 - 1|$ is of order unity. However, the above relation shows that at earlier epochs $|\Omega - 1|$ is far less than unity. If we assume that the initial conditions for the universe were 'set' at the Planck time, when $T \simeq 10^{19}$ GeV, then the a and $\dot{a}$ terms should have been matched at that epoch to an accuracy of $|(\Omega - 1)| \simeq 10^{-60}$ so as to result in the present values for a and $\dot{a}$. Had this fine-tuning not been resorted to, the universe would have contracted back to $a = 0$ (for $k = 1$) or diffused out to $a = \infty$ (for $k = -1$) *long before* the present epoch. In the absence of any physical mechanism, this fine-tuning has to be imposed in an ad hoc manner at some early epoch.

An inflationary phase in the early universe can resolve the above difficulties. The 'flatness problem' which concerns the unusually small value of the quantity $|\Omega - 1|$, or equivalently, the curvature term (k/a^2) can be tackled in the following manner. Inflation of $a(t)$ by a factor A decreases the value of this term by a factor $A^{-2} \cong 10^{-60}$. Thus one can start with moderate values of (k/a^2) before inflation and bring it down to a very small value at $t \gtrsim 10^{-33}$ s. This solves the 'flatness problem', interpreted as the smallness of (k/a^2). Notice that no classical process can change a $k \neq 1$ universe to a $k = 0$ universe, since it involves a change in topological properties. What inflation does is to decrease the value of (k/a^2) so much that, for all practical purposes, we can ignore the dynamical effect of the curvature term, thereby having the same effect as setting $k = 0$. This has the consequence that Ω at present must be unity to a high degree of accuracy. We have mentioned on several occasions that 'inflationary models predict that $\Omega = 1$'. The above analysis shows how this result arises.

Inflation also solves the horizon problem by bringing the entire observed region of the last scattering surface (LSS) into a causally connected patch. The coordinate size of the region in the LSS from which we receive signals today is

$$l(t_0, t_{\mathrm{dec}}) = \int_{t_{\mathrm{dec}}}^{t_0} \frac{dt}{a(t)} \cong \frac{3}{a_{\mathrm{dec}}}\left(t_{\mathrm{dec}}^{2/3} t_0^{1/3}\right), \tag{10.10}$$

while the coordinate size of the horizon at $t = t_{\mathrm{dec}}$ will be

$$l(t_{\mathrm{dec}}, 0) = \int_0^{t_{\mathrm{dec}}} \frac{dt}{a(t)} \cong \frac{4t_i}{a_{\mathrm{dec}}}\left(\frac{t_{\mathrm{dec}}}{t_f}\right)^{1/2} A. \tag{10.11}$$

(We have used the facts that $t_0 \gg t_{\rm dec}$, $A \gg 1$, $t_i \simeq H^{-1}$ and $a_{\rm dec} = a_i A(t_{\rm dec}/t_f)^{1/2}$.) The ratio

$$R = \frac{l(t_{\rm dec}, 0)}{l(t_0, t_{\rm dec})} \cong 2A \frac{t_i}{(t_f t_{\rm dec})^{1/2}} \left[\frac{2}{3} \left(\frac{t_{\rm dec}}{t_0} \right)^{1/3} \right] \approx 4 \times 10^4 \left(\frac{A}{10^{30}} \right) \tag{10.12}$$

is far larger than unity for $A \simeq 10^{30}$. Thus all the signals we receive today are from a causally connected domain in the LSS. Note that, in the absence of inflation, $l(t_{\rm dec}, 0) = (2t_{\rm dec}/a_{\rm dec})$ so that $R = (2/3)(t_{\rm dec}/t_0)^{1/3} \ll 1$. This value is amplified by the factor $2At_i (t_f t_{\rm dec})^{-1/2} \approx 10^7$ in the course of inflation.

Inflation can also reduce the density of any stable, relic particle – like the magnetic monopoles – by a dilution factor $A^{-3} \approx 10^{-90}$ provided these relics were produced before the onset of inflation. Likewise, any topological discontinuities like the domain wall etc. are expanded away from one another so that the chance of observing these structures by a typical observer is negligibly small. By the same token, any relic of the early universe which survives until today (like baryon number) must be generated *after* the end of inflation.

Notice that the above conclusions only depend on the scale factor growing rapidly (by a factor 10^{30} or so) in a short time. For example, if the energy density $\epsilon(t)$ varies slowly during $t_i < t < t_f$, then the expansion is almost exponential with

$$a(t_f) = a(t_i) \exp \left[\int_{t_i}^{t_f} H(t)\, dt \right] \equiv a(t_i) \exp N, \tag{10.13}$$

where $H^2(t) = (8\pi G\epsilon(t)/3)$; this provides a more general definition of N.

To avoid possible misunderstanding, we stress the following fact: Both the 'horizon' and 'flatness' problems deal with the 'initial conditions' for our universe. Since Einstein's equations permit the values $k = -1, 0$ or $+1$ in a Friedmann solution, the value of k needs to be supplied as an extra input in classical theory. But the choice of 'initial conditions' for the universe (i.e. the physics at $t < t_{\rm Pl}$) needs to be understood quantum mechanically rather than classically. It is quite possible that quantum gravitational effects make the $k = 0$ model highly probable[3]. Inflation tries to provide a *classical* solution to an inherently *quantum gravitational* question.

This anomaly is quite striking in the horizon problem. The horizon problem exists because the integral

$$r(t) = \int_0^t \frac{dx}{a(x)} \tag{10.14}$$

is thought to be finite. However, to make such a claim we have to assume that there was a singularity at $t = 0$ and that we know the behaviour of

$a(t)$ arbitrarily close to $t = 0$. For $t < t_{\rm Pl}$, quantum gravitational effects will modify the behaviour of $a(t)$ and will most probably eliminate the singularity problem. Then, for almost all $a(t)$ (except for a class of functions of measure zero), the above integral will diverge, automatically solving the horizon problem. In other words, flatness and horizon problems owe their existence to using classical physics beyond its domain of validity. There are several quantum gravitational models in which these problems are solved as an offshoot of elimination of the singularity problem[4].

Even within the context of inflationary models, the solutions to these problems work only for a limited stretch of time[5]. For example, it can be shown that the flatness problem will resurface in the late-time behaviour of a $k \neq 0$ universe. This should be clear from the fact that, at sufficiently late times, $k = +1$ and $k = -1$ models will behave very differently irrespective of the present value of the (k/a^2) term. In the case of horizon problem, it can be easily shown that LSS will appear homogeneous only for times $t < t_{\rm crit}$ where

$$t_{\rm crit} \cong \frac{t_i^2}{t_f} A^2 = \frac{t_i^2}{t_f} \exp[2H(t_f - t_i)]. \tag{10.15}$$

(For $t_i \approx 10^{-34}$ s, $t_f \approx 100 t_i$, $H \approx 10^9$ GeV, $t_{\rm crit} \approx 3 \times 10^{23}$ s which is far larger than $t_0 \approx 3 \times 10^{17}$ s; so, right now, $t_0 < t_{\rm crit}$.) This quantity $t_{\rm crit}$ is determined once and for all by microscopic physics at $t \approx 10^{-34}$ s. If we wait long enough t will be larger than $t_{\rm crit}$ and the horizon problem will resurface. Thus inflation offers only a temporary relief – albeit for a very long time – from these problems (see exercise 10.1). In contrast, solutions based on quantum gravitational models solve these problems permanently.

10.3 The epicycles of inflation

Since the inflationary idea seems to be quite attractive, several mechanisms were devised to implement this idea. Each of these models has some advantages and disadvantages and none of them is completely satisfactory. We will briefly summarize three different models.

For the universe to expand exponentially, the energy density should remain (at least approximately) constant. Various models of inflation differ in the process by which this is achieved. In most of them the 'quasi-constant' energy density $\epsilon(t)$ is derived from a phase transition which could have taken place at $T \simeq 10^{14}$ GeV; this is the energy scale above which strong and electroweak interactions are unified in several particle physics models. Although the specific details of such models (called the grand unified theories) may differ from one another, all of them are gauge theories which contain a scalar field ϕ (called the 'Higgs field'; see appendix B). The behaviour of this Higgs field is governed by a potential $V(\phi)$ which has a minimum at some non-zero value of ϕ, say, at $\phi = \sigma$. This value σ corresponds to the true ground state of the scalar field ϕ.

The dynamics of such a scalar field gets modified when the field is interacting with a system having an equilibrium temperature T. In that case the potential energy density $V(\phi)$ of the scalar field ϕ will also depend on the ambient temperature T of the universe[6]. The ground state of the theory (which is a state with minimum energy) also changes at finite temperatures. For temperatures higher than a critical temperature T_c, the minimum of $V(\phi; T > T_c)$ is at $\phi = 0$ and *not* at the zero temperature value of $\phi = \sigma$. We may term this minimum at $\phi = 0$ the 'high temperature ground state' of ϕ. In fact, for temperatures $T \gg T_c$, the potential V has only one minimum (at $\phi = 0$) with $V(0) \approx (10^{14}\,\mathrm{GeV})^4$. As the temperature is lowered to $T \simeq T_c$, a second minimum appears at $\phi = \sigma$. For $T \ll T_c$, the $\phi = \sigma$ minimum is the 'true' minimum (i.e. $V(\sigma) \approx 0 \ll V(0)$).

Now consider what happens in the early universe as matter cools through $T \approx T_c$. When this happens, the field may not instantaneously switch over from the value $\phi = 0$ to $\phi = \sigma$. The universe can get 'stuck' at the $\phi = 0$ configuration (called the 'false vacuum'), with $V = V(0)$, even at $T < T_c$. If that happens, then the energy density of the field will be dominated by the constant term $[V(0) - V(\sigma)] \approx V(0)$ and the universe will expand exponentially. Thermal fluctuations and quantum tunnelling will eventually induce a transition from the 'false' vacuum ($\phi = 0$) to the 'true' vacuum ($\phi = \sigma$) ending the inflation in localized regions (called 'bubbles'). The phase transition is expected to be completed by the expanding 'bubbles' colliding, coalescing and reheating the matter.

Detailed analysis, however, shows that this model does not work[7]. In order to have sufficient amount of inflation, it is necessary to keep the 'false' vacuum fairly stable. In such a case the bubble nucleation rate is small and even the resulting bubbles do not coalesce together efficiently. The final configuration is very inhomogeneous and quite different from the universe we observe.

The original model was soon replaced by a version based on a very special form for $V(\phi)$ called the Coleman–Weinberg potential[8]. At zero temperature this potential is given by

$$V(\phi) = \frac{1}{2}\lambda\sigma^4 + \lambda\phi^4\left[\ln\frac{\phi^2}{\sigma^2} - \frac{1}{2}\right]; \quad \lambda \approx 10^{-3}; \quad \sigma \approx 2 \times 10^{15}\,\mathrm{GeV}. \tag{10.16}$$

This potential is extremely flat for $\phi \lesssim \sigma$ and drops rapidly to zero near $\phi \approx \sigma$. At finite temperatures, the potential acquires a small barrier

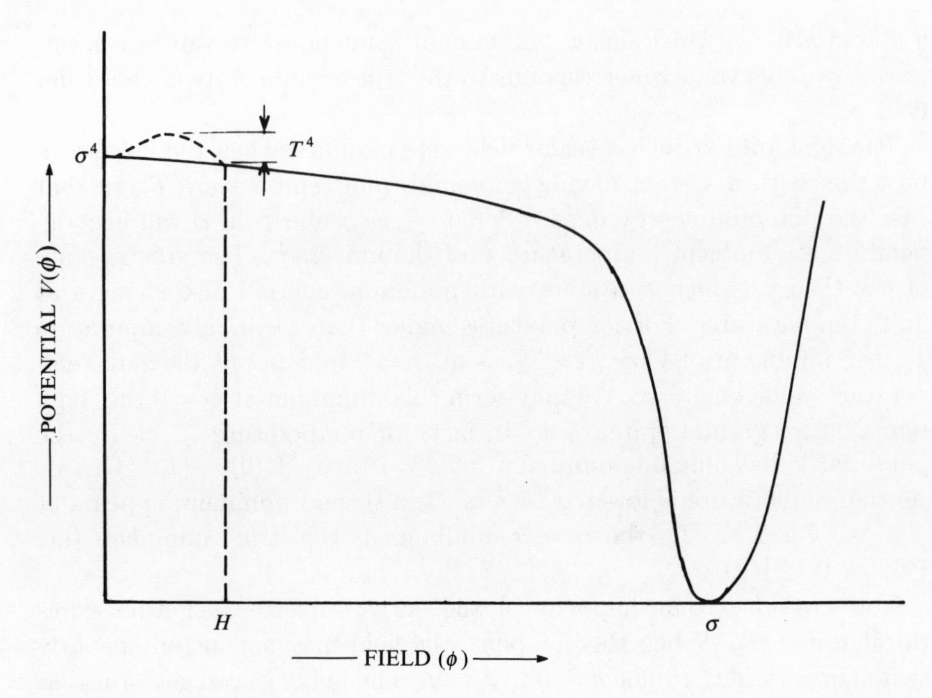

Fig. 10.2. This diagram shows the form of the Coleman–Weinberg potential. The unbroken line corresponds to the potential at zero temperature. The small bump indicated by the dotted lines (at the left end) denotes the change in the effective potential due to the finite temperature.

near the origin (at $\phi \simeq \mathcal{O}(T)$) with a height $\mathcal{O}(T^4)$, creating a local minimum at $\phi = 0$ (see figure 10.2). This 'false' vacuum, however, is quite unstable when the temperature becomes $\mathcal{O}(10^9\,\text{GeV})$. The scalar field then rapidly tunnels through the barrier to a value $\phi \approx \phi_0 \approx \mathcal{O}(H)$, and starts 'rolling down' the gently sloped potential towards $\phi = \sigma$. Since the potential is nearly flat in this region, the energy density driving the universe is approximately constant and about $V(0) \approx (3 \times 10^{14}\,\text{GeV})^4$. The evolution of the scalar field in this 'slow roll-over' phase can be described by the equation:

$$\Box\phi + V'(\phi) = \ddot{\phi} + 3H\dot{\phi} + V'(\phi) \approx 3H\dot{\phi} + V'(\phi) = 0, \tag{10.17}$$

where we have ignored the $\ddot{\phi}$ term and set $H = (4\pi\lambda G\sigma^4/3)^{1/2} \approx 2 \times 10^{10}\,\text{GeV}$. If the slow roll over lasts when ϕ varies from $\phi_{\text{start}} \simeq \mathcal{O}(H)$ to some $\phi_{\text{end}} \lesssim \mathcal{O}(\sigma)$ then

$$N \equiv \int_{t_i}^{t_f} H\,dt = H \int_{\phi_s}^{\phi_e} \frac{d\phi}{|\dot{\phi}|} \approx 3 \int_{\phi_s}^{\phi_e} \frac{H^2}{|V'(\phi)|} d\phi. \tag{10.18}$$

For the typical values of the Coleman–Weinberg potential, this number can easily be about 10^2, ensuring sufficient inflation.

As ϕ_0 approaches σ, the field 'falls down' the potential and oscillates around the minimum at $\phi = \sigma$ with the frequency $\omega^2 = V''(\sigma) \approx (2 \times 10^{14}\,\text{GeV})^2 \gg H^2$. These oscillations are damped by the decay of ϕ into other particles in some characteristic decay time, Γ^{-1}, and by the expansion of the universe. If $\Gamma^{-1} \ll H^{-1}$, the coherent field energy $[(1/2)\dot{\phi}^2 + V]$ will be converted into relativistic particles in a timescale $\Delta t_{\text{reheat}} \approx \Gamma^{-1} \ll H^{-1}$. This will allow the universe to be reheated to a temperature of about $T_{\text{reheat}} \approx \omega \approx 2 \times 10^{14}\,\text{GeV} \approx T_{\text{initial}}$. The decay width of several Coleman–Weinberg models can be about $\Gamma^{-1} \approx 10^{13}\,\text{GeV} \gg H$. This ensures good 'reheating' of the universe[9]. Since the field has already tunnelled out of the false vacuum before the onset of inflation, the problems which arose in the earlier version of the inflation are avoided. Instead of several bubbles having to collide, coalesce and make up the observed universe, we have one huge bubble encompassing the entire region which is observable today.

Though an improvement on the original version, this model is also not free from problems. It turns out that the field should start its slow 'roll over' from a value $\phi_s \approx H$ to ensure sufficient inflation. It can, however, be shown[10] that the quantum fluctuations in the scalar field have a magnitude of about $\Delta\phi \simeq (H/2\pi)$. Since $\phi_s \simeq \Delta\phi$, the entire analysis, based on semiclassical potential $V(\phi)$, is of doubtful validity. The second – and more serious – difficulty stems from the calculation of the density perturbations in this model: they turn out to be too large by a factor of about 10^6, unless the parameter λ is artificially reduced by a factor of 10^{-12} or so. We will discuss this point in detail later.

The original model for inflation was based on a strongly first order phase transition while the second model may be thought of as using a weakly first order (or even second order) phase transition. It is possible to construct inflationary scenarios in which no phase transition is involved. The idea of 'chaotic inflation', falls in this class[11]. In this model, the potential has a very simple form: $V(\phi) = \lambda\phi^4$. Inflation results because of the rather slow motion of ϕ from some initial value ϕ_0 towards the minimum. (The initial non-zero value of the ϕ_0 is supposed to arise due to 'chaotic' initial conditions). This model can also lead to sufficient inflation but suffers from two other difficulties: (i) To obtain the correct value for the density perturbation, it is necessary to fine-tune λ to very small values: $\lambda \approx 4 \times 10^{-14}$. (ii) In order for the inflation to occur, the kinetic energy of the scalar field has to be small relative to its potential energy. Detailed calculation shows that this requires the field to be uniform over sizes bigger than the Hubble radius. This is completely contrary to the original spirit of inflation.

The schemes and shortcomings discussed above are typical of several other models suggested in the literature. The most serious constraint on the inflationary scenarios arises from the study of density perturbations which we shall study in the next section. No single model for inflation, suggested so far, can be considered completely satisfactory.

10.4 Origin of density perturbations

We shall now discuss how the inflationary models can produce the seeds for density perturbations[12]. The most natural choice for the seed perturbations is the quantum fluctuations in the scalar field $\phi(t, \mathbf{x})$ driving the inflation. The computation of *classical* perturbations, generated by a *quantum* field is a difficult and technically involved issue. Several questions of principle are still unresolved in this calculation[13]. However, the main idea can be understood as follows.

During inflation, the universe was assumed to be described by a Friedmann model with small inhomogeneities. This implies that the source – which should be some *classical* scalar field $\Phi(t, \mathbf{x})$ – can be split as $[\phi_0(t) + f(t, \mathbf{x})]$ where $\phi_0(t)$ denotes the average, homogeneous part and $f(t, \mathbf{x})$ represents the space dependent, fluctuating part. Since the energy density due to a scalar field is $\rho \cong (1/2)\dot{\phi}^2$, we get,

$$\delta\rho(t, \mathbf{x}) = \rho(\mathbf{x}, t) - \overline{\rho}(t) \cong \dot{\phi}_0(t)\dot{f}(t, \mathbf{x}), \tag{10.19}$$

where $\overline{\rho}(t) = (1/2)\dot{\phi}_0^2(t)$ and we have assumed that $f \ll \phi_0$. Fourier transforming this equation, we get

$$\delta\rho(\mathbf{k}, t) \cong \dot{\phi}_0(t)\dot{Q}_{\mathbf{k}}(t), \tag{10.20}$$

with

$$f(t, \mathbf{x}) \equiv \int \frac{d^3\mathbf{k}}{(2\pi)^3} Q_{\mathbf{k}}(t) \mathrm{e}^{i\mathbf{k}\cdot\mathbf{x}}. \tag{10.21}$$

Since the average energy density during inflation is dominated by the constant term V_0, the density contrast will be

$$\delta(\mathbf{k}, t) \cong \frac{\delta\rho}{V_0} = \frac{\dot{\phi}_0(t)\dot{Q}_{\mathbf{k}}(t)}{V_0}. \tag{10.22}$$

The 'mean' value ϕ_0 and the fluctuating field $f(t, \mathbf{x})$ appearing in this equation are supposed to be some *classical* objects *mimicking* the quantum fluctuations. It is not easy to devise and justify such quantities. What is usually done is to choose some convenient quantum mechanical measure for fluctuations and *define* ϕ_0 and $Q_{\mathbf{k}}$ in terms of this quantity.

In quantum theory, the field $\hat{\phi}(t, \mathbf{x})$ and its Fourier coefficients $\hat{q}_{\mathbf{k}}(t)$ will become operators related by

$$\hat{\phi}(t, \mathbf{x}) = \int \frac{d^3\mathbf{k}}{(2\pi)^3} \hat{q}_{\mathbf{k}}(t) \mathrm{e}^{i\mathbf{k}\cdot\mathbf{x}}. \tag{10.23}$$

The quantum state of the field can be specified by giving the quantum state $\psi_{\mathbf{k}}(q_{\mathbf{k}}, t)$ of each of the modes $\hat{q}_{\mathbf{k}}$. One can think of $q_{\mathbf{k}}$ as coordinates of a particle and $\psi_{\mathbf{k}}(q_{\mathbf{k}}, t)$ as the wavefunction describing this particle. The fluctuations in $q_{\mathbf{k}}$ can be characterized by the dispersion

$$\sigma_{\mathbf{k}}^2(t) =< \psi|\hat{q}_{\mathbf{k}}^2(t)|\psi > - < \psi|\hat{q}_{\mathbf{k}}(t)|\psi >^2 =< \psi|\hat{q}_{\mathbf{k}}^2(t)|\psi > \tag{10.24}$$

in this quantum state. (The mean value of the scalar field operator $< \hat{\phi}(t, \mathbf{x}) >$ is zero in the inflationary phase. Therefore, we have set $< \hat{q}_{\mathbf{k}} >$ to zero in the above expression. Note that we are interested in only the $\mathbf{k} \neq 0$ modes). Expressing $\hat{q}_{\mathbf{k}}$ in terms of $\hat{\phi}(t, \mathbf{x})$ it is easy to see that

$$\sigma_{\mathbf{k}}^2(t) = \int d^3\mathbf{x} < \psi|\hat{\phi}(t, \mathbf{x} + \mathbf{y})\hat{\phi}(t, \mathbf{y})|\psi > \mathrm{e}^{i\mathbf{k}\cdot\mathbf{x}}. \tag{10.25}$$

In other words, the 'power spectrum' of fluctuations $\sigma_{\mathbf{k}}^2$ is related to the Fourier transform of the two-point-correlation function of the scalar field. Since $\sigma_{\mathbf{k}}^2(t)$ appears to be a good measure of quantum fluctuations, we may attempt to *define* $Q_{\mathbf{k}}(t)$ as

$$Q_{\mathbf{k}}(t) = \sigma_{\mathbf{k}}(t). \tag{10.26}$$

This is equivalent to *defining* the fluctuating classical field $f(t, \mathbf{x})$ to be

$$f(t, \mathbf{x}) \equiv \int \frac{d^3\mathbf{k}}{(2\pi)^3} \sigma_{\mathbf{k}}(t)\mathrm{e}^{i\mathbf{k}\cdot\mathbf{x}}. \tag{10.27}$$

The quantity $\phi_0(t)$ may be taken to be the classical solution to the slow-roll-over phase. For a potential which can be approximated as $V(\phi) \simeq V_0 - (\lambda/4)\phi^4$, we need to solve

$$3H\dot{\phi}_0 \simeq \lambda\phi_0^3, \tag{10.28}$$

with the boundary condition $\phi \simeq H$ at some $t = t_{\text{start}} \equiv t_s$. The $\lambda\phi^4$ approximation is valid for $H \ll \phi \lesssim \sigma$ which is the range we are concerned with; we have ignored the $\ddot{\phi}$ term because the roll over is 'slow'. The solution to this equation is

$$\phi_0^{-2}(t) \cong H^{-2}\left[1 - \frac{2}{3}\lambda H(t - t_s)\right] \tag{10.29}$$

in the initial stages. For late times, $\phi_0(t)$ should be determined by numerical integration of the equation. This leads to the result

$$\delta(\mathbf{k}, t) = \frac{\dot{\phi}_0(t)}{V_0}\dot{\sigma}_{\mathbf{k}}(t). \tag{10.30}$$

It should be stressed that the expressions chosen for $Q_{\mathbf{k}}$ and ϕ_0 are only two out of many possible choices available. Such an ambiguity cannot be avoided when semiclassical expressions have to be computed from quantum mechanical operators.

The expression derived above gives the value of $\delta(\mathbf{k}, t)$ in the inflationary phase: $t_i < t < t_f$. To compare this with observations, we need to know the value of $\delta(\mathbf{k}, t)$ at $t = t_{\text{enter}}(k)$ – that is when the perturbations enter the Hubble radius. This can be done by using the fact that, for modes which are outside the Hubble radius, the quantity $(1 + p/\rho)^{-1}\delta$ is approximately conserved (see chapter 4). This conservation relates the value $\delta(\mathbf{k}, t_{\text{enter}})$ with $\delta(\mathbf{k}, t_{\text{exit}})$, where $t_{\text{exit}}(k)$ is the time at which the relevant perturbation 'leaves' the Hubble radius in the inflationary epoch, by:

$$\frac{\delta(\mathbf{k}, t_{\text{exit}}(k))}{1 + w(t_{\text{exit}})} = \frac{\delta(\mathbf{k}, t_{\text{enter}}(\mathbf{k}))}{1 + w(t_{\text{enter}})}, \tag{10.31}$$

where $w(t)$ is the ratio between pressure $p(t)$ and density $\rho(t)$ of the background (mean) medium: $w(t) = p(t)/\rho(t)$. In the inflationary phase, dominated by the scalar field:

$$p(t) = \frac{1}{2}\dot{\phi}_0^2 - V_0; \quad \rho(t) = \frac{1}{2}\dot{\phi}_0^2 + V_0; \quad 1 + w(t) \cong \frac{\dot{\phi}_0^2}{V_0}, \tag{10.32}$$

where we have used the fact $\dot{\phi}_0^2 \ll V_0$. In the radiation dominated phase (at $t = t_{\text{enter}}$), $1 + w = 4/3$. Therefore

$$\delta(\mathbf{k}, t_{\text{enter}}) = \delta(\mathbf{k}, t_{\text{exit}}) \frac{4}{3}\left(\frac{V_0}{\dot{\phi}_0^2}\right). \tag{10.33}$$

This result may be understood, more directly, in the following manner: Let us suppose that the particular mode (corresponding to the galactic scale, say) we are interested in leaves the Hubble radius at $t = t_1$ during the inflationary phase, with the amplitude $\epsilon(t_1) \equiv \delta(\mathbf{k}, t_1(\mathbf{k}))$. We have suppressed the $\mathbf{k}$-dependence in ϵ to simplify the notation. Let $t = t_f$ be the instant at which the universe makes a transition from the inflationary phase to the radiation dominated phase. During the time $t_1 < t < t_f$, the mode is outside the Hubble radius in an inflating universe. For modes bigger than the Hubble radius, we have derived a growth law in chapter 4:

$$\left(\frac{\delta\rho}{\rho}\right) \propto \frac{1}{\rho_{\text{bg}} a^2}. \tag{10.34}$$

In the present context, ρ_{bg} =constant and we get

$$\epsilon(t) \propto \exp(-2Ht). \tag{10.35}$$

This result can also be expressed as

$$\frac{\epsilon(t_f)}{\epsilon(t_1)} = \left[\frac{H(t_1)a(t_1)}{H(t_f)a(t_f)}\right]^2, \tag{10.36}$$

since $H(t)$ is a constant during this interval. If the transition from the inflationary phase to the radiation dominated phase is approximated as instantaneous, then the radiation density ρ_R and the fluctuations in the radiation density $\delta\rho_R$ originate directly from the corresponding terms in the inflationary phase: $\rho_R \simeq \dot{\phi}^2$ and $\delta\rho$ (radiation phase) $\approx \delta\rho$ (inflationary phase). Further, since the background energy density ρ in the inflationary phase is dominated by V_0, we can write

$$\left(\frac{\delta\rho}{\rho}\right)_{\rm rad} \simeq \left(\frac{\delta\rho}{\rho}\right)_{\rm deSitter} \cdot \left(\frac{\rho_{\rm deSitter}}{\rho_{\rm rad}}\right) \simeq \left(\frac{\delta\rho}{\rho}\right)_{\rm deSitter} \cdot \left(\frac{V_0}{\dot{\phi}^2}\right). \quad (10.37)$$

During the time $t_f < t < t_{\rm enter}$, the mode is bigger than the Hubble radius and is evolving in the radiation dominated universe. We know from chapter 4 that such modes grow as $\delta \propto a^2 \propto t$, while the quantity $a(t)H(t) \propto t^{-1/2}$. Hence, at $t = t_{\rm enter} \equiv t_2$

$$\begin{aligned}\epsilon(t_2) &= \left(\frac{V_0}{\dot{\phi}^2}\right)\left(\frac{\delta\rho}{\rho}\right)_{\rm deSitter} \cdot \left[\frac{H(t_f)a(t_f)}{H(t_2)a(t_2)}\right]^2 \\ &= \epsilon(t_1)\left(\frac{V_0}{\dot{\phi}^2}\right)\left[\frac{H(t_1)a(t_1)}{H(t_f)a(t_f)}\right]^2\left[\frac{H(t_f)a(t_f)}{H(t_2)a(t_2)}\right]^2 \\ &= \epsilon(t_1)\left(\frac{V_0}{\dot{\phi}^2}\right)\left[\frac{H(t_1)a(t_1)}{H(t_2)a(t_2)}\right]^2 .\end{aligned} \quad (10.38)$$

But for a mode which is leaving the Hubble radius at t_1 and entering it again at t_2 $H(t_1)a(t_1) = H(t_2)a(t_2)$. So

$$\epsilon(t_2) = \epsilon(t_1)\left(\frac{V_0}{\dot{\phi}^2}\right). \quad (10.39)$$

More exact matching of energy densities at $t = t_f$ gives an additional factor of (4/3). On using (10.30), we get

$$\delta(\mathbf{k}, t_{\rm enter}) = \frac{4}{3}\left(\frac{\dot{\sigma}_{\mathbf{k}}}{\dot{\phi}_0}\right)_{t=t_{\rm exit}} \cong \left(\frac{\dot{\sigma}_{\mathbf{k}}}{\dot{\phi}_0}\right)_{t=t_{\rm exit}} . \quad (10.40)$$

This is the final result. Note that this expression gives the density contrast when the mode *enters* the Hubble radius even though the quantities on the right hand side should be evaluated at $t = t_{\rm exit}(k)$ for the mode labelled by k. This condition of $t = t_{\rm exit}$ implies that $(2\pi/k)a = H^{-1}$; or, rather, $(k/Ha) = 2\pi$ at $t = t_{\rm exit}$.

The problem now reduces to computing $\sigma_{\mathbf{k}}(t)$ and $\phi_0(t)$, which can be done once the potential $V(\phi)$ is known. For a Coleman–Weinberg potential, detailed calculations give the result:

$$\delta(\mathbf{k}, t_{\rm enter}) \approx \lambda^{1/2}N^{3/2}k^{-3/2} \approx 10^2 k^{-3/2}, \quad (10.41)$$

where we have taken $N \approx 50$ and $\lambda \approx 0.1$. We see that the density perturbations have the scale invariant spectrum (discussed in detail in the previous chapters) but too high an amplitude. Scales which are of astrophysical interest enter the Hubble radius during the radiation dominated era. From the bounds on the anisotropy of MBR, we know that $k^{3/2}|\delta_k| < 10^{-4}$ even at the time of decoupling. Hence the amplitude must necessarily be *smaller* at $t = t_{\rm enter}$ since $t_{\rm enter} < t_{\rm dec}$. To bring down the amplitude to the acceptable value of about 10^{-4}, we need to take the dimensionless parameter λ to be about 10^{-13}. This requires extreme fine-tuning of a dimensionless parameter, especially since we have no other motivation for such a value.

We shall now provide a simplified derivation of the above result. (The rest of the material in this section uses some elementary concepts from quantum field theory and may be omitted in the first reading.) To perform this analysis exactly, we need to use a suitable potential $V(\phi)$ for the scalar field. However the essential idea can be illustrated even by considering a toy model with constant $V(\phi) \cong V_0$. Since such a constant potential does not contribute to the quantum fluctuations, we may evaluate $\sigma_{\mathbf{k}}^2$ and other quantities using a free field theory.

Consider the action for a massless scalar field ϕ in a Friedmann universe. This field can be decomposed into a bunch of 'oscillators':

$$\begin{aligned}\mathcal{A} &= \frac{1}{2}\int d^3\mathbf{x}\, dt\, \sqrt{-g}(\phi^i\phi_i) = \frac{1}{2}\int dt\, d^3\mathbf{x}\, a^3(\dot\phi^2 - \frac{|\nabla\phi|^2}{a^2}) \\ &= \frac{1}{2}\int\frac{d^3\mathbf{k}}{(2\pi)^3}\int dt\, a^3\left\{|\dot q_{\mathbf{k}}|^2 - \frac{\mathbf{k}^2}{a^2}|q_{\mathbf{k}}|^2\right\}.\end{aligned} \tag{10.42}$$

To quantize this field, we have to quantize each independent oscillator $q_{\mathbf{k}}$. For the kth oscillator, we have the Schrödinger equation:

$$i\frac{\partial\psi_{\mathbf{k}}}{\partial t} = -\frac{1}{2a^3}\frac{\partial^2\psi_{\mathbf{k}}}{\partial q_{\mathbf{k}}^2} + \frac{1}{2}a^3\omega_{\mathbf{k}}^2 q_{\mathbf{k}}^2\psi_{\mathbf{k}}; \quad \omega = |\mathbf{k}|. \tag{10.43}$$

The quantum state of the field can be expressed by a wavefunction ψ which is the product of the $\psi_{\mathbf{k}}$ for all k. This Schrödinger equation can be solved by the ansatz:

$$\psi_{\mathbf{k}} = A_{\mathbf{k}}(t)\exp\left\{-B_{\mathbf{k}}(t)[q_{\mathbf{k}} - f_{\mathbf{k}}(t)]^2\right\}. \tag{10.44}$$

Substituting (10.44) into (10.43) and equating the coefficients of various powers of $q_{\mathbf{k}}$, we get the equations

$$i\dot B = \frac{2B^2}{a^3} - \frac{1}{2}ak^2, \tag{10.45}$$

$$i(\dot B f + B\dot f) = \frac{2B^2}{a^3}f, \tag{10.46}$$

$$i\frac{\dot{A}}{A} = i\dot{B}f^2 + 2iBf\dot{f} + \frac{B}{a^3} - \frac{2B^2f^2}{a^3}, \tag{10.47}$$

where we have suppressed the index k, for simplicity. These equations can be transformed to a simpler form by introducing a variable Q defined by the relation: $B = -(i/2)a^3(\dot{Q}/Q)$. Simple algebra then shows that

$$f = (\text{const})(a^3\dot{Q})^{-1}, \quad A = (\text{const})Q^{-1/2}\exp\left\{-\frac{i}{2}\int k^2 a f^2 dt\right\}, \tag{10.48}$$

while Q satisfies the linear equation:

$$\frac{1}{a^3}\frac{d}{dt}\left(a^3\frac{dQ}{dt}\right) + \frac{k^2}{a^2}Q = 0. \tag{10.49}$$

From (10.44) we also see that

$$|\psi_{\mathbf{k}}|^2 = N_{\mathbf{k}}\exp\left\{-\frac{(q_{\mathbf{k}} - R_{\mathbf{k}})^2}{2\sigma_{\mathbf{k}}^2}\right\}, \tag{10.50}$$

with

$$\sigma_{\mathbf{k}}^2 = \frac{1}{2}(B_{\mathbf{k}} + B_{\mathbf{k}}^*)^{-1}; \quad R_{\mathbf{k}} = \frac{B_{\mathbf{k}}f_{\mathbf{k}} + B_{\mathbf{k}}^*f_{\mathbf{k}}^*}{B_{\mathbf{k}} + B_{\mathbf{k}}^*}. \tag{10.51}$$

Equation (10.51) can be further simplified by noting that (10.49) implies the relation:

$$\frac{d}{dt}\left\{a^3(Q_{\mathbf{k}}^*\dot{Q}_{\mathbf{k}} - \dot{Q}_{\mathbf{k}}^*Q_{\mathbf{k}})\right\} = 0, \tag{10.52}$$

giving $Q_{\mathbf{k}}^*\dot{Q}_{\mathbf{k}} - \dot{Q}_{\mathbf{k}}^*Q_{\mathbf{k}} = i(\text{const})a^{-3}$. Using this result in (10.51), we get

$$\sigma_{\mathbf{k}}^2 = (\text{const})|Q_{\mathbf{k}}|^2; \quad R_{\mathbf{k}} = \text{Re}[(\text{const})\sigma_{\mathbf{k}}^2 B_{\mathbf{k}}f_{\mathbf{k}}]. \tag{10.53}$$

Thus all relevant quantities can be expressed in terms of Q. To determine $Q_{\mathbf{k}}(t)$ we have to solve (10.49) in the Friedmann universe with $a(t) = \exp(Ht)$. This can be done by transforming the independent variable from t to another time coordinate τ with $d\tau = [dt/a(t)]$. Then

$$t = -H^{-1}\ln|(1 - H\tau)|; \quad a(\tau) = (1 - H\tau)^{-1} \tag{10.54}$$

and the equation for Q becomes

$$\frac{d^2Q}{d\tau^2} - \frac{2H}{H\tau - 1}\frac{dQ}{d\tau} + k^2Q = 0. \tag{10.55}$$

Changing the variable to $x = kH^{-1}(1 - H\tau) = kH^{-1}e^{-Ht}$, (10.55) can be written as

$$\frac{d^2Q}{dx^2} - \frac{2}{x}\frac{dQ}{dx} + Q = 0. \tag{10.56}$$

The general solution to this equation is:

$$Q(x) = a(1 - ix)e^{ix} + b(1 + ix)e^{-ix}, \tag{10.57}$$

where a and b are two constants which depend on the initial conditions imposed on $Q(x)$. Since the quantum state of the oscillator $\psi_{\mathbf{k}}(q_{\mathbf{k}})$ is completely determined by the function $Q_{\mathbf{k}}(t)$, we now have a *set* of quantum states, parametrized by the constants $a_{\mathbf{k}}, b_{\mathbf{k}}$. Expectation values of physical variables will, of course, depend on the quantum state in which they are evaluated. Therefore, to proceed further, we need to make a specific choice for the quantum state of the field.

Since the minimum amount of fluctuations will arise from the 'ground state' of the system, it seems natural to choose the quantum state to be the ground state of the system. It is, however, not easy to define a ground state in the expanding background. One way of doing this is the following: We know that when $a = 1$, and $H = 0$ we should recover the ordinary, flat space, field theory. As $H \to 0$, equation (10.57) becomes

$$\begin{aligned} Q_{\text{flat}}(\tau, \mathbf{k}) = \lim_{H \to 0} \Big\{ & a\left(1 + \frac{ik}{H} - ik\tau\right) e^{i(k\tau - kH^{-1})} \\ & + b\left(1 - \frac{ik}{H} + ik\tau\right) e^{-i(k\tau - kH^{-1})} \Big\} \\ = {} & \alpha e^{ik\tau} + \beta e^{-ik\tau}, \end{aligned} \tag{10.58}$$

provided we keep $\alpha = ikaH^{-1}\exp(-ik/H)$ and $\beta = -ikbH^{-1}\exp(ik/H)$ as *finite* constants when the limit is taken. But to obtain the 'standard vacuum' state in the flat spacetime each oscillator must be described by the wavefunction

$$\begin{aligned} \psi &= N \exp\left(-\frac{i}{2}k\tau\right) \exp\left(-\frac{1}{2}kq^2\right) \\ &= (\text{const}) Q^{-1/2} \exp\left(-\frac{i}{2}k\tau\right) \exp\left(+\frac{i}{2}\frac{\dot{Q}}{Q}q^2\right). \end{aligned} \tag{10.59}$$

Therefore, we must satisfy three conditions: (i) $(\dot{Q}/Q) = ik$, (ii) $\beta = 0$ and (iii) α is independent of k. This implies that our solution for Q_k must have the form:

$$Q_k = a(1 - ix)e^{ix} = \alpha H e^{ik/H}(ik)^{-1}(1 - ix)e^{ix}. \tag{10.60}$$

(Notice that $\tau = t$ in the limit of $H \to 0$). We can now compute all the physical quantities: Direct calculation gives

$$B_k = \frac{k^3 H^{-2}}{2(1 + k^2/H^2a^2)}\left(1 - \frac{iHa}{k}\right); \quad \sigma_k^2 = \frac{1}{2}(B + B^*)^{-1} = \frac{H^2}{2k^3} + \frac{1}{2ka^2}. \tag{10.61}$$

To compute $\delta(k, t_{\text{enter}})$ we need the values of $\dot{\sigma}_k$ and $\dot{\phi}_0$ at $t = t_{\text{exit}}$. From the expression for $\sigma_k(t)$, we have

$$\begin{aligned}|\dot{\sigma}_k| &= \frac{H}{\sqrt{2}k^{3/2}}\frac{1}{2}\left(1+\frac{k^2}{H^2G^2}\right)^{-1/2}\frac{2k^2\dot{a}}{H^2a^3} \\ &= \frac{H^2}{\sqrt{2}k^{3/2}}\left(1+\frac{k^2}{H^2a^2}\right)^{-1/2}\left(\frac{k^2}{H^2a^2}\right)\end{aligned} \tag{10.62}$$

at any time t. At $t = t_{\text{exit}}$, $(k/Ha) = 2\pi$. So, at $t = t_{\text{exit}}$, we find

$$|\dot{\sigma}_k| = \frac{4\pi^2}{\sqrt{2}(1+4\pi^2)^{1/2}}\frac{H^2}{k^{3/2}} \simeq 4\left(\frac{H^2}{k^{3/2}}\right) \quad (\text{at } t = t_{\text{exit}}). \tag{10.63}$$

The quantity $\dot{\phi}_0$ (t_{exit}) can be computed directly from the classical solution to the field equation. It turns out that $\dot{\phi}_0(t_{\text{exit}}) \simeq [H^2/(\lambda^{1/2}N^{3/2})]$ where we have set $H(t_{\text{exit}} - t_{\text{start}}) = N$ and assumed $N\lambda \gg 1$. Substituting this result in the expression for δ_k we get

$$\delta(\mathbf{k}, t_{\text{enter}}) \simeq \frac{\dot{\sigma}_k}{\dot{\phi}_0} \simeq \frac{1}{k^{3/2}}\lambda^{1/2}N^{3/2}. \tag{10.64}$$

Thus the spectrum is scale invariant with $k^{3/2}|\delta_k| = \text{constant} \simeq \lambda^{1/2}N^{3/2}$. This is the result quoted above; to obtain an acceptable value (about 10^{-4}) we need to keep λ extremely small.

There is an alternative way of defining the quantity $\phi_0(t)$ which – though plagued with divergences – clearly shows the inevitability of the large amplitude. It is possible to provide an alternative definition for $\phi_0(t)$ purely from quantum theory, as the root-mean-square fluctuation in the ground state:

$$\phi_0(t) \equiv <\psi|\hat{\phi}^2(\mathbf{x}, t)|\psi> . \tag{10.65}$$

Since

$$<\psi|\hat{\phi}(t, \mathbf{x}+\mathbf{y})\hat{\phi}(t, \mathbf{y})|\psi> = \int \sigma_k^2(t)\mathrm{e}^{-i\mathbf{k}\cdot\mathbf{x}}\frac{d^3\mathbf{k}}{(2\pi)^3}, \tag{10.66}$$

we see that

$$\phi_0^2(t) = \int \sigma_k^2(t)\frac{d^3\mathbf{k}}{(2\pi)^3}. \tag{10.67}$$

So we need to compute

$$\sigma^2(0,t) = \int \frac{d^3k}{(2\pi)^3}\sigma_k^2 = \int_0^\infty \frac{k^2\,dk}{2\pi^2}\left\{\frac{H^2}{2k^3} + \frac{1}{2ka^2}\right\}, \tag{10.68}$$

which is divergent at both the limits. Such divergences arise in the field theory because of the existence of an infinite number of degrees of freedom in a field. To obtain a meaningful, finite result from this expression, we proceed as follows: First of all notice that the second term $(2ka^2)^{-1}$ exists even in flat space (i.e. even when $H = 0$) and represent the flat space vacuum fluctuations. Since this term will be present in *any* spacetime, we will drop this term. The first term, which can be written as

$$\sigma^2(0,t) = \frac{H^2}{4\pi^2}\int_0^\infty \frac{dk}{k} = \frac{H^2}{4\pi^2}\int_0^\infty \frac{d\lambda}{\lambda}, \tag{10.69}$$

is still divergent at both the limits. But notice that we are only interested in the modes which were smaller than the Hubble radius. Hence the upper limit on λ integration can be taken to be H^{-1}. Further, the semiclassical description must fail at extremely short distances. If the proper length scale at which this happens is L, then $\lambda_{\min} a(t) = L$, giving $\lambda_{\min} = (L/a)$. With these cutoffs we get

$$\sigma^2(0,t) = \frac{H^2}{4\pi^2}\int_{(L/a)}^{H^{-1}} \frac{d\lambda}{\lambda} = \frac{H^2}{4\pi^2}\ln\left(\frac{a}{HL}\right) = \frac{H^3 t}{4\pi^2} - \frac{H^2}{4\pi^2}\ln(HL). \tag{10.70}$$

We will now drop the second term because it again contributes only a constant 'zero-point-fluctuation'. Thus we are left with

$$[\sigma(0,t)]_{\text{regularized}} = \frac{H^{3/2}}{2\pi} t^{1/2}. \tag{10.71}$$

Taking $\phi_0(t) = \sigma(0,t) = (H^{3/2}/2\pi)t^{1/2}$, we find that $\dot\phi_0(t) = (H^2/4\pi)$ $(Ht)^{-1/2}$. At $t = t_{\text{exit}}$, we get, $= (H^2/4\pi)\ [\ln(k/2\pi H)]^{-1/2}$. This gives us the density contrast

$$\begin{aligned}\delta(\mathbf{k},t) &= \left(\frac{\dot\sigma_k}{\dot\phi_0}\right) \simeq \left(\frac{5H^2}{k^{3/2}}\right)\left(\frac{4\pi}{H^2}\right)\left[\ln\left(\frac{k}{2\pi H}\right)\right]^{1/2} \\ &\simeq \frac{50}{k^{3/2}}\left[\ln\left(\frac{k}{2\pi H}\right)\right]^{1/2}.\end{aligned} \tag{10.72}$$

This spectrum is scale invariant (except for a very weak $\ln k$ dependence of $k^{3/2}|\delta_k|$) but has an amplitude of about 50. This is too large by a factor of about 10^6.

This has been the most serious difficulty faced by all the realistic inflationary models: they produce too large an inhomogeneity. The qualitative reason for this result can be found from (10.40). To obtain *slow* roll-over and sufficient inflation we need to keep $\dot\phi_0$ small; this leads to an increase in the value of δ. The difficulty could have been avoided if it were possible to keep σ_k arbitrarily small; unfortunately, the inflationary phase induces a fluctuation of about $(H/2\pi)$ on any quantum field due to field theoretical reasons. This lower bound prevents us from getting acceptable values for δ unless we fine-tune the dimensionless parameters of $V(\phi)$. Several 'solutions' have been suggested in the literature to overcome this difficulty but none of them appear to be very compelling[14].

Finally, we mention a definite prediction which emerges from inflationary models[15]. It turns out that the same mechanism which produces the density inhomogeneities also produces gravitational wave perturbations (see exercise 10.3). These perturbations also have a scale invariant power spectrum and a root-mean-square amplitude of about $(H/10^{19}\,\text{GeV})$. The energy density of these gravitational waves contributes a fraction

$$\Omega_{\text{grav}} \approx 10^{-5}(H/m_P)^2 h^{-2} \tag{10.73}$$

to the critical density. Such perturbations can induce a quadrupole anisotropy in the MBR background (see chapter 6). The present bounds on the quadrupole anisotropy ($\lesssim 10^{-4}$) suggest that $H < 10^{15}$ GeV. The value of Ω_{grav} can be also restricted by the timing measurements of the millisecond pulsar; the present bound is $\Omega_{\text{grav}}(\lambda \sim 1\,\text{pc}) \leq 3 \times 10^{-7}$. A positive detection of the quadrupole anisotropy in MBR or a direct detection of relic gravitational radiation will certainly go a long way in boosting the confidence in inflation.

10.5 Cosmic strings

An entirely different mechanism for producing perturbations (and, consequently, structures) in the universe is based on certain class of objects, called 'cosmic strings'. This model has very few parameters and hence is aesthetically more appealing than the inflationary scenarios. We shall now describe the essential aspects of this model.

In a class of gauge theories, there arise solutions which are static, have finite energy and a linear structure. These solutions are stable because of topological reasons[16]. Consider, for example, a Lagrangian describing a complex scalar field interacting with a vector field

$$L[\phi, A_\alpha] = [(\partial_\alpha - igA_\alpha)\phi]^2 - \frac{1}{4}F_{\alpha\beta}F^{\alpha\beta} - \lambda[\phi^\dagger\phi - \sigma^2]^2, \tag{10.74}$$

with $F_{\alpha\beta} = \partial_\alpha A_\beta - \partial_\beta A_\alpha$. There exists a 'string like' solution to the field equations in this model which has the approximate form

$$A_\theta \cong \frac{1}{gr}[1 - \exp(-r/r_1)]^2; \quad \phi \cong \frac{\sigma}{\sqrt{2}}[1 - \exp(-r/r_2)]\exp(-\theta), \tag{10.75}$$

where r_1 and r_2 are of the same order as σ^{-1}. For this solution, the energy density is $\mu \simeq \sigma^2$ (i.e. energy per unit length) and size is $\omega \simeq \lambda^{1/2}\sigma^{-1}$. In the grand unified models, $\sigma \simeq 10^{16}$ GeV and $\lambda^{1/2} \simeq \mathcal{O}(1)$. Hence the thickness of the string is quite small and we can think of the string as a linear concentration of energy. Thus the string is characterized by a single dimensionless number $G\mu \simeq 10^{-6}\ (\sigma/10^{16}\,\text{GeV})^2$.

The stress-tensor for the string can be worked out from the above string solution[17]. It has the form $T^{\alpha\beta} = \mu\delta(x)\delta(y)$ dia $(1,0,0,-1)$ if we take the z-axis to be the direction of the string. (This expression shows that these strings have tension in z-axis, in addition to the energy density; in contrast, an 'ordinary string' will have a stress-tensor of the form $T^{\alpha\beta} =$ dia $(\rho,0,0,0)$.) The metric produced by such a stress tensor has a remarkable form. For $G\mu \ll 1$ the line interval is:

$$ds^2 = dt^2 - dr^2 - dz^2 - (1-4G\mu)r^2\, d\theta^2. \tag{10.76}$$

This space is flat; however, the range of θ is now only $0 \le \theta \le 2\pi(1-4G\mu)$ rather than the usual $0 \le \theta \le 2\pi$. The plane perpendicular to the string can be thought of as a flat, xy-plane with a wedge of angle $8\pi G\mu$ removed and the edges identified. Such a 'deficit angle' leads to three important consequences:[18]

(1) Consider a distant source (like a quasar) and an observer, both of which are located in a plane perpendicular to the string. Since the conical wedges are identified, the observer will see two images of the quasar with the angular separation $\Delta\theta \simeq 8\pi G\mu\ [l/(l+d)]$ where l is the distance between the string and the quasar and d is the distance between the string and the observer. Thus a series of 'double galaxies' or 'double quasars' will be a tell-tale sign for strings.

(2) Consider a string moving with a velocity v transverse to the line of sight. If the string was stationary, every point on the last scattering surface will be 'double-lensed' due to the effect discussed above. Since it happens to all points, the effects will be washed out. But if the string has a transverse velocity, v, the momentum vectors of the photons forming the two images will pick up small transverse components. This Doppler-like shift will lead to the temperature anisotropy $(\Delta T/T) \simeq 8\pi G\mu v$. Since $(\Delta T/T) \lesssim 10^{-5}$ we are immediately led to the bound $(G\mu) \lesssim 10^{-5}$.

(3) The most important consequence for structure formation is the following: Consider a long string moving through the universe with velocity v. As it moves past the particles, the particles are deflected transversely and will acquire a 'wake' velocity $u \simeq 4\pi G\mu v$ and will condense on the plane defined by the motion of the string. We will thus obtain planar regions (with $(\delta\rho/\rho) \simeq 1$) having an opening angle of $(8\pi G\mu)$ and thickness of about vH^{-1}. The mass contained in this wake can be a fraction $(8\pi G\mu v^3)$ of the total mass contained inside the Hubble radius and hence could be quite large.

The discussion so far was based on long, linear strings. It is also possible to have loops of cosmic strings with different radii. They are essentially produced by the interaction of strings, in the following manner: Numerical work suggests that[19] when two strings intersect, the further evolution proceeds as though the strings have exchanged ends. Hence, a self-intersecting infinite string will split off a closed loop. Therefore, in a realistic situation, we have to deal with both long strings and string loops.

The string loops are not stable because of the following reason[20]: Closed loops of radius R oscillate with period $\tau \simeq R$ and thus radiate gravitational waves. Since the quadrupole moment of a loop of radius R is $Q \simeq \mu R^3$ and the period of oscillation is $\tau \simeq R$, the power radiated as gravitational waves is about $P \simeq G\mu^2$; a more precise calculation gives $P = \gamma(G\mu)\mu$ with $\gamma \simeq 50$. The time taken by the string loop to decay through this process is $\tau_{\rm GW} \simeq (\mu R/P) \simeq (\gamma G\mu)^{-1} R$. Notice that small loops disappear faster. It can be easily shown (see exercise 10.4) that the energy density of the gravitational waves produced in this manner is $\rho_{\rm GW} \cong (32\pi/3)\,(G\mu/\gamma)^{1/2}\rho_R$ where ρ_R is the energy density of radiation.

So far we have considered strings as isolated entities. To work out the consequences for structure formation, we have to study the production and evolution of strings in an expanding universe, which we shall do next.

Strings are produced when the universe cools through the temperature $T_c \simeq \sigma$. In the ensuing phase transition, the field can 'get stuck' in the string configuration in regions of size $H^{-1}(T_c)$. As the universe expands, the strings evolve according to their equation of motion with the decay also taken into account (see exercise 10.5). Numerical experiments with the network of long evolving strings suggest that the energy density of long strings rapidly approaches that of a scaling solution[21], in the form:

$$\rho_s \simeq \nu \left(\frac{\mu}{t^2}\right) \quad \text{(long strings)} \tag{10.77}$$

where ν is a numerical coefficient of the order of 30 or so (see exercise 10.6).

The situation is much more complicated as regards string *loops.* Since ρ_s for long strings has a scale invariant form, we would have expected the loops also to have a similar spectrum. However, numerical simulations do not give a clear verdict on this point[22]. If $R_i(t)$ denotes the initial radius of a loop formed at time t, then the numerical simulations can provide us with a distribution of loops $n(R_i)\,dR_i$ giving the number of loops with radius between R_i and $R_i + dR_i$ at the time of production. Unfortunately the different simulations do not agree on the form of the 'radius spectrum' $n(R_i)$; however, they all seem to agree that the *mean* radius R_i of loops formed at time t is $\overline{R}_i \simeq \alpha t$ with $\alpha \simeq 10^{-2}$.

If we assume, for the sake of simplicity, that (i) a scaling solution does exist and (ii) the radius of strings produced at t is $R = \alpha t$, then $n(R,t)$ can be derived from the scaling solution for long strings (see exercise 10.7). We find that

$$n(R_i(t), t) = \frac{1}{2}\beta^{-1}\alpha^{-2}\nu t^{-4}. \tag{10.78}$$

At a later time, the number density will be

$$n(R,t) = \left[\frac{1+z(t)}{1+z[t_f(R)]}\right]^3 n(R, t_f(R)), \tag{10.79}$$

where $t_f(R)$ is the time when loops of radius R are formed. Clearly,

$$n(R,t) \propto R^{-4} z(R)^{-3}, \tag{10.80}$$

where $z(R)$ is the redshift at $t = R$. Using the cosmological model to find $z(R)$, we can write this result as

$$n(R,t) \simeq \begin{cases} R^{-5/2} t^{-3/2} & t < t_{\rm eq} \\ R^{-5/2} t_{\rm eq}^{1/2} t^{-2} & t > t_{\rm eq} > t_f(R) \\ R^{-2} t^{-2} & t_f(R) > t_{\rm eq} \end{cases} . \tag{10.81}$$

Decay of small loops due to gravitational radiation will give a lower cutoff to R at $R_{\rm min} \simeq \gamma G\mu t$; so we must add to the above scalings the result $n(R,t) = n(\gamma G\mu t, t)$ for $R < \gamma G\mu t$.

In forming structures, we can use these strings as seeds. This can be achieved through two different processes[23]. First of all, a string loop will accrete matter around it due to gravitational attraction. (For this purpose a loop of radius R is essentially equivalent to a mass of $M \simeq 2\pi\mu R$; see exercise 10.8.) Secondly, a long moving string produces a planar wake of overdensity behind it. Numerical simulations suggest that the second process is dominant over the first.

It should be stressed that the string-induced structure formation is conceptually quite different from the 'standard' picture of structure formation we have been studying so far. In the latter, the density field of the universe was taken to be a Gaussian random variable. The power spectrum was fixed at the initial instant by some physical process like, say, inflation. The process in which cosmic strings act as gravitational seeds is distinctly non-Gaussian. At some initial instant, we have a density contrast which is essentially a series of 'spikes'. Structures grow because of the gravitational influence of the seeds on the surrounding matter.

It is easy to understand that the wakes produced by the moving strings will lead to characteristic scale which is $l \simeq t_{\rm eq}$. To see this, notice that the wakes produced at $t < t_{\rm eq}$ will suffer damping between the time of formation and $t_{\rm eq}$. The perturbations which are formed at $t \geq t_i$ will grow according to the linear growth law:

$$\left(\frac{\delta\rho}{\rho}\right)(t, t_i) \simeq \left(\frac{t}{t_i}\right)^{2/3} \left(\frac{\delta\rho}{\rho}\right)(t_i). \tag{10.82}$$

Since the initial form of the wake perturbation is independent of t_i, it follows from this equation that the most dominant wakes are the ones produced at $t_i = t_{\rm eq}$. Since the characteristic size of wakes produced at time t is about $H^{-1}(t)$, we expect planar structures of the size $t_{\rm eq} \simeq$ 26 Mpc if $h = 0.5$.

The actual power spectrum produced at a later instant depends on the nature of the dark matter. Since cold dark matter has no characteristic scale associated with it, the only relevant scale in the problem is $t_{\rm eq}$, set by the strings. The accretion is, therefore, quite effective at all scales starting from the small scales. This leads[23] to a density profile around the loops which is somewhat steep: $\rho(r) \propto r^{-9/4}$. The situation is much better in models with hot dark matter because such models have a characteristic free streaming scale, comparable to $t_{\rm eq}$. Because of the free streaming, the accretion is suppressed at small scales. The density profile around each loop is now less steep and has an acceptable $\rho \propto r^{-2}$ profile. Also notice that the cosmic strings are unaffected by the free streaming of hot dark matter. Even though free streaming wipes out the power at small scales, strings can regenerate it at these scales. This process produces sufficient power at galactic scales even in a hot dark matter scenario. Larger, cluster size structures are formed as in the usual scenario.

It is also necessary to ensure that these perturbations grow to nonlinearity before the present epoch. This requires the strings to have some 'minimum' density; the calculations[23] suggest that this condition is satisfied provided $(G\mu) \gtrsim 5 \times 10^{-7}$. We saw before that the constraint from MBR anisotropy translates to the bound $G\mu \lesssim 10^{-5}$. Thus the parameter $G\mu$ is fairly well constrained in these scenarios.

Exercises

10.1 (a) Show that inflation can keep the observed universe homogeneous only for a time $t < t_{crit}$ where

$$t_{crit} \simeq \frac{t_i^2}{t_f} \exp 2H(t_f - t_i)$$

and t_i and t_f are the times at which the inflation started and ended.
(b) Consider a $k = 1$ universe which went through an inflationary phase. Since such a model is expected to recollapse eventually, observations will reveal that $\Omega \neq 1$ at some stage; say, at $t > T$. Estimate T.

10.2 The perturbation equations describing gravitational waves were studied in the exercises to chapter 4. These equations can be used to describe the generation of gravitation in inflationary models.
(a) Show that each of the two, transverse, traceless modes of the gravitational wave can be described by a scalar function $\phi(t_1, \mathbf{x})$ which satisfies the wave equation in the Friedmann universe. Scale the amplitude of h_{ik} such that $h_{ik} = \sqrt{16\pi G\phi}\ e_{ik}$ where e_{ik} is the unit tensor describing the polarization of the waves.

(b) Argue that the power in the wave modes is

$$P_{\rm grav}(k) \cong \frac{k^3|h_k|^2}{2\pi^2} = \frac{4}{\pi}\left(\frac{H}{m_{pl}}\right)^2,$$

where h_k is a typical Fourier amplitude at wave vector k. When this mode enters the Hubble radius, show that it carries the energy density

$$\frac{k}{\rho}\frac{d\rho_{\rm grav}}{dk} \simeq \frac{4}{3\pi}\left(\frac{H}{m_{pl}}\right)^2 \quad (\text{at } t = t_{\rm enter})$$

where ρ is the energy density of the background.

(c) Once the mode has entered the Hubble radius its energy density decreases as a^{-4}. Let $\Omega_{\rm grav}(k) = [k(d\rho_{\rm grav}/dk)/\rho_c]$ evaluated today. Show that

$$\Omega_{\rm grav}(k)h^2 \simeq 10^{-4}\left(\frac{M}{m_{pl}}\right)^3,$$

where h is the scaled Hubble constant and M is the energy scale at which inflation took place. What will be the spectrum of these waves?

10.3 For $t < t_{eq}$, *assuming* that a scaling solution exists, we can write the number density of loops as

$$n_{\rm loop}(E,t)dE \simeq \alpha\left(\frac{\mu t}{E}\right)^{3/2}\frac{dE}{Et^3},$$

where $E = 2\pi R\mu$ is the energy in the loop.

(a) Calculate the total energy density in the loops by evaluating

$$\rho_{\rm loop} = \int_{E_1}^{E_2} E n_{\rm loop}\, dE \simeq 3\alpha(\gamma G\mu)^{-1/2}(\mu t^{-2}),$$

where $E_1 \cong \gamma G\mu^2 t$ and $E_2 \simeq \mu t$. Why do we have these cutoffs?

(b) The loops which decay produce gravitational waves with energy $\rho_{\rm GW}(t)$. Show that the conservation of energy implies

$$\dot\rho_{\rm GW} + 4\left(\frac{\dot a}{a}\right)\rho_{\rm GW} \cong \left(\frac{\mu}{G\gamma}\right)^{1/2}\frac{1}{t^3}.$$

Solve this equation and show that

$$\left(\frac{\rho_{\rm GW}}{\rho_R}\right) \simeq \frac{32\pi}{3}\left(\frac{G\mu}{\gamma}\right)^{1/2}\ln\left(\frac{t}{t_i}\right),$$

where $t_i \simeq (\gamma G\mu)^{-1}t_{\rm form}$ with $t_{\rm form}$ being the time of formation of the string network.

10.4 The purpose of this exercise is to derive the action for a cosmic string starting from the field theory (see ref. 24). Let

$$\mathcal{A} = \int d^4y\, \mathcal{L}[\phi(y)]$$

be the action for a set of fields which does have a 'linear topological defect' as a solution. Such a defect is identified with the string. A propagating string can be expressed by the functions $X^\mu(\tau, \sigma)$ where τ is the time parameter and – at any given τ – $X^\mu(\sigma, \tau = \text{constant})$ defines a curve which is identified as the string. In flat spacetime, we can take $\tau = t$. Let w be the width of the string and R its radius of curvature.
(a) Change the coordinates from y^μ $(\mu = 0, 1, 2, 3)$ to a new set $\sigma^a = (\tau, \sigma, \rho^1, \rho^2)$ (where (ρ^1, ρ^2) are in the normal plane to $X^\mu(\sigma, \tau)$) via the equation

$$y^\mu(\sigma^a) = X^\mu(\sigma, \tau) + \rho^i n_i^\mu(\sigma, \tau)$$

where $i = 1, 2$ and n_i^μ are basis vectors in the normal plane to the string world sheet. Show that the volume element d^4y transforms to

$$d\mathcal{V} = d\sigma \, d\tau \, d\rho^1 \, d\rho^2 (\det M_a^\mu)$$

with

$$M_a^\mu = \frac{\partial y^\mu}{\partial \sigma^a} = \begin{pmatrix} \partial X^\mu / \partial(\sigma, \tau) \\ n_i^\nu \end{pmatrix} + \mathcal{O}(\rho).$$

(b) Evaluate the determinant using the fact

$$\det M_a^\mu = (-\det \eta_{\mu\nu} M_a^\mu M_b^\nu)^{1/2} = (-\det D_{ab})^{1/2},$$

where

$$D = \begin{pmatrix} \frac{\partial X^\mu}{\partial(\sigma,\tau)} \frac{\partial X^\nu}{\partial(\sigma,\tau)} \eta_{\mu\nu} & \frac{\partial X^\mu}{\partial(\sigma,\tau)} n_b^\nu \eta_{\mu\nu} \\ \frac{\partial X^\mu}{\partial(\sigma,\tau)} n_a^\nu \eta_{\mu\nu} & n_a^\mu n_b^\nu \eta_{\mu\nu}. \end{pmatrix}$$

Show that

$$D \simeq \begin{pmatrix} X_{,a}^\mu X_{,b}^\nu \eta_{\mu\nu} & 0 \\ 0 & \delta_{ab} \end{pmatrix} + \mathcal{O}\left(\frac{w}{R}\right).$$

(c) Hence show that the action can be written as

$$\begin{aligned} \mathcal{A} &= \int d\sigma \, d\tau (-\det g_{ab})^{1/2} \int d\rho^1 \, d\rho^2 \, \mathcal{L}[y(\sigma^a)] + \mathcal{O}\left(\frac{w}{R}\right) \\ &= -\mu \int d\sigma \, d\tau \, (-\det g_{ab})^{1/2} + \mathcal{O}\left(\frac{w}{R}\right), \end{aligned}$$

where $g_{ab} \equiv X_{,a}^\mu \, X_{,b}^\mu \eta_{\mu\nu}$ and μ is the energy density per unit length of the string. Vary this action, in flat spacetime, and obtain the equation of motion for the string

$$\frac{\partial^2 \mathbf{x}}{\partial \tau^2} - \frac{\partial^2 \mathbf{x}}{\partial \sigma^2} = 0.$$

10.5 There is a dynamical reason behind the existence of a scaling solution for long cosmic strings (see ref. 21). To see this, proceed as follows:
(a) Let $\nu(t)$ be the mean number of long strings per Hubble volume ($\propto t^3$). Show that the energy density from these strings is $\rho_{\text{long}}(t) \cong \mu\nu(t)t^{-2}$.

(b) Loops are produced during self intersections of strings. Show that

$$\frac{d}{dt}n_{\rm loop} \cong c\nu^2(t)t^{-4},$$

where $n_{\rm loop}(t)$ is the number density of loops and c is a constant of order unity. Show also that the conservation of energy of the strings implies

$$\frac{d}{dt}\rho_{\rm long}(t) + \frac{3}{2t}\rho_{\rm long}(t) + c'\mu t\frac{dn_{\rm loop}}{dt} = 0,$$

where c' is another constant.
(c) Convert this equation to the form

$$\dot{\nu} - \frac{\nu}{2t} = -cc'\nu^2 t^{-1}.$$

Argue why this equation will have a stable solution with ν of order unity, hence demonstrating the existence of the scaling solution. When does the above analysis break down?

10.6 Assuming that the energy density in long strings – in as much as it is not redshifted – must go into string loops and that about one loop of radius (αt) is produced per expansion time per Hubble volume, work out the number density of loops $n(R,t)$.

10.7 Cosmic strings are an example of 'seeds' which accrete matter gravitationally thereby producing structures. In this exercise, we will study the influence of a general distribution of seeds on collisionless matter.
(a) Consider a collisionless distribution of particles, described by $f(\mathbf{x},\mathbf{q},t)$ where $\mathbf{x}$ is the comoving coordinates and $\mathbf{q}$ is the corresponding momentum. Show that the correct form of collisionless equation, in the Newtonian limit, is

$$\frac{\partial f}{\partial t} + \frac{\mathbf{q}}{ma^2}\cdot\frac{\partial f}{\partial \mathbf{x}} - m\frac{\partial \phi}{\partial \mathbf{x}}\cdot\frac{\partial f}{\partial \mathbf{q}} - m\ddot{a}a\mathbf{x}\cdot\frac{\partial f}{\partial \mathbf{q}} = 0.$$

(b) Expand the solution in the form $\phi = \phi_0 + \phi_1, f = f_0 + \delta f$ with

$$\frac{\partial \phi_0}{\partial \mathbf{x}} = \frac{4\pi G}{3}\rho_b a^2\mathbf{x}; \quad \nabla^2\phi_1 = 4\pi G a^2(\delta\rho + \delta\rho_{\rm ext}),$$

where $f_0 = f_0(\mathbf{q},t)$ and $\delta\rho_{\rm ext}(\mathbf{x},t)$ is an externally specified mass density of 'seeds'. Show that the spatial Fourier transform $Q_{\mathbf{k}}(\mathbf{q},t)$ of $\delta f(\mathbf{x},\mathbf{q},t)$ satisfies the equation

$$\frac{\partial Q_{\mathbf{k}}}{\partial t} + \frac{i\mathbf{k}\cdot\mathbf{q}}{ma^2}Q_{\mathbf{k}} = -\frac{4\pi G m a^2}{k^2}\left[i\mathbf{k}\cdot\frac{\partial f_0}{\partial \mathbf{q}}\right]\left(\rho_{\mathbf{k}} + \rho_{\mathbf{k}}^{ext}\right),$$

where $\rho_{\mathbf{k}}$ and $\rho_{\mathbf{k}}^{\rm ext}$ are the Fourier transforms of $\delta\rho$ and $\delta\rho_{\rm ext}$ respectively.
(c) Use the fact that

$$\rho_{\mathbf{k}}(t) = \frac{m}{a^3}\int\frac{d^3\mathbf{q}}{(2\pi)^3}f_{\mathbf{k}}(\mathbf{q},t)$$

to convert the above equation into the integral equation:

$$\rho_k(\tau) = \frac{Gm^2}{ka^3(\tau)}\int_{\tau_0}^{\tau} d\tau'\, a^4(\tau')\left[\rho_k(\tau') + \rho_k^{ext}(\tau')\right] I_k(\tau - \tau'),$$

where $d\tau = (dt/a^2)$ and

$$I_k(z) = \frac{2}{\pi}\int_0^\infty q\,dq\,f_0(q)\sin\left[\frac{kq}{m}z\right]$$

and we have assumed $f_0(\mathbf{q}) = f_0(|\mathbf{q}|)$.

(d) Describe the evolution of $\rho_k(\tau)$ qualitatively for the case of wimps if $\rho_k^{\rm ext} = a^{-3}M$ where M is a seed mass.

11
Dark matter

11.1 Introduction

This chapter discusses the nature and distribution of dark matter in the universe. The observations which imply the existence of large quantities of dark matter in various systems are reviewed in section 11.2. The nature of dark matter is taken up in section 11.3; the theoretical constraints on baryonic dark matter and non-baryonic wimps are studied separately in two subsections. Sections 11.4 and 11.5 discuss the role of massive neutrinos and axions as dark matter candidates.

11.2 Observational evidence for dark matter

Any form of energy density which makes its presence felt only by its gravitational effects may be called 'dark matter'. For the purpose of this chapter, we shall not make any further distinction about the nature of the dark matter. (For example, planets like Jupiter which may exist around other stars will fall under the category of dark matter as far as the discussion in this chapter is concerned. This is in contrast to the discussion in earlier chapters where the term dark matter was used to denote the *non-baryonic* dark matter.) Observations suggest that there exists a significant amount of dark matter in our universe.

The quantity of dark matter in a system is usually determined in the following way: The gravitational field of the dark matter affects the motion of other objects (stars, clouds of gas ...) which are visible. If v is the typical velocity of test particles in a system with *total* mass $M(R)$ and size R, then

$$v^2 = k\frac{GM(R)}{R}, \tag{11.1}$$

where k is a numerical constant which depends on the geometry of the system and the density distribution. The velocities of test particles can be inferred from the measurement of the redshifts. The size R is determined by using different techniques for different objects; for the extragalactic objects, distances are usually determined from the redshift using Hubble's law. Once v and R are known from the observations and k from a theoretical model, M can be determined as $M = (v^2R/kG)$.

It is conventional to quote the results of such an analysis in terms of a quantity Q, called the 'mass-to-light' ratio: $Q = (M/L)$ where L is the luminosity of the system. This quantity can be expressed in units of the mass-to-light ratio for the sun $\mathrm{Q}_\odot = (\mathrm{M}_\odot/\mathrm{L}_\odot) \simeq 0.5\,\mathrm{g\,erg^{-1}\,s}$. The existence of dark matter is signalled if the observed value of Q, based on luminous matter, is smaller than the value of Q inferred from the motion of test particles. Quite often it is convenient to determine the mass-to-light ratio of the system from the ratio of mass *density* ρ to the luminosity *density* J.

The following scaling may be noted as regards the mass-to-light ratio of the extragalactic objects. If R is measured using the mean redshift and the angular size, then it will scale with the Hubble constant; i.e. $R \propto h^{-1}$, making $M \propto R \propto h^{-1}$. Similarly, the absolute luminosity of the system L will be proportional to the square of the distance to the object and hence will scale as h^{-2}. So $Q = (M/L) \propto (h^{-1}/h^{-2}) \propto h$. The inferred density ρ of the system (M/R^3) will scale as $(h^{-1}/h^{-3}) = h^2$; since the critical density ρ_c of the universe also varies as h^2, the estimated value of the density parameter $\Omega = (\rho/\rho_c)$ will be *independent* of h.

Since the determination of Q involves theoretical modelling, the actual procedure which is used to determine Q will be different for different systems. We shall now discuss a series of methods which are used in different contexts, from the smallest scales to the largest.

(a) Solar neighbourhood

The mass density of visible stars, near the sun, in our galaxy is about $0.044\,\mathrm{M}_\odot\,\mathrm{pc}^{-3}$. Almost an equal amount of density, $0.042\,\mathrm{M}_\odot\,\mathrm{pc}^{-3}$, is contributed by gas. Further, the standard theory of stellar evolution suggests that the mass density of stellar remnants with negligible luminosity should be about $0.028\,\mathrm{M}_\odot\,\mathrm{pc}^{-3}$. Thus the total mass density near the sun is about $0.114\,\mathrm{M}_\odot\,\mathrm{pc}^{-3}$. The luminosity is mainly contributed by stars and amounts to about $0.067\,\mathrm{L}_\odot\,\mathrm{pc}^{-3}$. Dividing the mass density by the luminosity density we find[1] that the mass-to-light ratio near the sun is

$$Q_{\mathrm{solar}}(\mathrm{observed}) \cong 1.7\,\mathrm{Q}_\odot. \tag{11.2}$$

This value is based on the *volume* density of mass and luminosity. Since different stellar populations are distributed differently in the direction perpendicular to the galactic plane, it is probably more meaningful to consider the projected *surface* density of mass and luminosity. The surface mass density near the sun (obtained by integrating the mass density within a size of 700 pc perpendicular to the galactic plane) is about $50\,\mathrm{M}_\odot\,\mathrm{pc}^{-2}$ while the surface brightness is $15\,\mathrm{L}_\odot\,\mathrm{pc}^{-2}$. This leads to a slightly larger value for Q:

$$Q_{\mathrm{solar}}(|z| < 700\,\mathrm{pc},\ \text{observed}) \cong 3.3\,\mathrm{Q}_\odot. \tag{11.3}$$

To estimate the mass-to-light ratio due to the total gravitating mass in the solar neighbourhood, we will first have to model the galaxy and obtain a relation between the mass density and the stellar velocities. The basic principle behind this estimate can be understood as follows: Consider a class of stars (say, the K-giants) near the sun. Let ρ be their density in the plane of the disc ($z \simeq 0$) and v_z be the mean random velocity in the z direction. If H is the mean scale height of the stars and g_z the z-component of gravitational force, then 'vertical equilibrium' requires $\rho g_z H = \rho v_z^2$; or $g_z \simeq (v_z^2/H)$. For a plane slab of gravitating matter with surface mass density μ, the Gauss theorem implies $\mu = (g_z/2\pi G)$. Hence $\mu = (v_z^2/2\pi GH)$. Measuring v_z and H we can determine the density of the gravitating mass.

To do the calculation more precisely, we shall use the model for our galaxy based on the collisionless Boltzmann equation (see chapter 1). Let the stellar distribution function for our galaxy be $f(\mathbf{x}, \mathbf{v})$. (In steady state, this function will be independent of time). We define the volume density of stars $s(\mathbf{x})$ and the mean velocity dispersion $\sigma_{ij}(\mathbf{x})$ by the relations

$$s(\mathbf{x}) \equiv \int f\, d^3\mathbf{v}; \quad \sigma_{ij}(\mathbf{x}) \equiv < v_i v_j > (\mathbf{x}) \equiv \frac{1}{s}\int f v_i v_j\, d^3\mathbf{v}. \tag{11.4}$$

Multiplying the collisionless, steady state equation by $v_i v_j$ and integrating over all $\mathbf{v}$, it is easy to show that (see exercise 11.1):

$$\frac{\partial}{\partial r}(s\sigma_{rz}) + \frac{\partial}{\partial z}(s < v_z^2 >) + \frac{s}{r}\sigma_{rz} + s\frac{\partial \phi}{\partial z} = 0. \tag{11.5}$$

Near the galactic disc, the terms which involve the gradient in the z-direction will dominate over other terms; hence we can approximate this equation by:

$$\frac{\partial}{\partial z}(s < v_z^2 >) + s\frac{\partial \phi}{\partial z} \simeq 0. \tag{11.6}$$

Similarly, the Poisson equation can be approximated as (see exercise 11.2):

$$\frac{\partial^2 \phi}{\partial z^2} \simeq 4\pi G\rho. \tag{11.7}$$

Combining (11.6) and (11.7) we get

$$\frac{\partial}{\partial z}\left[\frac{1}{s}\frac{\partial}{\partial z}(s < v_z^2 >)\right] = -4\pi G\rho. \tag{11.8}$$

If we can measure s and $< v_z^2 >$ as a function of z, then this equation can be used to estimate the mass density ρ. Unfortunately, such a procedure leads to large observational errors because the left hand side involves two differentiations. (In reality, we need three differentiations because the quantity which is usually observed is not s but the cumulative stellar density). To minimize the uncertainties one can try to obtain independent estimates of s for different stellar populations like K giants, F stars etc. and average the results. Such an analysis, using the above equation, leads to the estimate $\rho \cong 0.15\,\mathrm{M}_\odot\,\mathrm{pc}^{-3}$ for the volume density. The integrated (surface) mass density

$$S(z) = \int_{-z}^{+z} \rho(z')\,dz' \tag{11.9}$$

can be determined somewhat more accurately since it involves only two differentiations. A calculation similar to the above one gives $S(700\,\mathrm{pc}) \simeq 90\,\mathrm{M}_\odot\,\mathrm{pc}^{-2}$.

It is possible to increase the reliability of these measurements by more elaborate statistical techniques[2]. Such a study leads to the estimates $\rho \simeq 0.18 \pm 0.03\,\mathrm{M}_\odot\,\mathrm{pc}^{-3}$ and $S(700\,\mathrm{pc}) \simeq 75\,\mathrm{M}_\odot\,\mathrm{pc}^{-2}$. Since the luminosity density (near the sun) is about $0.067\,\mathrm{L}_\odot\,\mathrm{pc}^{-3}$, we get the mass-to-light ratio to be

$$Q = 2.7\,\mathrm{Q}_\odot. \tag{11.10}$$

Similarly, using the fact that the surface brightness is about $15\,\mathrm{L}_\odot\,\mathrm{pc}^{-2}$ we find the integrated mass-to-light ratio to be about $5\,\mathrm{Q}_\odot$. Both these values exceed the corresponding mass-to-light ratios, based on luminous matter, by about 50 per cent. Thus, nearly one third of the material in the solar neighbourhood must be considered to be dark matter.

The results quoted above are fairly sensitive to the statistical analysis and modelling which have been used. There have also been reports in the literature[2] claiming that the amount of dark matter in the solar neighbourhood is much lower than the figure quoted above. Hence the numbers quoted above should still be considered tentative.

We can estimate the amount of surface mass density of our galaxy from the rotational speed of stars in the disc. Let us assume, for simplicity, that the stars are moving in circular orbits with velocity $v(r)$. If the surface mass density of the disc is $S(r)$, then the balance between centrifugal force and gravitational force gives the relation:

$$\frac{v^2(r)}{r} = G\int_0^\infty dx\; xS(rx)\int_0^{2\pi} d\phi\frac{(1 - x\cos\phi)}{(1 + x^2 - 2x\cos\phi)^{3/2}}. \tag{11.11}$$

Observations of $v^2(r)$ in our galaxy suggest that it is constant for a wide range of values of r. A constant v can be obtained from the surface density $S(r) \propto r^{-1}$. The integrals can be easily evaluated for this case giving $v^2 = 2\pi G S(r) r$. Since $v \simeq 220\,\mathrm{km\,s^{-1}}$ near the location of the sun, $r = R_0 = 8.5\,\mathrm{kpc}$, we get:

$$S = \frac{v^2}{2\pi G R_o} \simeq 210\,\mathrm{M_\odot\,pc^{-2}}. \qquad (11.12)$$

Note that this value is nearly three times larger than the value $S(700\,\mathrm{pc})$ determined from the analysis of z-components of velocities of stars. Clearly a significant fraction of the dark matter must be distributed with a scale height much greater than 700 pc. The same result can be stated in a different manner. If all the matter was confined to the disc and had a surface density of only $75\,\mathrm{M_\odot\,pc^{-2}}$ (as determined before) then the rotational velocity of stars will be lower than $220\,\mathrm{km\,s^{-1}}$. Thus all the dark matter cannot be confined to the disc (also see exercise 11.3).

The existence of dark matter near the sun raises the question: How far does the dark matter halo extend? It is difficult to answer this question precisely because we do not have reliable observations at distances far greater than the distance of the sun from the galactic centre, R_0. However, a rough estimate[3] can be made along the following lines: Let us assume that the rotational velocity of our galaxy has a constant value V_0 up to a radius $r = L$ and falls as $r^{-1/2}$ beyond L. This is equivalent to assuming that the mass distribution of the halo is given by

$$M(r) = \begin{cases} (V_0^2 r/G) & (\text{for } r < L) \\ M_0 & (\text{for } r > L). \end{cases} \qquad (11.13)$$

Corresponding to this mass distribution, we get the gravitational potential ϕ:

$$\phi = \begin{cases} V_0^2\,[\ln(r/L) - 1] & (\text{for } r < L) \\ -(V_0^2 L/r) & (\text{for } r > L). \end{cases} \qquad (11.14)$$

All the stars in the solar neighbourhood with speeds significantly higher than the escape velocity in this potential would have escaped by now. So the maximum stellar velocity v_{max} we expect in solar neighbourhood will be given by the condition

$$\frac{1}{2} v_{\mathrm{max}}^2 + \phi(R_0) < 0. \qquad (11.15)$$

Observations in the solar neighbourhood suggest that the velocity distribution of stars shows a sharp cut-off around $v_{\mathrm{max}} = 500\,\mathrm{km\,s^{-1}}$. Substituting $V_0 = 220\,\mathrm{km\,s^{-1}}$ and $v_{\mathrm{max}} = 500\,\mathrm{km\,s^{-1}}$ in the above equation we find that $L \gtrsim 4.9 R_0$. Or, since $R_0 \simeq 8.5\,\mathrm{kpc}$, $L \gtrsim 41\,\mathrm{kpc}$, corresponding to a total mass of $M_0 \gtrsim 4.6 \times 10^{11}\,\mathrm{M}_\odot$. Since the total luminosity of our galaxy is about $L_{\mathrm{total}} \simeq 1.4 \times 10^{10}\,\mathrm{L}_\odot$ the mass-to-light ratio for our galaxy is bounded by the inequality $Q \gtrsim 33\,\mathrm{Q}_\odot$. This value is at least six times larger than the value in the solar neighbourhood, suggesting that Q increases with increasing scale.

An independent estimate of the extent of our galactic halo can be made[4] using the Magellanic clouds and the dynamics of the satellite galaxies which are gravitationally bound to the Milky Way (see exercise 11.4). These procedures lead to somewhat larger values of L, in the range of (50–80) kpc.

(b) Rotation curves of discs

One of the most striking and reliable pieces of evidence for the presence of dark matter in galactic systems comes from the study of rotation curves of disc galaxies[5]. A rotation curve is a plot of the rotational velocity $v(r)$ of some suitable test particle at a distance r from the centre of the spiral galaxy. The velocities are measured optically from the emission lines in the HII regions or from the radio measurements of the 21 cm line of neutral hydrogen. Since the neutral hydrogen clouds exist even at a large radius from the centre of the galaxy, they serve as good tracers of the mass distribution for distances much larger than the visible extent of the galaxy.

For a spiral galaxy with finite mass, we would expect the rotation curve to fall off at sufficiently large distances. Such a behaviour is not seen. We have now measurements for over 70 spiral galaxies and in almost all of them the rotation curve is either flat or slowly rising. The simplest interpretation of this result is that the spiral galaxies, like ours, contain massive spherical halos with the halo mass increasing linearly with radius.

If both the rotation curve and accurate photometry of the galaxy are available, then one can attempt to model the distribution of the dark halo[6]. This has been done, for example, for the Sc galaxy NGC 3198. The luminosity profile of this galaxy can be fitted by an exponential disc of scale height $2h^{-1}$ kpc and total luminosity $4 \times 10^9 h^{-2}\,\mathrm{L}_\odot$. The dark matter halo is well described by a density profile of the form $\rho(r) = \rho_0[1 + (r/a)^n]^{-1}$. If we assume a constant mass-to-light ratio throughout the galaxy, then the parameters which will provide the least mass-to-light ratio turn out to be $\rho_0 = 0.013h^2\,\mathrm{M}_\odot\,\mathrm{pc}^{-3}$, $a = 6.4h^{-1}$ kpc and $n = 2.1$. For this set of values $Q = 5.8h\,\mathrm{Q}_\odot$. The total mass inside a radius of $22h^{-1}$ kpc (which is the distance up to which the rotation curve has been determined) is about $1.1 \times 10^{11} h^{-1}\,\mathrm{M}_\odot$, corresponding to a total mass-to-light ratio of $28h\,\mathrm{Q}_\odot$. (While these results seem reasonable, they are unfortunately not unique. It is possible to model this system with mass distributed differently between the disc and the halo leading to considerable variation in these parameters.)

(c) Elliptical cores and dwarf spheroidals

There are two systems which can be modelled theoretically by an isothermal sphere (see chapter 1). These are the dwarf galaxies and the core regions of the elliptical galaxies. Using such a theoretical model, one can estimate the amount of dark matter in these systems.

The isothermal sphere is parametrized by two variables: the velocity dispersion, σ^2 and the central density ρ_0. Using these two variables, one can define a core radius $r_0 = (9\sigma^2/4\pi G\rho_0)^{1/2}$. Calculations show that the central *surface* density Σ_0 of the isothermal sphere is given by $\Sigma_0 \simeq 2.018\rho_0 r_0 \simeq 2\rho_0 r_0$. To determine the mass-to-light ratio for such a system – say, the core of an elliptical galaxy – we proceed as follows: The observed luminosity profile of the elliptical core is fitted to an isothermal sphere, allowing one to determine the best fit values for r_0 and central brightness I_0. The central emissivity is, therefore, $j_0 \cong (I_0/2r_0)$. On the other hand, the central density is $\rho_0 = (9\sigma^2/4\pi G r_0^2)$. Thus, the central mass-to-light ratio is

$$Q = \frac{M}{L} = \frac{\rho_0}{j_0} = \frac{9\sigma^2}{2\pi G I_0 r_0}. \tag{11.16}$$

Since all the quantities on the right hand side are known from observation, Q can be found.

This procedure has been applied to cores of elliptical galaxies, bulges of spiral galaxies and dwarf spheroidal galaxies[7]. The elliptical cores and spiral bulges give a mass-to-light ratio of about $12h\,\mathrm{Q}_\odot$. This is consistent with the mass-to-light ratio one would obtain in the solar neighbourhood if young stars and gas are excluded.

The results of the analysis for dwarf spheroidals have been somewhat controversial. The main uncertainty is in the determination of the velocity dispersion of the stars. The analysis gives the $(Q/\mathrm{Q}_\odot)$ to be 1.7, 6.8, 12, 71 and 220 for the satellite galaxies Fornax, Sculptor, Carina, Draco and Ursa Minor respectively. The wide variation in Q values, as well as the large value for Ursa Minor, make these observations somewhat suspect. An anomalous value for σ^2 will be obtained if the stars for which the observations are carried out are members of a binary system; in that case, part of the velocity due to orbital motion will be erroneously attributed to σ^2. Detailed statistical study[8], however, suggests that this effect is small and could possibly change the results at most by twenty per cent. Another possibility (which is difficult to test) is that Draco and Ursa Minor are not gravitationally bound systems. The situation regarding the dwarf galaxies is still somewhat unclear.

There exists one dwarf irregular galaxy (called DDO 154) which is very rich in gas. Gaseous hydrogen has been detected[7] in this galaxy up to a distance of 8 kpc while the visible extent is about 2 kpc. Up to this point, the rotation curve is flat, implying that ninety per cent of the matter is dark. The estimated mass-to-light ratio is greater than about 75. One of the simplest theoretical models gives $M_{\rm vis} \simeq 5 \times 10^7\,{\rm M}_\odot$ and $M_{\rm dark} \simeq 4 \times 10^9\,{\rm M}_\odot$.

In this connection, it is worth mentioning that star clusters, both open and globular, do not contain any significant amount of dark matter. This is of some importance because both globular clusters and dwarf spheroidals have comparable masses.

(d) Dark matter estimates from the dynamics of the Local Group

The Local Group consists of two dominant spiral galaxies, Milky Way and Andromeda, each surrounded by a number of smaller galaxies. The next nearest group of large galaxies are Sculptor and M81 which are about 3 Mpc away. So, to first approximation, we may consider the Local Group to be a gravitationally isolated system. By studying the dynamics of the Local Group, we can estimate the total gravitational mass in it and thus its mass-to-light ratio. The basic idea is extremely simple and can be understood as follows: The relative velocity v between Andromeda and the Milky Way, the most dominant members of the Local Group, is about $10^2\,{\rm km\,s^{-1}}$ and their relative separation r is about 10^3 kpc. Taking the age of the universe t_0 to be $(1\text{–}2) \times 10^{10}$ yr, we can form the dimensionless combination $(vt_0/r) \simeq (1\text{–}2)$. The fact that this ratio (which could have had any value, a priori) is close to unity suggests that the Andromeda–Milky Way system is gravitationally bound. In that case, the mass may be estimated to be $M \simeq (v^2 r/G) \simeq 10^{12}\,{\rm M}_\odot$.

To do a more precise job, we proceed as follows[9]: The line-of-sight velocity which will be measured by a hypothetical observer located on the sun (called the heliocentric velocity), is known for all the members of the Local Group. Since we know the motion of the sun, we can compute the line-of-sight velocity with respect to the galactic centre. If an external galaxy is at the galactic angular coordinates (l, b), then the line-of-sight velocity in a frame at rest with respect to the galaxy centre is given by

$$v_G = v_{\rm he} + v_{\rm LSR} + v_\odot, \tag{11.17}$$

where $v_{\rm he}$ is the heliocentric velocity which is observed, $v_{\rm LSR}$ is the velocity of a hypothetical star in a circular orbit at the location of the sun and $v_\odot$ is the deviation of the velocity of the sun from $v_{\rm LSR}$ (see chapter 7). The latter two quantities, when projected along the radial direction, have the values

$$v_{\rm LSR} = (220\,{\rm km\,s^{-1}}) \sin l \cos b, \tag{11.18}$$

and

$$v_{\odot} = (16.5\,\mathrm{km\,s^{-1}})[\cos b \cos 25^\circ \cos(l - 53^\circ) + \sin b \sin 25^\circ]. \qquad (11.19)$$

Since all the members of the Local Group are at distances much greater than R_0, the line-of-sight velocity is approximately radial from the centre of the galaxy. In particular, the line-of-sight velocity of Andromeda is $(-297)\,\mathrm{km\,s^{-1}}$. Since Andromeda is located at $l = 121.2, b = -21.6$, the above equation gives $v_G = -119\,\mathrm{km\,s^{-1}}$. The most natural explanation for this motion of the Milky Way and Andromeda towards each other (as indicated by the negative sign) is that the relative Hubble expansion between the two galaxies has been halted by their mutual gravitational attraction. In that case, we may treat Andromeda and the Milky Way as a gravitationally bound system, with each galaxy moving in a high eccentricity Keplerian orbit. For such an orbit, the separation of the galaxies, r, as a function of the cosmic time t, is given by the Kepler equation

$$r = \alpha(1 - e\cos\eta); \quad t = \left(\frac{\alpha^3}{GM}\right)^{1/2}(\eta - e\sin\eta) + T, \qquad (11.20)$$

where α is the semi-major axis, e is the eccentricity and M is the total mass of the Local Group. Using the conditions that at $t = 0$, $r = 0$ we may set $e = 1$ and $T = 0$ thereby obtaining a radial orbit. (The orbit may not be precisely radial because the tidal torques between the galaxies can lead to the exchange of orbital angular momentum and spin. However, such a transfer cannot produce *large* deviations from radial velocity.) From these equations we find the radial velocity to be

$$\frac{dr}{dt} = \frac{dr/d\theta}{dt/d\theta} = \left(\frac{GM}{\alpha}\right)^{1/2}\left(\frac{\sin\eta}{1-\cos\eta}\right) = \frac{r}{t}\frac{\sin\eta(\eta - \sin\eta)}{(1-\cos\eta)^2}. \qquad (11.21)$$

At the present epoch, $t_0 = (1\text{–}2)\times 10^{10}\,\mathrm{yr}$, we have $(dr/dt) = -119\,\mathrm{km\,s^{-1}}$ and $r = 730\,\mathrm{kpc}$. Substituting these values and solving the resulting equation, we find that η must be between 4.11 and 4.46 radians. Substituting this value in the $r(\eta)$ equation we find that α is between 0.47 and 0.58 Mpc. Finally, using this information on the $t(\eta)$ equation we get the mass of the Local Group to be between $3.2 \times 10^{12}\,\mathrm{M_{\odot}}$ and $5.5 \times 10^{12}\,\mathrm{M_{\odot}}$, with the larger age leading to smaller mass. The luminosity of the Milky Way is $1.4 \times 10^{10}\,\mathrm{L_{\odot}}$ while that of Andromeda is about $2.8 \times 10^{10}\,\mathrm{L_{\odot}}$. Thus we get the mass-to-light ratio of the Local Group to be between $76\,\mathrm{Q_{\odot}}$ and $130\,\mathrm{Q_{\odot}}$. This value is consistent with our earlier observation that the Milky Way halo must extend beyond 50 kpc.

(e) Groups of galaxies

Groups of galaxies contain typically 10–100 galaxies with a mean separation which is much smaller than the typical intergalactic separation in the universe. The masses of these groups can be estimated if we assume that they are gravitationally bound systems in steady state. For such systems, the virial theorem gives the relation: $2T + W = 0$, where T is the kinetic energy and W is the potential energy of the system. For a group of N galaxies, treated as point masses with masses m_i, positions $\mathbf{r}_i$ and velocities $\mathbf{v}_i$, the virial theorem can be written as

$$\sum_{i=1}^{N} m_i v_i^2 = \sum_{i \neq j}^{N} \frac{G m_i m_j}{|\mathbf{r}_i - \mathbf{r}_j|}. \tag{11.22}$$

Rigorously speaking, we should treat the quantities appearing in the above equation as time-averages; we shall assume that the instantaneous values will be close to the time-averaged values for systems with a sufficiently large number of galaxies. Observationally, we can only determine the line-of-sight velocity dispersion σ^2 and the projection $\mathbf{R}_i$ of $\mathbf{r}_i$ on to the plane of the sky. On the average, we expect $v_i^2 = 3\sigma^2$; the projected inverse separation can be related to the true inverse separation by averaging over all possible orientations of the group. It is easy to show that (see exercise 11.5):

$$< (|\mathbf{R}_i - \mathbf{R}_j|)^{-1} >= \frac{\pi}{2} (|\mathbf{r}_i - \mathbf{r}_j|)^{-1}. \tag{11.23}$$

Finally, we will also assume that the mass-to-light ratio Q of all the member galaxies is the same. Substituting $m_i = QL_i$ in the virial equation and using the above equation we get

$$Q_{\rm group} = \frac{3\pi}{2G} \frac{\sum_{i=1}^{N} L_i \sigma_i^2}{\sum_{i \neq j}^{N} L_i L_j \left(|\mathbf{R}_i - \mathbf{R}_j|^{-1}\right)}. \tag{11.24}$$

Since all the quantities on the right hand side can be determined from the observations, we can find Q for the group. This procedure will give reasonably reliable results for groups with a sufficiently large number of galaxies. Several groups have been studied[10] using the above equation. The median value of Q is about $260h\,Q_\odot$ but there is a wide spread in the values. It is believed that the spread is mostly statistical. In spite of the statistical uncertainty, it is obvious that the mass-to-light ratios in the groups are much larger than the values obtained in the luminous part of the galaxies. (The procedure needs to be modified if the groups are dominated by a single large galaxy or if the the dark matter halo forms a uniform background around all galaxies; these modifications are discussed in exercise 11.6.)

(f) Clusters of galaxies

For a rich cluster containing several hundred galaxies, like the Coma cluster, one can use a detailed model of the cluster for determining the mass-to-light ratio. One such approach uses the moments of the Boltzmann equation in spherical coordinates (see exercise 11.1). For a spherically symmetric system, the velocity moment of the collisionless Boltzmann equation becomes

$$\frac{d}{dr}(s < v_r^2 >) + \frac{2s}{r}[< v_r^2 > - < v_\phi^2 >] = -s\frac{GM(r)}{r^2}. \tag{11.25}$$

Introducing the anisotropy parameter $\beta(r) \equiv (1 - < v_\phi^2 > / < v_r^2 >)$ this equation can be rewritten as

$$\frac{d}{dr}(s < v_r^2 >) + \frac{2s}{r} < v_r^2 > \beta(r) = -s\frac{GM(r)}{r^2}. \tag{11.26}$$

Among the four quantities $\beta(r)$, $M(r)$, $< v_r^2 >$ and $s(r)$ we can observationally determine $s(r)$ and the line-of-sight velocity dispersion. These two functions, along with the moment equation, still constitute an underdetermined set. To proceed further, we have to make some assumption about the structure of the cluster. The mass which is calculated will depend on the nature of the assumption we make.

As an example of the procedures which are adopted, let us consider the Coma cluster. Among the several assumptions made in the literature[11] to determine the mass of the Coma cluster, the following three attempts seem to be reasonably reliable: (1) We may assume that s is proportional to ρ, i.e. the number density traces the actual mass density. This provides the extra relation needed to close the set of equations. The calculated mass for the Coma cluster turns out to be about $1.8 \times 10^{15} h^{-1}\,\mathrm{M}_\odot$. (2) It is possible to use suitable forms of the density profile $\rho(r)$, containing some free parameters, and choose these parameters so as to obtain minimum mass for the cluster. This often leads to fairly implausible density configurations but does constrain the mass from below. Such attempts have yielded a mass of about $0.7 \times 10^{15} h^{-1}\,\mathrm{M}_\odot$. (3) It is also possible to estimate the mass by choosing suitable functional forms for the anisotropy $\beta(r)$. In particular, the form $\beta(r) = [r^2/(r^2 + a^2)]$, where a is a free parameter, has several attractive features. (One can show, for example, that this anisotropy parameter arises from a distribution function of the form $f = f(E + L^2/2a^2)$.) Such a procedure also leads to masses slightly greater than $2 \times 10^{15} h^{-1}\,\mathrm{M}_\odot$.

The total luminosity of the Coma cluster is about $5 \times 10^{12} h^{-2}\,\mathrm{L}_\odot$. Our estimates therefore suggest a mass-to-light ratio of about $400h\,\mathrm{Q}_\odot$ for the Coma cluster. Similar analysis of the Perseus cluster gives a value of $600h\,\mathrm{Q}_\odot$. Quite clearly, these clusters contain a large amount of dark matter.

An indirect way of determining the amount of dark matter around the clusters is by using clusters as gravitational lenses for distant objects[12]. In one such study, the lensing of faint blue galaxies (with redshifts between 0.8 and 3) by 23 foreground clusters (in the redshift range of 0.2 to 0.4) is investigated. The images of the faint galaxies are used to determine the mass distribution around the clusters. Such an analysis indicates that the dark matter content of the clusters is (10–25) times larger than the visible mass. The central density of dark matter in these clusters is about $2 \times 10^5 \rho_c$ and the core radius of the dark matter distribution is about (30–50) kpc. These observations also suggest that the dark matter distribution in these clusters is extremely smooth to the level of about $(10^{10}\text{–}10^{11})\,\mathrm{M}_\odot\,\mathrm{kpc}^{-1}$.

(g) Virgo-centric flow and velocity fields

We have seen in chapter 7 that the velocity field of the nearby galaxies is extremely complicated. At least part of the velocity of the Local Group must be due to the gravitational attraction of the Virgo cluster. This fact can be used to estimate the amount of mass contained in a sphere, centred at Virgo and having a radius equal to the distance between Virgo and the Local Group[13]. Though this method has now become somewhat obsolete, we will discuss it because it illustrates an interesting method for determining the Ω of the universe.

The recession velocity between Virgo and the Local Group will be $H_0 r_{V_0}$ (where r_{V_0} is the distance to Virgo) if we ignore the gravitational influence of Virgo on the Local Group. To take into account the gravitational influence, we may proceed as follows: We assume that the mass concentration of the Virgo supercluster is spherically symmetric and that, at any time t, the total mass contained inside a sphere of radius $r_V(t)$ is M. We have seen in chapter 2 that any spherical region in a Friedmann universe behaves as another isolated Friedmann universe with its own parameters. Thus a sphere, centred on Virgo with the Local Group at its periphery can be treated as a Friedmann universe with density $\rho_V(t) = (3M/4\pi r_V^3(t))$ and Hubble parameter $H_V(t) = (\dot{r}_V/r_V)$. This sphere can also be assigned a density parameter $\Omega_V = (8\pi G\rho_V/H_V^2)$. We shall assume that Ω_V is less than unity. Using the standard equations for the $k = -1$ Friedmann universe, we can show that (see exercise 11.7):

$$H_{V_0} t_0 = \frac{\sinh \eta_{V_0} (\sinh \eta_{V_0} - \eta_{V_0})}{(\cosh \eta_{V_0} - 1)^2} \equiv F_1(\eta_{V_0}), \tag{11.27}$$

$$\rho_{V_0} = \left(\frac{3}{4\pi G t_0^2}\right) \frac{(\sinh \eta_{V_0} - \eta_{V_0})^2}{(\cosh \eta_{V_0} - 1)^3} \equiv \frac{3}{4\pi G t_0^2} F_2(\eta_{V_0}); \tag{11.28}$$

$$\Omega_{V_0} = 2\,(1 + \cosh \eta_{V_0})^{-1}. \tag{11.29}$$

To proceed further, we shall compare these values with that of a background universe which, we shall assume, has $\Omega < 1$. Since the spherical region should match the background universe at the initial singularity, the constant t_0 can be eliminated by such a comparison. It follows that

$$\frac{H_{V_0}}{H_0} = \frac{F_1(\eta_{V_0})}{F_1(\eta_0)}; \quad \frac{\rho_{V_0}}{\rho_0} = \frac{F_2(\eta_{V_0})}{F_2(\eta_0)}. \tag{11.30}$$

These equations give two relations between the four variables: η_{V_0}, η_0, and the ratios (ρ_{V_0}/ρ_0) and (H_{V_0}/H_0). If two of these quantities, viz. (ρ_{V_0}/ρ_0) and (H_{V_0}/H_0), can be determined observationally, then these equations can be used to fix η_0; once η_0 is known, Ω_{V_0} can be determined.

The quantities (H_{V_0}/H_0) and (ρ_{V_0}/ρ_0) can be measured in the following way. Suppose the mean velocity of the galaxies in the core of the Virgo cluster is u_1 and the mean velocity of the galaxies in some very distant cluster (located at distance l) is u_2. Let the observed brightness of the Virgo cluster and the distant cluster be L_1 and L_2 respectively. Since $u_1 = H_{V_0} r_{V_0}, u_2 = H_0 l$ and $L_1/L_2 = \left(l^2/r_{V_0}^2\right)$ it follows that

$$\frac{H_{V_0}}{H_0} = \frac{u_1}{u_2}\left(\frac{L_1}{L_2}\right)^{1/2}. \tag{11.31}$$

Since all quantities on the right hand side are measurable, we can determine this ratio. Similarly, we can determine (ρ_{V_0}/ρ_0) by estimating the ratio of luminosity distances, provided we assume that the average mass-to-light ratio on a sphere centred on Virgo is the same as that of the universe as a whole. Given these ratios and the two equations derived above, Ω can be determined.

Estimates of (H_{V_0}/H_0) are in the range of (0.7–0.8) with most of the scatter originating due to: (i) uncertainties in the velocity of the Virgo cluster and (ii) the differences in the definition used for the volume on which the sampling is made. The estimates for (ρ_{V_0}/ρ_0) range between 3 and 4. Using these values one obtains $\Omega = 0.25 \pm 0.1$. The mean value of Q is about $400h\,Q_\odot$.

It must be noted that this procedure is plagued with several uncertainties. The procedure adopted for measuring (ρ_{V_0}/ρ_0) requires the knowledge of the absolute luminosities of nearby galaxies, which in turn, requires the knowledge of the distance to these galaxies. The distance indicators are calibrated against a set of galaxies whose distances are usually determined using Hubble's law. If the Virgo cluster affects these galaxies (as it will) this primary calibration needs to be corrected in an iterative manner. This can be done by using more sophisticated models for the velocity flow in the local region. For example, we described in chapter 7 several attempts to map the velocity field of the universe and compare it with the observed mass distribution. Each of these comparisons depends on the combination $(b^{-1}\Omega^{0.6})$, where b is the biasing factor, as a free parameter and can be used to determine the best-fit value for this parameter. The QDOT-IRAS galaxy survey, for example, suggests that $b^{-1}\Omega^{0.6} = (0.81 \pm 0.15)$. Assuming $b = 1$ gives Ω to be about $0.7(\pm 0.3, -0.2)$. On the other hand $\Omega = 1$ would correspond to a biasing factor of $b = 1.23 \pm 0.23$. This is the largest scale at which we have a direct dynamical measurement of Ω.

The results discussed so far are summarized in table 11.1 in which we have given both the mass-to-light ratio and the corresponding value of Ω. We have also provided a rough 'length scale' at which each of the observations is relevant.

For comparison, note that the density parameter due to luminous matter in the universe is about $\Omega_{\rm lum} \lesssim 0.01$, while that due to baryons is constrained by the primordial nucleosynthesis to the range $0.011 < \Omega_b < 0.12$.

11.3 Nature of dark matter

It is clear that the density parameter Ω is at least about 0.2 over scales at which reasonably accurate dynamical measurements are possible. There is also some evidence which suggests that Ω increases with the scale and is possibly as high as unity at the largest scales. We have seen in the last chapter that inflationary models predict Ω to be unity. The luminous matter contributes only about one hundredth of this value.

It is, therefore, important to ask what the dark matter is composed of. *There is no a priori reason for the dark matter in different objects*

Table 11.1. Mass density in the universe

System	Method	Scale	$Q/Q_\odot$	Ω
Milky Way	Near sun	100 pc	5 (?)	$0.003h^{-1}$ (?)
	Escape velocity	20 kpc	30	$0.018h^{-1}$
	Satellites	100 kpc	30	$0.018h^{-1}$
	Magellanic stream	100 kpc	> 80	$> 0.05h^{-1}$
Ellipticals	Core-fitting	2 kpc	$12h$	0.007
	X-ray halo	100 kpc	> 750	$> 0.46h^{-1}$
Spirals	Rotation curves	50 kpc	$> 30h$	> 0.018
Groups	Local group	800 kpc	100	$0.06h^{-1}$
	Other groups	1 Mpc	$260h$	0.16
Clusters	Coma	2 Mpc	$400h$	0.25
Velocity fields	IRAS survey	$100h^{-1}$ Mpc	–	0.6

to be made of the same constituent. For example, our analysis of dark matter near the sun suggests that roughly half the mass in the solar neighbourhood is not visible. This component of dark matter is *confined* to the galactic disc and hence must have undergone dissipation. It is, therefore, reasonable to conclude that this component must be baryonic. In fact, there *must* be baryonic dark matter even at a larger scale. This is because the dark matter in the solar neighbourhood contributes only about $\Omega \simeq 0.003$ which is at least a factor four less than the *lower* limit on the baryonic density arising from the nucleosynthesis bounds.

At the same time, it is unlikely that *all* the dark matter in the universe is contributed by baryons. There are two reasons for this conclusion: (i) We have seen in chapter 6 that baryonic universes with $\Omega_b \simeq 0.2$ (and adiabatic initial fluctuations) violate the MBR anisotropy bounds. It is possible to get out of this constraint by invoking non-adiabatic fluctuations and re-ionization. However, these models appear to be rather contrived at present. (ii) We saw in the last chapter that $\Omega = 1$ is the only stable value in the standard Friedmann cosmology. Universes with $\Omega \neq 1$ evolve to $\Omega = 0$ or $\Omega = \infty$ within a very short time, unless we resort to extreme fine tuning. This argument favours the value $\Omega = 1$, which cannot be contributed entirely by baryons, because such a scenario would violate the upper bounds on Ω_b arising from the nucleosynthesis.

The above arguments, taken together, suggest that *both* baryonic and non-baryonic dark matter exist in the universe with non-baryonic matter being dominant. We will now discuss these components separately.

(a) Baryonic dark matter

Baryonic dark matter can exist in several forms. Interstellar clouds with masses less than about $0.08\,M_\odot$ will not be able to reach central temperatures which are needed for nuclear ignition. Such clouds can end up as 'brown dwarfs'. Even if *all* the dark matter is in the form of brown dwarfs, they would still be too sparsely distributed to allow detection. (The dwarf nearest to the sun will be too faint to be seen.) However, the standard theory of star formation predicts very few stars with masses less than $0.08\,M_\odot$. So, in order for this model to be viable, the initial mass function governing the star formation must have a non-standard shape. Since the process of star formation is not fully understood, such a possibility cannot be ruled out at this stage[14].

The baryonic dark matter could also be in the form of white dwarfs, neutron stars or black holes, all of which are remnants of the stellar evolution. There is a constraint on the density which can be contributed by such remnants which arises from the fact that stellar evolution should not contribute too much to the background radiation. This constraint gives $\Omega \lesssim 0.03$. In this case also, one needs to modify the standard initial mass function so that the enhanced production of high mass stars (which will turn into the remnants mentioned above) does not affect observations of low and intermediate mass stars[15].

Another possibility is for the baryons to exist in the form of primordial black holes. This can happen in scenarios in which there existed an epoch of star formation *before* the onset of galaxy formation. In such a case, primordial stars heavier than a few hundred solar masses can collapse directly, forming a black hole. This is to be contrasted with the black holes produced in the standard stellar evolution, in which much of the stellar mass is expelled during the black hole formation. The latter scenario is severely constrained by the fact that the interstellar medium should not be contaminated by too much heavy element material. (The resulting bound is $\Omega < 10^{-4}$). The bounds are much less severe on the formation of black holes directly from supermassive stars with masses in the range $10^2\,\mathrm{M}_\odot < M < 10^6\,\mathrm{M}_\odot$. (The upper bound of $10^6\,\mathrm{M}_\odot$ comes from the fact that gravitational perturbation from the black holes should not heat up the disc stars too much.) There are, however, several other indirect constraints on the existence of such supermassive black holes[16].

It may be possible to detect some of the compact objects discussed above (usually called 'Massive Astrophysical Compact Halo Objects' or 'MACHOS' for short) by the gravitational microlensing effect of these objects on the stars in the Large Magellanic Cloud. This process operates as follows: when we look at the stars in the LMC through our galactic halo, every once in a while, a MACHO will cross the line of sight. This MACHO will act a lens for the starlight and will enhance the intensity of the star by some amplification factor which can be calculated. By continuously monitoring about a million stars in the LMC for a period of about four months (10^7 s) one hopes to either find or impose constraints on the MACHOS in our galactic halo. For a MACHO with mass $10^{-2}\,\mathrm{M}_\odot$, one expects about 5 events in such a search, each lasting about 9 days. For objects with a mass of $10^{-6}\,\mathrm{M}_\odot$, there will be about 500 events each lasting about 2 hours. A wide span of MACHOS can be constrained by this observation which is currently underway[17].

Many other forms of baryonic dark matter can be either ruled out or severely constrained. For example, small objects like dust grains, asteroids, comets etc. (which are dominated by molecular forces rather than by gravitational forces) cannot contribute much to dark matter. This is because such bodies will be composed of heavy elements like silicon, carbon etc. which are always much less abundant (lower by a factor of at least 100) than hydrogen. Hydrogen in the form of solid snowballs is also not viable since these would evaporate too fast[18].

Elliptical galaxies and clusters do contain hot gas, made of baryons, which is detectable from X-ray emission. However, the mass in the gaseous component is not dynamically significant. For example, in M87, the total mass within 100 kpc is about $10^{12}\,M_\odot$ but the mass of gas is only about $3 \times 10^{11}\,M_\odot$. Similarly, observations from the 21 cm line of neutral hydrogen clouds show that the neutral gas present around galaxies is also not dynamically significant.

Neutral hydrogen at extragalactic distances can be constrained from the observations of the quasar absorption spectra. These bounds were discussed in chapter 9, where it was shown that the baryonic density, in either uniformly distributed neutral hydrogen or in compact Lyman-α clouds, should be much smaller than the critical density. Ionized matter in the intergalactic medium can be constrained by indirect methods similar to the ones discussed in chapter 6. These constraints show that the intergalactic medium cannot have too much baryonic dark matter.

(b) Non-baryonic dark matter

Any massive neutral fermion with weak interactions provides an a priori candidate for dark matter. Any such particle would have existed in large numbers in the early universe. As the universe cools, the number density of such particles will change depending on the details of their interaction. Given a specific particle physics model, one can compute, in a rather straightforward manner, the relic abundance of any such particle today. Knowing their mass we can work out the density contributed by these particles.

Such particles are of primary interest only if they could contribute significantly to the closure density. We saw in chapter 3 that, for fermions with weak interactions, three mass ranges are of interest. Particles which decouple when they are relativistic will contribute closure density if their total mass is about 100 eV; more precisely $\Omega h^2 \simeq (m/92\,\mathrm{eV})$. Particles which are in the mass range $100\,\mathrm{eV} \ll m < 100\,\mathrm{GeV}$ and decouple after becoming non-relativistic can contribute closure density if their masses are a few GeV; $\Omega h^2 \simeq 3(m/1\,\mathrm{GeV})^{-2}$. This result depends on the estimate of annihilation cross sections which is valid provided the mass of the particle is less than that of Z_0. For particles with m greater than 100 GeV the annihilation cross section decreases as m^{-2} and Ωh^2 can again increase to about unity for $m \simeq 1\,\mathrm{TeV}$. In this connection, it must be mentioned that the annihilation cross section for any such structureless fermionic particle is bounded from above by certain field-theoretical considerations[19]. This bound implies that particles with masses higher than about 340 TeV need not be considered at all. Thus the cosmologically interesting mass ranges are: $m < 100\,\mathrm{eV}$, $m = (1\text{–}3)\,\mathrm{GeV}$ and $m \simeq 1\,\mathrm{TeV}$.

The minimal particle physics model, based on $SU(3)_C \otimes SU(2)_L \otimes U(1)_Y$ provides *no* such dark matter candidate. The only neutral fermion in such a theory – the neutrino – is modelled to have zero mass. However, almost all extensions of the standard model provide several massive candidates. These involve generalizations of the standard model with massive neutrinos, grand unified theories, theories incorporating supersymmetry etc. We shall now discuss some of these candidates and the constraints on their masses.

There exist several simple generalizations of the standard model in which the neutrino becomes massive with a low mass: $m_\nu < 100\,\text{eV}$. All such models can provide a hot dark matter candidate. Such a light neutrino is the most difficult candidate to constrain experimentally. The current upper bound on the neutrino mass is about 12 eV. Considering the importance of this candidate, we will discuss it separately in a later section.

Much better bounds are available on heavier wimps, including heavy neutrinos, from the decay width of the Z_0-boson at SLC and LEP. The essential idea behind these experiments is as follows[20]: The decay width Γ of the Z boson can be written in the form

$$\Gamma = \Gamma_{\text{hadron}} + 3\Gamma_{\text{lepton}} + N_\nu \Gamma_{\nu\overline{\nu}}, \tag{11.32}$$

where the first two terms represent the width due to the hadronic channel and due to $e\overline{e}$, $\mu\overline{\mu}$ and $\tau\overline{\tau}$ channels, while the last term incorporates contributions due to all 'neutrino-like' particles which couple to Z. In the standard model, the first two terms can be computed quite reliably; one can also compute $\Gamma_{\nu\overline{\nu}}$ from first principles. Since all quantities on the right hand side except N_ν are known, the experimentally observed value of Γ can be used to put bounds on N_ν. (It turns out that a change of N_ν by unity will change Γ by about 13 per cent; thus this is a fairly sensitive test.) The LEP measurements, for example, give $N_\nu = 2.95 \pm 0.11$; all other results are also consistent with the value $N_\nu = 3$. Since 3 neutrino species are known to exist, this does not allow for any other neutral wimp to which Z_0 can decay. Any other wimp, if it exists, must be very massive – so that Z_0 cannot decay by that channel; or they must be coupled to the Z_0 much more weakly than the 'standard' neutrinos. For example, a heavy neutrino, if it exists, must have a mass greater than about 35 GeV.

Further constraints are available from semiconductor ionization detectors designed to observe the wimps in our halo[21]. As the earth moves through the halo, the wimp strikes the nucleus of the detector, depositing a small recoil energy (about 10 keV). A fraction of this energy (0.3 to 0.2) goes into ionization giving a signal which is typically one count per keV per kg per day. The best bounds[22] currently available are from UCSB/LBL/UCB groups. Assuming that the halo density of wimps on the location of earth is about $0.3\,\text{GeV}\,\text{cm}^{-3}$ (i.e. $\rho_{\text{DM}} \simeq 5 \times 10^{-25}\,\text{g}\,\text{cm}^{-3}$) and that the root mean square velocity of halo particles is about

$300\,\mathrm{km\,s^{-1}}$, these groups find that Dirac fermions in the mass range 10 GeV to 3 TeV can be excluded. Since SLC–LEP results rule out the (4–30) GeV range and the cosmological constraints eliminate the (30 eV–4 GeV) range, we have excluded Dirac fermions in the entire range of interest (30 eV–1 TeV). This bound rules out *all* dark matter candidates which have coherent interaction with the nuclei and a coupling strength which is comparable to that of standard weak interaction.

These results also constrain several other particles. In the models incorporating supersymmetry, every particle will have a supersymmetric partner (called 'sparticle'). The lightest of these supersymmetric particles (LSP, for short) is stable in *most* models. If the stable LSP with a mass of $(1\text{–}10^4)$ GeV has an electric charge or strong interaction, then it would be detectable as an anamolous heavy isotope in geological studies[23]. The observed bounds rule out this possibility, suggesting that LSP must be a neutral, weakly interacting particle; and hence, LSP is a viable candidate for dark matter. Some of these LSPs are constrained by the LEP–SLC result while the others are not. For example, if the LSP is the sneutrino, superpartner to the neutrino, then the decay width of Z_0 constrains its mass to be higher than about 35 GeV. (Note that the sneutrino is a boson.) Moreover, sneutrinos in the halo (with $m \gtrsim 3\,\mathrm{GeV}$) would have been captured by the sun. The annihilation of solar sneutrinos will lead to high energy neutrinos from the sun which should have been noticed in the experiments designed to detect the proton decay[24]. The absence of such signals rules out the range $m > 3\,\mathrm{GeV}$. These two results, together, rule out the sneutrino as an LSP, and as a dark matter candidate.

On the other hand, if the LSP is a neutralino – denoted by χ, which is a fermionic superpartner to the gauge and Higgs bosons in the theory – the situation is more complicated. An analysis similar to the one performed in chapter 3 shows that $\Omega_\chi \simeq 1$ if $15\,\mathrm{GeV} < m_\chi < 1\,\mathrm{TeV}$. The coupling of χ to Z_0 is complicated and somewhat model dependent. However, in most scenarios, much of the parameter space for $m_\chi < 30\,\mathrm{GeV}$ is excluded by current observations[25]. The range $30\,\mathrm{GeV} < m_\chi < 1\,\mathrm{TeV}$ is still open.

Another particle which has been suggested[26] as a dark matter candidate is a 'cosmion'. Such a hypothetical particle should have a mass in the range of (4–10) GeV and a coupling to matter which is about 100 times stronger than the weak interaction. This particle was postulated as a possible solution to the solar neutrino problem, which we will discuss in the next section. The lower bound on the mass comes from the constraint that the cosmion should not evaporate too quickly while the upper bound arises from the constraint that too many of these particles should not 'sink' to the core of the sun. The detection experiments mentioned above have ruled out the range $m > 7\,\mathrm{GeV}$ for these particles. This result is based on the assumption that the root-mean-square velocity of cosmions in the halo is about 260 km s^{-1}. If this velocity is taken to be larger than about 300 km s^{-1}, then the current data will exclude the full range (4–10 GeV) of (interesting) values for m.

The above discussion shows that a wide class of cold dark matter candidates are now ruled out, though not all of them. One of the remaining cold dark matter candidates is an axion, which we will discuss separately in section 11.5, after discussing massive neutrinos in the next section.

11.4 Massive neutrinos

We mentioned before that the minimal standard model of electroweak interactions has only zero mass neutrinos. This result, in some sense, is put in by hand in the standard model. Since all known experimental results are consistent with a left-handed neutrino (ν_L) and a right handed antineutrino ($\overline{\nu}_R$), the *minimal* model for electroweak interaction does not use ν_R and $\overline{\nu}_L$, and sets the neutrino masses to zero. In particular, note that the vanishing of neutrino mass does not arise from any fundamental physical symmetry (unlike, for example, the gauge invariance which makes the photon massless).

There are several theoretical and observational motivations for generalizing the electroweak model to include massive neutrinos. On the theoretical side, the absence of any guiding principle which makes neutrinos massless, coupled with the fact that the experimental bounds[27] on the masses of the neutrinos (12 eV, 250 keV and 35 MeV on ν_e, ν_μ and ν_τ respectively) are too weak, suggests that it is worth examining models with massive neutrinos. The impetus on the observational side comes from two different issues. The first is the possibility that a massive neutrino will be cosmologically important. The second – and probably more important factor – is the fact that the most natural solution for the 'solar neutrino problem' would require the neutrinos to be massive. We shall now summarize briefly the main features of the solar neutrino problem[28].

We saw in chapter 1 that energy is produced in the stars through a series of nuclear reactions. In several of these chains, neutrinos are also produced. For example, the cumulative effect of a *p–p* chain will be to convert four protons into a helium nucleus releasing two positrons, two ν_e and about 28 MeV of energy. Similarly, several other cycles release neutrinos. A typical flux of neutrinos in the above reaction can be easily computed. Since two neutrinos come out with the release of 28 MeV of energy, the total number of neutrinos released per second in this reaction will be $2(L_\odot/28\,\mathrm{MeV})$. Hence the neutrino flux on earth will be about

$$F = \frac{2\,L_\odot}{4\pi R^2(28\,\mathrm{MeV})} = \frac{2\times 4\times 10^{33}\,\mathrm{erg\,s^{-1}}}{4\pi(1.5\times 10^{13}\,\mathrm{cm})^2\times 28\,\mathrm{MeV}} \\ = 6\times 10^{10}\,\mathrm{cm^{-2}\,s^{-1}}. \tag{11.33}$$

(Since the density of the stellar core is about $100\,\mathrm{g\,cm^{-3}}$ and the e–ν_e scattering cross section is about $10^{-43}\,\mathrm{cm}^2$, the mean free path of neutrinos is about 10^{17} cm; this is much larger than the radius of the sun, and hence the neutrinos travel to earth unhindered, carrying about 3 per cent of the total energy emitted by the sun). Given a specific solar model one can compute the flux of neutrinos arising from various reactions and their characteristic energy ranges which are expected on the earth.

There have been several attempts to detect and measure these solar neutrinos. The ^{37}Cl experiment (in which the neutrino is allowed to convert ^{37}Cl in the detector to ^{37}Ar, which is measured) is expected to measure a solar neutrino flux of 7.9 ± 2.6 snu where 1 snu is a unit which counts 10^{-36} atoms of Argon per Chlorine atom per day. This is to be compared with the observed value of 2.1 ± 0.9 snu which is only a fraction of about 0.27 ± 0.04. Two other experiments are doing no better. The Kamiokande results are nearly 0.46 ± 0.05 times the expected one and the Gallex experiment gives a preliminary result which is about one-third the predicted value. This discrepancy has been called the 'solar neutrino problem'.

The discrepancy clearly indicates that either the neutrino physics or the solar physics is in error. For example, if there exist a specific kind of particles – called cosmions, invented specially for this purpose – which have been captured by the sun, then the energy transport in the sun's interior can be modified. This, in turn, could reduce the core temperature and consequently the flux of neutrinos. However, we saw in the last section that the models involving cosmions are severely constrained by observations.

There are several other possible ways of getting out of this difficulty, all of which require the neutrino to be massive. The most natural one involves an 'oscillation'; that is, the conversion of some of the ν_e to ν_μ or ν_τ which will reduce the effective flux of ν_e on earth. The simplest such model will involve masses of 10^{-6} eV, 10^{-3} eV and $(0.1$–$10^2)$ eV for ν_e, ν_μ and ν_τ respectively. In such a case, ν_τ can possibly act as a hot dark matter candidate. The probability for conversion between two types of neutrinos, say ν_μ and ν_τ, is usually denoted by $\sin^2 2\theta_{\mu\tau}$ where $\theta_{\mu\tau}$ is called the mixing angle. Accelerator experiments[29] can be used to exclude regions in the m–θ plane. For example, if $\nu_\tau \simeq 30\,\mathrm{eV}$, then

$\sin^2\theta_{\mu\tau}$ must be less than about 4×10^{-3}; and the range $m[\nu_\tau] \geq 3\,\text{eV}$ is excluded if $\sin^2 2\theta_{\mu\tau}$ is greater than about 0.008.

To confirm (or exclude) the existence of these particles we need a further handle on the parameter space. One of the approaches is to look for the effects of non-zero mass of the neutrino in the beta decay: $(A, Z) \to (A, Z+1) + e + \overline{\nu}_e$. The mass measurement from this experiment is quite difficult and has led to controversial results in the past. At present, we only have an upper bound on the ν_e mass: $m[\nu_e] \leq 12\,\text{eV}$.

The beta decay experiments, of course, cannot hope to detect masses like $10^{-6}\,\text{eV}$ discussed before. However, any *positive* detection of massive neutrinos in a β decay experiment will create serious problems for the scenario described above. As an example, consider the reported detection of a 17 keV neutrino which has a mixing probability of $\sin^2\theta = 0.0085 \pm 0.001$ for being an electron neutrino. Such a neutrino, if it is a Majorona-ν, will violate the bounds on neutrinoless double beta decay in most models. If it is a Dirac-ν, then its properties are constrained by the laboratory bounds on the mixing between ν_e and ν_μ and the observed pulse profiles of supernova 1987A. These constraints[30] suggest that the 17 keV state is mostly contributed by ν_τ. In order to escape the cosmological bound on Ωh^2, the ν_τ must be unstable with a lifetime of 10^4–10^6 years. Even such scenarios with unstable neutrinos are severely constrained and quite contrived[31]. Neither the dark matter problem nor the solar-neutrino problem will be solved in this scenario. Thus beta decay experiments are important in testing the models, though in a negative manner.

Finally, we mention a general constraint which can be imposed on the mass of any collisionless wimp based on the scale at which it is clustering[32,33]. Consider a fermionic wimp which was a relic from the early universe. At some time in the past, when the universe was homogeneous, the wimps were described by the Fermi–Dirac distribution $f_{\rm FD}(\mathbf{p})$. The maximum phase space density (i.e. the maximum value for number $\mathcal{N}$ of neutrinos per unit phase space volume) is $\mathcal{N}_{\rm max} = f_{\rm FD}$ $(\mathbf{p} = 0) = 2g/(2\pi\hbar)^3$ for each neutrino species. The phase space density, of course, decreases with larger $|\mathbf{p}|$. During the subsequent evolution, the distribution function evolves to some fairly complicated, space-dependent, form $f(\mathbf{x}, \mathbf{p})$. Let the *average* phase space density of wimps in some system (galaxy, cluster, ...) today be $\mathcal{N}$. Since collisionless evolution *only* mixes up regions of high phase density with those of low phase density, the final, average phase density $\mathcal{N}$ cannot be larger than the initial maximum phase density $\mathcal{N}_{\rm max}$. (This is a special case of a more general result; see exercise 11.8.) We can use this fact to obtain a lower bound on the mass m of the wimp. Suppose N wimps, each of mass m, have clustered in a scale R. The phase volume is about $\mathcal{V} = (4\pi/3)R^3(4\pi/3)(mv_{\rm max})^3$ where $v_{\rm max}^2 = (2GM/R) \cong 2\sigma^2$; so $\mathcal{V} \propto R^3 m^3 \sigma^3$. The average phase space density is $\mathcal{N} = (N/\mathcal{V}) \propto (M/m)\mathcal{V}^{-1} \propto (\sigma^2 R/m)\mathcal{V}^{-1} \propto \sigma^{-1}R^{-2}m^{-4}$. The condition $\mathcal{N} < \mathcal{N}_{\rm max}$, on putting in the numbers, leads to the constraint:

$$m > 30\,{\rm eV}\left(\frac{\sigma}{220\,{\rm km\,s^{-1}}}\right)^{-1/4}\left(\frac{R}{10\,{\rm kpc}}\right)^{-1/2}. \tag{11.34}$$

Clearly, the wimps should have higher mass to cluster at smaller scales. If, for example, it is confirmed that the wimps cluster at the scale of dwarf galaxies which have masses of $M \simeq 10^6\,{\rm M}_\odot$ and radius $R \simeq 1\,{\rm kpc}$, then $m > 10^2\,{\rm eV}$, ruling out the possibility of hot dark matter in these systems.

11.5 Axions

Another possible candidate for dark matter is a particle called the axion. Since this candidate is somewhat peculiar compared with the rest, we shall discuss it in some detail.

Axions arise in the theory of quantum chromodynamics, which is a very successful model for explaining the strong interactions. The basic constituents of this model are quarks (which make up the baryons) and gluons (which mediate the strong interactions). The gluons are described by a vector field $A^i_\mu(x)$ where μ denotes the usual spacetime index and i denotes an 'internal' symmetry space index. This field A^i_μ is analogous to the vector potential in electrodynamics. One can also construct a field tensor $F^i_{\mu\nu}$ from A^i_μ in a well defined manner; $F^i_{\mu\nu}$ will denote the field strength of the gluon field and is analogous to the electromagnetic field.

In electromagnetism, the ground state of the theory is characterized by the vanishing of the electromagnetic field tensor. Any vector potential which leads to a vanishing electromagnetic field can itself be made to vanish by a suitable gauge transformation. It turns out that the corresponding transformation is *not* possible in quantum chromodynamics. There can exist different configurations $A^i_\mu(x)$, all of which lead to the vanishing of $F^i_{\mu\nu}(x)$ but are *not* connected by gauge transformation. Any one such configuration of the gauge field is a valid vacuum state in quantum chromodynamics. It can be shown that the inequivalent vacuum states can be classified by the set of integers[34]. The most general ground state can, therefore, be expressed as a superposition:

$$|\theta> = \sum_{n=0}^{\infty} \exp\left(-in\theta\right)|n>, \tag{11.35}$$

where θ is a free parameter and $|n>$ is a ground state labelled by the integer n. In other words, the non-trivial nature of the vacuum introduces an extra parameter θ into the theory, whose value needs to be determined by experiments.

The non-zero value of θ leads to an electric dipole moment[35] for the neutron of the order of $d_n \simeq 10^{-15}\theta e\,{\rm cm}$, comparing this value with the experimental bound that $d_n < 10^{-25} e\,\mathrm{cm}$ we see that $\theta < 10^{-10}$. (To be precise, the parameter θ which is constrained in the above manner is related to the θ in (11.35) by the addition of a quark mass determinant. We have ignored this complication, for simplicity.) We are, therefore, led to the existence of a dimensionless parameter which is extremely small; and the question arises as to why θ is so small.

The simplest solution could be to ignore the non-trivial structure of the vacuum state. Unfortunately, this will lead to contradictions with the observed mass spectrum of mesons[36]. It is, therefore, necessary to explain the smallness of θ by some other means. This can be done by introducing a new dynamical field (called the axion field) and arranging the potential for the axion field in such a way that the effective value of θ is zero at the minimum of the potential. In such a model, the axion field will acquire a mass which depends on the energy scale f at which the symmetry of the new potential is broken. Taking f to be anywhere between 100 GeV and 10^{19} GeV, the corresponding axion mass m_a will range between 10^{-12} eV and 1 MeV, with the larger f leading to smaller m_a.

Given a specific model for the axion field one can also work out the coupling of axions to other matter fields[37]. It turns out that, in general, the axion couples to the electrons, nucleons and photons. The strengths of these couplings are somewhat model dependent; however, the smaller the axion mass, the weaker is the coupling. In particular, the axion can decay into two photons, with the lifetime

$$\tau \cong 6 \times 10^{24}\,\mathrm{s}\, q^{-2} \left(\frac{m_a}{1\,\mathrm{eV}}\right)^{-5}, \tag{11.36}$$

where the value of q ranges between 0.05 and 0.7 and depends on the model.

The interaction between the axion and other matter fields can be used to put severe constraints on the possible range of m_a. To begin with, laboratory observations[38] of the decay of K-mesons and quarkonium allow one to constrain the coupling constant by $f > 10^3$ GeV giving $m_a \lesssim 6$ keV. The remaining window (10^{-12} eV to 6 keV) can be constrained from the study of stellar evolution.

The key idea behind such constraints can be understood as follows: Consider a star like the sun. The nuclear energy produced in the solar core is transported to the surface by photon diffusion. Since photons are very strongly coupled to the solar material, the energy transport to the surface is a very slow process (i.e. 'photon cooling' is quite inefficient); this is the key reason for the long lifetime of main sequence stars. The situation changes considerably when the axion is introduced into the picture. Since axions couple very weakly to matter, they can transport energy rapidly from the core. Such an efficient cooling will accelerate several phases of the stellar evolution. From the observed properties of the stars, we can put bounds on the axion coupling and hence on the axion mass.

The constraints from the evolution of the main sequence stars give a maximum bound for the axion mass in the range of (1–20) eV, depending on the model. A similar analysis of evolution of the red giant phase gives a bound of $(10^{-2}$–$2)$ eV. Thus the possible mass range for the axion is reduced[39] to the range 10^{-12} eV to (about) 1 eV.

Further constraints on the mass of the axion can be imposed from cosmological considerations. At very early phases of the universe, axions will be in thermal equilibrium with other particles. Depending on the relative values of the interaction rate and expansion rate, the axions will decouple from the rest of the matter at some temperature. Axions with masses greater than 10^{-3} eV decouple while they are relativistic (just like neutrinos) and hence have a relic number density which is comparable to photons. From our earlier discussion, we know that, for such particles:

$$\Omega_a h^2 \simeq 10^{-2}(m_a/1\,\text{eV}). \tag{11.37}$$

So these axions can close the universe only if their mass is about $100h^2$ eV. However, we see from (11.36) that such massive axions have a lifetime which is much shorter than the age of the universe. The lifetime can be larger for axions with masses in the range of a few eV. However, the decay of such axions in the halo of our galaxy will produce a detectable line signal of photons at a frequency which depends on the mass of the axion[40]. The absence of such signals excludes axions of mass greater than about 4 eV. Thus the relic axions, even if they exist, cannot contribute significantly to the mass density of the universe.

There is, however, another way by which axions can be produced. We saw that axions arise when a symmetry breaking drives the value of the parameter θ to zero. When this symmetry breaking occurs, θ does not instantaneously relax to zero. Instead, it oscillates around the value $\theta = 0$. These oscillations of a scalar field behave[41] like a concentration of energy density with the equation of state $p = 0$. The *effective* mass of these condensates made of coherent oscillations of the scalar field will be quite different from the mass of the axion. For $m_a \simeq 10^{-5}$ eV the effective mass can be as high as a few GeV, making these axions an ideal cold dark matter candidate. A careful analysis shows that this process contributes to the density parameter

$$\Omega_a h^2 = \mathcal{O}(1)\,(\Lambda/200\,\text{MeV})^{-0.7}\left(m_a/10^{-5}\,\text{eV}\right)^{-1.18}, \tag{11.38}$$

where Λ is a parameter which arises in the QCD models. Notice that Ω_a decreases with increasing mass of the axions. Since Ωh^2 must be less than about unity it follows that $m_a \gtrsim 10^{-6}$ eV. Thus we are left with a small window in the range 10^{-6} eV to 10^{-3} eV for the mass of the axion.

An axion produced at the symmetry breaking, having a mass of about 10^{-5} eV, is thus a viable candidate for cold dark matter.

It is possible to detect axions in this mass range using their electromagnetic coupling[42]. For example, an axion can be converted into a microwave photon in the presence of a strong magnetic field. In a properly tuned microwave cavity, one may be able to detect this signal. Such experiments are now in progress.

11.6 Cosmological constant as dark matter

One important feature which must be stressed in the determination of dark matter densities is the following: Gravitational effects acting on a system of size l are usually insensitive to energy densities which are distributed smoothly over scales much larger than l. If the universe contains such a smooth component of energy density it may not be detectable by the conventional probing of dark matter via gravitational effects. Nevertheless, it will contribute to the expansion of the universe and in particular can close the universe. By introducing such a dark matter candidate we can reconcile the observational result of $\Omega \simeq 0.2$ with the theoretical prejudice for $\Omega = 1$, in a natural fashion.

For this idea to be viable one must prevent the smooth component from condensing on smaller scales. One way of doing this will be to use a smooth component made of particles which are relativistic at present. This idea, however, runs into several difficulties; for example, the age of a radiation dominated universe is smaller than that of a matter dominated universe and can lead to difficulties with the age of globular clusters.

Another alternative for a smooth energy density is a cosmological constant. The source term for Einstein's equations can be any conserved stress-tensor. If this stress-tensor has the form $T^i_k = \Lambda\delta^i_k$ with a constant Λ (corresponding to an 'equation of state' $p = -\rho = -\Lambda$, implying either pressure or density to be negative), then the quantity Λ is called the 'cosmological constant.' Such a term was originally postulated by Einstein and has an interesting history[43]. We shall first discuss some general features of the cosmological constant before turning to models which use non-zero Λ.

The value of such a constant term is severely constrained by cosmological consideration. Since Λ contributes $\Omega_\Lambda = (8\pi G\Lambda/3H^2)$ to the critical density, one can safely conclude (in spite of any astronomical uncertainties) that

$$|\Lambda| \lesssim 10^{-29}\mathrm{g\,cm^{-3}} \approx 10^{-47}(\mathrm{GeV})^4. \tag{11.39}$$

As we shall discuss below, the smallness of this value is a deep mystery.

The surprise in the smallness of $|\Lambda|$ is mainly due to the following fact: We do not know of any symmetry mechanism which requires it to be zero. In fact, we know of several independent, unrelated phenomena which contribute to Λ. To produce such a small Λ, these terms have to be fine-tuned to a bizarre accuracy.

To begin with, nothing prevents the existence of a Λ-term (say, Λ_0) in Einstein's equations. This will make the gravitational part of the Lagrangian dependent on two fundamental constants Λ_0 and G, *which differ widely in scale*; the dimensionless combination made of fundamental constants $(G\hbar/c^3)^2\Lambda_0$ has a value less than 10^{-124} which appears to be very unaesthetic.

Quantum field theory provides a wide variety of contributions to Λ. For example, consider a scalar field with a potential $V(\phi)$. The particle physics predictions do not change (except possibly in theories with exact supersymmetry, which is anyway not realized in nature) if we add a constant term V_0 to this potential. A potential like $V_1(\phi) = (1/2)\mu^2\phi^2 + (\lambda/4)\phi^4$ and $V_2 = (\lambda/4)(\phi^2 - \mu^2/\lambda)^2$ will lead to the same particle physics even though they differ by the constant term $(\mu^4/4\lambda)$. But such a shift in the energy-density will contribute to Λ. According to currently accepted scenarios, the value of the constant term in $V(\phi)$ changes in every phase transition by the amount E^4 where E is the energy scale at which the phase transition occurs: at GUTS transition this change is $10^{56}(\mathrm{GeV})^4$; at Salam–Weinberg transition it changes by $10^{10}(\mathrm{GeV})^4$. These are enormous numbers relative to the present value of $10^{-47}(\mathrm{GeV})^4$. It is not clear how a physical quantity can change by such a large magnitude and finally adjust itself to be zero to such a fantastic accuracy.

Lastly, one should not forget that the 'zero-point energy' of quantum fields will also contribute to gravity[44]. Each degree of freedom contributes an amount

$$\Lambda \cong \int_0^{k_{\max}} \frac{4\pi k^2\, dk}{(2\pi)^3} \sqrt{k^2 + m^2} \cong \frac{k_{\max}^4}{8\pi^2}, \tag{11.40}$$

where $k_{\max}$ is an ultra-violet cut-off. If we take general relativity to be valid up to the Planck energy, then we may take $k_{\max} \approx 10^{19}\,\mathrm{GeV}$ and the contribution to Λ will be $10^{74}(\mathrm{GeV})^4$.

If we assume that all the contributions are indeed there, then they have to be fine-tuned to cancel each other, for no good reason. Before the entry of GUTS into cosmology, one needed to worry only about the first and last contribution, both of which could be tackled in an ad hoc manner. One arbitrarily sets $\Lambda_0 = 0$ in the Lagrangian defining gravity and tries to remove the zero-point contribution by complicated regularization schemes. (Neither argument is completely satisfactory, but appears plausible). But with the introduction of GUTS and inflationary scenarios, the cosmological constant becomes a dynamical entity and the situation has become more serious. Notice that it is precisely the large change in the $V(\phi)$ which leads to a successful inflation; it has to be large to inflate the universe and change to a small value in the end for a graceful exit from the inflationary phase.

Several mechanisms have been suggested in the literature to make the cosmological constant zero[45]: Supersymmetry, complicated dynamical mechanisms and probabilistic arguments from quantum gravity are only a few of them. None of these seems to provide an entirely satisfactory solution. The smallness of the cosmological constant is probably *the* most important problem that needs to be settled in cosmology. We have no idea as to why this happens; but if it is because of some general symmetry consideration, it may be necessary for Λ to vanish identically at all epochs. This can, for example, wipe out the entire inflationary picture.

Let us now consider cosmological models based on non-zero value of Λ. These models invoke a universe with $\Omega_{\rm total} = 1$ with $\Omega_{\rm vac} \neq 0$; a typical set of values could be $\Omega_b = 0.03$, $\Omega_{\rm CDM} \simeq 0.17$ and $\Omega_{\rm vac} \simeq 0.8$. In such a model[46] the cosmological constant (or equivalently the vacuum energy density) dominates in the present epoch. Being perfectly smooth, no *local* gravitational field measurement will allow us to detect this energy density.

It is obvious that such a model is aesthetically quite ugly. As we mentioned before the only two natural values for the vacuum energy density are 0 and $(10^{19}\,{\rm GeV})^4$. Of these the second value is clearly ruled out by observations leaving us with the natural choice of a vanishing cosmological constant. Any non-zero value for this parameter requires extreme fine-tuning of the model so that just around the present epoch vacuum energy begins to dominate over matter. In spite of this fact, this cosmological model is worth investigating purely because it seems to alleviate several difficulties faced in the conventional models. We shall now briefly review how a non-zero vacuum density changes the dynamics of the universe.

For the values suggested above the universe would have been dominated by the cosmological constant at $z < z_{\rm vac} \simeq 0.6$. For $z_{\rm eq} > z > z_{\rm vac}$ the universe would have been matter dominated where $z_{\rm eq} \simeq 2.3 \times 10^3$. Einstein's equations can be easily integrated to give $a(t)$ for all t. For $z \ll z_{\rm eq}$ the scale factor evolves as

$$\frac{a(t)}{a_0} = \left(\frac{\Omega_{\rm NR}}{\Omega_{\rm vac}}\right)^{1/3} \sinh^{2/3}\left(\frac{3\sqrt{\Omega_{\rm vac}}H_0 t}{2}\right). \tag{11.41}$$

It can be easily verified from the above expression that the age of such a universe is given by

$$t_0 = \frac{2H_0^{-1}}{3\sqrt{\Omega_{\rm vac}}} \ln\left[\frac{1+\sqrt{\Omega_{\rm vac}}}{\sqrt{\Omega_{\rm NR}}}\right]. \tag{11.42}$$

Clearly the addition of a cosmological constant increases the age of the flat universe; the above expression is always greater than $(2H_0^{-1}/3)$, which is the age of the matter-dominated universe with $\Omega = 1$. For the parameters we have assumed $t_0 \simeq 15.5\,\mathrm{Gyr}$, if $h = 0.7$ (say). This is quite a comfortable value and is consistent with the age of globular clusters. In fact, this model is older than a model with zero cosmological constant at *any* given epoch. This means that objects at any given redshift get more time to evolve. For $z \gg z_{\rm vac}$, $t(z) \simeq [2H_0^{-1}/3\sqrt{\Omega_{\rm NR}}(1+z)^{3/2}]$ which is older than a flat matter dominated model by a factor $(\Omega_{\rm NR})^{-1/2}$.

A model with non-zero cosmological constant can also help to enhance power at large scales. The power spectrum $P(k) \propto k^{3/2}|\delta_k|$ depends on the value of Ωh^2 for its overall scale. Let us suppose that the spectrum is normalized at $\lambda = 8h^{-1}\,\mathrm{Mpc}$; decreasing the value of Ωh^2 will now increase the power on all scales greater than the normalization scale. (Another way of seeing this result is as follows: The ratio between the characteristic scale in the spectrum, $\lambda_{\rm eq} = 13(\Omega h^2)^{-1}\,\mathrm{Mpc}$ and the normalization scale $\lambda_N = 8h^{-1}\,\mathrm{Mpc}$ is $(1.6/\Omega h)$. For the parameter values we are using, this ratio is about 3.5 times larger than the corresponding ratio in standard CDM models.) This certainly helps to enhance power at large scale.

The negative features of such a model are the following: First of all, notice that a model with cosmological constant allows for lesser growth of perturbations than the standard model. (For the parameters used here the growth is lower by a factor of about 3). Secondly, models with cosmological constant, since they have more power at larger scales, predict larger MBR anisotropy at large angles. (This comparison is not quite straightforward because the relation between λ and θ also gets modified in the presence of a cosmological constant.) For our parameters, the $(\delta T/T)$ is increased by about 20 per cent at $\theta \simeq 1^\circ$. Though marginally consistent with the current bounds, the situation is potentially hazardous if the observational bounds improve. Lastly, the increased power is not dynamically significant as far as the velocity streaming results are concerned; at $50h^{-1}\,\mathrm{Mpc}$ scale, the model predicts $v_{\rm rms} \simeq 200\,\mathrm{km\,s^{-1}}$ (marginally better than the CDM value of $160\,\mathrm{km\,s^{-1}}$) which is still far short of the observed $600\,\mathrm{km\,s^{-1}}$. Hence it is not clear whether we have really made any substantial progress by invoking the cosmological constant.

Exercises

11.1 (a) Show that the collisionless Boltzmann equation (CBE), in the cylindrical (r, ϕ, z) coordinate system is

$$\frac{\partial f}{\partial t} + \left[v_r \frac{\partial f}{\partial r} + \frac{v_\phi}{r}\frac{\partial f}{\partial \phi} + v_z \frac{\partial f}{\partial z}\right]$$
$$+ \left(\frac{v_\phi^2}{r} - \frac{\partial \Phi}{\partial r}\right)\frac{\partial f}{\partial v_r} - \frac{1}{r}\left(v_r v_\phi + \frac{\partial \Phi}{\partial \phi}\right)\frac{\partial f}{\partial v_\phi} - \frac{\partial \Phi}{\partial z}\frac{\partial f}{\partial v_z} = 0,$$

where Φ is the gravitational potential.
(b) Consider an axisymmetric system in which all derivatives with respect to ϕ vanish. Integrate the equation given above with respect to velocities to obtain

$$\frac{\partial s}{\partial t} + \frac{1}{r}\frac{\partial}{\partial r}(rs < v_r >) + \frac{\partial}{\partial z}(s < v_z >) = 0,$$

where

$$s = \int f\, d^3\mathbf{v}; \quad < v_i > = \frac{1}{s}\int f v_i\, d^3\mathbf{v}.$$

(c) Multiply the CBE by velocity components and integrate over velocities to obtain

$$\frac{\partial}{\partial t}(s < v_r >) + \frac{\partial}{\partial r}(s\sigma_{rr}) + \frac{\partial}{\partial z}(s\sigma_{rz}) + s\left(\frac{\sigma_{rr} - \sigma_{\phi\phi}}{r} + \frac{\partial \Phi}{\partial r}\right) = 0;$$

$$\frac{\partial}{\partial t}(s < v_\phi >) + \frac{\partial}{\partial r}(s\sigma_{r\phi}) + \frac{\partial}{\partial z}(s\sigma_{\phi z}) + \frac{2s}{r}\sigma_{\phi r} = 0;$$

$$\frac{\partial}{\partial t}(s < v_z >) + \frac{\partial}{\partial r}(s\sigma_{rz}) + \frac{\partial}{\partial z}(s\sigma_{zz}) + \frac{s}{r}\sigma_{rz} + \nu\frac{\partial \Phi}{\partial z} = 0;$$

where

$$\sigma_{ij}(\mathbf{x}, t) = \frac{1}{s}\int v_i v_j f(\mathbf{x}, \mathbf{v}, t)\, d^3\mathbf{v}.$$

(d) Observations show that, in our galaxy, $s \propto \exp(-r/L)$ with $(R_0/L) \simeq 2.4$ where R_0 is the distance to the sun from the centre of the galaxy. Argue that, in a steady state, $[\partial(s\sigma_{rz})/\partial r]$ and $[s\sigma_{rz}/r]$ are unlikely to be larger than $K = (\sigma_{rr} - \sigma_{zz})\,(z/rL)$; this quantity K is a factor (z^2/rL) smaller than the other two terms in the equations. (Under what conditions can this assumption be invalid?) When this approximation is valid, we get the result used in the text:

$$\frac{1}{s}\frac{\partial}{\partial z}(s\sigma_{zz}) = -\frac{\partial \Phi}{\partial z}.$$

(e) In the case of a *spherically symmetric*, steady state system, show that

$$\frac{d}{dr}(s < v_r^2 >) + \frac{2s}{r}\left[< v_r^2 > - < v_\phi^2 >\right] = -s\frac{d\Phi}{dr}.$$

This equation was used in the text in determining (M/L) values for clusters.

11.2 (a) For an axisymmetric system, Poisson's equation can be written as

$$\frac{\partial^2 \Phi}{\partial z^2} = -\frac{1}{r}\frac{\partial}{\partial r}\left(r\frac{\partial \Phi}{\partial r}\right) + 4\pi G\rho(r, z).$$

Show that, as the system becomes more and more flattened – that is, when $\rho(r, z)$ tends to a delta function $\delta(z)$ – the second term on the right hand side dominates over the first. Then we can write $(\partial^2\Phi/\partial z^2) \cong 4\pi G\rho$.

(b) Suppose the visible (disc) part of our galaxy is embedded in a large, spherical, concentric halo with density distribution $\rho_{\rm halo}(r)$. This density distribution is such that the rotation velocity $v_c(r)$ near the sun (at a location of $R_0 \simeq 8.5\,{\rm kpc}$) is constant at $220\,{\rm km\,s^{-1}}$. Show that the approximation $(\partial^2\Phi/\partial z^2) \simeq 4\pi G\rho$ is still valid. What is the density of the dark matter halo near the sun?

11.3 Consider a system with the density distribution $\rho(r,\theta) = \rho_0 S(\theta)(r_0/r)^2$ which is axisymmetric and falls as r^{-2} (see ref. 47). Let $S(\theta)$ be normalized such that $\int_0^\pi S(\theta)\sin\theta\, d\theta = 2$. In general, we would have expected the radial and angular components of the force

$$F_r = -\frac{\partial\Phi}{\partial r}; \quad F_\theta = -\frac{1}{r}\frac{\partial\Phi}{\partial\theta}$$

to depend on θ as well as r.

(a) Show that $(\partial F_r/\partial\theta) = 0$; i.e. the radial force is independent of the latitude. Hence show that $F_r(r,\theta) = F_r(r) = -4\pi G\rho_0 r_0^2/r \equiv -v_0^2/r$.

(b) Show that the potential must have the form

$$\Phi(r,\theta) = v_0^2[\ln(r/r_0) + P(\theta)].$$

Determine the equation connecting $P(\theta)$ and $S(\theta)$.

(c) What is the potential, if $S(\theta) = a\delta_{\rm Dirac}(\theta - \pi/2) + b$, which corresponds to the disc embedded in a spherical halo?

11.4 (a) The Large Magellanic Cloud (LMC; $d_{\rm LMC-MW} \simeq 52\,{\rm kpc}$, $M_{LMC} \simeq 2\times 10^{10}\,{\rm M}_\odot$) and Small Magellanic Cloud (SMC; $d \simeq 63\,{\rm kpc}$; $M_{\rm SMC} \simeq 2\times 10^{9}\,{\rm M}_\odot$) are the nearest satellite galaxies to us. There exists a long, circular, trail of neutral hydrogen (the 'Magellanic stream') connecting the clouds and our galaxy. The material at the tip of the stream is falling on the Milky Way with a speed of about 220 km s^{-1}. The stream may be interpreted as due to the material torn out of the clouds in the past encounter with our galaxy. Make a reasonable, physical model for the observations and argue that they suggest an extended halo for the Milky Way.

(b) Our galaxy is surrounded by several satellite galaxies and globular clusters. Assume that they are outside our halo and feel the potential $\phi = -GM/r$. Let the radial velocities and galactocentric distances of these objects be (v_i, r_i) respectively; $i = 1, 2\cdots N$. Show that the time average of $v^2 r$ for any one satellite in Kepler orbit is

$$< v^2 r >= \frac{GM}{4}.$$

Assuming time average is the same as 'ensemble average', we can estimate the mass of our galaxy to be

$$M = \frac{4}{GN}\sum_{i=1}^{N} v_i^2 r_i.$$

This gives a value of about $M \simeq 3.8 \times 10^{11}\,\mathrm{M}_\odot$.

11.5 Let $\mathbf{x}_i$ be the position of a galaxy and let $\mathbf{R}_i$ be the projection of this vector on the sky. Show that

$$< |\mathbf{R}_i - \mathbf{R}_j|^{-1} > = \frac{\pi}{2} |\mathbf{r}_i - \mathbf{r}_j|^{-1}$$

where the average is taken over all possible orientations of the members of a set.

11.6 In estimating Q for a group of galaxies in the text, we have treated all the members of the group on the same footing. The equation for Q needs to be modified when this is not the case.

(a) Suppose that the group is dominated by a massive galaxy of luminosity L. Let v_i and R_i be the line-of-sight velocity and projected distance relative to the dominant galaxy. Show that

$$Q \simeq \frac{3\pi}{2GL} \left(\sum_{i=1}^{N} v_i^2 \right) \left(\sum_{i=1}^{N} R_i^{-1} \right)^{-1}.$$

(b) Suppose the dark matter is distributed over the size of the group, with the galaxies embedded in it. Further, assume that the spatial distribution of galaxies is similar to that of dark matter. Show that, in this case,

$$Q \simeq \frac{3\pi N}{2G} \left(\sum_{i=1}^{N} v_i^2 \right) \left(\sum_{i=1}^{N} L_i \right)^{-1} \left(\sum_{i=1}^{N} \sum_{j<i} |\mathbf{R}_i - \mathbf{R}_j|^{-1} \right)^{-1}.$$

If the dark matter distribution is more extended than that of galaxies, will Q be larger than $Q_{\rm true}$ or smaller?

11.7 Integrate Einstein's equations $(\dot{a}^2/a^2) + (k/a^2) = (8\pi G/3)\rho$ for the case of $k = -1$ with $\rho \propto a^{-3}$. From this result, deduce the expressions for F_1 and F_2 used in the text in section 11.2(g).

11.8 Consider a collisionless system described by a distribution function $f(\mathbf{x}, \mathbf{v}, t)$. We define a 'coarse-grained' distribution function $f_c(\mathbf{x}, \mathbf{v}, t)$ by using a weighting function $K(\mathbf{x}, \mathbf{v}; \mathbf{x}', \mathbf{v}')$:

$$f_c(\mathbf{x}, \mathbf{v}, t) = \int K(\mathbf{x}, \mathbf{v}; \mathbf{x}', \mathbf{v}') f(\mathbf{x}', \mathbf{v}', t)\, d^3\mathbf{x}'\, d^3\mathbf{v}'.$$

(a) Define an 'H-function' by the relation (see ref. 33)

$$H[f_c] = -\int C(f_c)\, d\mathbf{x}\, d\mathbf{v}$$

where $C(z)$ is a convex function of its argument. Suppose, $f_c = f$ at $t = t_1$. Show that $H(t_2) > H(t_1)$ for all $t_2 > t_1$. (Note that the condition $H(t_2) > H(t_3)$ need *not* hold for $t_2 > t_3 > t_1$.)

(b) Given an f_c, we can define the volume of the phase space with the phase density larger than q as

$$V(q) = \int d^3\mathbf{x}\, d^3\mathbf{v}\, \theta[f_c(\mathbf{x}, \mathbf{v}) - q]$$

where $\theta(z) = 1$ for $z > 0$ and zero otherwise; we have omitted the time dependence, for simplicity. Similarly, the mass contained in a region of the phase space which has phase density larger than q is

$$M(q) = \int d^3\mathbf{x}\, d^3\mathbf{v}\, f_c(\mathbf{x}, \mathbf{v})\theta[f_c(\mathbf{x}, \mathbf{v}) - q].$$

From these functions, we can construct $M(V)$. Show that

$$M(V) = \int_0^V q(V')\, dV'$$

where $q(V)$ is the inverse function of $V(q)$.
(c) Show that a distribution function f_1 can evolve to another f_2 only if $M_2(V) \leq M_1(V)$. Also show that, near $v \simeq 0$, $M(V) \simeq f_{\max} V$; hence $f_{\max}$ must decrease during evolution.

12
Epilogue

12.1 Introduction

This chapter summarizes the major ideas discussed in the book and provides a critical evaluation of the various concepts and prejudices.

12.2 Structure formation – an appraisal

The discussion in the previous chapters shows clearly that we do not yet have a satisfactory theory to explain the observed structures in the universe. Some of the ideas and beliefs, on which the models are based, must be incorrect. It is, therefore, worth examining these ideas with a view to identify those which are most suspect.

The cornerstones of all the models suggested in the literature are the following two assumptions: (i) At sufficiently large scales, the universe is smooth and homogeneous. (ii) The universe is evolving and, in the past, the universe was smoother. These assumptions allow us to describe the observed universe as a smooth Friedmann background with small-scale inhomogeneities. The observational evidence which supports the above assumptions is the following:

First of all, the two-dimensional angular distribution of galaxies in the sky appears to be statistically homogeneous, when averaged over large angular sizes. Further, root-mean-square fluctuation $(\delta N/N)_R$ in the number of galaxies contained in a sphere of radius R seems to decrease as R increases. This conclusion, of course, is based indirectly on the assumptions (i) and (ii) above, in the sense that the Friedmann model is usually used in the estimate of $(\delta N/N)$; but, it does show that the model is internally consistent. Similar consistency arguments can be given based on the isotropy of the MBR; any model in which (i) and (ii) are violated will produce significant anisotropy in the MBR.

The above arguments, strictly speaking, can only apply to the smoothness of the source term $T_{\alpha\beta}$ in Einstein's equations. Since these equations are nonlinear in $g_{\alpha\beta}$, it is not obvious that the large scale averages of the metric $< g_{\alpha\beta} >$ and the stress-tensor $< T_{\alpha\beta} >$ will be related in the same way as $g_{\alpha\beta}$ and $T_{\alpha\beta}$. This question is difficult to answer reliably but certainly deserves more attention than it has received.

Finally, it must also be noted that we do not have today any alternative cosmological model which does not invoke (i) or (ii). We are, therefore, forced to discuss various models only in the backdrop of these assumptions.

Within this framework, one can divide the discussion into two class of questions: (a) How well do we understand the parameters of the smooth universe? and (b) What are the successes and failures of the models for structure formation? Let us begin with the first question.

Among the various parameters which describe the smooth universe, the most important ones are H and Ω and Λ. Of these three, the Hubble constant is not constrained by any theory directly and needs to be introduced as an observational input. Recent trends suggest that the lower value of $H = 50\,\mathrm{km\,s^{-1}\,Mpc^{-1}}$ is quite likely. The situation regarding Ω is more complex. It seems to be well-established that $\Omega > 0.3$ (at least) at the largest scales and that $0.038(2h)^{-2} < \Omega_B < 0.064(2h)^{-2}$. Further, the luminous matter in the universe contributes only $\Omega_{\mathrm{lum}} \approx 0.007$. Taken together, these observations lead to two important conclusions: (1) Most of the matter in the universe, which is non-luminous, is non-baryonic. (2) Part of the dark matter must be baryonic.

Observations are quite inconclusive as regards the cosmological constant, Λ. But, theoretically, the most natural value for Λ is zero. Any cosmologically interesting value for Λ will require extreme fine-tuning of parameters and is quite ugly.

Since dark matter is the gravitationally dominant component, the scenarios for structure formation depend crucially on the amount and nature of dark matter. Since observations imply $0.3 < \Omega < 3$ (say), the most natural value for Ω is unity. Once again, any other value will require a fine-tuning of the parameters. (Inflationary models do predict $\Omega = 1$; however, the argument based on naturalness is independent of inflation and seems to be more robust.) It may also be noted that most of the low-Ω models produce too much anisotropy in the MBR.

Dark matter candidates which lead to $\Omega \gtrsim 0.3$ can be classified as 'hot' or 'cold'. Direct laboratory experiments unfortunately do not provide us with any suitable candidate. Instead, laboratory results have imposed rather stringent bounds on the allowed mass range for wimps. At the same time, these bounds have not ruled out the possible existence of either hot or cold wimps. Thus, it is not possible to decide about the nature of dark matter on the basis of laboratory results.

In a universe containing wimps and baryons, it is possible to amplify any initial perturbation through the process of gravitational instability. As long as $(\delta\rho/\rho) \ll 1$, the theoretical formulation is unambiguous and can be used with complete confidence to relate $(\delta\rho/\rho)$ at $t = t_{\rm dec}$ to $(\delta\rho/\rho)$ at $t = t_{\rm enter}$. The form of the spectrum at $t = t_{\rm dec}$ will depend on the primordial spectrum and the nature of the dark matter. It is at this stage that observations play a crucial role in constraining the theory.

The various restrictions which arise from the observations are shown in figures 12.1 and 12.2. In figure 12.1 we have chosen the y-axis to be the root-mean-square fluctuation, $(\delta M/M)$, in the mass and the x-axis to be the average background mass, M. The y-axis is marked in terms of the value of the density contrast $\delta_{\rm dec} = (\delta M/M)_{\rm dec}$ at decoupling as well as the corresponding (hypothetical) value today, evaluated using the linear theory. (It is the latter quantity which was called δ_0 in chapter 8; $(\delta M/M)_{\rm now} \equiv \delta_0 = (3/5)\delta_{\rm dec}(1 + z_{\rm dec}) \simeq 660\delta_{\rm dec}$. As we have seen in chapter 8, δ_0 is a convenient parameter to use even though the real density contrast today will be much higher than δ_0.) The length scale in

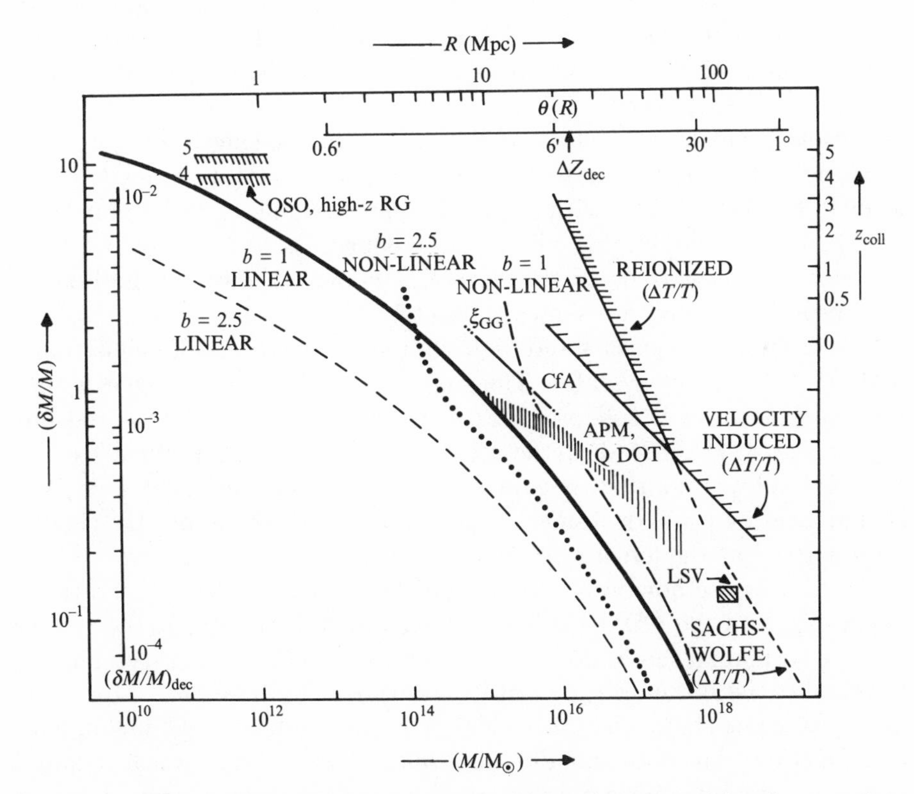

Fig. 12.1. The constraints on the density inhomogeneities arising from various observations are summarized. See text for discussion.

the smooth universe which will contain the mass M is shown along the top horizontal axis. This mass is computed using the Gaussian window function:

$$M(R) = (2\pi^2)^{3/2} R^3 \overline{\rho} \simeq 10^{12}\,\mathrm{M}_\odot\, h_{0.5}^2 (R/1\,\mathrm{Mpc})^3. \qquad (12.1)$$

The top axis also indicates the relevant angle $\theta(R)$ in the MBR anisotropy measurement which will probe the length scale R.

Let us now consider the constraints on $\delta_0(M) = (\delta M/M)$ arising from various observations. (We will assume that $\Omega = 1$ and $h = 0.5$): (a) The large scale streaming velocities of $v_{\mathrm{pec}} \simeq 600\,\mathrm{km\,s}^{-1}$ at a distance of $v_H \simeq 5000\,\mathrm{km\,s}^{-1}$ – which corresponds to 100 Mpc for $h = 0.5$ – requires a density contrast of $\delta \simeq (v_{\mathrm{pec}}/v_H) \simeq 0.12$ at this scale; this point is marked 'LSV' in the figure. (One can, of course, make these calculations more precise; however it is not necessary for obtaining an overall picture.) (b) The IRAS-QDOT survey and the APM survey give information about the power spectrum in the range of, say, $(10\text{–}10^2)$ Mpc. From the original data, one can estimate[1] the value of $(\delta M/M)$ in this range; this result is shown as a shaded band marked 'APM, QDOT'. (The curvature of this shaded region seems to be a real effect.) (c) The observed galaxy–galaxy correlation function appears as a line with a slope of $(-1.88/2 = -0.94)$ in this figure. The CfA survey result, that $(\delta N/N)$ is unity at $8h^{-1}$ Mpc, is also shown as a point.

Anisotropy measurements of MBR at 1° probe the length scale of about 200 Mpc. The bound due to the Sachs–Wolfe effect (at angles $\theta \geq 1^\circ$) leads to the dashed line (with slope -2) marked at the right hand bottom corner. Unfortunately, this effect can be wiped out at $\theta \lesssim 8^\circ$ or so if the universe was re-ionized. In the case of a re-ionized universe the bounds are much weaker and are shown separately.

The existence of quasars and high-z radio galaxies provides constraints on $(\delta M/M)$ at smaller scales. The spherical top-hat model suggests that a mass scale M will collapse (and form a bound system) at a redshift $z_{\mathrm{coll}} = (\delta_0(M)/1.687) - 1$; the z_{coll} is marked on the vertical axis at the right side. Assuming that quasars etc., with masses $(10^{11}\text{–}10^{12})\,\mathrm{M}_\odot$ should form by $z = 4$ or 5 shows that $(\delta M/M)$ must be above the shaded lines marked at the left as 'QSO, high-z RG'.

Let us now see how well various models account for the observations, beginning with the CDM models. As explained in chapter 9, the proper way to normalize these models is to set the *nonlinear* density contrast to unity at the normalization scale, which we take to be $8h^{-1}$ Mpc. The thick black line shows the *linear* density evaluated using a Gaussian window function. The corresponding *nonlinear* density contrast is calculated using the top-hat approximation up to the epoch of turn-around and is matched with the results of CDM numerical simulation by polynomial fit.

(Such an analysis allows one to obtain a function $\delta_{\rm nl} = F(\delta_l)$ giving the nonlinear density contrast for any given linear density contrast.) This nonlinear density contrast is shown by the dash-dot curve labelled 'b = 1, nonlinear'. The overall scale is adjusted so that the *nonlinear* density contrast $\delta_{\rm nl} = 1$ at $8h^{-1}$ Mpc. (This corresponds to setting $\delta \simeq 0.58$ at $8h^{-1}$ Mpc, as discussed in chapter 9.)

The mismatch of slopes between ξ_{GG} and the nonlinear curve at $8h^{-1}$ Mpc shows that this model predicts too steep a correlation function, ξ_{GG}. It is also obvious that the model has less power than that which the observations demand at *both* large and small scales; the curve is consistently below observations.

To obtain the correct value of ξ_{GG} we have to resort to some form of biasing. For example, we may assume that the mass fluctuation $(\delta M/M)$ is a factor b lower than the fluctuations in the number of galaxies $(\delta N/N)$. The linear and nonlinear curves for $b = 2.5$ are also shown in the figure. Since $\delta_l(b) = [\delta_l(1)/b]$ the linear curve for $b = 2.5$ is obtained by merely translating the curve for $b = 1$ downwards. The nonlinear curve has to be generated afresh using $\delta_{\rm nl}(b) = F[\delta_l(1)/b]$. The slope of the nonlinear curve at $8h^{-1}$ Mpc now matches with the slope of the ξ_{GG}. In the process, of course, we have reduced the power at all scales. The discrepancy at large and small scales is now higher by a factor $b = 2.5$.

It may seem from the above discussion that the problem arises essentially due to normalization. This is not quite true. Suppose we ignore the CfA results and try to normalize the spectrum at very large scales – say, at 200 Mpc. This will enhance small scale power considerably and of course will lead to disparities with ξ_{GG}. But what is more important, the theory will again disagree with QDOT-APM results, now at small scales. The CDM power spectrum does not curve rapidly enough to account for all the observations in the (10–100) Mpc range. Thus there seems to be some difficulty with the shape of the spectrum as well[1].

Figure 12.2 shows two other aspects of the spectrum. In the diagram on the left hand side we have plotted the depth of the gravitational potential well – measured by the velocity dispersion $v = (GM/R)^{1/2}$ – at different scales. The horizontal axis on top shows the *physical* size of the virialized structures, calculated using the spherical top-hat model. In the range (10–100) kpc we get the depths of the potential wells to be about (80–200) km s^{-1}. This is reasonable, though slightly on the lower side. (The rough analysis we are performing cannot be used to decide this issue.) In the diagram on the right we have extended the CDM spectra right up to the scale of the present Hubble radius, H_0^{-1}. The lower left side of the diagram indicates the large scale peculiar velocities calculated for the $b = 1$ (dotted curve) and $b = 2.5$ (dash-dot curve) CDM spectra.

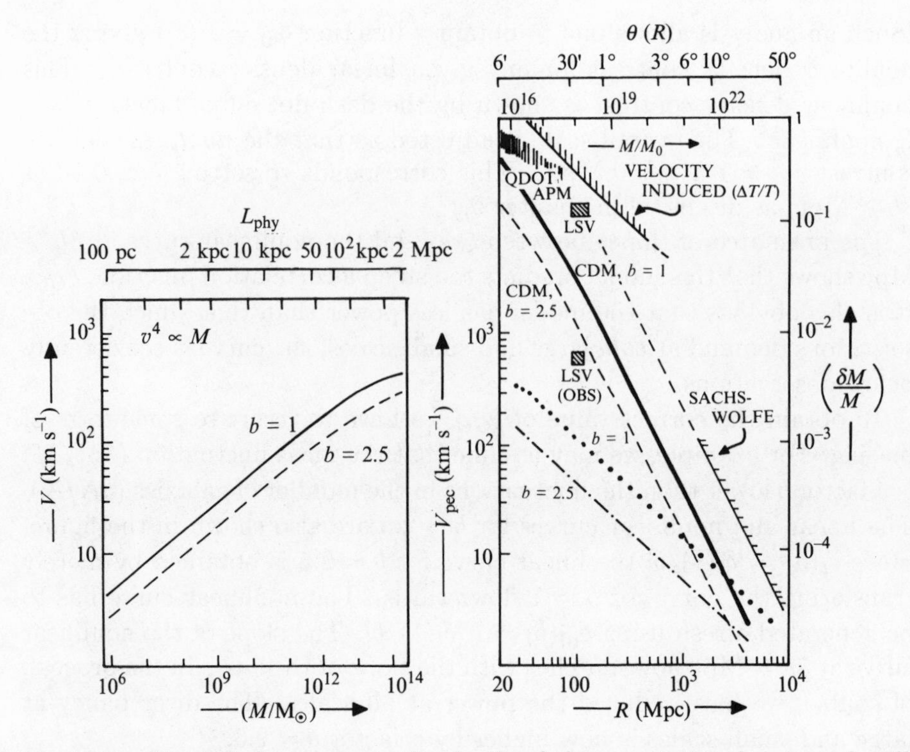

Fig. 12.2. The left figure shows the depth of the gravitational potential well as a function of mass scale. The figure on the right shows the peculiar velocity at large scales and the corresponding density contrasts. See text for detailed discussion.

The observed result (marked 'LSV (OBS)') is way above the theoretical values. The figure also shows the density contrast, $(\delta M/M)$, which is required to account for the observed large scale streaming velocities (marked 'LSV'; this is also shown in figure 12.1).

The very large scales shown in this diagram correspond to MBR anisotropy measurements in the range of (1–50) degrees. Beyond 8 degrees or so, the Sachs–Wolfe effect cannot be wiped out by re-ionization. The Sachs–Wolfe bound is shown by a line with slope R^{-2}. It is now clear that the problems cannot be solved by a simple scaling up of the power at all scales; that would produce MBR anisotropies which are (marginally) too large.

The HDM models do not work any better. Since the first structures which form in the HDM scenarios are cluster-size objects, the normalization of the spectrum is more complicated in this case. This allows one to increase the power slightly at large scales compared to the CDM but the resulting spectrum still falls short of the observations.

What could be a possible solution to these problems? There are, broadly speaking, three possibilities.

The first possibility is that we are missing some crucial physical phenomenon which either accelerates the growth of amplitude or produces a scale dependent biasing factor. It should be stressed that this possibility is not as remote as one might think. Since we are dealing with a complicated N-body problem, with very little analytical insight, it is easy to miss out such phenomena. (Violent relaxation, for example, could have been discovered at least a century earlier!) Signals for such processes should exist in the numerical simulations provided all the relevant ingredients are included in the code. The difficulty, of course, is in deciding what is relevant, since all codes need to use some approximations. If this is the cause for the discrepancies between theory and observation, then it is only a question of time before the problem is solved. We are, at least, on the right track.

Secondly, there is always the chance that the observations may be wrong; or, at least, misleading. With more observations and analysis of data, the character of the results might be hoped to change in the right direction. Though this possibility cannot be ruled out, one must admit that it appears somewhat far-fetched at the moment. In this context, it must be noted that there exist two crucial experimental results which – if obtained – can change the entire complexion of this subject: (i) Detection of an anisotropy in the MBR and (ii) Detection of a viable dark matter candidate in the laboratory. It is gratifying that experiments in these directions are in full swing.

Lastly, it is possible that the difficulties signal a far deeper problem. The absence of anisotropies in the MBR is probably as significant as the null result of the Michelson–Morley experiment. Notice that, eventually, a cosmological model has to address the question of how the universe originated. (It is this question, which is inevitable in any evolving model of the universe, that is so neatly avoided in the steady state models.) Any model for the 'creation' of the universe, based on some viable theory of quantum gravity, will have implications for the parameters describing our universe. It is hard to predict how such a model will affect the present conclusions.

In the final analysis, the first possibility seems the strongest and, probably, the most tractable path. It is hoped that we will have a less contradictory picture of structure formation by the end of the 1990s.

"bhramasya jagatasya 'sya jatasya 'kasavarnavat
apunah smaranam manye sadho vismaranam varam."
(The world-appearance is a confusion, even as the
blueness of the sky is an optical illusion. I think it
is better not to let the mind dwell on it, but to ignore it.)
– Yoga Vasistha

Appendix A
Aspects of general relativity

A.1 Introduction

This appendix introduces the key concepts of general relativity. Detailed derivations, which can be found in standard text books[1], are omitted. The emphasis is on the physical concepts and the logical structure of general relativity rather than on the mathematical details.

A.2 Principle of equivalence

The motion of a particle in a Newtonian gravitational field is described by the action,

$$A = \int dt\, \frac{1}{2} m\dot{\mathbf{x}}^2 - \int dt\, m\phi(\mathbf{x}, t) = m \int dt\, \left(\frac{1}{2}\dot{\mathbf{x}}^2 - \phi\right) \qquad \text{(A.1)}$$

where m is the mass of the particle and ϕ is the Newtonian gravitational potential. The kinetic energy term in the action is independent of the interaction and is *always* proportional to the mass m. It is, however, remarkable that the second term – which represents the influence of gravitational field on the mass – is *also* proportional to the mass, allowing us to pull m out of the integral. In contrast, note that the coupling to an electromagnetic field is not dependent on m; instead, it is proportional to some other parameter (q) of the particle called the electric charge.

Since the multiplication of the action functional by a constant does not change the equation of motion, the trajectory of a particle in a *given* gravitational field is completely independent of *any* of the parameters characterizing the particle. In other words, the gravitational potential $\phi(\mathbf{x}, t)$ itself defines a set of curves, which will be trajectories of the material particles in that field. In contrast, one cannot obtain a similar set of curves from the electromagnetic potentials; the trajectory of a particle will depend not only on the potentials but also on the parameter (q/m) of the particle which is moving.

This fact leads to the following peculiar property for the gravitational field: Consider a small volume $\mathcal{V}$ around some point $\mathbf{x}_0$ in space. It is always possible to choose a coordinate system in $\mathcal{V}$, such that *all* the particles in it move with no acceleration in this coordinate system. This can be achieved by using the coordinates

$$\xi^i = x^i - \frac{1}{2}g^i t^2 \tag{A.2}$$

where x^i $(i = 1, 2, 3)$ are the original inertial coordinates and $g^i = -(\partial\phi/\partial x^i)_0$ is the gravitational acceleration at $\mathbf{x}_0$. Notice that the coordinates $\{\xi^i\}$ represent a *non-inertial* frame of reference. It follows that the mechanical phenomena in a non-inertial frame of reference are indistinguishable from those in a suitably chosen gravitational field, as far as the local behaviour is concerned.

The above results also hold for a special relativistic particle interacting with a gravitational field described by a scalar potential ϕ. In this case, the action will be

$$A = -m\int dt\sqrt{1-\mathbf{v}^2} - m\int dt\,\phi(\mathbf{x},t) = -m\int dt\left[\sqrt{1-\mathbf{v}^2} + \phi(\mathbf{x},t)\right]. \tag{A.3}$$

The trajectories $\mathbf{x}(t)$ are specified by ϕ alone and are independent of m. In this case, we can derive some more important consequences. Consider two frames of references S and S', with S' moving with an *acceleration* g with respect to S along the direction of positive x-axis. An observer in S' will feel that all particles in his vicinity are accelerated towards the $(-\hat{\mathbf{x}})$ direction with the magnitude of the acceleration being g. Hence he can attribute a gravitational potential $\phi = gx$ in the region where he is moving and ignore the fact that he is accelerating. Further suppose that there exists a series of clocks along the x-axis in S. The observer in S' can compare the reading in his clock with each of the clocks in S as he crosses it. It can be easily shown using special relativity that the clock readings in the two frames are connected by the relation $t = g^{-1}\sinh gt'$. So, for $gt \ll c$, we have

$$\Delta t' \cong \Delta t\left(1 - \frac{1}{2}\frac{g^2t^2}{c^2}\right). \tag{A.4}$$

But $x \cong (1/2)gt^2$ is the location of the accelerated observer at time t. Hence

$$\Delta t' \cong \Delta t\left(1 - \frac{gx}{c^2}\right) = \Delta t\left(1 - \frac{\phi}{c^2}\right). \tag{A.5}$$

Since the accelerated observer believes that he is at rest in a gravitational field he has to interpret this result by saying that the gravitational potential ϕ present in his frame is slowing down the clocks. We see that, in special relativity, there arises a possibility of gravitational field influencing the rate of clocks. Similar analysis will show that gravitational field can affect the frequency of radiation and length measurements.

As long as clocks are made of normal material particles, their rate will be influenced by the gravitational field. In an arbitrary gravitational field, this influence can be quite complicated. In principle, we can work out this effect by studying in detail the influence of gravitational field on the rods and clocks used to measure the spacetime intervals. However, since this effect seems to be fundamental and universal, it would be desirable to have an alternative mathematical description of the gravitational field in which these effects are incorporated in a natural fashion.

Since non-inertial frames mimic a gravitational field locally, it is clear that a better description of gravitational field should use the same apparatus as that used in the description of a non-inertial frame. At the same time, one must realize a crucial difference between the gravitational field and a non-inertial frame: It will be impossible to make the gravitational field vanish over an *extended* region of spacetime by the coordinate choice. In other words, a gravitational field is 'more general' than a non-inertial frame, though *locally* equivalent to it.

Thus we need to use a mathematical description which is more general than that used to describe a non-inertial frame, but uses the same language. To discover such a description, we should first determine the correct mathematical description of a non-inertial frame. We shall now turn to this question.

A.3 Curved spacetime

Let $X^\alpha = (T, X, Y, Z)$ be the special relativistic, inertial frame of reference and let $x^\alpha = x^\alpha(X)$ be some set of non-inertial coordinates, related to X^α by some specified coordinate transformation. Since $dX^\beta = (\partial X^\beta/\partial x^\alpha)\, dx^\alpha$, the interval ds^2 between two events can be written in terms of dx^μ as:

$$ds^2 \equiv \eta_{\alpha\beta}\, dX^\alpha\, dX^\beta = \eta_{\alpha\beta} \left(\frac{\partial X^\alpha}{\partial x^\mu}\right) \left(\frac{\partial X^\beta}{\partial x^\nu}\right) dx^\mu\, dx^\nu \equiv g_{\mu\nu}(x)\, dx^\mu\, dx^\nu . \tag{A.6}$$

Here $\eta_{\alpha\beta} = \mathrm{dia}\,(1, -1, -1, -1)$ and we are using the 'summation convention': Summation is implied over any index which occurs twice in a given term; for example, $(\partial X^\alpha/\partial x^\mu)\, dx^\mu$ will actually stand for the sum $\sum_{\mu=0}^{3} (\partial X^\alpha/\partial x^\mu)\, dx^\mu$, etc. The last equality defines the set of ten functions:

$$g_{\mu\nu}(x) \equiv \eta_{\alpha\beta} \left(\frac{\partial X^\alpha}{\partial x^\mu}\right) \left(\frac{\partial X^\beta}{\partial x^\nu}\right) \tag{A.7}$$

called the 'metric tensor'. The central difference between an inertial frame and non-inertial frame is apparent in (A.6). The interval retains the same form – with the same numerical value for $\eta_{\alpha\beta}$ – in all inertial frames connected by the Lorentz transformations; in contrast, the transformation to a non-inertial frame introduces a metric $g_{\alpha\beta}$ which depends on the coordinates.

This immediately suggests the possibility that we may be able to describe the gravitational field by the functions $g_{\alpha\beta}(x)$. In other words, we would like to assume that the interval between two events in a general spacetime is given by

$$ds^2 = g_{\mu\nu}(x)\, dx^\mu\, dx^\nu \tag{A.8}$$

where the form of $g_{\mu\nu}(x)$ depends on the nature of the gravitational field present on the spacetime.

Such a hypothesis satisfies the two properties discussed before: (i) Since non-inertial frames are described by a metric which depends on the spacetime coordinates, it is possible that we can set up a local equivalence between gravitational field and the non-inertial frame. We will make this correspondence more precise a little later. (ii) More importantly, note that (A.8) represents a much wider class of spacetimes than those which can be obtained by coordinate transformation from the inertial frame. An intuitive way of understanding this result is the following: A coordinate transformation is characterized by four independent functions; hence any metric obtained by such a coordinate transformation really has only four-function-degrees of freedom (i.e. four numbers per event). But a *general* metric $g_{\mu\nu}(x)$ has ten-function-degrees of freedom (ten numbers at each event). Therefore, (A.8) with a general $g_{\mu\nu}(x)$ represents a much wider class of metrics than those obtainable by coordinate transformation. Since we *do* expect the gravitational field to be more general than a non-inertial frame, this situation is completely acceptable.

The above conclusion can be stated in a different manner: Consider the spacetime in (A.8), with some arbitrary $g_{\mu\nu}(x)$. If we change the coordinate system from x^μ to $\overline{x}^\mu$, the form of the metric will change. However, in general, it will *not* be possible to reduce $g_{\mu\nu}$ to the form dia $(1, -1, -1, -1)$ by a coordinate choice. Such a spacetime is said to be 'curved'. Our original hypothesis can, therefore, be stated as follows: 'Gravitational field is equivalent to the curvature of the spacetime.' Let us consider some of the consequences of this hypothesis.

First of all, notice that this assumption is consistent with the known properties of the weak ($\phi \ll c^2$) gravitational field. Consider, for example, the action for a special relativistic particle in a gravitational field:

$$\begin{aligned} A &= -\int mc^2\sqrt{1-v^2/c^2}\,dt - m\int \phi\,dt \\ &= -mc\int (c\,dt)\left[\sqrt{1-v^2/c^2}+\frac{\phi}{c^2}\right] \\ &\cong -mc\int \left[(1+\frac{2\phi}{c^2})c^2\,dt^2 - d\mathbf{x}^2\right]^{1/2} \end{aligned} \tag{A.9}$$

where the last equality is valid for $\phi \ll c^2$. This expression is identical to the form of the special relativistic action

$$A = -mc\int ds = -mc\int \sqrt{g_{\alpha\beta}\,dx^\alpha\,dx^\beta} \tag{A.10}$$

provided we identify

$$g_{\alpha\beta}\,dx^\alpha\,dx^\beta \cong \left(1+\frac{2\phi(\mathbf{x},t)}{c^2}\right)c^2\,dt^2 - dx^2 - dy^2 - dz^2. \tag{A.11}$$

Thus, weak Newtonian ($\phi \ll c^2$) gravitational fields can indeed be interpreted as a non-trivial change in the metric; with $g_{00} = (1+2\phi/c^2)$, $g_{ik} = -\delta_{ik}$.

This metric also shows the influence of gravity on clock readings. Suppose – as is usually the case – the potential ϕ vanishes as $|\mathbf{x}| \to \infty$. The time interval dT measured by a clock at infinity is related to dt measured at finite $\mathbf{x}$ by $dT \cong (1+\phi/c^2)\,dt$. So the reading of clocks in a gravitational potential is $dt \cong dT(1-\phi/c^2)$ where dT may be thought of as the reading in a clock unaffected by gravity. This is consistent with the result derived earlier.

It is also easy to verify the earlier claim that our hypothesis respects the local equivalence between gravity and non-inertial frame. If we change the coordinates from x^μ to $\overline{x}^\mu$, then the metric $g_{\mu\nu}$ will change to $\overline{g}_{\mu\nu}$ with

$$g_{\mu\nu}\,dx^\mu\,dx^\nu = g_{\mu\nu}\frac{\partial x^\mu}{\partial \overline{x}^\alpha}\frac{\partial x^\nu}{\partial \overline{x}^\rho}\,d\overline{x}^\alpha\,d\overline{x}^\rho \equiv \overline{g}_{\alpha\rho}\,d\overline{x}^\alpha\,d\overline{x}^\rho \tag{A.12}$$

so that

$$\overline{g}_{\alpha\rho} = g_{\mu\nu}\frac{\partial x^\mu}{\partial \overline{x}^\alpha}\frac{\partial x^\nu}{\partial \overline{x}^\rho}. \tag{A.13}$$

We would like to choose the transformation from x^μ to $\overline{x}^\mu$ in such a way that, at some chosen event $\mathcal{P}$ in the spacetime, $\overline{g}_{\alpha\rho} = \eta_{\alpha\rho}$ (ten conditions) *and* $(\partial\overline{g}_{\alpha\rho}/\partial\overline{x}^\beta) = 0$ (forty conditions). Near the event $\mathcal{P}$, let us expand $\overline{x}^\mu$ as

$$\overline{x}^\mu = \alpha^\mu + \beta^\mu_\nu x^\nu + \gamma^\mu_{\nu\sigma}x^\nu x^\sigma + \cdots. \tag{A.14}$$

Simple counting shows that we have enough coefficients in this expansion to ensure the fifty conditions. Thus we can always choose a coordinate system such that, in an infinitesimal region around an event, the effects of the gravitational field can be made to vanish. Such a coordinate system is called 'locally inertial'. Notice that, *even to linear order*, we can maintain $g_{\alpha\beta} = \eta_{\alpha\beta}$; the deviation is quadratic in the distance from the chosen point $\mathcal{P}$. In any such local region, we can use the results of special relativity.

Treating $g_{\alpha\beta}$ as a matrix, we can define the quantities $g^{\mu\nu}$ as the elements of the inverse matrix g^{-1}; that is, $g^{\mu\nu} g_{\nu\sigma} = \delta^{\mu}_{\sigma}$. The transformation law for $g^{\mu\nu}$ is easy to deduce from the one for $g_{\alpha\beta}$; we get

$$g'^{\mu\nu} = \left(\frac{\partial x'^{\mu}}{\partial x^{\alpha}}\right)\left(\frac{\partial x'^{\nu}}{\partial x^{\rho}}\right) g^{\alpha\rho}. \tag{A.15}$$

By taking the determinant of the transformation equation for $g_{\alpha\beta}$, we see that $g' = J^2 g$ where J is the Jacobian of the transformation from x to x'. Since the metric tensor has the signature $(+ - - -)$, g is negative and we may write this relation as $\sqrt{-g'} = J\sqrt{-g}$. But notice that, under the transformation $x \to x'$, the infinitesimal volume elements are related by $d^4x = J\, d^4x'$. Therefore we see that $\sqrt{-g'}\, d^4x' = \sqrt{-g}\, d^4x$. Thus the *invariant* volume element in a spacetime with metric $g_{\alpha\beta}$ is $\sqrt{-g}\, d^4x$.

For future use, we will also derive an expression for the derivative of the determinant: $(\partial g/\partial x^{\alpha})$. The quantity dg is obtained by taking the differential $dg_{\alpha\beta}$ of each component of $g_{\alpha\beta}$ and multiplying by the corresponding minor $\mathcal{M}^{\alpha\beta}$. But, by definition of the inverse, $g^{\alpha\beta} = (\mathcal{M}^{\alpha\beta}/g)$. So $dg = dg_{\alpha\beta}\,(g g^{\alpha\beta}) = g g^{\alpha\beta}\, dg_{\alpha\beta} = -g g_{\alpha\beta}\, dg^{\alpha\beta}$, giving

$$\left(\frac{\partial g}{\partial x^{\alpha}}\right) = g g^{\mu\nu}\left(\frac{\partial g_{\mu\nu}}{\partial x^{\alpha}}\right) = -g g_{\mu\nu}\left(\frac{\partial g^{\mu\nu}}{\partial x^{\alpha}}\right). \tag{A.16}$$

A.4 Physics in curved spacetime

In the flat spacetime of special relativity, there exists a natural set of coordinates X^{α}, in which the metric is $\eta_{\alpha\beta} = \mathrm{dia}\,(1, -1, -1, -1)$. Any other coordinate system can be studied with reference to this inertial frame; laws of physics need only to be stated in such a form that they remain valid in the inertial frame. The form of the laws in the non-inertial frame can always be found by transformation.

The situation is different in curved spacetime. Since it is impossible to choose coordinates in such a way that an arbitrary $g_{\mu\nu}$ will be reduced to a pre-assigned form, no choice of coordinates is special. The laws of physics, therefore, have to be written in such a way that they retain the same form in all coordinates. (We say that the laws must be 'covariant'). In other words, the form of the equations should be the same after a coordinate transformation. This can be easily achieved by using tensors.

The simplest covariant equation is a scalar (tensor of zero rank) equation of the form $f(x^\alpha) = g(x^\alpha)$; under a coordinate transformation $x^\alpha \to x'^\alpha$, scalars 'transform' as $f'(x'^\alpha) = f(x^\alpha)$ and any scalar equation retains its form.

Let us consider vectors next. Given some curve $x^\alpha(\lambda)$ in the spacetime, the quantity $v^\alpha = (dx^\alpha/d\lambda)$ clearly has the interpretation of a tangent vector to the curve. It is trivial to verify that v^α transforms as

$$v'^\alpha = \left(\frac{\partial x'^\alpha}{\partial x^\mu}\right) v^\mu \tag{A.17}$$

under a change of coordinates. Since any vector is a tangent vector to its integral curve, this definition is quite general. By assigning a vector to each event of spacetime, we can define a vector field $v^\alpha(x)$. Addition of vectors and multiplication of vectors by scalars are also, clearly, covariant operations provided they are performed on the same location.

Let us next define a quantity $T^{\alpha\beta} = v^\alpha v^\beta$. To maintain the form of this equation in another coordinate frame, both sides must transform in the same way. Since we know the transformation property of the right hand side, we can find the transformation law of $T^{\alpha\beta}$:

$$T'^{\alpha\beta} = \left(\frac{\partial x'^\alpha}{\partial x^\mu}\right)\left(\frac{\partial x'^\beta}{\partial x^\nu}\right) T^{\mu\nu}. \tag{A.18}$$

We shall *define* this to be the transformation law for an arbitrary 'second rank' tensor. One such special, second rank tensor is $g^{\mu\nu}$, the inverse of the metric $g_{\mu\nu}$ defined by the relation $g^{\mu\nu}g_{\nu\sigma} = \delta^\mu_\sigma$; it can be easily verified that $g^{\mu\nu}$ transforms as above. Higher order tensors like $A^{\alpha\rho\beta}, B^{\alpha\rho\beta\cdots}$ etc. can be assigned similar transformation properties.

It is convenient to define two operations – 'lowering' and 'raising' of indices – using $g_{\mu\nu}$ and $g^{\mu\nu}$. For example, given a tensor $T^{\alpha\beta}$ we define the quantities T^α_β, $T_{\alpha\beta}$ etc. by:

$$T^\alpha_\beta \equiv g_{\mu\beta}T^{\alpha\mu}; \quad T_{\alpha\beta} \equiv g_{\alpha\mu}g_{\beta\nu}T^{\mu\nu}. \tag{A.19}$$

Similarly, $T^\alpha_\beta = g^{\alpha\mu}T_{\mu\beta}$ etc. It is easy to verify that: (i) these equations are covariant and (ii) every upper index (called a 'contravariant' index) transforms with a factor $(\partial x'^\alpha/\ \partial x^\mu)$ while every lower index (called a 'covariant' index) transforms with a factor $(\partial x^\mu/\partial x'^\alpha)$. With these definitions, all local operations – like $u_\alpha T^{\alpha\beta}$, $u^\mu T^{\nu\sigma}$, $g_{\mu\nu}u^\mu u^\nu$ etc. – are covariant, and all local algebraic laws of physics can be cast in a form which is independent of the coordinates. The transformation laws also show that a tensor which vanishes in one coordinate system will vanish in all coordinate systems.

An interesting difficulty arises when we have to deal with the derivatives of the tensorial quantities. To see this, consider the motion of a free particle in a flat spacetime. The trajectory $X^\alpha(\lambda)$, obtained by varying the special relativistic action

$$A = -m \int ds = -m \int \left(\eta_{\alpha\beta} \frac{dX^\alpha}{d\lambda} \frac{dX^\beta}{d\lambda} \right)^{1/2} d\lambda, \tag{A.20}$$

is a straight line, satisfying the equation: $(d^2X^\alpha/d\lambda^2) = (du^\alpha/d\lambda) = 0$. Given the tangent vector $u^\alpha = (dX^\alpha/d\lambda)$ of a curve $X^\alpha(\lambda)$, we can express u^α in terms of the coordinates X^α along the curve and obtain the vector field $u^\alpha(X)$ along the curve. The equation of motion can now be stated as:

$$\frac{du^\alpha}{d\lambda} = \left(\frac{dX^\beta}{d\lambda} \right) \left(\frac{\partial u^\alpha}{\partial X^\beta} \right) = u^\beta \left(\frac{\partial u^\alpha}{\partial X^\beta} \right) \equiv u^\beta (u^\alpha{}_{,\beta}) = 0. \tag{A.21}$$

Suppose we now choose a non-inertial coordinate system x^μ to describe the motion. Since the particle will *not* be moving in a straight line (that is, since, $d^2x^\mu/d\lambda^2 \neq 0$) in these coordinates, (A.21) *cannot* be valid in an arbitrary frame. We have to modify this law into a form which will be valid in any frame.

The mathematical reason for this difficulty is easy to understand: The derivative of the vector field $u^\alpha{}_{,\beta}(x)$ does not transform like a second rank tensor; by explicit transformation, it is easy to see that

$$\begin{aligned} u'^\alpha{}_{,\beta} &= \frac{\partial u'^\alpha}{\partial x'^\beta} = \frac{\partial x^\mu}{\partial x'^\beta} \frac{\partial}{\partial x^\mu} \left\{ \frac{\partial x'^\alpha}{\partial x^\nu} u^\nu \right\} \\ &= \left(\frac{\partial x^\mu}{\partial x'^\beta} \frac{\partial x'^\alpha}{\partial x^\nu} \right) u^\nu{}_{,\mu} + \left[\left(\frac{\partial x^\mu}{\partial x'^\beta} \right) \frac{\partial^2 x'^\alpha}{\partial x^\mu \partial x^\nu} \right] u^\nu \end{aligned} \tag{A.22}$$

for *any* vector field. The presence of the second term shows that $u^\alpha{}_{,\beta}$ does not transform like a tensor.

To find the correct equation which should replace (A.21) in a curved spacetime, we should start with the correct action principle describing the motion of a particle in a curved spacetime. Since the trajectory of the particle should be independent of the properties of the particle and depend only on $g_{\mu\nu}(x)$, the action functional should be constructed out of $g_{\mu\nu}$ except for an overall multiplicative constant. Further, in the absence of a genuine gravitational field, the action must reduce to the special relativistic form. Lastly, the numerical value of the action should not depend on the coordinate system used. The only choice which will satisfy these conditions is

$$A = -m \int ds = -m \int \sqrt{g_{\mu\nu}\, dx^\mu\, dx^\nu} = -m \int \left(g_{\mu\nu} \frac{dx^\mu}{d\lambda} \frac{dx^\nu}{d\lambda} \right)^{1/2} d\lambda. \tag{A.23}$$

The trajectory $x^\mu(\lambda)$ can be found by varying this action. Since we can invoke special relativity at a local region around any event, it follows that u^μ is timelike for the massive particles and zero for massless particles. Also, since the Lagrangian multiplying $d\lambda$ in (A.23) is independent of λ, it follows that $(g_{\mu\nu}u^\mu u^\nu)$ is constant along the trajectory. For massive particles we will choose this constant to be unity, which is equivalent to choosing $\lambda = s$, the proper time. The Euler–Lagrange equations then give

$$\frac{d}{ds}\left[\frac{1}{2}(g_{\mu\nu}u^\mu u^\nu)^{-1/2} 2 g_{\alpha\beta}u^\beta\right] = \frac{1}{2}(g_{\mu\nu}u^\mu u^\nu)^{-1/2}\frac{\partial g_{\mu\nu}}{\partial x^\alpha}u^\mu u^\nu. \tag{A.24}$$

Setting $g_{\mu\nu}u^\mu u^\nu = 1$ and writing $(dg_{\alpha\beta}/ds) = (\partial g_{\alpha\beta}/\partial x^\mu)u^\mu$ we find that

$$g_{\alpha\beta}\frac{du^\beta}{ds} = \frac{1}{2}\frac{\partial g_{\mu\nu}}{\partial x^\alpha}u^\mu u^\nu - \frac{\partial g_{\alpha\beta}}{\partial x^\mu}u^\mu u^\beta = \frac{1}{2}\left(\frac{\partial g_{\mu\nu}}{\partial x^\alpha} - \frac{\partial g_{\alpha\nu}}{\partial x^\mu} - \frac{\partial g_{\alpha\mu}}{\partial x^\nu}\right)u^\mu u^\nu. \tag{A.25}$$

In arriving at the last equality, we have used the symmetry of $u^\mu u^\nu$ to write the second term in a symmetric manner. This equation may be written as

$$\frac{du^\mu}{d\lambda} = -\frac{1}{2}g^{\mu\nu}\left(-\frac{\partial g_{\sigma\delta}}{\partial x^\nu} + \frac{\partial g_{\sigma\nu}}{\partial x^\delta} + \frac{\partial g_{\nu\delta}}{\partial x^\sigma}\right)u^\sigma u^\delta \equiv -\Gamma^\mu_{\sigma\delta}u^\sigma u^\delta, \tag{A.26}$$

where the second equality defines the quantities $\Gamma^\mu_{\nu\sigma}$ (called 'affine connection' or 'Christoffel symbols') in terms of the derivatives of the metric. Notice that $\Gamma^\mu_{\nu\sigma}$ is symmetric in the lower indices; $\Gamma^\mu_{\nu\sigma} = \Gamma^\mu_{\sigma\nu}$. This equation (called a 'geodesic equation') governs the motion of material particles in a given gravitational field. We can rewrite (A.26) as

$$\frac{du^\mu}{d\lambda} = \left(\frac{dx^\alpha}{d\lambda}\right)\left(\frac{\partial u^\mu}{\partial x^\alpha}\right) = u^\alpha\left(\frac{\partial u^\mu}{\partial x^\alpha}\right) = -\Gamma^\mu_{\rho\alpha}u^\rho u^\alpha \tag{A.27}$$

or,

$$u^\alpha\left[\frac{\partial u^\mu}{\partial x^\alpha} + \Gamma^\mu_{\rho\alpha}u^\rho\right] \equiv u^\alpha u^\mu_{;\alpha} = 0, \tag{A.28}$$

where the first equality defines the symbol $u^\mu_{;\alpha}$. Since our action principle is covariant, this equation must be valid in any frame of reference. It follows that the quantity

$$u^\alpha_{;\beta} \equiv u^\alpha_{,\beta} + \Gamma^\alpha_{\mu\beta}u^\mu \tag{A.29}$$

must transform as a tensor. From the transformation law for $g_{\mu\nu}$, it is easy to verify that $\Gamma^\alpha_{\beta\epsilon}$ transforms as

$$\Gamma'^\alpha_{\beta\epsilon} = \frac{\partial x'^\alpha}{\partial x^\mu}\frac{\partial x^\nu}{\partial x'^\beta}\frac{\partial x^\sigma}{\partial x'^\epsilon}\Gamma^\mu_{\nu\sigma} + \frac{\partial x'^\alpha}{\partial x^\mu}\frac{\partial^2 x^\mu}{\partial x'^\beta \partial x'^\epsilon}. \tag{A.30}$$

Comparing the second term in (A.30) with the second term in (A.22), it is easy to see that $u^\alpha_{;\beta}$ does transform as a tensor even though neither $\Gamma^\alpha_{\beta\epsilon}$ nor $u^\alpha_{,\beta}$ is a tensor.

The quantity $u^\alpha_{;\beta}$, called the 'covariant derivative', generalizes the notion of differentiation in a useful manner. We can define the covariant derivatives of tensors of different rank by invoking the principle that covariant differentiation must obey the usual 'chain rule' of differentiation. We are, then, led to the following conclusions: (1) Since the ordinary derivative of a scalar $\phi_{,\alpha} = (\partial\phi/\partial x^\alpha)$ does transform like a covariant vector, we define $\phi_{;\alpha} = \phi_{,\alpha}$ for scalars. (2) Since $(u^\mu u_\mu)$ is scalar, $(u^\mu u_\mu)_{,\beta} = (u^\mu u_\mu)_{;\beta}$. Using the chain rule on both sides and the known form of $u^\mu_{;\beta}$, it is easy to verify that

$$u_{\mu;\beta} = u_{\mu,\beta} - \Gamma^\nu_{\mu\beta} u_\nu. \tag{A.31}$$

This defines the covariant derivative of a covariant vector. (3) The derivatives of higher rank tensors are defined by using these results and the chain rule. Consider, for example, the tensor $T^\alpha_\beta = u^\alpha u_\beta$. Since $T^\alpha_{\beta;\mu} = (u^\alpha u_\beta)_{;\mu} = (u^\alpha_{;\mu})u_\beta + u^\alpha(u_{\beta;\mu})$, it follows that

$$T^\alpha_{\beta;\mu} = T^\alpha_{\beta,\mu} + \Gamma^\alpha_{\nu\mu} T^\nu_\beta - \Gamma^\nu_{\beta\mu} T^\alpha_\nu. \tag{A.32}$$

This will be the definition for $T^\alpha_{\beta;\mu}$ for an arbitrary T^α_β. A similar procedure is adopted for other tensors. (4) It can be easily verified by direct computation that $g_{\mu\nu;\beta} = 0$. This fact allows us to interchange the order of covariant differentiation and the process of raising and lowering the indices. For example,

$$u_{\alpha;\beta} = (g_{\alpha\mu} u^\mu)_{;\beta} = (g_{\alpha\mu;\beta})u^\mu + g_{\alpha\mu}(u^\mu_{;\beta}) = g_{\alpha\mu}(u^\mu_{;\beta}). \tag{A.33}$$

The covariant derivative can be given a geometrical interpretation along the following lines: Let $v^\alpha(x)$ be a vector field and let $x^\alpha(\lambda)$ be some curve. Then the quantity

$$v^\alpha_{;\beta}\left(\frac{dx^\beta}{d\lambda}\right) = v^\alpha_{,\beta}\left(\frac{dx^\beta}{d\lambda}\right) + \Gamma^\alpha_{\mu\beta} v^\mu \left(\frac{dx^\beta}{d\lambda}\right) \tag{A.34}$$

can be thought of as the generalization of the 'directional derivative' of the vector field along a particular direction specified by the tangent vector $(dx^\beta/d\lambda)$. Suppose we are given a vector k^α at one event $\mathcal{P}$ and some curve $x^\alpha(\lambda)$ passing through $\mathcal{P} = x^\alpha(0)$. We can then solve the differential equation

$$\frac{dv^\alpha}{d\lambda} + \Gamma^\alpha_{\mu\nu}(\lambda)\frac{dx^\mu}{d\lambda} v^\nu = 0 \tag{A.35}$$

where $\Gamma^\alpha_{\mu\nu}(\lambda) = \Gamma^\alpha_{\mu\nu}[x(\lambda)]$ etc., with the boundary condition $v^\alpha(\lambda = 0) = k^\alpha$, thereby obtaining the function $v^\alpha(\lambda)$. This allows us to *define* a vector *field* along the curve $x^\alpha(\lambda)$, given the vector k^α at only one point. If the space time is actually flat, then this construction is equivalent to 'moving' the vector from event to event maintaining the same Cartesian components; i.e., the vector is moved 'parallel' to itself. The above construction generalizes the notion of 'parallel transport' to curved spacetime.

The following point should be noted. Let $x^\alpha_A(\lambda)$ and $x^\alpha_B(\lambda)$ be two arbitrary curves intersecting at two events $\mathcal{P}$ and $\mathcal{M}$. Given a vector k^α at $\mathcal{P}$ we can parallel transport it along either of the curves to obtain a vector at $\mathcal{M}$. In a flat spacetime, we will get the same vector at $\mathcal{M}$ whichever curve we choose. (This result is obvious in the Cartesian coordinates; since parallel transport is a covariant procedure, the result must hold in any frame.) However, this result is *not* true in a curved spacetime. The value of the vector at $\mathcal{M}$ will depend on the curve along which the vector is transported. We will discuss this feature in detail in the next section.

The calculation of covariant derivatives can be simplified whenever repeated indices occur. To obtain more convenient formulas, we begin by proving two simple identities. From the definition of $\Gamma^\mu_{\nu\sigma}$ it follows that

$$\Gamma^\alpha_{\beta\alpha} = \frac{1}{2} g^{\alpha\gamma} \frac{\partial g_{\alpha\gamma}}{\partial x^\beta} = \frac{1}{2g} \frac{\partial g}{\partial x^\beta} = \frac{\partial}{\partial x^\beta} (\ln \sqrt{-g}), \tag{A.36}$$

where we have used (A.16). Similarly,

$$g^{\beta\epsilon}\Gamma^\alpha_{\beta\epsilon} = g^{\beta\epsilon} g^{\alpha\gamma} \left(\frac{\partial g_{\gamma\beta}}{\partial x^\epsilon} - \frac{1}{2} \frac{\partial g_{\beta\epsilon}}{\partial x^\gamma} \right) = -\frac{1}{\sqrt{-g}} \frac{\partial(\sqrt{-g} g^{\alpha\beta})}{\partial x^\beta}. \tag{A.37}$$

Consider now the covariant divergence of a vector: $A^\alpha_{;\alpha}$. From the definition of the covariant derivative we get

$$A^\alpha_{;\alpha} = \frac{\partial A^\alpha}{\partial x^\alpha} + \Gamma^\alpha_{\epsilon\alpha} A^\epsilon = \frac{\partial A^\alpha}{\partial x^\alpha} + A^\epsilon \frac{\partial (\ln \sqrt{-g})}{\partial x^\epsilon} = \frac{1}{\sqrt{-g}} \frac{\partial(\sqrt{-g} A^\alpha)}{\partial x^\alpha} \tag{A.38}$$

where we have used (A.36). Further, if A^α was a gradient of some function ϕ, $A^\alpha = \partial^\alpha \phi$, then $A^\alpha_{;\alpha}$ will represent the covariant Laplacian $\phi^{;\alpha}_{;\alpha}$ of the scalar ϕ. We see that

$$\phi^{;\mu}_{;\mu} = \frac{1}{\sqrt{-g}} \frac{\partial}{\partial x^\mu} \left(\sqrt{-g} g^{\mu\nu} \frac{\partial \phi}{\partial x^\nu} \right). \tag{A.39}$$

In particular, note that $\phi^{;\mu}_{;\mu}$ is *not* given by $g^{\mu\nu}\partial_\mu \partial_\nu \phi$. A similar result can be derived for the covariant derivative of an antisymmetric tensor $A^{\alpha\beta}$. We have

$$A^{\alpha\beta}_{;\beta} = \frac{\partial A^{\alpha\beta}}{\partial x^\beta} + \Gamma^\alpha_{\gamma\beta} A^{\gamma\beta} + \Gamma^\beta_{\gamma\beta} A^{\alpha\gamma} = \frac{\partial A^{\alpha\beta}}{\partial x^\beta} + \Gamma^\beta_{\gamma\beta} A^{\alpha\gamma} \tag{A.40}$$

since $\Gamma^\alpha_{\gamma\beta} A^{\gamma\beta} = 0$ for an antisymmetric $A^{\gamma\beta}$. Using (A.36), we get

$$A^{\alpha\beta}_{;\beta} = \frac{1}{\sqrt{-g}} \frac{\partial(\sqrt{-g} A^{\alpha\beta})}{\partial x^\beta}. \tag{A.41}$$

Also notice that, $A_{\alpha;\beta} - A_{\beta;\alpha} = (\partial A_\alpha/\partial x^\beta) - (\partial A_\beta/\partial x^\alpha)$ since the term involving Γ cancels out.

With the notions developed above, we can express any special relativistic law in a form which is valid in an arbitrary coordinate system. Consider, for example, the action for the electromagnetic field in flat spacetime, which is

$$A_{\rm em} = -\frac{1}{16\pi}\int F_{\alpha\beta}F^{\alpha\beta}\, d^4X; \quad F_{\alpha\beta} = A_{\beta,\alpha} - A_{\alpha,\beta}. \tag{A.42}$$

This action can be rewritten in a form which is generally covariant. The covariant volume element in arbitrary coordinates is $\sqrt{-g}\, d^4x$; we also have to change the ordinary derivative $A_{\beta,\alpha}$ to the covariant derivative $A_{\beta;\alpha}$ etc. This leads to the action

$$A_{\rm em} = -\frac{1}{16\pi}\int F_{\alpha\beta}F^{\alpha\beta}\sqrt{-g}\, d^4x; \quad F_{\alpha\beta} = A_{\beta;\alpha} - A_{\alpha;\beta}. \tag{A.43}$$

The electromagnetic field equation in a curved spacetime can be obtained by varying this action; we will find that $F^{\alpha\beta}_{\ \ ;\beta} = 0$. Using the identities derived above, we can write these equations as

$$\frac{1}{\sqrt{-g}}\partial_\alpha(\sqrt{-g}F^{\alpha\beta}) = 0; \quad F_{\alpha\beta} = \frac{\partial A_\beta}{\partial x^\alpha} - \frac{\partial A_\alpha}{\partial x^\beta}. \tag{A.44}$$

These equations describe the influence of gravitational field on the electromagnetic phenomenon.

The following point must be stressed: By changing ordinary derivatives to covariant derivatives, $\eta_{\alpha\beta}$ to $g_{\alpha\beta}$ and d^4X to $\sqrt{-g}\, d^4x$ we can express any (special relativistic) action in a form which is valid in an arbitrary coordinate system. Such a translation is purely kinematic and has no physical content. The physical content of the theory rests in the principle of equivalence which we invoke at this stage to claim that this action (or, more precisely, the Lagrangian which is a local quantity) must continue to be valid even in a genuinely curved spacetime. This hypothesis is a dynamical principle and immediately specifies the effect of gravitational field on the physical system in question.

A.5 Characterization of the gravitational field

We mentioned earlier that, in curved spacetime, the parallel transport of a vector along two different curves will not result in the same final value. The mathematical reason for this fact is the following: The parallel transport equation is equivalent to $v^\alpha_{\ ;\epsilon} = 0$; that is

$$\frac{\partial v^\alpha}{\partial x^\epsilon} = -\Gamma^\alpha_{\beta\epsilon}v^\beta. \tag{A.45}$$

Such an equation will have an acceptable, unique solution only if the integrability conditions are satisfied. (Note that we are essentially demanding $dv^\alpha = -\Gamma^\alpha_{\beta\epsilon}(x)v^\beta(x)\,dx^\epsilon$ to be an exact differential.) This condition is easy to obtain. Differentiating (A.45) with respect to x^γ and using (A.45) again, we find

$$\frac{\partial^2 v^\alpha}{\partial x^\gamma \partial x^\epsilon} = -\left[\frac{\partial \Gamma^\alpha_{\theta\epsilon}}{\partial x^\gamma} - \Gamma^\alpha_{\beta\epsilon}\Gamma^\beta_{\gamma\theta}\right] v^\theta. \tag{A.46}$$

The integrability condition $(\partial^2 v^\alpha/\partial x^\gamma \partial x^\epsilon) = (\partial^2 v^\alpha/\partial x^\epsilon \partial x^\gamma)$ is equivalent to $R^\alpha_{\theta\gamma\epsilon}v^\theta = 0$ where

$$R^\alpha_{\theta\gamma\epsilon} = \frac{\partial \Gamma^\alpha_{\theta\epsilon}}{\partial x^\gamma} - \frac{\partial \Gamma^\alpha_{\theta\gamma}}{\partial x^\epsilon} + \Gamma^\alpha_{\beta\gamma}\Gamma^\beta_{\epsilon\theta} - \Gamma^\alpha_{\beta\epsilon}\Gamma^\beta_{\gamma\theta}. \tag{A.47}$$

If this result should hold for an arbitrary v^α, then $R^\alpha_{\theta\gamma\epsilon}$ has to vanish. In general, this quantity will *not* vanish and hence parallel transport will not give a unique vector.

This important result can also be interpreted in a different way. The covariant derivative differs from the ordinary derivative in one significant detail. For ordinary differentiation, we know that

$$A^\alpha{}_{,\beta,\epsilon} - A^\alpha{}_{,\epsilon,\beta} = \frac{\partial^2 A^\alpha}{\partial x^\beta \partial x^\epsilon} - \frac{\partial^2 A^\alpha}{\partial x^\epsilon \partial x^\beta} = 0. \tag{A.48}$$

This is not true as regards covariant differentiation; direct calculation shows that

$$A^\alpha{}_{;\beta;\epsilon} - A^\alpha{}_{;\epsilon;\beta} = -R^\alpha{}_{\rho\beta\epsilon}A^\rho \tag{A.49}$$

where $R^\alpha{}_{\rho\beta\epsilon}$ is as defined above. This result has several interesting consequences.

First of all, notice that the left hand side of (A.49) is a tensor. It follows that $R^\alpha{}_{\beta\epsilon\gamma}$ is a fourth rank tensor, even though the $\Gamma^\alpha_{\beta\epsilon}$, out of which it is constructed, is not a tensor. Consider now the flat spacetime, described in the inertial coordinate system X^α, in which $\Gamma^\alpha_{\beta\epsilon} = 0$; it follows that $R^\alpha{}_{\beta\epsilon\gamma} = 0$ in this frame. If we now transform to a non-inertial coordinate system x^μ, then $\Gamma^\alpha_{\beta\epsilon}$ will become nonzero. But a tensorial quantity like $R^\alpha{}_{\beta\epsilon\gamma}$, which vanishes in one frame, must remain equal to zero in all frames. It follows that $R^\alpha{}_{\beta\epsilon\gamma} = 0$ in the flat spacetime, in *any* coordinate system.

On the other hand, suppose we are given a spacetime with some metric $g_{\mu\nu}(x)$ which leads to nonzero values of $R^\alpha{}_{\beta\epsilon\gamma}$; it is then *not* possible to reduce the metric to the form dia $(1, -1, -1, -1)$. The spacetime *is* curved. The converse is also true: If $R^\alpha{}_{\beta\epsilon\gamma} = 0$, then it is always possible to choose a set of coordinates in which $g_{\alpha\beta}$ reduce to the form dia $(1, -1, -1, -1)$. For these reasons, $R^\alpha{}_{\beta\epsilon\gamma}$ is called the curvature tensor.

A curved spacetime is characterized by the condition that $R^{\alpha}_{\beta\epsilon\gamma}$ should be nonzero; and the flat spacetime by the condition $R^{\alpha}_{\beta\epsilon\gamma} = 0$. Note that both the curved spacetime and the flat spacetime in non-inertial coordinates will have space dependent $g_{\mu\nu}$ and nonzero $\Gamma^{\alpha}_{\beta\epsilon}$. Only the curvature tensor can distinguish between the flat spacetime in curvilinear coordinates and a genuinely curved spacetime.

From the definition of $R_{\alpha\beta\epsilon\gamma} = g_{\alpha\mu} R^{\mu}{}_{\beta\epsilon\gamma}$, it is easy to verify that $R_{\alpha\beta\epsilon\gamma}$ is anti-symmetric in $\alpha\beta$, antisymmetric in $\epsilon\gamma$ and symmetric under pair exchange. It is also possible to verify directly the identity, called the 'Bianchi identity',

$$R^{\theta}{}_{\alpha\epsilon\gamma;\beta} + R^{\theta}{}_{\alpha\beta\epsilon;\gamma} + R^{\theta}{}_{\alpha\gamma\beta;\epsilon} = 0 \tag{A.50}$$

by evaluating the expression in locally inertial coordinates.

We can define a second rank tensor $R_{\alpha\beta} = R^{\rho}{}_{\alpha\rho\beta}$ and a scalar $R = g^{\alpha\beta} R_{\alpha\beta}$ from the curvature tensor. No other independent, lower rank tensors can be constructed from $R^{\alpha}_{\beta\epsilon\gamma}$. A useful identity involving R^{α}_{γ} and R can be obtained by contracting the Bianchi identity on θ and ϵ and α and β. This will give

$$R^{\theta\alpha}{}_{\theta\gamma;\alpha} + R^{\theta\alpha}{}_{\alpha\theta;\gamma} + R^{\theta\alpha}{}_{\gamma\alpha;\theta} = 0. \tag{A.51}$$

Using the symmetries of $R^{\alpha\beta}{}_{\gamma\delta}$ and the definitions of R^{α}_{γ} and R this can be written as

$$\left(R^{\alpha}_{\gamma} - \frac{1}{2}\delta^{\alpha}_{\gamma} R \right)_{;\alpha} = 0. \tag{A.52}$$

In general, the curvature tensor $R_{\alpha\beta\epsilon\gamma}$ depends on $g_{\alpha\beta}$ and its first and second derivatives. It is, however, possible to construct a tensor with the symmetries of $R_{\alpha\beta\epsilon\gamma}$ purely from the metric tensor. Such a tensor will have the form

$$R_{\alpha\beta\epsilon\gamma} = f(g_{\alpha\epsilon} g_{\beta\gamma} - g_{\alpha\gamma} g_{\beta\epsilon}); \quad f = \text{constant}. \tag{A.53}$$

Such spaces – with any dimension – are called 'maximally symmetric.'

So far, we have treated the gravitational field as an externally specified quantity described by the metric $g_{\alpha\beta}$. We now consider the dynamical equations connecting the metric to the source. In the non-relativistic limit, these equations should reduce to the Newtonian equations $\nabla^2\phi = 4\pi G\rho$. Further – in the same limit – we found that $g_{00} \simeq (1 + 2\phi)$. It, therefore, follows that the correct dynamical equations will be second order differential equations with $g_{\alpha\beta}$ as the independent variables.

To obtain such equations from an action principle, we would have normally started with a Lagrangian which depends on $g_{\alpha\beta}$ and its first derivatives. But, it is obvious that no (nor trivial) scalar can be constructed from the metric and its first derivatives alone since the derivatives of the metric tensor vanish in the local inertial frame. Thus any non-trivial Lagrangian must depend on the second derivatives of $g_{\alpha\beta}$ as well. If the variation of the action should lead to second order equations in $g_{\alpha\beta}$ – rather than third order – then this Lagrangian should be linear in the second derivatives.

The only scalar which can be constructed from $g_{\alpha\beta}$ and its first and second derivatives and is linear in the second derivative is R. Hence we take the Lagrangian for the gravitational field to be

$$A_g = -\frac{1}{16\pi\lambda}\int R\sqrt{-g}\,d^4x \tag{A.54}$$

where the form of the constant $(-1/16\pi\lambda)$ is so chosen for future convenience. The total action for the gravitational field interacting with matter will be $A_{\text{total}} = A_g + A_{\text{matter}}\ (\phi, g_{\alpha\beta})$ where $A_{\text{matter}}(\phi, g_{\alpha\beta})$ depends on the matter variables – denoted by ϕ, symbolically – *and* the metric $g_{\alpha\beta}$. It can be shown that[1] the variation of this action gives the field equations

$$R_{\alpha\beta} - \frac{1}{2}g_{\alpha\beta}R = 8\pi\lambda T_{\alpha\beta} \tag{A.55}$$

where $T_{\alpha\beta}$ is *defined* by the variation of the matter term:

$$\delta A_{\text{matter}} \equiv \frac{1}{2}\int T_{\alpha\beta}\delta g^{\alpha\beta}\sqrt{-g}\,d^4x. \tag{A.56}$$

The quantity $T_{\alpha\beta}$ is called the energy-momentum tensor of matter; its explicit form depends on the matter action A_m. The above equation, derived by Einstein, determines $g_{\alpha\beta}$ in terms of the matter variables $T_{\alpha\beta}$. Notice that the (contracted) Bianchi identity $(R^{\alpha}_{\ \beta} - (1/2)\delta^{\alpha}_{\beta}R)_{;\alpha} = 0$ implies the condition $T^{\ \alpha}_{\beta;\alpha} = 0$ on the energy–momentum tensor.

To see the physical nature of $T_{\alpha\beta}$ let us consider the result of the above operation when applied to the action for the electromagnetic field:

$$A = -\frac{1}{16\pi}\int F_{\alpha\beta}F^{\alpha\beta}\sqrt{-g}\,d^4x = -\frac{1}{16\pi}\int F_{\alpha\beta}F_{\epsilon\gamma}g^{\alpha\epsilon}g^{\gamma\beta}\sqrt{-g}\,d^4x. \tag{A.57}$$

On varying $g_{\alpha\beta}$, we get

$$\begin{aligned}\delta A &= -\frac{1}{16\pi}\int d^4x\left[2F_{\alpha\beta}F_{\epsilon\gamma}g^{\alpha\epsilon}\delta g^{\beta\gamma}\sqrt{-g} + F_{\mu\nu}F^{\mu\nu}\delta(\sqrt{-g})\right]\\ &= -\frac{1}{16\pi}\int d^4x\left[2F_{\alpha\beta}F^{\alpha}_{\gamma}\delta g^{\beta\gamma}\sqrt{-g} + F_{\mu\nu}F^{\mu\nu}\frac{1}{2}\frac{(-1)}{\sqrt{-g}}(-g)g_{\beta\gamma}\delta g^{\beta\gamma}\right]\\ &= \frac{1}{16\pi}\int d^4x\sqrt{-g}\delta g^{\beta\gamma}\left[-F_{\alpha\beta}F^{\alpha}_{\gamma} + \frac{1}{4}F_{\mu\nu}F^{\mu\nu}g_{\beta\gamma}\right]\end{aligned} \tag{A.58}$$

so that

$$4\pi T_{\beta\gamma} = -F_{\alpha\beta}F^{\alpha}_{\gamma} + \frac{1}{4}F_{\mu\nu}F^{\mu\nu}g_{\beta\gamma} \tag{A.59}$$

which is the standard expression for the energy-momentum tensor of the electromagnetic field; T^0_0 denotes the energy density and T^μ_0 denotes the energy flux along the μ-direction, etc. The conditions $T^\alpha_{\beta;\alpha} = 0$ can be easily shown to give the equations $F^\alpha_{\beta;\alpha} = 0$ which are the Maxwell equations in curved spacetime derived before. In this sense, Einstein's equations imply the equations of motion for the source term.

As a second example, consider the action for a system of particles

$$A = \sum_A m_A \int ds_A. \tag{A.60}$$

Similar analysis will now show that

$$T_{\alpha\beta} = \rho u_\alpha u_\beta, \tag{A.61}$$

where ρ is the mass energy density and u^α is the four velocity. In the rest frame of the particles, $u^\alpha = (1, 0)$ and $T^\alpha_\beta = \text{dia}\ (\rho, 0, 0, 0)$, which is to be expected. The condition $T^\alpha_{\beta;\alpha} = 0$ now gives the geodesic equation for the particles $u^\beta u^\alpha_{;\beta} = 0$.

It remains to determine the value of λ. This can be done by computing the form of (A.55) in the non-relativistic limit. Taking the trace of (A.55), we find that $R - 2R = 8\pi\lambda T$ or $R = -8\pi\lambda T$. Substituting for R, we can rewrite Einstein's equation as

$$R_{\alpha\beta} = 8\pi\lambda \left(T_{\alpha\beta} - \frac{1}{2} g_{\alpha\beta} T \right). \tag{A.62}$$

In the non-relativistic limit, we have seen that $g_{00} = (1 + 2\phi)$ and $g_{\mu\nu} = -\delta_{\mu\nu}$. Computing $R_{\alpha\beta}$ from this metric, one finds that the dominant term is $R^0_0 \cong \nabla^2\phi$. For a system of particles with $T^\alpha_\beta = \rho u^\alpha u_\beta$, the zero–zero component of the right hand side of (A.62) is $(4\pi\lambda\rho)$. Thus the equation (A.62) becomes $\nabla^2\phi = 4\pi\lambda\rho$; comparing with Newton's law of gravitation, we find that $\lambda = G$.

The general relativistic description of the gravitational field may be summarized as follows: Given the form of the matter stress-tensor, Einstein's equations can be solved, in principle, to obtain the metric tensor. Once the metric is known, we can compute the effects of the gravitational field on various physical systems in a straightforward manner. The behaviour of measuring rods and clocks can be directly read off from the metric. The motion of the material particles and light rays can be obtained from the geodesic equation. Finally, the effect of gravity on any arbitrary physical system can be studied, by replacing $\eta_{\alpha\beta}$ by $g_{\alpha\beta}$, d^4X by $\sqrt{-g}\, d^4x$, and the ordinary derivatives by the covariant derivatives in the action for that system.

Appendix B
Aspects of field theory

B.1 Introduction

This appendix describes some essential features of particle physics and field theory, especially those which are used in the attempts to unify the forces of nature[1]. The discussion is not rigorous and is confined to a few selected topics.

B.2 Quarks, leptons and gauge bosons

At sufficiently low energies, the interactions between the elementary particles can be classified as of four different kinds: gravitational, electromagnetic, weak and strong. All particles are affected by gravity and all charged particles are affected by the electromagnetic forces. However, not all particles feel the strong interaction force; those which are affected by the strong interactions as well are called hadrons while those which participate only in the other three interactions are called leptons. Observations suggest that there are six different leptons: electron (e), muon (μ), tau-lepton (τ), electron-neutrino (ν_e), muon-neutrino (ν_μ) and tau-neutrino (ν_τ). The most prominent members of the hadrons are the proton (p) and the neutron (n) though there exist many other unstable particles. For each of these particles there exists an anti-particle with the same mass and opposite charge.

The hadrons are in turn made of quarks which are spin-half particles that participate in all the interactions. There are also six quarks called up (u), down (d), charm (c), strange (s), top (t) and bottom (b). Most of the models for the particle interactions pair the quarks and the leptons into 'families': $(e-\nu_e)$, $(\mu-\nu_\mu)$, $(\tau-\nu_\tau)$ and $(u-d)$, $(c-s)$, $(t-b)$. The u, c and t quarks have electric charge $(+2/3)$ and the d, s and b quarks have the charge $(-1/3)$. Baryons are made of three quarks (for example, $p = uud$ and $n = udd$) while the mesons are made of quark–antiquark states (for example, $\pi^+ = u\bar{d}$). Each quark also carries an internal quantum number called 'colour'. The force between coloured particles is so large that the stable states seen in nature (like baryons and mesons) are colour singlets.

Each quark–lepton pair is called a 'family'. At present there is experimental evidence for the existence of three families. Five out of the six particles in this quark–lepton set of three families have been discovered; the top quark is yet to be discovered but it is believed to exist. Some recent experiments also suggest that it is unlikely that there are more than three families.

The interactions between these particles are mediated by the spin-one bosons called the gauge bosons. (The term 'gauge' refers to the particular way in which the interactions are introduced into the theory.) The simplest example of a gauge boson is a photon (γ) which mediates the electromagnetic interaction. The weak interactions are mediated by three gauge bosons ($W^{\pm}, Z^0$). The photon is massless while the W and Z bosons have masses of about 80 GeV and 90 GeV respectively. Since the range of an interaction is inversely related to the mass of the boson mediating the interaction, the electromagnetic interaction has infinite range while the weak interaction has an extremely short range (less than 10^{-16} cm). In the same way, the strong interactions are mediated by a set of vector bosons containing at least eight gluons. The properties of any extra bosons in the theory (other than gluons) depend on the models; in some of the simplest models there will be 12 more bosons called collectively the X and Y bosons.

There have been several attempts in the last two decades to produce a unified description of all these interactions – a task in which partial success has been achieved. We shall now briefly summarize the essential ingredients of these attempts.

In quantum theory, we associate a field with every particle. How a field transforms under the Lorentz transformation depends on the spin of the particle described by the field. A spin-zero particle can be described by a scalar field, a spin-half particle by a spinor field, a spin-one particle by a vector field etc. The Lagrangian describing the field will carry information about the mass of the particle and its interactions. Given the Lagrangian, one should be able to identify the 'coordinates' and their 'conjugate momenta' and thus arrive at a quantum theory. The physical states of a field should correspond to a ground state, a state with one particle, a state with two particles etc.

There arises a peculiar difficulty in implementing this programme. A free field (like the electromagnetic field) is mathematically equivalent to an infinite number of independent harmonic oscillators. The lowest energy state of such a field will correspond to the state in which all the oscillators are in their respective ground states. Since each oscillator has a finite zero-point energy but there are an infinite number of oscillators, the total energy of the ground state diverges. This difficulty arises directly from the basic concepts of Lorentz invariance and quantum theory and no satisfactory solution to this problem has been found. Similar difficulties arise when one tries to describe the interaction between the particles in a perturbative fashion, by separating the Hamiltonian describing the system into a part describing a free field and a part describing the interaction, and attempt to incorporate the effects of the latter by a systematic perturbation on the theory based on the former. It is not known whether these difficulties will disappear if the full interaction is treated non-perturbatively. However, since we lack the necessary mathematical machinery to do so, this question is still unsettled.

Because of the difficulty mentioned above, physicists have concentrated on a very special class of Lagrangians to describe particle interactions. These Lagrangians allow for a reinterpretation of the divergent results which arise in the perturbation theory. The method works in the following manner: If the theory produces divergences of a certain kind which can be handled by redefining the coupling constants of the free Hamiltonian, then it is possible to obtain finite predictions from the perturbation theory. The Lagrangians which have the special form allowing the above procedure to work are called 'perturbatively renormalizable'.

For example, the Lagrangian in quantum electrodynamics (QED) describing the interaction of charged massive fermions with photons belongs to this class. The renormalizability of the QED Lagrangian is intimately related to a property called 'local gauge invariance': There exists a transformation of the spinor and electromagnetic fields, involving one arbitrary function, which leaves the Lagrangian invariant. The class of all such transformations form a group called $U(1)$ group. The connection between the gauge invariance and renormalizability is quite deep. It is possible to produce more complicated gauge theories with Lagrangians which are invariant under some other group of transformations.

It is, for example, possible to construct a model describing the electromagnetic and the weak interactions starting with a Lagrangian which possesses invariance under two groups $U(1)$ and $SU(2)$. (Unitary 2×2 matrices with unit determinant under the operation of matrix multiplication form a group which is in one-to-one correspondence with the group $SU(2)$.) Some properties of the strong interactions, which arise due to the gluons, can be understood from a gauge theory based on a group called $SU(3)$. However, before such models can describe the real world we have to tackle another important difference between the electromagnetic interaction and the other two interactions.

We mentioned before that the electromagnetic interaction has a long range because the photon is massless; the weak interaction has a short range since W and Z bosons are massive. But the gauge invariance is intimately linked with the zero-mass character of the mediating particle; electromagnetism is gauge invariant only because the photon is massless. A mass term for the photon in the electromagnetic Lagrangian will break the gauge invariance. Thus it appears impossible to produce a gauge invariant theory for an interaction (like the weak interaction) which needs the mediating particles to be massive.

To solve this difficulty, one has to invoke a procedure called 'spontaneous symmetry breaking'. This is done by introducing into the theory a set of scalar fields with a specific kind of interaction. The basic Lagrangian will then contain, in addition to the leptons and gauge bosons (which will all be massless) some scalar bosons. The *physical* fields will not be these basic fields which appear in the Lagrangian but certain combinations of them. These combinations will act as though they are massive even if the original (basic) fields are massless. The symmetry of the theory will be manifest in terms of the original fields while it will be *hidden* when expressed in terms of the physical fields. By using this method one can produce a theory which is gauge invariant and at the same time contains massive mediating particles.

The scalar fields (called 'Higgs fields') have a Lagrangian with a potential $V(\phi)$ which has non-trivial minima. If such a system comes into contact with matter fields which are in thermal equilibrium at some temperature T, then the effective potential energy will acquire a temperature dependence. That is, $V(\phi)$ will become $V(\phi, T)$. Such a temperature dependence can lead to several non-trivial effects – like phase transitions – in the early universe[2].

Even though the transformation group underlying the theory can be determined from some general principles, the detailed transformation properties of the fields representing specific particles cannot be derived from any fundamental considerations. These details are fixed using the known laboratory properties of these particles. For example, consider the fields describing the leptons. Given a spinor field ψ one can construct its 'right-handed' and 'left-handed' components by the decomposition

$$\psi_L = \frac{1}{2}(1 - \gamma_5)\psi; \quad \psi_R = \frac{1}{2}(1 + \gamma_5)\psi. \tag{B.1}$$

where γ_5 is the 4×4 matrix

$$\gamma_5 = \begin{pmatrix} 0 & I \\ I & 0 \end{pmatrix}. \tag{B.2}$$

In the standard electro-weak theory, the right-handed components behave as singlets (that is, they do not change) while the left-handed components transform as a doublet (that is, under an $SU(2)$ transformation these fields are changed into linear components of themselves). It is a consequence of this feature that in the simplest electro-weak theory there is no necessity for the right-handed neutrino, ν_R, and that the left-handed neutrino ν_L is massless.

Since the transformation properties of the fields are put in by hand into the theory, it is possible to generalize these models in many ways. In particular, it is possible – though not necessary – to have massive neutrinos in the theory.

There is another complication regarding the fermions which deserves to be mentioned. As we have seen above, a spinor field (ψ) can be split up into right-handed (ψ_R) and left-handed (ψ_L) states. Suppose that, in some frame of reference, an electron was moving towards the z direction with z-component of spin $(-1/2)$. Since spin and momentum are antiparallel, we call this state left-handed and describe it by ψ_L. Consider another observer, moving along $+z$, faster than the electron. This observer will see the momentum *and* spin of the electron to be both along the $(-z)$ direction and will describe it by the state (ψ_R). Notice that we also have the states $\overline{\psi}_L$ and $\overline{\psi}_R$, which represent the left-handed and right-handed states of a positron. Lorentz transformations cannot change the electric charge of a state and hence cannot convert an electron state to a positron state. So the moving observer obviously cannot use $\overline{\psi}_R$ to describe the electron.

The situation is different as regards neutrinos because they do not have any electric charge. Consider first the massless neutrinos. They can be described by just two states ν_L and $\overline{\nu}_R$. A left-handed neutrino will remain left-handed in all Lorentz frames because no observer can overtake a *massless* particle. When we add mass to the neutrino one may think that we will require all the four states ν_L, ν_R, $\overline{\nu}_L$ and $\overline{\nu}_R$ to describe a neutrino–antineutrino system. This can indeed be done; in that case, the structure of the states is similar to that of the electron and such a neutrino is called a Dirac neutrino. However, for neutrinos, there exists another possibility. We may say that the neutrino is its own antiparticle and restrict ourselves to the states ν_L and $\overline{\nu}_R$ even when the neutrino is massive. An observer travelling faster than a left-handed neutrino will indeed see it as right-handed; but we now arrange matters so that under Lorentz transformations the state ν_L gets mapped to $\overline{\nu}_R$. This is now possible because neutrinos do not carry electric charge or any such distinguishing quantum numbers making the particle and antiparticle different. (To be precise, there does exist a quantum number, called the lepton number, which distinguishes ν from $\overline{\nu}$. But the conservation of lepton number is not as sacred as that of the electric charge; it is more a consequence of specific model than a result of dynamical significance.) A massive neutrino which is its own antiparticle is called a Majorana neutrino.

The unification scheme mentioned above can be extended to include quarks and gluons by using the $SU(3)$ group. The full theory will then be invariant under the combined group $SU(3) \otimes SU(2) \otimes U(1)$ which is a product of three separate groups. Such a feature allows the theory to have three independent coupling constants. Further, the interaction of the Higgs field with the fermionic fields introduces another host of coupling constants. The minimal model based on the above group contains nearly 20 free parameters. Since such features appear to be rather unaesthetic, there have been several attempts to produce a theory describing leptons, quarks and gluons from one single gauge group. Most of these models, since they are based on larger groups, also bring additional particles and gauge bosons. Such models go under the collective name 'grand unified theories'. The simplest of such models unify the strong interactions with the electro-weak interactions at energy scales which are above 10^{15} GeV. Since these scales are far beyond the reach of accelerators, the early epochs of the universe are often used to test the predictions of the theory.

The symmetry transformations discussed above take place in an abstract internal space. It is, however, possible to construct theories in which the symmetry transformation also acts on the real spacetime. Such theories – of which supergravity is one example – imply a symmetry between bosons and fermions which exist in the theory. In other words, these models assign one fermionic partner to every boson and vice-versa.

Appendix C
COBE results and implications

C.1 Introduction

The analysis of data collected by the satellite Cosmic Background Explorer (COBE) has now shown that the microwave background does have temperature anisotropy of $(\Delta T/T) \simeq 10^{-5}$. This result (which was announced when the processing of this book was well under way!) has several important cosmological implications. We will discuss these implications in this appendix.

C.2 COBE results and cosmology

The COBE instruments have a resolution of about 7° and are designed to detect the large angle anisotropies in the MBR. The data analysis has now given[1] three crucial numbers: (i) The dominant term is a dipole anisotropy of magnitude $(\Delta T/T)_{\rm dipole} \cong 1.23 \times 10^{-3}$, towards the direction $l = 284.7^\circ \pm 0.8^\circ$, $b = 48.2^\circ \pm 0.5^\circ$. We saw in chapter 6 that such an effect arises due to purely kinematic reasons. (ii) The root-mean-square fluctuation in temperature, in the angular range of 15°–165°, is found to be

$$\left(\frac{\Delta T}{T}\right)_{\rm rms} = (1.1 \pm 0.2) \times 10^{-5}. \tag{C.1}$$

(iii) The quadrupole contribution to the anisotropy is

$$\left(\frac{\Delta T}{T}\right)_{Q} = (0.48 \pm 0.15) \times 10^{-5}. \tag{C.2}$$

These results have important implications for the models of structure formation which we shall now discuss.

Given a power spectrum $P(k)$ in some model, one can predict the values of $(\Delta T/T)_{\rm rms}$ and $(\Delta T/T)_Q$. This can be done using the formulas derived in chapter 6. All the information about these anisotropies is contained in the coefficients a_{lm} which arise in the expansion of the temperature anisotropy in spherical harmonics. We saw in chapter 6 that

$$< |a_{lm}|^2 >= C_l = \frac{H_0^4}{2\pi} \int_0^\infty dk \, \frac{|\delta_k|^2}{k^2} |j_l(k\eta)|^2, \tag{C.3}$$

where j_l is the spherical Bessel function of order l and

$$\eta = \int_{t_{\rm rec}}^{t_0} \frac{dt}{a(t)} \cong \int_0^{t_0} \left(\frac{t_0}{t}\right)^{2/3} dt = 3t_0 = 2H_0^{-1} \equiv 2R_H, \tag{C.4}$$

with $R_H = H_0^{-1} = 3000h^{-1}$ Mpc. Since the lth mode has a degeneracy of $(2l+1)$, it contributes an amount $(2l+1)(C_l/4\pi)$ to the mean square fluctuation. We therefore get

$$\left(\frac{\Delta T}{T}\right)^2_{\rm rms} = \frac{1}{4\pi} \sum_{l=2}^\infty (2l+1) C_l \tag{C.5}$$

and

$$\left(\frac{\Delta T}{T}\right)^2_Q = \frac{1}{4\pi}(5C_2) = \frac{5}{4\pi} C_2. \tag{C.6}$$

There is, however, one minor complication which needs to be taken into account. The above formulas, especially (C.5), are correct provided the instrument can probe all the modes with equal sensitivity. But, in practice, any instrument will have a finite angular resolution, θ_c. This implies that the response of the instrument will decrease significantly for modes with $l > l_c$ where $l_c \simeq \theta_c^{-1}$. This effect can be taken into account by introducing a gaussian response profile for the detector. The modified formulas for the anisotropy will be:

$$\left(\frac{\Delta T}{T}\right)^2_{\rm rms} = \frac{1}{4\pi} \sum_{l=2}^\infty (2l+1) C_l \exp\left(-\frac{l^2\theta_c^2}{2}\right), \tag{C.7}$$

$$\left(\frac{\Delta T}{T}\right)^2_Q = \frac{5}{4\pi} C_2 \exp(-2\theta_c^2). \tag{C.8}$$

The quantity θ_c can be determined from the response function of the detector, which is usually quoted in terms of a quantity called 'full-width-at-half-maximum', $\theta_{\rm FWHM}$. If the response function is $\mathcal{R}(\theta) = \exp(-\theta^2/2\theta_c^2)$ then, $\theta_{\rm FWHM}$ is determined by the condition $\mathcal{R}(\theta_{\rm FWHM}) = 0.5$. This gives $\theta_c^2 = (\theta_{\rm FWHM}^2/8\ln 2)$ or $\theta_c \cong 0.425\ \theta_{\rm FWHM}$. For COBE, $\theta_{\rm FWHM} \simeq 7°$ giving $\theta_c \cong 0.052$ radian.

To compute C_l, we need the power spectrum $P(k) = |\delta_k|^2$. We saw in chapter 4 that most models of structure formation predict a power spectrum which has the Harrison–Zeldovich form, $P(k) \cong Ak$, for small k. Since COBE is essentially probing large angular scales, it is legitimate to use this asymptotic form in computing C_l. Setting $n = 1$ in equation (6.47) we find that

$$C_2 = \frac{AH_0^4}{24\pi}; \quad C_l = \frac{6C_2}{l(l+1)}. \tag{C.9}$$

Substituting these relations into (C.7) and (C.8) we get

$$\left(\frac{\Delta T}{T}\right)_Q^2 \cong \left(\frac{5}{96\pi^2}\right)(AH_0^4) = (5.28 \times 10^{-3})(AH_0^4) \tag{C.10}$$

and

$$\begin{aligned}\left(\frac{\Delta T}{T}\right)^2 &= \frac{1}{4\pi}\sum_{l=2}^{\infty}(2l+1)\left[\frac{6C_2}{l(l+1)}\right]\exp\left(-\frac{1}{2}l^2\theta_c^2\right) \\ &= \frac{C_2}{4\pi} \times 28.45 = 0.03(AH_0^4).\end{aligned} \tag{C.11}$$

In arriving at the final result we have used $\theta_c \cong 0.052$ and evaluated the sum over l numerically. (Note, however, that the exponential factor suppresses the contribution from $l \gtrsim l_c = 28$. For $l < 28$, the sum is reasonably well approximated by the integral over l, giving $I = 6\ln[l_c(l_c+1)/6] \cong 29$.) The quantity (AH_0^4) is directly related to the fluctuations in the gravitational potential $\Phi^2 = (k^3|\phi_k|^2/2\pi^2)$ at large scales. Since $\phi_k = (4\pi G\rho_b)(\delta_k/k^2) = (3/2)H^2(\delta_k/k^2)$ we find that

$$\Phi^2(k) = \frac{k^3|\phi_k|^2}{2\pi^2} = \frac{9}{4}\left(\frac{H}{k}\right)^4\left(\frac{k^3|\delta_k|^2}{2\pi^2}\right) = \frac{9}{8\pi^2}(AH_0^4) \tag{C.12}$$

if $|\delta_k|^2 \cong Ak$. Therefore, $AH_0^4 = (8\pi^2/9)\Phi^2$ and we can re-express (C.10) and (C.11) as,

$$\left(\frac{\Delta T}{T}\right)_Q \cong 0.22\Phi; \quad \left(\frac{\Delta T}{T}\right)_{\rm rms} \cong 0.51\Phi. \tag{C.13}$$

We can now compare the theoretical results with the COBE observations. To begin with, notice that $(\Delta T_{\rm rms}/\Delta T_Q) \cong 2.3$ if the spectrum has $n = 1$. The COBE results allow this ratio to fall between 1.43 and 3.94 with a mean value of 2.29. This is quite consistent with the assumption of $n = 1$. (It turns out that a least square fit to the COBE data gives $n = 1.1 \pm 0.5$.) We shall, hereafter, assume that $P(k) \propto k$ for small k.

The parameter Φ and the amplitude A can now be determined by comparing (C.13) with the COBE result. Ideally, of course, both $(\Delta T/T)_{\rm rms}$ and $(\Delta T/T)_Q$ should lead to the same value for Φ and A. However, because of instrumental and systematic errors, we get slightly different values. Comparing the root-mean-square values we find

$$\begin{aligned}\Phi_{\rm rms} &= 1.96(\Delta T/T)_{\rm rms} = (2.16 \pm 0.39) \times 10^{-5},\\ A_{\rm rms} &= (8\pi^2/9)\Phi^2 R_H^4 = (4.09 \pm 1.5) \times 10^{-9} R_H^4 \\ &= (24 \pm 2\, h^{-1}\,{\rm Mpc})^4.\end{aligned} \tag{C.14}$$

Similarly, by comparing the quadrupole anisotropy, we get

$$\begin{aligned}\Phi_Q &= 4.55(\Delta T/T)_Q = (2.2 \pm 0.68) \times 10^{-5},\\ A_Q &= (8\pi^2/9)\Phi^2 R_H^4 = (4.2 \pm 2.6) \times 10^{-9} R_H^4 \\ &= (24 \pm 4\, h^{-1}\,{\rm Mpc})^4.\end{aligned} \tag{C.15}$$

Within the error bars, we may take $\Phi \cong 2.2 \times 10^{-5}$ and $A = (24h^{-1}\,{\rm Mpc})^4$. Also note that the quadrupole result gives $C_2^{1/2} = (4\pi/5)^{1/2}\,(\Delta T/T)_Q = (0.76 \pm 0.24) \times 10^{-5}$; thus the maximum permitted value for $C_2^{1/2}$ is about 10^{-5}.

These results impose rather stringent conditions on the models for structure formation. Consider, for example, pure HDM models with a power spectrum of the form:

$$P(k) = Ak\exp[-4.61(k/k_{\rm FS})^{3/2}], \tag{C.16}$$

with $k_{\rm FS} = 0.16\,{\rm Mpc}^{-1}\,(m_\nu/30\,{\rm eV})$ (see equation (4.229); we have set $\alpha = 3/2$ to get $n = 1$). For this spectrum, the density contrast, $\Delta_k = [k^3 P(k)/2\pi^2]^{1/2}$ reaches maximum at $k_m = 0.69 k_{\rm FS}$. The maximum value is

$$\Delta_{\rm max} = \left[\frac{k_m^3 P(k_m)}{2\pi^2}\right]^{1/2} = 7.3 \times 10^{-4} \left(\frac{A}{1\,{\rm Mpc}^4}\right)^{1/2} \left(\frac{m_\nu}{30\,{\rm eV}}\right)^2. \tag{C.17}$$

Using the COBE result, $A \simeq (24h^{-1})^4\,{\rm Mpc}^4$ we get

$$\Delta_{\rm max} \cong 0.42 h^{-2} \left(\frac{m_\nu}{30{\rm eV}}\right)^2. \tag{C.18}$$

This value is far too small to produce any nonlinear structures by today. We saw in chapter 8 that for a mass scale to go nonlinear at a redshift $z_{\rm coll}$, the density contrast should be about $\delta_0 \cong 1.69\,(1 + z_{\rm coll})$ (see equation (8.41)). Even with $h = 0.5$, (C.18) gives only $z_{\rm coll} \simeq 0$; that is, structures are beginning to go nonlinear only at the present epoch. Thus any model in which free-streaming wipes out small scale power will face difficulties to account for COBE results.

The models like the ones based on CDM have a $\Delta(k)$ which flattens at large k. In such models, small scales go nonlinear first and structures form hierarchically. Taken in isolation, those models have no difficulty in accommodating the COBE results. However, they could also run into some trouble when we combine the COBE results with other constraints on the density contrast. To see how this comes about, it is best to compare the density contrast predicted by various observations. Most of the galaxy surveys suggest that $\sigma(50h^{-1}\,\mathrm{Mpc}) \simeq 0.2$ where $\sigma^2(R) =< (\delta M/M)^2_R >$ is the mean-square-fluctuation in the mass contained within a sphere of radius R. On the other hand, the COBE result which suggests that $P(k) \simeq Ak$ for small k, leads to $\sigma(R) \cong (R_0/R)^2$ for $R \gtrsim 10^3 h^{-1}\,\mathrm{Mpc}$, with $R_0 \simeq A^{1/4} \simeq 24h^{-1}\,\mathrm{Mpc}$. This relation can give $\sigma(50h^{-1}\,\mathrm{Mpc}) \simeq (24/50)^2 \simeq 0.2$ *provided we assume that the Harrison–Zeldovich form* ($P \simeq Ak$) *is extended all the way down to* $50h^{-1}\,\mathrm{Mpc}$. In most models, the power spectrum will flatten much before this and hence will *not* give $\sigma(50h^{-1}\,\mathrm{Mpc}) \simeq 0.2$. Even if we arrange matters so that the Harrison–Zeldovich spectrum is extended all the way up to $50h^{-1}\,\mathrm{Mpc}$ or so, the shape of the spectrum will not be quite right. The galaxy survey data requires a rather sharp bend in the spectrum which scale-free models cannot produce.

The situation is illustrated in figure C.1 using a series of model power spectra of the form

$$P(k) = \frac{Ak}{1 + (k/k_c)^n}. \tag{C.19}$$

The amplitude A is fixed using the COBE result of $(C_2)_{\mathrm{max}} = 10^{-5}$. That leaves two free parameters: (i) an index n which determines the slope of the spectra at large k and (ii) a length scale k_c^{-1} which fixes the location at which the spectrum bends. The density contrast $\sigma(R) =< (\delta M/M)^2_R >^{1/2}$ calculated from (C.19) is plotted for five different values in figure C.1. Also plotted on the same curve are the $\sigma(R)$ determined from a few galaxy surveys.[2] A good fit is achieved for $n = 2.4$, $k_c = 0.047h\,\mathrm{Mpc}^{-1}$ (curve E). We have also shown two curves ($n = 2.2$, $k_c = 0.057h\,\mathrm{Mpc}^{-1}$ and $n = 2.4$, $k_c = 0.084h\,\mathrm{Mpc}^{-1}$) which overshoot the data and two curves ($n = 2.2$, $k_c = 0.025h\,\mathrm{Mpc}^{-1}$ and $n = 2.4$, $k_c = 0.025h\,\mathrm{Mpc}^{-1}$) which undershoot the data. These fits suggest that a sharply bending power spectrum with a length scale of about $L \simeq k_c^{-1} \simeq 21h^{-1}\,\mathrm{Mpc}$ may be required to explain all the observations. The models which have been suggested so far do not have such a power spectrum.

How reliable is the above conclusion? If the galaxy survey results, indicating $\sigma(50h^{-1}\,\mathrm{Mpc}) \simeq 0.2$ are correct, then it may be difficult to escape this conclusion. There are, however, a couple of other effects which could affect the reasoning at $R \lesssim 10h^{-1}\,\mathrm{Mpc}$. (i) The first effect is biasing. If biasing is dependent on scale, then the true shape of the spectrum could be quite different from the one based on galaxy counts. (ii) In general, nonlinear effects tend to steepen the spectrum. Thus the

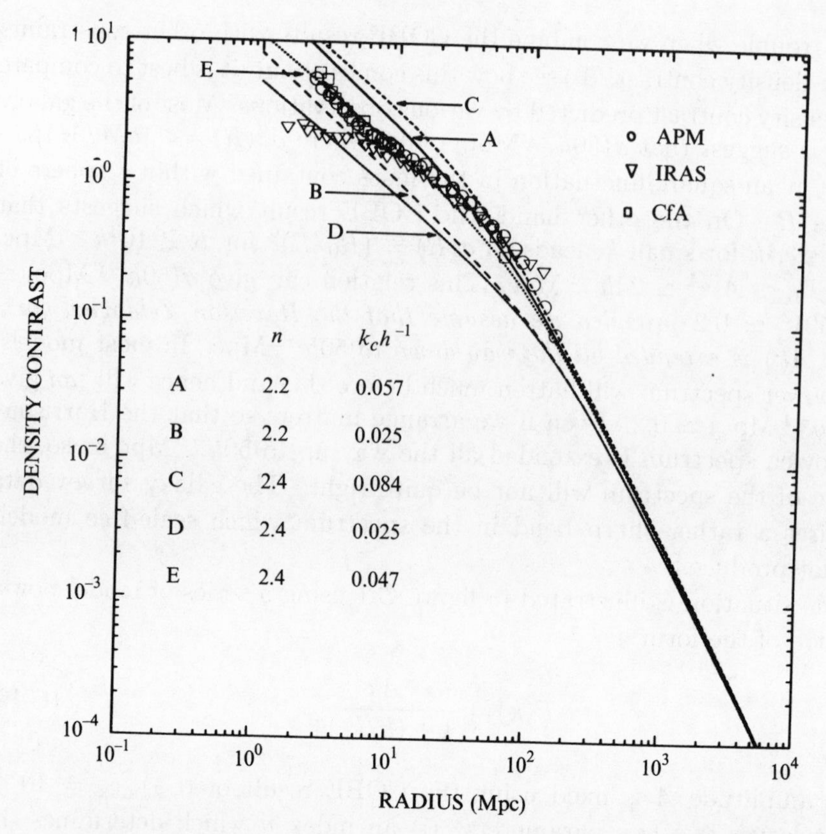

Fig. C.1. The root-mean-square fluctuation in mass $\sigma(R) =< (\delta M/M)_R^2 >^{1/2}$, determined from galaxy surveys and the COBE results, is fitted using a class of test spectra of the form $P(k) = Ak[1 + (k/k_c)^n]^{-1}$. All the spectra are normalized using the maximum value of $C_2^{1/2} = 10^{-5}$ permitted by COBE. There is reasonable agreement for $n = 2.4$, $k_c = 0.047h\,\mathrm{Mpc}^{-1}$ suggesting a length scale of $k_c^{-1} = 21h^{-1}$ Mpc in the theory. We have set $h = 0.5$.

slope of the spectrum calculated from linear theory will be smaller than that in actual galaxy surveys. Even though these effects are important at small scales, it seems unlikely that they can modify the results at, say, $R \gtrsim 40h^{-1}$ Mpc. Thus the galaxy surveys, together with COBE results, seem to impose rather stringent constraints of the shape of the power spectrum.

Notes and references

Chapter 1

1. D. Clayton, (1968), *Principles of Stellar Evolution and Nucleosynthesis*, (McGraw-Hill, New York).
2. The derivation of the masses and sizes of stars, galaxies etc. from fundamental considerations has been attempted by several authors in the context of the anthropic principle. See, for example, B. J. Carr and M. J. Rees, (1979), Nature, **278**, 605; J. D. Barrow and F. J. Tipler, (1982), *The Anthropic Principle*, (Oxford University Press, Oxford) and the references cited therein. Note that, in our discussion, we have accepted the values of the fundamental constants as inputs from physics and have merely worked out the resulting parameters for various structures. Such a derivation, of course, is quite independent of the anthropic principle. Also see L. M. Celnikier, (1989), *Basics of Cosmic Structures*, (Editions Frontiers, Paris).
3. Stellar evolution is discussed at an elementary level in the following texts: F. H. Shu, (1982), *The Physical Universe: An Introduction to Astronomy*, (University Science Books, California); M. Harwit, (1988), *Astrophysical Concepts*, (Springer-Verlag, New York).
4. An extensive discussion of the stellar evolution can be found in G. W. Collins, (1989), *The Fundamentals of Stellar Astrophysics*, (W. H. Freeman, New York); also see S. Chandrashekhar, (1939/1957), *An Introduction to the Study of Stellar Structure*, (Dover, New York); A. S. Eddington, (1926/1959), *The Internal Constitution of the Stars*, (Dover, New York); J. P. Cox and R. T. Giuli, (1968), *Principles of Stellar Structure*, (Gordon and Breach, New York); C. Rolfs and W. S. Rodney, (1986), *Cauldrons in the Cosmos*, (University of Chicago, Chicago); M. Schwarzschild, (1958), *The Structure and Evolution of the Stars*, (Princeton University Press, Princeton, N.J.).
5. For a description of the properties of galaxies, see D. Mihalas and J. J. Binney, (1981), *Galactic Astronomy*, 2nd Ed., (Freeman, San Fran-

cisco); S. Faber, (1986), (ed.) *Nearly Normal Galaxies: from the Planck Time to the Present*, (Springer-Verlag, New York). A popular account can be found in P. W. Hodge, (1986), *Galaxies*, (Harvard University Press, Harvard).

6. An excellent discussion of the solutions to the collisionless Boltzmann equation, in the context of modelling galaxies, can be found in: J. Binney and S. Tremaine, (1987), *Galactic Dynamics*, (Princeton University Press, Princeton); The Lane–Emden equation is discussed in S. Chandrashekhar, (1939/1957), cited in ref. 4 above.
7. For a visual display of galaxy distribution in the nearby region see, R. B. Tully and J. R. Fisher, (1987), *Nearby Galaxies Atlas*, (Cambridge University Press, Cambridge). Detailed description of various structures can also be found in *The Cambridge Atlas of Astronomy (Second edition)*, (1988), eds. J. Audouze and G. Israel, (Cambridge University Press, Cambridge).
8. A non-technical description of the Local Supercluster can be found in: R. B. Tully, (1982), *Sky and Telescope*, p. 550.
9. Some of the galaxy surveys are described in: V. de Lapparent, M. J. Geller and J. Huchra, (1986), Ap. J., **302**, L1; (1988), Ap. J., ibid, **332**, 44; M. J. Geller, J. Huchra and V. de Lapparent, (1987), in *Observational Cosmology (IAU Symp. 124)*, eds. A. Hewitt, G, Burbidge and Li-Zhi Fang, (Reidel, Dordrecht), p. 301; M. Seldner et al., (1977), Astron. J., **82**, 249; R. P. Kirshner et al., (1981), Ap. J., **248**, L57; M. Rowan-Robinson et al., (1990), MNRAS, **247**, 1; W. Saunders et al., (1991), Nature, **349**, 32; S. J. Maddox et al., (1990), MNRAS, **242**, 43p.
10. M. Davis and P. J. E. Peebles, (1983), Ap. J., **267**, 437; 465; P. J. E. Peebles, (1980), *Large Scale Structure of the Universe*, (Princeton University Press, Princeton).
11. N. A. Bahcall, (1988), Ann. Rev. Astron. Ap., **26**, 631.
12. V. de Lapparent, M. J. Geller and J. Huchra, (1986), Ap. J., **302**, L1. For a good popular account, see, M. J. Geller and J. P. Huchra, (1989), Science, **246**, 897.
13. See, for example, H. Spinrad and S. Djorgovski, (1987) in *Observational Cosmology* cited in ref. 9 above, p. 129; S. J. Lilly, (1989), *The Epoch of Galaxy Formation*, eds. C. S. Frenk et al., (Kluwer, Dordrecht), p. 63.
14. See, for example, A. L. Sandage, (1988), Ann. Rev. Astron. Ap., **26**, 561.
15. For a discussion of the properties of the quasars see D. W. Weedman, (1986), *Quasar Astronomy*, (Cambridge University Press, Cambridge); D. W. Weedman, (1986), in *Structure and Evolution of Active Galactic nuclei*, eds. G. Giuricin et al., (Reidel, Dordrecht); P. Veron,

(1983), in *Proc. 24th Liège Astrophysics Colloq: Quasars and Gravitational Lenses*, ed. J. P. Swings, (Institute d'Astrophysique, Liège); S. J. Warren and P. C. Hewett, (1990), Rep. Prog. Phys., **53**, 1095; D. E. Osterbrock, (1991), Rep. Prog. Phys., **54**, 579.
16. Detailed references can be found in the review by S. J. Warren and P. C. Hewett cited in ref. 15 above.
17. See, for example, M. Ted Ressell and M. S. Turner, (1990), Comm. Ap., **14**, 323.
18. D. T. Wilkinson, (1986), in *Inner Space/Outer Space*, eds. E. W. Kolb et al., (University of Chicago Press, Chicage), p. 126; D. T. Wilkinson, (1987), in *13th Texas Symposium on Relativistic Astrophysics*, ed. M. P. Ulmer, (World Scientific, Singapore), p. 209; R. B. Patridge, (1988), Rep. Prog. Phys., **51**, 647; R. D. Davis, et al., (1987), Nature, **326**, 462; A. C. S. Readhead et al., (1989), Ap. J., **346**, 566; J. B. Peterson, et al., (1986), in *Inner Space/Outer Space*, op. cit., p. 119; M. Bersanelli, et al., (1989), Ap. J., **339**, 632.
19. For an excellent discussion of the techniques used for the determination of distances in cosmology, see M. Rowan-Robinson, (1986), *The Cosmological Distance Ladder*, (Freeman, San Francisco).
20. See, for example, J. E. Hesser et al., (1987), P. A. S. P., **99**, 739; D. Winget, et al., (1987), Ap. J. Lett., **315**, L77; W. A. Fowler, (1987), Q. J. R. A. S., **28**, 37.

Chapter 2

1. Most books on general relativity discuss Friedmann models briefly. More detailed discussion can be found in: Ya. B. Zeldovich and I. D. Novikov, (1983), *Relativistic Astrophysics Vol. II*, (Chicago University Press, Chicago); S. Weinberg, (1972), *Gravitation and Cosmology*, (Wiley, New York); J. V. Narlikar, (1983), *Introduction to Cosmology*, (Jones and Bartlett, Boston) and E. W. Kolb and M. S. Turner, (1990), *The Early Universe*, (Addision-Wesley, California).
2. The determination of H_0, q_0 etc. is discussed in A. Sandage and G. A. Tammann, (1986), in *Inner Space/Outer Space*, eds. E. W. Kolb et al., (University of Chicago Press, Chicago), p. 41; P. W. Hodge, (1981), Ann. Rev. Astron. Ap., **19**, 357; J. P. Huchra, (1987), in *13th Texas Symposium on Relativistic Astrophysics*, ed. M. P. Ulmer (World Scientific, Singapore); J. Kristian, A. Sandage and J. Westphal, (1978), Ap. J., **221**, 383; H. Spinrad and S. Djorgovski, (1987), in *Observational Cosmology (IAU Symposium 124)*, eds. A. Hewitt, G. Burbidge and Li-Zhi Fang (Reidel, Dordrecht), p. 129; also see M. Rowan-Robinson, (1985), *The Cosmological Distance Ladder*, (Freeman, San Francisco).

Chapter 3

1. The calculation of $g(T)$ proceeds as follows: For $T < m_e$ only γ and ν contribute giving $g = (29/4)$. For $m_e < T < m_\mu$, an $e\bar{e}$ pair is added giving $(43/4)$; for $m_\mu < T < m_\pi$, $g = (57/4)$ due to addition of muons etc. In this manner $g(T)$ can be computed from the known mass spectrum of the particles.
2. General rules for the calculation of such cross-sections can be found in standard books on quantum field theory. See, for example, C. Quigg, (1983), *Gange Theories of the Strong, Weak and Electromagnetic Interactions*, (Benjamin/Cummings, California); J. D. Bjorken and S. D. Drell, (1964), *Relativistic Quantum Mechanics*, (McGraw-Hill, New York).
3. The freezing-out of wimps and relic abundance of massive particles have been discussed by several authors. A sample of references will be : Ya. B. Zeldovich, (1965), Zh. Eksp. Teor. Fiz, **48**, 986; Ya. B. Zeldovich, L. B. Okun and S. B. Pikelner, (1965), Usp. Fiz. Nauk., **84**, 113; H. - Y. Chiu, (1966), Phys. Rev. Lett., **17**, 712; B. W. Lee and S. Weinberg, (1977), Phys. Rev. Lett., **39**, 165; P. Hut, (1977), Phys. Lett., **69B**, 85; K. Sato and H. Kobayashi, (1977), Prog. Theo. Phys., **58**, 1775; M. I. Vysotskii, A. D. Dolgov and Ya. B. Zeldovich, (1977), JETP Lett., **26**, 188; R. Cowsik and J. McClelland, (1972), Phys. Rev. Lett., **29**, 669; G. Gerstein and Ya. B. Zeldovich, (1966), Zh. Eksp. Teor. Fiz. Pisma Red., **4**, 174; G. Marx and A. Szalay, (1972), in *Neutrino 72*, eds. A. Frenkel and G. Marx, (OMKDT-Technoinform, Budapest), p. 123; J. Bernstein, L. Brown and G. Feinberg, (1985), Phys. Rev., **D32**, 3261; R. Scherrer and M. S. Turner, (1986), Phys. Rev., **D33**, 1585; **D34**, 3263; (E); also see the review: A. Dolgov and Ya. B. Zeldovich, (1981), Rev. Mod. Phys., **53**, 1.
4. The nucleosynthesis calculations are performed using numerical codes which take into account all the complications discussed in the text. A description of such codes can be found in R. V. Wagoner, W. A. Fowler and F. Hoyle, (1967), Ap. J., **148**, 3; A. Yahil and G. Beandet, (1967), Ap. J., **206**, 26; B. V. Vainer et al., (1976), Sov. Astron., **22**, 1; Y. David and H. Reeves, (1980) in *Physical Cosmology*, eds. R. Balian, J. Audouze and D. N. Schramm, (North-Holland, Amsterdam); N. Terasawa and K. Sato, (1985), Ap. J., **294**, 9; R. V. Wagoner, (1973), Ap. J., **179**, 343; D. A. Dicus, et al., (1982), Phys. Rev., **D26**, 2694; For a general review of the nucleosynthesis in the early universe see, A. M. Boesgaard and G. Steigman, (1985), Ann. Rev. Astron. Ap., **23**, 319; D. N. Schramm and R. V. Wagoner, (1979), Ann. Rev. Nucl. Part. Sci., **27**, 37; S. M. Austin, (1981), Prog. Part. Nucl. Phys., **7**, 1.
5. Some of the recent determinations of the neutron lifetime can be found in P. Bopp et al., (1986), Phys. Rev. Lett., **56**, 919; 1986 ibid, **57**, 1192

E; J. Last et al., (1988), Phys. Rev., Lett., **60**, 995; E. Klemt et al., (1988), Z. Phys,, **C37**, 179.

6. Observational status of the primordial nucleosynthesis is reviewed by B. E. J. Pagel, (1991) Phys. Scripta, **T36**, 7. Models for inhomogeneous nucleosynthesis are reviewed by D. N. Schramm (1991) in *After the First Three Minutes*, eds. S. S. Holt, C. L. Bennett and V. Trimble (AIP, New York), p.12.
7. B. Adeva et al., (1989), Phys. Lett., **B231**, 509; ibid, (1990), **B237**, 136; D. Decamp et al., (1989), Phys. Lett., **B231**, 519; M. Z. Akrawy et al., (1989), Phys. Lett., **B231**, 530; P. Aarnio, et al., (1989), Phys. Lett., **B231**, 539; G. S. Abrams, et al., (1989), Phys. Rev. Lett., **63**, 2173.
8. Most of the radiative processes needed in this section are discussed in G. B. Rybicki and A. P. Lightman, (1979), *Radiative Processes in Astrophysics* (John Wiley, New York); the double-Compton process is studied in L. Danese and De Zotti, Astron. Ap., (1982), **107**, 39.
9. See, for example, E. M. Lifshitz and L. P. Pitaevskii, (1981), *Physical Kinetics*, (Pergamon Press, Oxford).
10. B. J. T. Jones and R. F. G. Wyse, (1985), Astron. Ap., **149**, 144; also see P. J. E. Peebles, (1968), Ap. J., **153**, 1; Ya. B. Zeldovich et al., (1969), Sov. Phys. JETP, **28**, 146; R. A. Sunyaev, Ya. B. Zeldovich, (1970), Astrophy. Sp. Sci, **7**, 1; T. Matsuda, H. Sato and H. Takeda, (1971), Prog. Theo. Phys., **46**, 416; S. A. Bonometto, et al., (1983), Astron. Ap., **126**, 377.
11. This Gaussian fitting is discussed in B. J. T. Jones and R. F. G. Wyse, (1985), Astron. Ap. **149**, 144.

Chapter 4

1. Perturbation theory is discussed in several standard texts and reviews; see, for example, P. J. E. Peebles, (1980), *The Large Scale Structure of the Universe*, (Princeton University Press, Princeton); L. D. Landau and E. M. Lifshitz, (1975), *Classical Theory of Fields*, (Pergamon Press, New York); G. Efstathiou and J. Silk, (1983), Fund. Cosmic Phys., **9**, 1; G. Efstathiou, (1990), in *Physics of the Early Universe*, eds. J. A. Peacock, A. F. Heavens and A. T. Davies, (Sussp Publications, New York); Also see E. M. Lifshitz and I.M. Khalatnikov, (1963), Adv. Phys., **12**, 185; W. Press and E. T. Vishniac, (1980), Ap. J., **239**, 1.
2. For a good discussion of the gauge invariant formalism see J. M. Bardeen, (1980), Phys. Rev., **D22**, 1882; J. M. Bardeen, P. J. Steinhardt and M. S. Turner, (1983), Phys. Rev., **D28**, 679; D. Salopek, J. R. Bond and J. M. Bardeen, (1989), Phys. Rev., **D40**, 1753; for a review, see H. Kodoma and M. Sasaki, (1982), Prog. Theo. Phys. Suppl., **68**, 1398.

3. Most of the standard books provide a discussion of the 'Newtonian' perturbation theory. See, for example, S. Weinberg, (1972), *Gravitation and Cosmology*, (Wiley, New York); P. J. E. Peebles, (1980), *The Large Scale Structure of the Universe*, cited in ref. 1 above.
4. The derivation given here is based on T. Padmanabhan (1990), TIFR preprint (unpublished); T. Padmanabhan and K. Subramanian, (1992), Bull. Astron. Soc. In., **20**, 1; identical results have been obtained in D. H. Lyth and E. D. Stewart, (1990), Ap. J., **361**, 343; also see A. Kodoma and M. Sasaki, (1987), IJMP, **A2**, 491.
5. See P. J. E. Peebles, (1980), cited in ref. 1 above.
6. P. Mészaros, (1975), Astron. Ap., **38**, 5.
7. See, for example, J. R. Bond, G. Efstathiou and J. Silk, (1980), Phys. Rev. Lett., **45**, 1980; A. G. Doroshkevich et al., (1981), in *Proceedings of the Tenth Texas Symposium*, Ann. N. Y. Acad. Sci., **375**, 32; P. J. E. Peebles, (1982), Ap. J., **258**, 415.
8. J. Silk, (1968), Ap. J., **151**, 459; P. J. E. Peebles, (1968), Ap. J., **153**, 1; M. L. Wilson and J. Silk, (1981), Ap. J., **243**, 14; J. R. Bond and G. Efstathiou, (1984), Ap. J., **285**, L45; also see, P. J. E. Peebles, (1981), Ap. J., **248**, 885; N. Kaiser, (1983), MNRAS, **202**, 1169.
9. E. R. Harrison, (1970), Phys. Rev. **D1**, 2726; Ya. B. Zeldovich, (1972), MNRAS, **160**, 1p.
10. The analytic formula for the HDM spectrum is given in J. R. Bond and A. S. Szalay, (1983), Ap. J., **276**, 443. The analytic formula for the CDM spectrum is given by P. J. E. Peebles, (1983), Ap. J., **263**, L1; M. Davis et al., (1985), Ap. J., **292**, 371. Another analytic approximation is suggested by A. A. Starobinsky and V. Sahni, (1986), University of Newcastle preprint NCL-TP12 and is discussed in S. F. Shandarin and Ya. B. Zeldovich, (1989), Rev. Mod. Phys., **61**, 185.

Chapter 5

1. An excellent discussion of the statistical properties of the fluctuations can be found in: Ya. B. Zeldovich and I. D. Novikov, (1983), *Relativistic Astrophysics (Vol. 2): The Structure and Evolution of the Universe*, (University of Chicago Press, Chicago); a more advanced treatment is available in: P. J. E. Peebles, (1980), *The Large Scale Structure of the Universe*, (Princeton University Press, Princeton).
2. M. Davis and P. J. E. Peebles, (1983), Ap. J., **267**, 465; also see P. J. E. Peebles, (1974), Astron. Ap., **32**, 391; P. J. E. Peebles and E. J. Groth, (1976), Astron. Ap., **53**, 131; G. Efstathiou, (1979), MNRAS, **187**, 117.
3. J. Huchra et al., (1983), Ap. J. Supp. Ser., **53**, 89; M. Davis and P. J. E. Peebles, (1983), Ap. J., **267**, 465.

4. For a discussion and examples of 'biasing' see M. J. Rees, (1985), MNRAS, **213**, 75p; J. Silk, (1985), Ap. J., **297**, 1; A. Dekel and J. Silk, (1986), Ap. J., **303**, 39. Biasing will be further discussed in chapter 8.
5. M. Davis and P. J. E. Peebles, (1977), Ap. J. Suppl. Ser., **34**, 425.
6. N. Kaiser, (1985), Ap. J., **284**, L9; (1986); in *Inner Space/Outer Space*, eds. E. W. Kolb et al., (University of Chicago Press, Chicago), p. 285; J. M. Bardeen, p. 212 of the same book; N. Kaiser, (1985), Ap. J., **284**, L9; J. M. Bardeen et al., (1986), Ap. J., **304**, 15.
7. W. H. Press and P. L. Schecter, (1974), Ap. J., **187**, 425; J. A. Peacock and A. F. Heavens, (1990), MNRAS, **243**, 133; S. Cole, (1989), Ph. D. Thesis, University of Cambridge.
8. S. Chandrashekhar, (1943), Rev. Mod. Phys., **15**, 1.

Chapter 6

1. For an overview of the theoretical aspects, see N. Kaiser and J. Silk, (1986), Nature, **324**, 529. The observational aspects of MBR anisotropies are discussed in, for example, R. B. Patridge, (1988), Rep. Prog. Phys., **51**, 647; D. T. Wilkinsion, (1987), in *Texas Symposium on Relativistic Astrophysics*, ed. M. P. Ulmer, (World Scientific, Singapore); An excellent summary of both theory and observation are available in: G. Efstathiou, (1990), in *Physics of the Early Universe*, eds. J. A. Peacock, A. F. Heavens and A. T. Davies, (SUSSP Publications, Adam Hilger, New York); Also see the papers in *The Galactic and Extragalactic Background Radiation*, (1990), eds. S. Bowyer and C. Leinert (Kluwer, Dordrecht).
2. P. M. Lubin and T. Villela, (1985), in *The Cosmic Background Radiation and Fundamental Physics*, ed. F. Melchiorri, (Editrice Compositore, Bologna), p. 65; P. M. Lubin et al., (1985), Ap. J. Lett., **298**, L1; D. J. Fixen, E. S. Cheng and D. T. Wilkinson, (1983), Phys. Rev. Lett., **277**, L23; I. A. Strukov, D. P. Skulachev and A. A. Klypin, (1987), in *Large Scale Structures of the Universe*, (IAU Symp. 130), eds. J. Audouze, M-C. Pelletar and H. Szalay, (Kluwer, Dordrecht), p. 27.
3. G. F. Smoot et al., (1991), Ap. J., **371**, L1.
4. F. Melchiorri et al., (1981), Ap. J., **250**, L1; R. D. Davies et al., (1987), Nature, **326**, 462; J. M. Uson and D. T. Wilkinson in Ap. J., (1984), **277**, L1; Nature, (1985), **312**, 427; and Ap. J., (1984), **283**, 471; A. C. S. Readhead et al., (1989), Ap. J., **346**, 566; K. I. Kellerman, et al., (1986), *Highlights of Astronomy – Vol. 6*, (Reidel, Dordrecht), p. 367.
5. R. K. Sachs and A. M. Wolfe, (1967), Ap. J., **147**, 73; Also see Ya. B. Zeldovich and M. V. Sazhin, (1987), Sov. Astron. Lett., **13**, 145; E. Martinez-Gonzalez, J. L. Sanz and J. Silk, (1990), Ap. J., **355**, L5; A. M. Anile and S. Motta, (1976), Ap. J., **207**, 685.

6. J. R. Bond and G. Efstathiou, (1984), Ap. J., **285**, L45; J. R. Bond and G. Efstathiou, (1987), MNRAS, **226**, 655; J. R. Bond, (1990), in *Frontiers in Physics - From Colliders to Cosmology*, Proceedings of the Lake Louise Winter Institute, eds. B. Campbell and F. Khanna, (World Scientific, Singapore).
7. J. Silk, (1986), in *Inner Space/Outer Space*, eds. E. W. Kolb et al., (University of Chicago Press, Chicago), p. 143; J. R. Bond, (1988) in *Large Scale Structures of the Universe*, cited in ref. 2 above, p. 93.
8. M. S. Turner and L. M. Widrow, (1988), Phys. Rev., **D37**, 3428; V. A. Rubakov, M. Sazhin and A. Veryasken, (1982), Phys. Lett., **115B**, 189; R. Fabbri and M. Pollock, (1983), Phys. Lett., **125B**, 445; B. Allen, (1988), Phys. Rev., **D37**, 2078; L. Abbot and M. Wise, (1984), Nucl. Phys., **B244**, 541; G. Dautcourt, (1969), MNRAS, **144**, 255; E. V. Linder, (1988), Ap. J., **326**, 517.
9. See G. F. Smoot et al., (1991), ref. 3 above; I. A. Strukov, D. P. Skulachev and A. A. Klypin, (1987), cited in ref. 2 above.
10. See J. R. Bond and G. Efstathiou, (1984, 1987), cited in ref. 6 above; G. Efstathiou and J. R. Bond, (1986), MNRAS, **218**, 103; P. J. E. Peebles and J. T. Yu, (1970), Ap. J., **162**, 815; M. L. Wilson and J. Silk, (1981), Ap. J., **243**, 14; S. A. Bonometto, A. Caldara and F. Lucchin, (1983), Astron. Ap., **126**, 377.
11. For an excellent discussion of the entire subject and a more detailed comparison of theory and observation, see: J. R. Bond (1988), in *The Early Universe*, eds. W. G. Unruh and G. W. Semenoff, (Reidel, Dordrecht); and in *Large scale structures of the universe*, cited in ref. 2 above, p.93. Table 6.1 is adapted from the bounds discussed in this reference.
12. See, for example, G. Efstathiou and J. R. Bond, (1986), MNRAS, **218**, 103.
13. See, for example, E. T. Vishniac, (1987), Ap. J., **322**, 597.
14. Ya. B. Zeldovich and R. A. Sunyaev, (1969), Astrophy. Sp. Sci., **4**, 301; P. J. E. Peebles, (1971), *Physical Cosmology*, (Princeton University Press, Princeton), Chap. 7; K. L. Chan and B. J. T. Jones, (1975), Ap. J., **195**, 1; J. G. Bartlett and A. Stebbins, (1991), Ap. J., **371**, 8.
15. See Ya. B. Zeldovich and I. Novikov, (1983), *Relativistic Astrophysics – Vol II*, (University of Chicago Press, Chicago).
16. See, for example, R. Cowsik and E. J. Kobetich, (1972), Ap. J., **177**, 585. For a more recent discussion (which also contains references to earlier work) see R. Subrahmanyan and R. Cowsik, (1989), Ap. J., **347**, 1.
17. For a review of this effect, see R. A. Sunyaev and Ya. B. Zeldovich, (1980), Ann. Rev. Astron. Ap., **18**, 537 and references cited therein.

18. M. Birkinshaw and S. F. Gull, (1984), MNRAS, **206**, 359; Nature, (1984), **309**, 34.

Chapter 7

1. For an overview of this subject, see D. Burstein, (1990), Rep. Prog. Phys., **53**, 421; G.E. Gunn, (1989), in *The Extragalactic Distance Scale*, eds. S. Van den Bergh and C.J. Pritchet (Provo: Astron. Soc. of Pacific), p. 344; V.C. Rubin, (1988), *Gerard and Antoinette de Vaucouleurs: A Life for Astronomy*, eds. M. Capaccioli and H.G. Corwin (Singapore: World Scientific); A. Dressler, (1989), Proc. 14th Texas Relativistic Symposium; M. Davis and P.J.E. Peebles, (1983), Ann. Rev. Astron. Ap., **21**, 109; also see the contributions in *Large Scale Motions in the Universe*, (1989) eds., V.C. Rubin and G. Coyne (Princeton; Princeton University Press).
2. B.E. Corey and D.T. Wilkinson, (1976), Bull. Am. Astron. Soc., **8**, 351; G.F. Smoot, M.V. Gerenstin, R.A. Muller, (1977), Phys. Rev. Lett., **39**, 898; [also see E.K. Conklin, (1969), Nature **222**, 971]; G.F. Smoot et al., (1991), Ap. J., 371, L1.
3. D. Mihalas and J.J. Binney, (1981), *Galactic Astronomy*, (Freeman: San Francisco); M. Fich, L. Blitz and A. Stark, (1989), Ap. J., **342**, 272; F.J. Kerr, D. Lynden-Bell, (1986), MNRAS, **221**, 1023.
4. A. Yahil, G. Tamman and A. Sandage, (1977), Ap. J., **217**, 903. For a review of different measurements, see: D. Lynden-Bell and O. Lahav in *Large Scale Motions in the Universe*, (1989), cited in ref. 1 above. G. de Vaucouleurs, A. de Vaucouleurs and H.G. Corwin, (1976), *Second Reference Catalog of Bright Galaxies*, (University of Texas Press, Austin).
5. For a discussion of the results from various 'dipoles', see: O. Lahav, (1987), MNRAS, **225**, 213; D. Lynden-Bell, O. Lahav and D. Burstein, (1989), MNRAS, **241**, 325; A. Yahil, D. Walter and M. Rowan-Robinson, (1986), Ap. J. Lett., **301**, 1; O. Lahav, M. Rowan-Robinson and D. Lynden-Bell, (1988), MNRAS, **234**, 677; M. Rowan-Robinson et al., (1990), MNRAS, **247**.
6. R.B. Tully and J.R. Fisher, (1977), Astron. Ap., **54**, 661; A. Dressler et al., (1987), Ap. J., **313**, 42; S. Djorgovski and M. Davis, (1987), Ap. J., **313**, 59. The estimates for the scatter in IRTF quoted in the literature vary quite a bit. For a discussion of this issue, see, for example, D. Burstein and S. Raychaudhury, (1989), Ap. J., **343**, 18; M. Han and J. Mould, (1990), Ap. J., **360**, 448.
7. S. Djorgovski, R. de Carvalho and M.S. Hans, (1989) in *The Extragalactic Distance Scale*, cited in ref. 1 above, p. 329; J. Silk, (1989), Ap. J., **345**, L1. Also see G. Giuricin et al., (1989), Ap. J., **345**, 101; D. Burstein, (1990), cited in ref. 1 above; For a similar discussion of the

Tully–Fisher relation see J. Mould, M. Han and G. Bothun, (1989), Ap. J., **347**, 112; These aspects are reviewed by E. Bertschinger, (1990), in *Particle Astrophysics – XXVth Rencontre de Moriond*, eds. J.M. Alimi et al., (Editions Frontiers, Paris).

8. For a discussion, see: D. Lynden-Bell et al., (1988), Ap. J., **326**, 19; A. Dekel and E. Bertschinger, (1990), *Large Scale Structure and Peculiar Motions in the Universe*, Eds. D.W. Latham and L.N. da Costa (Astronomical Society of Pacific, San Francisco) and the references cited therein.
9. D. Burstein et al., (1986), *Galaxy Distances and Deviations from Universal Expansion*, eds. B.F. Madore and R.B. Tully, (Reidel: Boston), p. 123; Also see A. Dressler, (1987), Ap. J., **317**, 1; P.B. Lilje, A. Yahil and B.J.T. Jones, (1986), Ap. J., **307**, 91; D. Lynden-Bell et al.,(1988), Ap. J., **326**, 16.
10. This is discussed in detail in the review by D. Burstein, (1990) cited in ref. 1 above.
11. M. Aaronson et al.,(1989), Ap. J., **338**, 654; S.M. Faber and D. Burstein, (1989), in *Large Scale Motions in the Universe*, cited in ref. 1 above; J.R. Lucey and D. Carter, (1988), MNRAS, **235**, 1177.
12. The 'Great Attractor' region has been surveyed extensively in recent years; see A. Dressler, (1988), Ap. J., **329**, 519; A. Dressler and S.M. Faber, (1990), Ap. J., **354**, 13 and **354**, L 45; R. Scaramella et al., (1989), Nature, **338**, 562; S. Raychaudhury, (1989), Nature, **342**, 251; An earlier survey is reported in H. Shapley, (1930), Harvard Coll. Obs. Bull., **374**, 9.
13. E. Bertschinger and A. Dekel, (1989), Ap. J., **336**, L5; A. Dekel, E. Bertschinger and S.M. Faber, (1990), Ap. J., **364**, 349; E. Bertschinger et al.,(1990), Ap. J., **364**, 370.
14. See the contributions by M. Strauss and M. Davis and by A. Yahil, (1988), in *Large Scale Motions in the Universe*, cited in ref. 1 above, p. 255; M. Strauss and M. Davis, 1988, in *The Large Scale Structure of the Universe, (IAU Symposium 130)*, ed. J. Audouze.
15. M. Rowan-Robinson et al.,(1990), MNRAS, **247**, 1; G. Efstathiou et al., (1990), MNRAS, **247**, 10p; W. Saunders et al., (1991), Nature, **349**, 32.
16. Whether the cold dark matter models are consistent with the existence of a Great Attractor is a much debated question. For a selection of references see: N. Vittoria, R. Juszkiewicz and M. Davis, (1986), Nature **323**, 132; E. Bertschinger and R. Juszkiewicz, (1988), Ap. J., **334**, L59; N. Kaiser, (1988), MNRAS, **231**, 149; N. Kaiser and O. Lahav, (1988), in *Large Scale Motions in the Universe (Proceedings of the Vatican Study Week 27)*, (Pontifical Academy of Sciences, Vatican). E. Groth,

R. Juszkiewicz and J. P Ostriker, (1989), Ap. J., **346**, 558; M. Aaronson et al., (1986), Ap. J., **302**, 536.
17. R. Juszkiewicz, K.M. Gorski and J. Silk, (1987), Ap. J., **323**, L1; K.M. Gorski, (1991), Ap. J., **370**, L5.
18. Y. Suto et al.,(1988), Nature, **332**, 328.
19. E. Bertschinger, K.M. Gorski and A. Dekel, (1990), Nature, **345**, 507.
20. J.P. Ostriker and Y. Suto, (1990), Ap. J., **348**, 378.
21. R. Juszkiewicz, N. Vittoro and R.F.G. Wyse, (1990), Ap. J., **349**, 408.

Chapter 8

1. See, for example, P.J.E. Peebles, (1980), *Large Scale Structure of the Universe*, (Princeton University Press, Princeton).
2. D. Lyndel-Bell, (1967), MNRAS, **136**, 101.
3. P.J.E Peebles, (1980), *Large Scale Structure of the Universe*, ref. 1 above. G. Lemaitre, (1931), MNRAS, **91**, 490; (1933), Comp. Rend., **196**, 903; 1085; R. C. Tolman, (1934), Proc. Nat. Acad. Sci., **20**, 169.
4. S.D.M. White and M.J. Rees, (1978), MNRAS, **183**, 341.
5. F. Hoyle, (1953), Ap. J., **118**, 513; M.J. Rees and J.P. Ostriker, (1977), MNRAS, **179**, 541; J. Silk, (1977), Ap. J., **211**, 638; J. Binney, (1977), Ap. J., **215**, 483; S.D.M. White and M.J. Rees, (1978), (ref. 4 above).
6. J.C. Raymond, D.P. Cox and B.W. Smith, (1976), Ap. J., **204**, 290; J.A. Peacock and A.F. Heavens, (1990), MNRAS, **243**, 133.
7. Cooling diagrams have been used by several people to analyse this problem. See, for example, J. Silk, (1977); M. J. Rees and J. P. Ostriker (1977); S. D. M. White and M. J. Rees, (1978) cited in ref. 5, 4 above. Also see, for a very detailed discussion, G.R. Blumenthal et al.,(1984), Nature, **341**, 517.
8. A. Dekel and J. Silk, (1986), Ap. J., **303**, 39; S.D.M. White and C.S. Frenk, (1990), Arizona Theoretical Astrophysics preprint No. 90-38.
9. Ya B. Zeldovich, (1970), Astron. and Ap., **5**, 84.
10. A.G. Doroshkevich, (1970), Astrofisika, **6**, 581; G. Efstathiou and J. Silk, (1983), Fund. Cosmic. Phys., **9**, 1.
11. S.N. Gurbatov, A.I. Saichev and S.F. Shandarin, (1989), MNRAS, **236**, 385; (1983), Sov. Phys. Usp., **26**, 857; (1985), Sov. Phys. Dok., **30**, 921; S.F. Shandarin and Ya. B. Zeldovich, (1990), Rev. Mod. Phys., **61**, 185.
12. J.M. Burgers, (1940), Proc. Roy. Neth. Acad. Sci., **43**, 2; J.M. Burgers, (1974), *The Nonlinear Diffusion Equation*, (Reidel, Dordrecht).
13. See, for example, D. H. Weinberg and J. E. Gunn, (1990), Ap. J., **352**, L25; MNRAS, **247**, 260; L. A. Kofman, D. Pogosian and S. F. Shandarin, (1990) MNRAS, **242**, 200.
14. T. Padmanabhan, (1991), TIFR-TAP-preprint (unpublished).

15. S.N. Gurbatov, A.I. Saichev and S.F. Shandarin, (1989), cited in ref. 11 above. Also see E.V. Kotok and S.F. Shandarin, (1989), Sov. Astron., **32**, 351; B.G. Williams et al.,(1991), MNRAS, **250**, 458.
16. F. Hoyle, (1949) in *Problems of Cosmical Aerodynamics*, (Central Air Documents Office, Ohio), p. 195; P.J.E. Peebles, (1969), Ap. J., **155**, 393; A.G. Doroshkevich, (1970), Astrofisika, **6**, 581; S.D.M. White, (1984), Ap. J., **286**, 34.
17. See, for example, J. Barnes and G. Efstathiou, (1987), Ap. J., **319**, 575.
18. S.D.M. White and M.J. Rees, (1978), ref. 4 above; S.M. Faber, (1982), in *Astrophysical Cosmology*, eds. H.A. Bruck, G.V. Coyne and M.S. Longair (Pontificia Acad. Sci., Vatican), p. 219.
19. For a sample of papers dealing with this issue, see: S.M. Fall and G. Efstathiou, (1980), MNRAS, **193**, 189; J.E. Gunn, (1982), in *Astrophysical Cosmology*, cited in ref. 18 above, p. 233.
20. J. Binney, (1978), MNRAS, **183**, 501.
21. R. B. Larson, (1975), MNRAS, **173**, 671; J.R. Gott, (1977), Ann. Rev. Astron. Ap., **15**, 235.
22. Several aspects of this idea are discussed in: A. Toomre, (1977), in *Evolution of Galaxies and Stellar Populations*, eds., B.M. Tinsley and R.B. Larson, (Yale University Observatory, New Haven), p. 401; F. Schweizer, (1986), in *Nearly Normal Galaxies*, eds., S.M. Faber, (Springer Verlag, Berlin), p. 18; F. Schweizer, (1982), Ap. J., **252**, 455; S.D.M. White, (1979), MNRAS, **189**, 831; J. Negroponte and S.D.M. White, (1983), MNRAS, **205**, 1009; S.M. Fall, (1979), Nature, **281**, 200.
23. S. D. M. White, (1982), in *Morphology and Dynamics of Galaxies*, ed., L. Martinet, M. Mayer, (Geneva Observatory), p. 291; J. Barnes, (1989), Nature, **338**, 123.
24. J. Kormendy, (1989), Ap. J. Lett., **342**, L63.
25. S. Van den Bergh, (1990), in *Dynamics and Interactions of Galaxies*, ed., R. Weilen, (Springer Verlag, Berlin), p. 492.
26. J. Barnes and G. Efstathiou, (1987), Ap. J., **319**, 575; C.S. Frenk et al.,(1988), Ap. J., **327**, 507.
27. R.W. Hockney and J.W. Eastwood, (1981), *Computer Simulation Using Particles*, (McGraw Hill, New York); S.J. Aarseth, (1984), in *Methods of Computational Physics*, eds. J.U. Blackbill and B.I. Cohen (Academic, New York); G. Efstathiou et al.,(1985), Ap. J. Suppl., **57**, 241.
28. M. Davis et al.,(1985), Ap. J., **292**, 371; C.S. Frenk et al.,(1985), Nature, **317**, 595; C.S. Frenk et al., (1988), Ap. J., **327**, 507; S.D.M. White et al.,(1987), Nature, **330**, 451; S.D.M. White et al.,(1987), Ap. J., **313**, 505.

29. See, for example, M.S. Turner, G. Steigman and L. Krauss, (1984), Phys. Rev. Lett., **52**, 2090; K. Olive, D. Sekel and E. Vishniac, (1985), Ap. J., **292**, 1; T. Padmanabhan and M.M. Vasanthi, (1987), Ap. J., **315**, 411.
30. M.J. Rees, (1985), MNRAS, **213**, 75p; J. Silk, (1985), Ap. J., **297**, 1; A. Dekel and M.J. Rees, (1987), Nature, **326**, 455; S.D.M. White et al.,(1987), Nature, **330**, 451.
31. R.G. Carlberg, H.M.P. Couchman and P.A. Thomas, (1990), Ap. J., **352**, L29; A. Evrad, (1986), Ap. J., **310**, 1; M.J. West and D.O. Richstone, (1988), Ap. J., **335**, 532.
32. C.S. Frenk et al.,(1988), Ap. J., **327**, 507.
33. Some of the large-scale observations are described in: P.F. Teague, D. Carter and P.M. Gray, (1990), Ap. J. Suppl., **72**, 715; N.A. Bahcall and R. Soneira, (1983), Ap. J., **270**, 20; W. Sutherland, (1988), MNRAS, **234**, 159; D. Lynden-Bell et al., (1988), Ap. J., **326**, 19. The following papers discuss the observations in the light of CDM model: S.D.M. White et al., (1987), Ap. J., **313**, 505; C.S. Frenk et al., (1990), Ap. J., **351**, 10; however, see P.J.E. Peebles, (1991), in *Observational Tests of Inflation*, (Institute of Advanced Study, Princeton), preprint; N. Kaiser and O. Lahav, (1989), MNRAS, **231**, 635; however, see J.P. Ostriker and Y. Suto, (1990), Ap. J., **348**, 378.
34. S.J. Maddox et al., (1990), MNRAS, **242**, 43p.
35. W. Saunders et al., (1991), Nature, **349**, 32.
36. J. Centralla and A. Mellot, (1982), Nature, **305**, 196; S.D.M. White, C.S. Frenk and M. Davis, (1983), Ap. J., **274**, L1.
37. S.D.M. White, (1986) in *Inner Space/Outer space*, eds. E.W. Kolb, et al.,(Chicago University Press, Chicago), p. 228.
38. E. Braun, A. Dekel and P.R. Shapiro, (1988), Ap. J., **328**, 34.
39. J.M. Centralla et al., (1988), Ap. J., **333**, 24.

Chapter 9

1. Several constraints on the epoch of galaxy formation are discussed in: P.J.E. Peebles, (1989), in *The Epoch of Galaxy Formation*, eds., C.S. Frenk, R.S. Ellis, T. Shanks, A.F. Heavens and J.A. Peacock, (Kluwer, Dordrecht), p. 1.
2. See, for example, J. Binney and S. Tremaine, (1987), *Galactic Dynamics*, (Princeton University Press, Princeton).
3. E. Baron and S.D.M. White, (1987), Ap. J., **322**, 585.
4. D. Koo, (1986), in *Spectral Evolution of Galaxies*, eds. C. Chiosi and A. Renzini, (Reidel, Dordrecht), p. 419.
5. J.A. Tyson, (1988), Astron. J. **96**, 1; L.L. Cowie et al.,(1990), Ap. J. Lett., **360**, L1; L.L. Cowie, (1991), Phys. Scripta, **36**, 102 and the

references cited therein; D. Koo, (1990) in *Evolution of the Universe of Galaxies*, ed. R.G. Kron, A.S.P. Conference Series, **10**, 268.
6. Y. Yoshi and F. Takahara, (1988), Ap. J., **326**, 1.
7. T.J. Broadhurst, R.S. Ellis and T. Shanks, (1988), MNRAS, **235**, 827; M. Colless et al., (1990), MNRAS, **244**, 408; L.L. Cowie and J. P. Gardner, (1990),
8. See the review by L.L. Cowie, (1991), cited in ref. 5 above.
9. A. Songanla, L.L. Cowie and S.J. Lilly, (1990), Ap. J., **348**, 371.
10. B. Rocca-Volmerange and B. Guiderdoni, (1990), MNRAS, **247**, 166.
11. M. Colless et al., (1990), MNRAS, **244**, 408; R.S. Ellis, (1990), in *Evolution of the Universe of Galaxies*, cited in ref. 5 above.
12. S.J. Lilly, L.L. Cowie and J.P. Gardner, (1991), Ap. J., **369**, 79; J.A. Peacock, (1991), Nature, **349**, 190.
13. D. P. Schneider, M. Schmidt and J. E. Gunn, (1991), Astron. J., (in press).
14. K. C. Chambers, G.K. Miley and W.J.M. Van Breugel, (1990), Ap. J., **363**, 21.
15. J.S. Dunlop and J.A. Peacock, (1990), MNRAS, **247**, 19; B.J. Boyle et al, (1987), MNRAS, **227**, 717; R.F. Green, (1989), in *The Epoch of Galaxy Formation*, cited in ref. 1 above, p. 121; B.J. Boyle, (1990), Talk at Texas/ESO-CERN Symposium on Relativistic Astrophysics, Cosmology and Fundamental Physics, Brighton, U.K.
16. G. Efstathiou and M.J. Rees, (1988), MNRAS, **230**, 5p.
17. A. Dekel and J. Silk, (1986), Ap. J., **303**, 39.
18. G. Efstathiou and M.J. Rees, (1988), ref. 16 above; E.L. Turner, (1991), Astron. J., **101**, 5; A. Kashlinsky and B.J.T. Jones, (1991), Nature, **349**, 753.
19. P. J. McCarthy et al., (1990), Astron. J., **100**, 1014.
20. J.E. Gunn and B.A. Peterson, (1965), Ap. J., **142**, 1633.
21. C.C. Steidel and W.L.W. Sargent, (1987), Ap. J. Lett., **318**, L11. Also see D. P. Schneider et al., (1989), Astron. J., **98**, 1951.
22. See, for example, P.R. Shapiro and M.L. Giroux, (1987), Ap. J. Lett., **321**, L107; P.R. Shapiro and M.L. Giroux, (1989) in *The Epoch of Galaxy Formation*, cited in ref. 1 above, p. 153; P. R. Shapiro, M. L. Giroux and A. Babul in *After the First Three Minutes*, eds., S. S. Holt, C. L. Bennet and V. Trimble (AIP, New York, 1991).
23. H. M. P. Couchman and M. J. Rees, (1986), MNRAS, **221**, 53.
24. See R.F. Carswell, (1989), in *The Epoch of Galaxy Formation*, cited in ref. 1 above, p. 89, for a review; R.W. Hunstead et al., (1988), Ap. J., **329**, 527; J. Bechtold, (1987), in *High Redshift and Primeval Galaxies*, eds. J. Bergeron, D. Kunth, B. Rocca-Volmerange and J. Tran Thanh Van (Frontiers, Paris), p. 397.

25. D.W. Weedman et al.,(1982), Ap. J. Lett., **255**, L5; K. Subramaniam and S.M. Chitre, (1984), Ap. J., **276**, 440; C.B. Foltz et al., (1984), Ap. J., **281**, L1.
26. S. Bajtlik, R. Duncan and J.P. Ostriker, (1987), Ap. J., **327**, 570.
27. D.E. Osterbrok, (1974), *Astrophysics of Gaseous Nebulae*, (Freeman, San Francisco); J.P. Ostriker and S. Ikeuchi, (1983), Ap. J. Lett., **268**, L63.
28. W.L.W. Sargent et al.,(1980), Ap. J. Suppl., **42**, 41; J.P. Ostriker, (1988) in *QSO Absorption Lines*, eds. J.C. Blades, D. Turnshek, C.A. Norman, (Cambridge University Press, Cambridge), p. 319; J.P. Ostriker and L.L. Cowie, (1981), Ap. J. Lett., **243**, L127; J.P. Ostriker and S. Ikeuchi, (1983), Ap. J. Lett., **268**, L63; S. Ikeuchi and J.P. Ostriker, (1986), Ap. J., **301**, 552; E. Baron et al.,(1989), Ap. J., **337**, 609; R.F. Carswell and M.J. Rees, (1987), MNRAS, **224**, 13.
29. M.J. Rees, (1986), MNRAS, **218**, 25; M.J. Rees, (1988), in *QSO Absorption Lines*, cited in ref. 28 above, p. 107; S. Ikeuchi, (1986), Ap. Sp. Sci., **118**, 509; S. Ikeuchi, I. Murakami and M.J. Rees, (1988), MNRAS, **236**, 21.
30. A.M. Wolfe (1989) in *The Epoch of Galaxy Formation*, cited in ref. 1 above, p. 101; A.M. Wolfe et al., (1986), Ap. J. Suppl., **61**, 249; D.A. Turnshek et al., (1989), Ap. J., **344**, 567; A.M. Wolfe, (1988) in *QSO Absorption Lines*, cited in ref. 28 above, p. 297; J.H. Black, F.H. Chaffe and C.B. Foltze, (1987), Ap. J., **317**, 442; K.M. Lanzetta, A.M. Wolfe and D.A. Turnshek, (1989), Ap. J., **344**, 277; S.M. Fall, Y.C. Pei and R.G. McMahon, (1989), Ap. J. Lett., **341**, L5; M. Pettini, A. Boksenberg and R.W. Hunstead, (1989), Ap. J., **348**, 48.
31. M.J. Rees, (1988) in *QSO Absorption Lines*, cited in ref. 28 above, p. 107; C.J. Hogan, (1987), Ap. J. Lett., **316**, L59; K. Subramanian, (1988), MNRAS, **234**, 459. For some other models, see J.A. Tyson, (1988), Astron. J., **96**, 1; M. Pettini et al., (1990), MNRAS, **246**, 545.
32. R.J. Weymann, R.F. Carswell and M.J. Smith, (1981), Ann. Rev. Astron. Ap., **19**, 41; W.L.W. Sargent, (1988), in *QSO Absorption Lines*, cited in ref. 28 above, p. 1; J. Bergeron, (1988), in *QSO Absorption Lines*, cited in ref. 28 above, p. 127.
33. W.L.W. Sargent, A. Boksenberg and C.C. Steidel, (1988), Ap. J. Suppl., **68**, 539.
34. P.J. McCarthy et al., (1987), in *Cooling Flows in Clusters and Galaxies*, ed. A.C. Fabian (Kluwer, Dordrecht), p. 325; P.J. McCarthy et al., (1987), Ap. J., **321**, L29; K.C. Chambers, G.K. Miley and W.J.M. van Breugel, (1987), Nature, **329**, 604; P.J. McCarthy, (1989), Ph.D. thesis, University of California, Berkeley.
35. For a review of the mechanisms suggested in this connection see, K.C. Chambers and G.K. Miley, (1990), in *Evolution of the Universe of*

Galaxies, cited in ref. 5 above, p. 373; W.J.M. van Breugel et al., (1985), Ap. J., **293**, 83; J.P. Brodie, S. Bowyer and P.J. McCarthy, (1985), Ap. J. Lett., **293**, L59; K.C. Chambers, G.K. Miley and W.J.M. Van Breugel, (1987), Nature, **329**, 604; P.J. McCarthy et al., (1987), in *Cooling Flows in Clusters and Galaxies*, cited in ref. 34 above; De Young, (1989), Ap. J. Lett., **342**, L59; M.C. Begelman and D.F. Cioffi, (1989), Ap. J. Lett., **345**, L21; R. Daly, (1990), Ap. J., **355**, 416. Also see, W.J.M. Van Breugel and P. J. McCarthy, (1990), in *Evolution of the Universe of Galaxies*, cited in ref. 5 above, p. 359; S.M. Fall and M.J. Rees, (1985), Ap. J., **298**, 18.

36. S.J. Lilly, (1989), in *The Epoch of Galaxy Formation*, cited in ref. 1 above, p. 63; S.J. Lilly, (1989), Ap. J., **340**, 77.
37. K.C. Chambers, G.K. Miley and R.R. Joyce, (1988), Ap. J., **329**, L75; M. Bithell and M.J. Rees, (1990), MNRAS, **242**, 570; K.C. Chambers and S. Charlot, (1989), Ap. J. Lett., **348**, L1.
38. S.J. Lilly and M.S. Longair, (1984), MNRAS, **211**, 833; S.J. Lilly, (1989), in *The Epoch of Galaxy Formation*, cited in ref. 1 above, p. 63; S.J. Lilly, (1989), Ap. J., **340**, 77; also see K.C. Chambers and S. Charlot, (1989), Ap. J. Lett., **348**, L1.
39. K.C. Chambers and S. Charlot, (1989), cited in ref. 38 above.
40. S.J. Lilly, (1989), in *The Epoch of Galaxy Formation*, cited in ref. 1 above, p. 63; S.J. Lilly, (1989), Ap. J., **340**, 77.

Chapter 10

1. D. Kazanas, (1980), Ap. J., **241**, L59; K. Sato, (1981), MNRAS, **195**, 467; A.H. Guth, (1981), Phys. Rev., **D23**, 347; For a review, see R.H. Brandenberger, (1985), Rev. Mod. Phys., **57**, 1; J. V. Narlikar and T. Padmanabhan, (1991), Ann. Rev. Astron. Astrophys., **29**, 325.
2. This aspect is emphasised in E.W. Kolb and M.S. Turner, (1990), *The Early Universe*, (Addison-Wesley, California), chap. 8; J.V. Narlikar and T. Padmanabhan, (1991), cited in ref. 1 above.
3. T. Padmanabhan, (1983), Phys. Lett., **A96**, 110.
4. See, for example, J.V. Narlikar, (1984), Found. Phys., **14**, 443; J.V. Narlikar and T. Padmanabhan, (1983), Ann. Phys., **150**, 289; T. Padmanabhan, (1983), Phys. Rev., **D28**, 756.
5. T. Padmanabhan and T. R. Seshadri, (1988), Class. Q. Grav., **5**, 221; J.F.R. Ellis, (1988), Class, Q. Grav., **5**, 891; J.F.R. Ellis and W. Stoger, (1988), Class. Q. Grav., **5**, 207.
6. These effects are discussed in: A. D. Linde, (1979), Rep. Prog. Phys., **42**, 389; R. Brandenberger, (1985), cited in ref. 1 above; L. Dolan and R. Jackiw, (1974), Phys. Rev., **D9**, 3320.
7. A. Guth and E. Weinberg, (1983), Nucl. Phys., **212**, 321.

8. A. Albrecht and P.J. Steinhardt, (1982), Phys. Rev. Lett., **48**, 1220; A.D. Linde, (1982), Phys. Lett., **B108**, 389; A.D. Linde, (1982), Phys. Lett., **B116**, 335.
9. L. Abbott, E. Farhi and M. Wise, (1982), Phys. Lett., **B117**, 29; A. Albrecht et al., (1982), Phys. Rev. Lett., **48**, 1437; A. Dolgov and A.D. Linde, (1982), Phys. Lett., **B116**, 329.
10. A.D. Linde, (1982), Phys. Lett., **116**, 335; A. Vilenkin and L. Ford, (1982), Phys. Rev., **D26**, 1231.
11. A.D. Linde, (1983), Phys. Lett., **B129**, 177.
12. J. Bardeen, P. Steinhardt and M.S. Turner, (1983), Phys. Rev., **D28**, 679; A. Guth and S.Y. Pi, (1982), Phys. Rev. Lett., **49**, 1110; S.W. Hawking, (1982), Phys. Lett., **B115**, 295; A.A. Starobinsky, (1982), Phys. Lett., **B117**, 175; also see references in ref. 13.
13. R.H. Brandenberger, (1985), cited in ref. 1 above; T. Padmanabhan, T.R. Seshadri and T.P. Singh, (1989), Phys. Rev. **D39**, 2100.
14. J. Ellis et al., (1985), Phys. Lett., **B152**, 175; R. Holman, P. Ramond and G.G. Ross, (1984), Phys. Lett., **B137**, 343; L. Jensen and K. Olive, (1986), Nucl. Phys., **B263**, 731; B. Ovrut and P.J. Steinhardt, (1983), Phys. Lett., **B133**, 161; T. Padmanabhan, (1988), Phys, Rev. Lett., **60**, 2229; T. Padmanabhan, T.R. Seshadri and T.P. Singh, (1989), Phys. Rev., **D39**, 2100; Q. Shafi and A. Vilenkin, (1984), Phys. Rev. Lett., **52**, 691.
15. L. Abbott and M. Wise, (1984), Nucl. Phys., **B244**, 541; B. Allen, (1988), Phys. Rev., **D37**, 2078; R. Fabri and M. Pollock, (1983), Phys. Lett., **B125**, 445; V.A. Rubakov, M. Sazhin and A. Veryaskin, (1982), Phys. Lett., **B115**, 189.
16. H.B. Nielson and P. Olesen, (1973), Nucl. Phys., **B61**, 45; T.W.B. Kibble, (1976), J. Phys., **A9**, 1387.
17. D. Forster, (1974), Nucl. Phys., **B81**, 84; A. Vilenkin, (1981), Phys. Rev., **D23**, 852. Also see R. Gregory, (1987), Phys. Rev. Lett., **59**, 740.
18. A. Stebbins, (1988), Ap. J., **327**, 584; F. Bouchet, D. Bennett and A. Stebbins, (1988), Nature, **355**, 410; T. Vachaspati, (1986), Phys. Rev. Lett., **57**, 1655; J. Charlton, (1988), Ap. J., **325**, 521; A. Stebbins, (1987), Ap. J., **322**, 1.
19. E.P.S. Shellard, (1987), Nucl. Phys., **B283**, 624; R. Matzner, (1988), Computers in Physics, **1**, 51; K. Moriarty, E. Myers and C. Rebbi, (1988), Phys. Lett., **B207**, 411; E.P.S. Shellard and P. Ruback, (1988), Phys. Lett., **B209**, 262; P. Ruback, (1988), Nucl. Phys., **B296**, 669.
20. T. Vachaspati and A. Vilenkin, (1985), Phys. Rev., **D31**, 3052.
21. A. Vilenkin, (1985), Phys. Rep., **121**, 263.
22. D. Bennett and F. Bouchet, (1988), Phys. Rev. Lett., **60**, 257; D. Bennett and F. Bouchet, (1990), Phys. Rev., **D41**, 2408; A. Albrecht and

N. Turok, (1989), Phys. Rev., **D40**, 973; B. Allen and E.P.S. Shellard, (1990), Phys. Rev. Lett., **64**, 119; Ya. B. Zeldovich, (1980), MNRAS, **192**, 663; A. Vilenkin, (1981), Phys. Rev. Lett., **46**, 1169; A. Vilenkin and Q. Shafi, (1983), Phys. Rev. Lett., **51**, 1716; N. Turok, (1983), Phys. Lett., **B123**, 387; N. Turok, (1984), Nucl. Phys., **B242**, 520; N. Turok and R. Brandenberger, (1986), Phys. Rev., **D33**, 2175; A. Stebbins, (1986), Ap. J. Lett., **303**, L21; H. Sato, (1986), Prog. Theo. Phys., **75**, 1342; J. Kung and R. Brandenberger, (1990), in *Symposium on the Formation and Evolution of Cosmic strings*, eds. G. Gibbons, S. Hawking and T. Vachaspati (Cambridge University Press, Cambridge).

23. J. Silk and A. Vilenkin, (1984), Phys. Rev. Lett., **53**, 1700; T. Vachaspati, (1986), Phys. Rev. Lett., **57**, 1655; A. Stebbins, et al.,(1987), Ap. J., **322**, 1; Ya. B. Zeldovich, (1980), MNRAS, **192**, 663; A. Vilenkin, (1981), Phys. Rev. Lett., **46**, 1169; A. Vilenkin and Q. Shafi, (1983), Phys. Rev. Lett., **51**, 1716; N. Turok, (1983), Phys. Lett., **B123**, 387; N. Turok, (1984), Nucl. Phys., **B242**, 520; N. Turok and R. Brandenberger, (1986), Phys. Rev., **D33**, 2175; A. Stebbins, (1986), Ap. J. Lett., **303**, L21; H. Sato, (1986), Prog. Theo. Phys., **75**, 1342; R. Brandenberger, L. Perivolaropoulos and A. Stebbins, (1990), Int. J. Mod. Phys., **A5**, 1633; L. Perivolaropoulos, R. Brandenberger and A. Stebbins, (1990), Phys. Rev., **D41**, 1764; Also see R. Brandenberger , (1990), Phys. Scripta, **T36**, 114.
24. D. Forster, (1974), cited in ref. 17 above.

Chapter 11

1. J.H. Oort, (1932), Bull. Astr. Inst. Netherlands, **6**, 349; J.H. Oort, (1965), in *Galactic Structure*, eds., A. Blaauw and M. Schmidt, p. 455 (Chicago University Press, Chicago); J. N. Bahcall, (1984), Ap. J., **276**, 169; **287**, 926; D. Mihilas and J.J. Binney, (1981), *Galactic Astronomy*, (2nd edition), (Freeman, San Francisco); J.N. Bahcall and R.M. Soneira, (1980), Ap. J. Suppl., **44**, 73.
2. These results are from the two papers by J.N. Bahcall, (1984), cited above. The negative result, claiming that these is no dark matter in the disc, is reported in K. Kuijken and G. Gilmore, (1989), MNRAS, **239**, 605, 651.
3. B.W. Carney and D.W. Latham, (1987), in *Dark Matter in the Universe, (IAU Symposium No. 117)*, p. 39, eds., J. Kormendy and G.R. Knapp (Reidel, Dordrecht).
4. T. Murai and M. Fujimoto, (1980), Publ. Astro. Soc. Japan, **32**, 581; D.N.C. Lin and D. Lynden-Bell, (1982), MNRAS, **198**, 707; D. Lynden-Bell, R.D. Cannon and P.J. Godwin, (1983), MNRAS, **204**, 87p; B. Little and S. Tremaine, (1987), Ap. J., **320**, 493.

5. S.M. Faber and J.S. Gallagher, (1979), Ann. Rev. Astr. Ap., **17**, 135; V.C. Rubin, W.K. Ford and N. Thonnard, (1980), Ap. J., **238**, 471; V.C. Rubin et al., (1982), Ap. J., **261**, 439; V.C. Rubin et al.,(1985), Ap. J., **289**, 81.
6. T.S. Van Albada et al.,(1985), Ap. J., **295**, 305; also see C. Carignan and K.C. Freeman, (1985), Ap. J., **294**, 494 for other examples. See the contributions by K.C. Freeman, (1987), and by R. Sancisi and T.S. Van Albada, (1987), in *Dark Matter in the Universe*, cited in ref. 3 above, p. 119 and p. 67. For a discussion of different models for NGC 3198 see V. C. Rubin in *After the First Three Minutes*, eds. S. S. Holt, C. L. Bennett and V. Trimble (AIP, New York, 1991).
7. J. Kormendy, (1987), in *Structure and Dynamics of Elliptical Galaxies, (IAU symposium No. 125)*, eds. de Zeeuw, T. (Reidel, Dordrecht); M. Aaronson and E. Olszewski, (1987), in ref. 3 above, p. 153; S. M. Faber and D. N. C. Lin, (1983), Ap. J.; **266**, 17; D. N. C. Lin and S. M. Faber, (1983), Ap. J. **266**, 21. For a discussion of DDO154 see, C. Carignan and S. Beanlieu, (1989), Ap. J., **347**, 760 and reference cited therein.
8. S. Tremaine and H. M. Lee, (1987) in *Dark Matter in the Universe*, (World Scientific, Singapore), p. 103.
9. F. D. Kahn and L. Woltjer, (1959), Ap. J., **130**, 705; J. E. Gunn, (1975), Comm. Ap. Sp. Phys. **6**, 7; D. Lynden-Bell, (1982), in *Astrophysical Cosmology*, eds. H. A. Bruck, G. V. Coyne and M. S. Longair, (Pont. Acad. Scient., Vatican), p. 85; R. Mishra, (1985), MNRAS, **212**, 163; also see, J. R. Gott and T. X. Thaun, (1978), Ap. J., **223**, 426; J. Einasto and D. Lynden-Bell, (1982), MNRAS, **199**, 67.
10. D. N. Limber and W. G. Mathews, (1960), Ap. J., **132**, 286; J. N. Bahcall and S. Tremaine, (1981), Ap. J., **244**, 805; J. Heisler, S. Tremaine and J. N. Bahcall, Ap. J., **298**, 8; J. P. Huchra and M. J. Geller, (1982), Ap. J., **257**, 423 (also see the correction in M.J. Geller, 1984, *Clusters and Groups of Galaxies*, eds., F. Mardirossian, G. Giuricin and M. Mezzetti, (Reidel, Dordrecht)), p. 353.
11. D. Merritt, (1987), Ap. J., **313**, 121; Also see S. M. Kent and W. L. W. Sargent, (1983), Astron. J., **88**, 692; C. Jones and W. Forman, (1984), Ap. J., **276**, 38; C. Sazarin, (1986), Rev. Mod. Phys., **58**, 1; H. J. Rood et al., (1972), Ap. J., **224**, 724.
12. J. A. Tyson in *After the First Three Minutes*, cited in ref. 6 above, p. 437.
13. M. Davis and P. J. E. Pebles, (1983), Ann. Rev. Astron. Ap., **21.**, 109; P. L. Schechter, (1980), Astron. J., **85**, 801; M. Aaronson et al., (1982), Ap. J., **258**, 64; J. L. Tonry and M. Davis, (1981), Ap. J., **246**, 680; M. Davis and J. Huchra, (1982), Ap. J., **254**, 437; A. Yahil, A. Sandage and G. A. Tammann, (1980), Ap. J., **242**, 448.

14. D. W. McCarthy, R. G. Probst and F. J. Low, (1985), Ap. J., **290**, L9; B. J. Carr, J. R. Bond and W. D. Arnett, (1984), Ap. J., **277**, 445; R. B. Larson, (1986), MNRAS, **218**, 409.
15. B. J. Carr, J. R. Bond and W. D. Arnett, (1984), Ap. J., **277**, 445.
16. C. G. Lacey and J. P. Ostriker, (1985), Ap. J., **299**, 633; M. Gorenstein et al.,(1984), Ap. J., **287**, 538; C. Canizares, (1982), Ap. J., **263**, 508.
17. B. Paczynski, (1986), Ap. J., **304**, 1; for a review, see, for example, the contribution by D. P. Bennett in *After the First Three Minutes*, cited in ref. 6 above, p. 446.
18. D. J. Hegyi and K. A. Olive, (1983), Phys. Lett., **B126**, 28.
19. K. Griest and M. Kamionkowski, (1990), Phys. Rev. Lett., **64**, 615.
20. L. M. Krauss, (1990), Phys. Rev. Lett., **64**, 999; K. Griest and J. Silk, (1990), Nature, **343**, 261.
21. For a general overview, see, for example, G. Gelmini, (1990), in *Dark Matter in the Universe*, eds. P. Galeotti and D. N. Schramm (Kluwer, Denmark) p.25; D. O. Caldwell, (1991) UCSB-HEP-preprint 91-04; J. Ellis, (1991), Phys. Scripta, **T36**, 142.
22. S. P. Ahlen et al., (1987), Phys. Lett., **B195**, 603; F. Boehm et al., (1991), Phys. Lett., **B255**, 143; O. Caldwell et al.,(1988), Phys. Rev. Lett., **61**, 510.
23. S. Wolfram, (1979), Phys. Lett., **B82**, 65; C. B. Dover, T. K. Gaisser and G. Steigman, (1979), Phys. Rev. Lett., **42**, 1117.
24. J. Silk, K. A. Olive and M. Srednicki, (1987), Nucl. Phys., **B279**, 804.
25. J. S. Ellis, J. S. Hagelin and D. V. Nanopoulos, (1985), Phys. Lett., **B159**, 26; M. Oreglia, (1990), Talk given at the *14th International Conference on Neutrino Physics and Astrophysics "Neutrino 90"* (CERN, Geneva); J. Silk and M. Srednicki, (1984), Phys. Rev. Lett., **53**, 624; K. Griest, (1988), Phys. Rev. Lett., **61**, 666; K. Griest, (1988), Phys. Rev., **D38**, 2357; J. Ellis, et al., (1984), Nucl. Phys., **B238**, 453; K. Griest, M. Kamiokowski and M. S. Turner, (1990), Phys. Rev., **D41**, 3565; K. Olive and M. Srednicki, (1989), Phys. Lett., **B230**, 78.
26. S. P. Ahlen et al., (1987), Phys. Lett., **195**, 603; F. Boehm et al.,(1991), Phys. Lett., **B255**, 143; J. Faulkner and R. L. Gilliland, (1985), Ap. J., **299**, 994; L. M. Krauss, (1985), Harvard Univ. Rep. No. HUTP-85/A0008 (unpublished); W. H. Press and D. N. Spergel, (1985), Ap. J., **296**, 679; R. L. Gilliland et al., (1986), Ap. J., **306**, 703; D. O. Caldwell et al.,(1990), Phys. Rev. Lett., **65**, 1305; A. Gould, (1990), Ap. J., **356**, 302; Y. Giraud-Herand et al.,(1990), Solar Phys., **128**, 21; A. Gould and R. Raffelt, (1990), Ap.J., **352**, 654; D. Dearborn, K. Griest and G. Raffelt, (1990), University of California Report No. CfPA-TH-90-012.

27. Review of Particle Properties, (1990), Phys. Lett., **B239**, 158; see the contributions by E. Holzschuh, (1990) and J. F. Wilkerson, (1990), in "Neutrino 90", cited in ref. 25 above.
28. For an excellent discussion of solar neutrino problem, see J. N. Bahcall, (1989), *Neutrino astrophysics*, (Cambridge University Press, UK).
29. These bounds are reviewed in ref. 21. Also see D. N. Schramm, (1991), in *After the First Three Minutes*, cited in ref. 6 above, p. 12.
30. A. Hime and N. A. Jelley, (1990), Oxford Univ. preprint OUNP-91-01; D. O. Caldwell et al.,(1990), Nucl. Phys. (Proc. Suppl. **B13**), 547 and other references therein; M. Dugan, A. Manohar and A. E. Nelson, (1985), Phys. Rev. Lett., **55**, 170; M. Fukugita and T. Yanagida, (1991), Kyoto Univ. Preprint No. YITP/K-906; L. A. Ahrens et al., (1985), Phys. Rev., **D31**, 2732; J. A. Grifols and E. Masso, (1990), Phys. Lett., **B242**, 77; R. Gandhi and A. Burrows, (1990), Phys. Lett. **B242**, 149.
31. See, for example, T. Padmanabhan and M. M. Vasanthi, (1987), Ap. J., **315**, 411; (1985), Nature, **317**, 335 and references cited therein.
32. S. Tremaine and J. E. Gunn, (1979), Phys. Rev. Lett., **42**, 407.
33. S. Tremaine, M. Henon and D. Lynden-Bell, (1986), MNRAS, **219**, 285.
34. See, for example, P. Ramond, (1981), *Field Theory: A Modern Primer*, (Benjamin-Cummings, Reading, MA).
35. V. Baluni, (1979), Phys. Rev., **D19**, 2227; R. Crewther et al., (1979), Phys. Lett., **B88**, 123; N. Ramsey, (1977), Phys. Rep., **43**, 409; I. S. Alterev, (1984), Phys. Lett., **B136**, 327.
36. S. Weinberg, (1975), Phys. Rev., **D11**, 3583 and references cited therein.
37. R. D. Peccei and H. R. Quinn, (1977), Phys. Rev. Lett., **38**, 1440; (1977), Phys. Rev. **D16**, 1791; S. Weinberg, (1978), Phys. Rev. Lett., **40**, 223; F. Wilezek, (1978), Phys. Rev. Lett., **40**, 279; K. Choi, K. Kangand, J. E. Kim, (1989), Phys. Rev. Lett., **62**, 849; D. Kaplan, (1985), Nucl. Phys., **B260**, 215; R. Mayle et al., (1988), Phys. Lett. **203B**, 188; G. G. Raffelt and D. Seckel, (1988), Phys. Rev. Lett., **60**, 1793.
38. For a more complete review of laboratory searches for axions, see, for example, J. E. Kim, (1987), Phys. Rep., **150**, 1.
39. S. Dimopoulos et al.,(1986), Phys. Lett., **B176**, 223; J. Frieman, S. Dimopoulos and M. S. Turner, (1987), Phys. Rev., **D36**, 2201; G. G. Raffelt and D. S. P. Dearborn, (1987), Phys. Rev., **D36**, 2211; D. S. P. Dearborn, D. N. Schramm and G. Steigman, (1986), Phys. Rev. Lett., **56**, 26; N. Iwamoto, (1984), Phys. Rev. Lett., **53**, 1198; S. Truruta and K. Nomoto (1987), in *Observational Cosmology: Prceedings of IAU symposium 124*, eds. A. Hewitt, G. Burbridge and Li Zhi Fang, (Kluwer, Dordrecht), p. 713; D. E. Morris (1986), Phys. Rev., **D34**, 843; G.

G. Raffelt and L. Stodolsky, (1988), Phys. Rev., **D37**, 1237; R. P. Brinkmann and M. S. Turner, (1988), Phys. Rev., **D38**, 2338.

40. T. Kephart and T. Weler, (1987), Phys. Rev. Lett., **58**, 171; M. S. Turner, (1987), Phys. Rev. Lett., **59**, 2489; A. L. Broadfoot and K. R. Kendall, (1968), J. Geophys. Res. Sp. Phys., **73**, 426.
41. M. S. Turner, (1986), Phys. Rev., **D33**, 889.
42. S. De Panfills et al., (1987), Phys. Rev. Lett., **59**, 839; P. Sikivie, N. Sullivan and D. Tanner, experiment in progress.
43. S. Weinberg, (1989), Rev. Mod. Phys., **61**, 1.
44. Ya. B. Zeldovich, (1967), JETP Lett., **6**, 316; T. Padmanabhan, (1989), Int. J. Mod. Phys., **A4**, 4735.
45. E. Gemmer et al., (1983), Phys. Lett., **B133**, 61; G. Moore, (1987) preprint; G. Moore, (1987), Nucl. Phys., **B293**, 139; B. Zumino, (1975), Nucl. Phys., **B89**, 535; A. D. Dolgov, (1982), in *The Very Early Universe*, eds., G. W. Gibbons, S. W. Hawking and S. T. C. Siklos, (Cambridge University Press, Cambridge); R. D. Peccei, J. Sola, C. Wetterich, (1987), Phys. Lett., **B195**, 183; T. P. Singh and T. Padmanabhan, (1988), Int. J. Mod. Phys., **A3**, 1593; E. Baum, (1983), Phys. Lett., **B133**, 185; S. W. Hawking, (1984), Phys. Lett., **B134**, 403; T. Padmanabhan, (1984), Phys. Lett., **A104**, 196; S. Coleman, (1988), Nucl. Phys., **B310**, 643; S. Weinberg, (1987), Phys. Rev. Lett., **59**, 2607.
46. M. S. Turner, (1990), Physica Scripta, **T36**, 167.
47. A. Toomre, (1982), Ap. J., **259**, 535.

Chapter 12

1. T. Padmanabhan, (1991), talk given at the 57th annual meeting of the Indian Academy of Sciences, Pune, India; (1992), Curr. Sci., **63**, 379; J. A. Peacock, (1991), MNRAS, **253**, 1p. Also see the analysis in C. Park, (1991), Caltech theoretical astrophysics preprint.

Appendix A

1. For a detailed discussion of the principles of general retalivity, see L. D. Landau and E. M. Lifshitz, (1975), *Classical Theory of Fields*, (Pergamon Press, Oxford) or C. W. Misner, K. Thorne and J. A. Wheeler, (1973), *Gravitation*, (Freeman, San Francisco).

Appendix B

1. There are several text books which discuss these ideas in great detail. See, for example, C. Quigg, (1983), *Gauge Theories of Strong, Weak and Electromagnetic Interactions*, (Addison-Wesley, California); G. G. Ross, (1984), *Grand Unified Theories*, (Addison-Wesley, California); J.

C. Taylor, (1976), *Gauge Theories of Weak Interactions*, (Cambridge University Press, Cambridge); I. J. R. Aitchison and A. J. G. Hey, (1982), *Gauge Theories in Particle Physics*, (Adam Hilger Ltd., Bristol). M. B. Green, J. H. Schwarz and E. Witten, (1987), *Superstring Theory, Vols. I and II*, (Cambridge University Press, Cambridge).

2. Phase transitions and thermal effects in the early universe are discussed in: A. D. Linde, (1979), Rep. Prog. Phys., **42**, 389; R. Brandenberger, (1984), Rev. Mod. Phys., **57**, 1; L. Dolan and R. Jackiew, (1974), Phys. Rev., **D9**, 3320.

Appendix C

1. Smoot, G.F. et al. (1992), Ap. J., **396**, L1.
2. T. Padmanabhan and D. Narasimha, 1992 preprint TIFR-TAP-3/92. The CfA, IRAS and APM data shown here is based on Huchra et al. (1983), Ap.J. Suppl. **52**, 89; Davis, M and Peebles, P.J.E. (1983), Ap.J., **267**, 465; Strauss, M.A. et al., (1990), Ap.J., **361**, 49; Davis, M et al. (1988), Ap.J., **333**, L9; Maddox, S.J. et al. (1990), MNRAS, **242**, 43P. The $\sigma(R)$ was computed from these survey results in Hamilton, A.J.S. et al. (1991) Ap.J. Lett., **374**, L1.

Some useful numbers

1 Conversions

1 GeV	=	1.16×10^{13} K
1 GeV^{-1}	=	1.97×10^{-14} cm
	=	6.58×10^{-25} s
λ (photon wavelength)	=	$1.24 \times 10^4\,\text{Å}\,(E/1\,\mathrm{eV})^{-1}$
1 radian	=	57.3 degrees
1 steradian	=	$3.28 \times 10^3\,\mathrm{deg}^2$
	=	$4.25 \times 10^{10} (\mathrm{arc\,sec})^2$
1 parsec	=	3.1×10^{18} cm
1 kiloparsec	=	3.1×10^{21} cm
1 megaparsec	=	3.1×10^{24} cm
1 year	=	3.16×10^7 s
1 Jansky	=	$10^{-23}\,\mathrm{erg\,cm^{-2}\,s^{-1}\,Hz^{-1}}$

2 Fundamental constants

G	=	$6.67 \times 10^{-8}\,\mathrm{cm^3\,g^{-1}\,s^{-2}}$
c	=	$2.998 \times 10^{10}\,\mathrm{cm\,s^{-1}}$
$\hbar$	=	1.05×10^{-27} erg s
G_F	=	$1.17 \times 10^{-5}\,\mathrm{GeV^{-2}} = (293\,\mathrm{GeV})^{-2}$
m_e	=	0.51 MeV
m_p	=	938.3 MeV
m_n	=	939.6 MeV

3 Astrophysical constants

Solar mass ($\mathrm{M}_\odot$)	=	$1.99 \times 10^{33}\,\mathrm{g} = 1.19 \times 10^{57}\,m_p$
Solar radius ($\mathrm{R}_\odot$)	=	6.96×10^{10} cm
Solar potential ($\mathrm{GM}_\odot/\mathrm{R}_\odot$)	=	$(437\,\mathrm{km\,s^{-1}})^2$
Solar luminosity ($\mathrm{L}_\odot$)	=	$3.9 \times 10^{33}\,\mathrm{erg\,s^{-1}}$
Hubble constant (H_0)	=	$100h\,\mathrm{km\,s^{-1}\,Mpc^{-1}}$
H_0^{-1}	=	$3.1 \times 10^{17} h^{-1}\,\mathrm{s} = 9.78 \times 10^9 h^{-1}$ yr
	=	$3000 h^{-1}$ Mpc
Critical density (ρ_c)	=	$1.88 \times 10^{-29} h^2\,\mathrm{g\,cm^{-3}}$
	=	$1.05 \times 10^4 h^2\,\mathrm{eV\,cm^{-3}}$
MBR photon density	=	$n_\gamma = 422 (T/2.75\,\mathrm{K})^3\,\mathrm{cm^{-3}}$

Index